MICRO- AND NANOENGINEERING OF THE CELL SURFACE

The *cover image* shows an artistic representation of self-assembled DNA arrays bound to the surfaces of cells through specific antibody—cell surface interactions. The versatile cargo-carrying ability of arrays is explored for directing cell-surface interactions, cell—cell bridging, and positioning multiple cells onto a DNA fabric. SEM of a cell containing a DNA patch is shown in the top left corner.

Illustration by Peter Allen using Autodesk Maya. Concept design by Alexey Koyfman and Norbert Reich. Please see Chapter 8 for details.

MICRO- AND NANOENGINEERING OF THE CELL SURFACE

Edited by

JEFFREY M. KARP

Department of Medicine, Brigham and Women's Hospital, Harvard Medical School, Boston, MA, USA

WEIAN ZHAO

Sue and Bill Gross Stem Cell Research Center, Department of Pharmaceutical Sciences, University of California, Irvine, CA, USA

ELSEVIER

AMSTERDAM • BOSTON • HEIDELBERG • LONDON
NEW YORK • OXFORD • PARIS • SAN DIEGO
SAN FRANCISCO • SINGAPORE • SYDNEY • TOKYO

William Andrew is an imprint of Elsevier

William Andrew is an imprint of Elsevier
The Boulevard, Langford Lane, Kidlington, Oxford, OX5 1GB, UK
525 B Street, Suite 1800, San Diego, CA 92101-4495, USA
225 Wyman Street, Waltham, MA 02451, USA

British Library Cataloguing-in-Publication Data
A catalogue record for this book is available from the British Library

Library of Congress Cataloging-in-Publication Data
A catalog record for this book is available from the Library of Congress

ISBN: 978-1-4557-3146-6

For information on all William Andrew publications visit
our website at http://store.elsevier.com/

14 15 16 17 18 10 9 8 7 6 5 4 3 2 1

Contents

11. Cell Engineering with Nanoparticles for Cell Imaging

DAVID C. YEO, RIDDHIMA MINOCHA AND CHENJIE XU

12. Micro/Nano-Engineering of Cells for Delivery of Therapeutics

OREN LEVY, EDWARD HAN, JESSICA NGAI, PRIYA ANANDAKUMARAN, ZHIXIANG TONG, KELVIN S. NG AND JEFFREY M. KARP

13. Molecular Engineering of Cell and Tissue Surfaces with Polymer Thin Films

JOHN T. WILSON AND ELLIOT L. CHAIKOF

14. Biofunctionalization of Hydrogels for Engineering the Cellular Microenvironment

MANIRAJ BHAGAWATI AND SANJAY KUMAR

15. Probe and Control of Cell–Cell Interactions Using Bioengineered Tools

FENG LI, RANGOLI AERAN, EGEST J. PONE, DONG-KU KANG, LINAN LIU, X. CHRIS LE AND WEIAN ZHAO

Foreword

The domestication of independently living cells to serve human needs is a major goal of biotechnology. Over millions of years of evolution, cells from archaeal, bacterial, and eukaryotic origin have evolved sophisticated capabilities that, if harnessed, could advance the treatment of human disease, produce carbon neutral fuels, and detect environmental agents of risk to human health. Already, cultured cells have been integrated into high-scale production systems for biotherapeutics. They are the workhorses of high-throughput assays that search for new drug leads. Much excitement surrounds the prospect of using human cells themselves as therapeutic agents, particularly stem cells with their potential to replace damaged or defective counterparts in human patients. Cells are uniquely capable of transforming simple molecular inputs, such as sugars, amino acids, and lipids, into complex biopolymers or fuel-related compounds with high efficiency. Accordingly, harnessing cells as intermediates in the conversion of solar energy to combustible compounds is a major goal of the biofuels industry. And cells have proven to be highly sensitive biosensors of pathogens and toxins. They are naturally endowed with receptors that recognize analytes and trigger auto-amplifying signaling events. These capabilities have stimulated much interest in the use of cells as components of clinical and environmental biosensing devices.

Central to realizing the goal of cell domestication is the necessity of controlling interactions of the cell with its surrounding, be it a synthetic matrix or tissue microenvironment. The cell surface, a complex landscape comprising membrane lipids, proteins, and polysaccharides, is the key translator of information between the cell's outside and inside worlds. Cell surface components make direct contacts with the surrounding matrix allowing cells to attach and respond to their setting. Some of these interactions trigger cell responses that are critical for tissue integration or biomolecule production. At the same time, cell surface molecules comprise a dynamic data set that reports on the cell's physiology; decoding cell surface information can reveal underlying disease states, such as cancer, viral infection, and immune activation. There is clearly a bidirectional flow of information across the cell surface, where the cell is influenced by interactions with its matrix and, reciprocally, the cell can affect its neighbors through surface contacts.

The ability to control the physical and biochemical properties of the cell surface, therefore, has powerful implications. Engineering the presentation of specific ligands on cells can direct their attachment or response to a particular matrix structure, for example, or engender new functions not possessed by native cells. Some cell surface engineering challenges can be met using conventional tools of molecular biology and genetic engineering. However, modern developments from the physical sciences—bioconjugate chemistry, lipid biophysics, polymer science, nanomaterials, and microfabrication, in particular—have created exciting new avenues for engineering cell surface properties far outside the realm of natural biology.

This book outlines the challenges and opportunities inherent to the emerging discipline of cell surface engineering and provides a detailed view of state-of-the-art methods to modulate cell surface architecture. The book concludes with the status of engineered cells in the commercial sector and previews challenges that lie ahead as cell surface engineering techniques converge with clinical efforts. A theme that pervades this book is the critical role of interdisciplinary science in achieving cell domestication. Success has involved integration of concepts and techniques from cell biology, chemistry, physics, materials science, and nanoscience. Likewise, the book will appeal to and inspire scientists and engineers from many disciplinary backgrounds. I anticipate that this book will catalyze new conversations across the aisles of the life and physical sciences, within industry and academia. It is a must-have for the bookshelves of seasoned practitioners of cell surface engineering as well as those newly learning about this exciting area of much future opportunity.

Carolyn Bertozzi
T. Z. and Irmgard Chu
Distinguished Professor of Chemistry and
Professor of Molecular and Cell Biology,
University of California, Berkeley, CA, and
Howard Hughes Medical Institute, Chevy
Chase, MD, USA.

List of Contributors

Ayman F. Abuelela Biological and Environmental Sciences and Engineering, King Abdullah University of Science and Technology, Thuwal, Kingdom of Saudi Arabia

Rangoli Aeran Department of Pharmaceutical Sciences, Chao and Family Comprehensive Cancer Center, Sue and Bill Gross Stem Cell Research Center, Edwards Lifesciences Center for Advanced Cardiovascular Technology, Department of Biomedical Engineering, University of California, Irvine, CA, USA

Kawther K. Ahmed College of Pharmacy, University of Iowa, Iowa City, IA, USA

Laura Alberch Department of Biomedical Engineering and the Translational Tissue Engineering Center, The Johns Hopkins University, Baltimore, MD, USA

Priya Anandakumaran Division of Biomedical Engineering, Department of Medicine, Center for Regenerative Therapeutics, Brigham and Women's Hospital, Boston, MA; Harvard Medical School, Harvard University, Boston, MA; Harvard Stem Cell Institute, Cambridge, MA; Harvard-MIT Division of Health Sciences and Technology, Cambridge, MA, USA

Maniraj Bhagawati Department of Bioengineering, University of California, Berkeley, CA, USA

Elliot L. Chaikof Department of Surgery, Beth Israel Deaconess Medical Center, Boston, MA; Harvard Medical School, Boston, MA; Wyss Institute for Biological Inspired Engineering of Harvard University, Cambridge, MA, USA

Jian Du Department of Biomedical Engineering and the Translational Tissue Engineering Center, The Johns Hopkins University, Baltimore, MA, USA

Debjit Dutta Institute for Stem Cell Biology and Regenerative Medicine, National Centre for Biological Sciences-Campus, UAS-GKVK, Bangalore, India

Wade M. Fox Department of Chemical and Biological Engineering, University at Buffalo, The State University of New York, Buffalo, NY, USA

Samir Gautam Department of Cell Biology, Yale School of Medicine, New Haven, CT, USA

Sean M. Geary College of Pharmacy, University of Iowa, Iowa City, IA, USA

Edward Han Division of Biomedical Engineering, Department of Medicine, Center for Regenerative Therapeutics, Brigham and Women's Hospital, Boston, MA; Harvard Medical School, Harvard University, Boston, MA; Harvard Stem Cell Institute, Cambridge, MA; Harvard-MIT Division of Health Sciences and Technology, Cambridge, MA, USA

Dong-Ku Kang Department of Pharmaceutical Sciences, Chao and Family Comprehensive Cancer Center, Sue and Bill Gross Stem Cell Research Center, Edwards Lifesciences Center for Advanced Cardiovascular Technology, Department of Biomedical Engineering, University of California, Irvine, CA, USA

Jeffrey M. Karp Division of Biomedical Engineering, Department of Medicine, Center for Regenerative Therapeutics, Brigham and Women's Hospital, Boston, MA; Harvard Medical School, Harvard University, Boston, MA; Harvard Stem Cell Institute, Cambridge, MA; Harvard-MIT Division of Health Sciences and Technology, Cambridge, MA, USA

Alexey Y. Koyfman Department of Biochemistry and Molecular Biology, and Sealy Center for Structural Biology and Molecular Biophysics,

University of Texas Medical Branch, Galveston, TX, USA

Sanjay Kumar Department of Bioengineering, University of California, Berkeley, CA, USA

X. Chris Le Department of Laboratory Medicine and Pathology, University of Alberta, Edmonton, AB, Canada

Oren Levy Division of Biomedical Engineering, Department of Medicine, Center for Regenerative Therapeutics, Brigham and Women's Hospital, Boston, MA; Harvard Medical School, Harvard University, Boston, MA; Harvard Stem Cell Institute, Cambridge, MA; Harvard-MIT Division of Health Sciences and Technology, Cambridge, MA, USA

Feng Li Department of Laboratory Medicine and Pathology, University of Alberta, Edmonton, AB, Canada; Department of Pharmaceutical Sciences, Chao and Family Comprehensive Cancer Center, Sue and Bill Gross Stem Cell Research Center, Edwards Lifesciences Center for Advanced Cardiovascular Technology, Department of Biomedical Engineering, University of California, Irvine, CA, USA

Linan Liu Department of Pharmaceutical Sciences, Chao and Family Comprehensive Cancer Center, Sue and Bill Gross Stem Cell Research Center, Edwards Lifesciences Center for Advanced Cardiovascular Technology, Department of Biomedical Engineering, University of California, Irvine, CA, USA

Srujan K. Marepally Institute for Stem Cell Biology and Regenerative Medicine, National Centre for Biological Sciences-Campus, UAS-GKVK, Bangalore, India

Jasmeen S. Merzaban Biological and Environmental Sciences and Engineering, King Abdullah University of Science and Technology, Thuwal, Kingdom of Saudi Arabia

Riddhima Minocha Division of Bioengineering, School of Chemical and Biomedical Engineering, Nanyang Technological University, Singapore

Kelvin S. Ng Division of Biomedical Engineering, Department of Medicine, Center for Regenerative Therapeutics, Brigham and Women's Hospital, Boston, MA; Harvard Medical School, Harvard University, Boston, MA; Harvard Stem Cell Institute, Cambridge, MA; Harvard-MIT Division of Health Sciences and Technology, Cambridge, MA, USA

Jessica Ngai Division of Biomedical Engineering, Department of Medicine, Center for Regenerative Therapeutics, Brigham and Women's Hospital, Boston, MA; Harvard Medical School, Harvard University, Boston, MA; Harvard Stem Cell Institute, Cambridge, MA; Harvard-MIT Division of Health Sciences and Technology, Cambridge, MA, USA

Jaina M. Patel Department of Pathology and Laboratory Medicine, Emory University School of Medicine, Atlanta, GA, USA

Egest J. Pone Department of Pharmaceutical Sciences, Chao and Family Comprehensive Cancer Center, Sue and Bill Gross Stem Cell Research Center, Edwards Lifesciences Center for Advanced Cardiovascular Technology, Department of Biomedical Engineering, University of California, Irvine, CA, USA

Norbert O. Reich Biomolecular Science and Engineering Program, and Department of Chemistry and Biochemistry, University of California Santa Barbara, Santa Barbara, CA, USA

Kosuke Sakashita Biological and Environmental Sciences and Engineering, King Abdullah University of Science and Technology, Thuwal, Kingdom of Saudi Arabia

Aliasger K. Salem College of Pharmacy, University of Iowa, Iowa City, IA, USA

Debanjan Sarkar Department of Biomedical Engineering, University at Buffalo, The State University of New York, Buffalo, NY USA

Periasamy Selvaraj Department of Pathology and Laboratory Medicine, Emory University School of Medicine, Atlanta, GA, USA

David A. Spiegel Department of Chemistry, Yale University, New Haven, CT, USA; Department of Pharmacology, Yale University, New Haven, CT, USA

Zhixiang Tong Division of Biomedical Engineering, Department of Medicine, Center for

Regenerative Therapeutics, Brigham and Women's Hospital, Boston, MA; Harvard Medical School, Harvard University, Boston, MA; Harvard Stem Cell Institute, Cambridge, MA; Harvard-MIT Division of Health Sciences and Technology, Cambridge, MA, USA

Mark L. Tykocinski Department of Pathology, Anatomy and Cell Biology, Jefferson Medical College, Thomas Jefferson University, Philadelphia, PA, USA

Vincent F. Vartabedian Department of Pathology and Laboratory Medicine, Emory University School of Medicine, Atlanta, GA, USA

Praveen Kumar Vemula Institute for Stem Cell Biology and Regenerative Medicine, National Centre for Biological Sciences-Campus, UAS-GKVK, Bangalore, India

John T. Wilson Department of Chemical and Biomolecular Engineering, Vanderbilt University, Nashville, TN, USA

Chenjie Xu Division of Bioengineering, School of Chemical and Biomedical Engineering, Nanyang Technological University, Singapore

Kevin J. Yarema Department of Biomedical Engineering and the Translational Tissue Engineering Center, The Johns Hopkins University, Baltimore, MD, USA

David C. Yeo Division of Bioengineering, School of Chemical and Biomedical Engineering, Nanyang Technological University, Singapore

Weian Zhao Department of Pharmaceutical Sciences, Chao and Family Comprehensive Cancer Center, Sue and Bill Gross Stem Cell Research Center, Edwards Lifesciences Center for Advanced Cardiovascular Technology, Department of Biomedical Engineering, University of California, Irvine, CA, USA

CHAPTER

1

Cell Membrane Biology and Juxtacrine Signal Conversion

Mark L. Tykocinski

Department of Pathology, Anatomy and Cell Biology, Jefferson Medical College, Thomas Jefferson University, Philadelphia, PA, USA

OUTLINE

1.1 INTRODUCTION

The cell membrane, bulwark to the exterior and gateway to the interior, has emerged as a most remarkable biological structure. Complex in composition, organization, and biogenesis, it now reveals itself to be a dynamic, plastic, and functionally rich entity. As an ever more sophisticated experimental toolkit is deployed to probe the cell membrane's complexities, our view of it is being enriched by the capacity to tease things apart—from single cell down to single molecule analytics—coupled to a growing ability to start to piece things back together again through system perspectives.

J.M. Karp & W. Zhao (Eds): Micro- and Nanoengineering of the Cell Surface.
DOI: http://dx.doi.org/10.1016/B978-1-45-573146-6.00001-5

We are still a long way off from a "grand unification" that brings together the various perspectives and narratives of the biochemists, the biophysicists, the cellular physiologists, and the systems and computational biologists into a comprehensive, structurally, and functionally integrated view of the cell membrane. Cell surface proteomics, lipidomics, and glycomics are still in their infancy, and sophisticated tools to dissect and visualize static and dynamic macromolecular interplay are only now coming into play. Yet, though the cell membrane's intricacies are yet to be unraveled, it has uncannily proven itself to be amenable to artificial manipulation. Quite remarkably, reductionism seems to work time and again, as simplistic interventions and modulatory inputs lead to predictable and utilitarian outputs. Somehow membrane engineering works despite the experimentalist's indulgence in tunnel vision and conscious ignorance of the wider membrane biology panorama in its spatiotemporal complexity. Further, beyond the engineering of cell membranes, there are early successes in artificially mimicking them and recapitulating their functions in diverse and creative ways.

This chapter offers some historical highlights and perspectives on membrane biology, as a prelude to showcasing one early cell membrane engineering story line, directed toward the use of paintable and signal converter proteins to modulate cell membrane protein repertoires and their juxtacrine and autocrine signaling properties. Thinking about cell membranes is framed within a membrane emergence paradigm, with membrane engineering cast as a way to perturb and craft emergent cell membrane states for defined ends.

1.2 CELL MEMBRANE BIOLOGY—EARLY MILESTONES

The history of cell membrane biology features a progression from simple lipid bilayer to complex macromolecular mixture, and from there to the current paradigm of a physiochemical entity that is intricate in substructure and dynamically molded by intrinsic and extrinsic factors. In the process, the cell membrane's primitive "boundary" role, minimally tasked to envelop the cell's constituents, has given way over time to a far richer constellation of functions, both outward and inward looking.

Our understanding of membrane biology has unfolded through several experimental narratives that have played out in parallel over the past century. Each narrative encompasses a series of historical milestones. These narratives are highly interwoven and relate to membrane composition and structure, membrane model systems, membrane protein structure and function, membrane biogenesis, and emergent properties of membranes.

A seminal insight into membrane composition and structure came in 1925, when Gorter and Grendel [1] proposed a lipid bilayer structure for cell membranes. This was built upon an interesting historical sequence of studies on the interaction of oil films with water, starting with Benjamin Franklin and furthered by Raleigh and then Langmuir [2]. The Gorter–Grendel lipid bilayer model was bolstered by a simple experimental observation: when lipids are extracted from erythrocytes, the area they cover at an air–water interface is twice as large as the original surface area of the cells. It became clear that the membrane bilayer results from the aggregation behavior of amphipathic phospholipids, in accordance with the hydrophobic effect, with the polar ends of the lipid molecules oriented toward the aqueous phase and the hydrophobic hydrocarbon regions organizing to prevent contact with the aqueous phase. The product is a planar biomolecular film separating two aqueous compartments. Proteins entered the picture in 1935 when Danielli and Davison offered a membrane model incorporating globular proteins [3]. Their image of a trilamellar sandwich, with globular proteins coating both surfaces of a lipid bilayer, was prompted by their attempt to accommodate the surface hydrophilicity of globular proteins. Their model was reinforced by the discovery back then of protein beta-sheet structure, which seemed to preclude protein penetration of the lipid bilayer. It was not until 1972 that the Danielli–Davison model was finally supplanted. In an alternative *fluid mosaic* model of cell membranes, Singer and Nicolson explicitly postulated two distinct classes of membrane proteins, peripheral and integral. Peripheral membrane proteins were defined as those removable with salt treatment or pH changes, whereas integral membrane proteins require detergents for removal, and importantly, can even span the entire membrane [4]. Various experimental findings had laid the groundwork for the Singer–Nicolson model, including the first high-resolution electron micrograph of a biological membrane by Palade [5], subsequently framed by Robertson into a "unit membrane" paradigm [6,7]; the elegant demonstration by Frye and Edidin [8], via experimental cell–cell fusion and fluorescein-labeled protein tracking, that membrane proteins can diffuse laterally in cell membranes; and the confirmation, by Hladky and Haydon [9], of the unit channel structure for Gramicidin A peptide.

Experiments with artificial membranes catalyzed this evolving understanding of membrane structure. A seminal milestone was the 1962 report by Mueller and Rudin [10] of the first artificial lipid bilayer. Such membrane modeling was taken to the functional level in 1969, when Huang [11] demonstrated that unilamellar lipid vesicles reconstituted with membrane proteins could be used to study ion flux. Since single membrane vesicles can be readily produced in large quantities, this technical achievement revolutionized transport studies, allowing insights into the kinetics of many channels, pumps, and transporters without having to know their structure. Next came the monolayer-derived planar bilayers of Montal and Mueller in 1972, virtually solvent-free synthetic membranes which allowed for the study of intrinsic properties of ion channels [12]. This technical advance set the stage for studying how lipid composition and distribution within cell membranes affects the activities of membrane proteins embedded in bilayers. In 1979, Schindler [13] described vesicle-supported planar bilayers, as synthetic membranes that are completely solvent free. The armamentarium of model membrane systems has continued to grow, including categories such as liposomes, sonicated vesicles, large unilamellar vesicles, black lipid membranes, lipid monolayers, high-density lipoprotein particles (reviewed in Ref. [14]), and dendrimersomes [15].

Another narrative that has in parallel shaped the membrane biology story revolves around membrane protein structure. Glycophorin was the first integral membrane protein to be sequenced, back in 1975 [16]. That same year, Henderson and Unwin [17] reported the first electron microscopy-derived structure for a membrane protein, bacteriorhodopsin. Its remarkable structure, with seven transmembrane α-helices, supported the notion that α-helices are the secondary structure of choice for transmembrane proteins. The first X-ray high-resolution structure of a membrane protein, by Diesenhofer, Huber and Michel in 1985, reinforced the image of an α-helical transmembrane segment [18]. The β-barrel model for bacterial porin and the β-helix structure of Gramicidin A, which came later, came to be seen as exceptions. Yet another landmark was the high-resolution structure of a bacterial ion channel, solved in 1999 by MacKinnon, which linked protein backbone structure, as opposed to the expected amino acid residues, to ion selectivity [19].

Then there are the functional narratives, diverse and multifaceted. Back in 1855, Naegeli and Cramer ascribed to plant cell membranes an essential role in osmosis. In the last decade of the nineteenth century, Ernest Overton [20] framed the theory that "lipoid" membranes enclosing animal and plant cells control their osmotic properties and permeation of molecules, and posited fundamental membrane processes such as ion transport. A multitude of signaling insights and milestones have ensued, along with breakthroughs touching on diverse membrane functions that extend well beyond signaling.

1.3 MEMBRANE MICRODOMAINS

The notion that there may be spatial separation of signaling reactions within membranes was first proposed for the cAMP system in 1981 [21]. Early on, the Singer–Nicolson fluid mosaic model stood in the way of the idea that cells might spatially and functionally compartmentalize signaling machinery, with skepticism over whether laterally diffusible molecules could be segregated effectively. However, this compartmentalization concept, instantiated in the paradigm of *membrane microdomains*, has by now entered the mainstream, taking the fluid mosaic model in a different direction. While proteins move laterally within membranes, membrane surfaces are in fact nonhomogeneous, instead displaying heterogeneity in substructure, molecular distribution, and inner and outer connections. Thus the "fluid mosaic" can be recast as a "mosaic of microdomains" [22].

Membrane microdomains have been defined as functional regions, of micron/submicron dimensions, that compartmentalize proteins, lipids, and signaling components into multimolecular assemblies [23,24]. Over the years, microdomains have been categorized in different ways, with the delineation of lipid rafts, tetraspanin webs (TEMs), and caveolae (reviewed in Ref. [25]). Lipid rafts, with an estimated size of 12–200 nm, showcase bilayer asymmetry with outer leaflets composed of primarily cholesterol that binds to glycosphingolipids and inner leaflets composed of saturated phospholipids [26,27]. TEMs are a distinct class of microdomains in which tetraspanins, each with a signature four transmembrane domain structure, serve as organizers of multimolecular complexes, organized through tetraspanin–tetraspanin interactions and selective associations with various other membrane proteins [28–30]. Caveolae are cholesterol-rich membrane invaginations comprising caveolin and cavin proteins, which themselves have significant functions. These various types of microdomains are not only dynamic in space and time, but they can also exchange molecular constituents amongst themselves [31–33].

Along with converging evidence for the actual physical existence of these evanescent structures [23,34,35], there have been early attempts to mathematically model them [36], keying in on relative rates of production, diffusion, and destruction of microdomain-localized signaling components. Such models attempt to factor in space and time, for instance, considering two defining spatiotemporal characteristics—the point at which the concentration of the activated signaling component is maximal and the concentration gradient down to the level in the surrounding milieu, with the gradient slope defining the boundaries of a given microdomain. With both parameters in constant flux, views of membrane microdomains are momentary snapshots, precluding simplistic steady-state assumptions. Factors underlying the formation and dissipation of these microdomains are complex and interdependent, and include cell shape, topology of the reaction network, diffusion coefficients, and intrinsic properties of the components involved.

The architecture of membrane microdomains, along with their mode of assembly, is starting to come into focus. This has been propelled by significant advances in the experimental tools available for tracking them, including the advent of super-resolution microscopy techniques, such as near-field scanning optical microscopy (NSOM), stimulated emission depletion (STED), structured illumination microscopy (SIM), total internal reflection fluorescence microscopy (TIRFM), Pointillism microscopy methods, and photoactivation localization microscopy coupled to pair correlation analysis, which are capable of nanoscale mapping of the cell surface landscape [37,38] (reviewed in Ref. [25]). These have been complemented by advances in single molecule imaging, such as fluorescence recovery after photobleaching-fluorescence lifetime imaging microscopy (FRET-FLIM) and multicolor single particle tracking (SPT), which provide insights into protein dynamics at the molecular level [39].

Formation of specialized membrane islands depends on an intricate array of protein–protein, lipid–protein, and lipid–lipid interactions. Beyond intramembranous molecular interplay, recent evidence points to a role for the cytoskeleton in creating membrane microdomain boundaries, with the intriguing implication of one particular integral membrane protein, CD317/tetherin, as a critical microdomain organizer. Unique structural features of CD317/tetherin, including its dual transmembrane and glycosyl-phosphatidylinositol (GPI) anchors and abilities to oligomerize and form disulfide-bonded parallel coiled coils, endow it with the potential to link integral membrane protein "pickets" to an actin cytoskeletal "fence," at once delimiting lipid rafts and enabling their linkage into larger microdomains [22]. This provides an elegant protein structural underpinning to the actin cytoskeleton-based "picket-fence model" of microdomain organization, that proposes a hierarchical, three-tiered, mesoscale (larger than a nanometer, smaller than a micron)-domain architecture [40].

Much remains to be learned about the rules governing which surface proteins enter and exit the various kinds of microdomains. Typical lipid raft constituents, anchored to the outer leaflet, are GPI-anchored proteins [41]. Some proteins constitutively reside in rafts, whereas others are only transiently associated with them. Processes such as protein oligomerization can influence transit time, and coalescence of rafts impacts the dynamics. Special structural domains or motifs have been implicated in microdomain localization [42].

What emerges is a new image of the cell membrane, as a mosaic of structurally and functionally differentiated islands, with special abilities to connect to the cell interior and extracellular entities, all the while dissolving in and out. Out of this dynamism, discrete functionalities somehow emerge. Microdomains have been implicated in roles that go well beyond receptor signal acquisition. Outbound signals are empowered through microdomains as well. One of the more elegant microdomain stories has indeed come together around the "immunological synapse," envisioning microdomain-based signaling elements on both the antigen-presenting cell (APC) signal sender and the T-cell signal receiver sides (reviewed in Ref. [25]). In addition to signaling, microdomains have been implicated in a variety of other aspects of cell dynamics and function such as membrane trafficking and regulated exocytosis. Membrane microdomains have also been implicated in pathogenic processes, the most straightforward being pathogen entry/egress and toxin entry. Human deficiencies in tetraspanin and other microdomain-associated proteins have emphasized the profound implications of perturbations in microdomain composition and biology (reviewed in Ref. [25]).

Looking beyond microdomain physiology and pathophysiology, one can begin to ponder theoretical opportunities to harness microdomain biology in the engineering of cell membranes for therapeutic purposes. Possibilities include purposeful perturbation of microdomain dynamics and selective alteration of microdomain-based protein and lipid repertoires. The therapeutic efficacy of currently available antibodies that are known to impact protein transit to microdomains hints at the future promise of microdomain targeting (reviewed in Ref. [25]).

1.4 CELL MEMBRANE EMERGENCE

It has been suggested that membrane microdomains, as locally restricted signaling reaction systems, constitute an emergent property of cells [36]. This notion can be generalized, beyond microdomains *per se*, to encompass the full array of membrane properties. Indeed, emergence offers a robust way to conceptually frame the panoply of structural and functional dimensions of cell membranes, as they play out in their spatiotemporal complexity. To capture this perspective, one could speak of *cell membrane emergence*. The phenomenon of emergence has been defined more generally as the way complex systems and patterns arise out of a multiplicity of relatively simple interactions. Emergence speaks to how novel and coherent structures, patterns, and properties arise during a process of self-organization in complex systems, with the emergent state being greater than the sum of its parts. Building upon this global concept, a cell membrane emergence paradigm would posit that membrane properties and states are malleable, in constant flux, and arise, sometimes in unpredictable ways, out of a dynamic interplay of a collection of intrinsic and extrinsic components. Further, the properties of cell membranes might reflect real-time "tuning" around quasi-steady emergent states, which can be shifted in incremental or quantum fashion by extrinsic perturbations, natural or artificial. Higher order complexity emerges, featuring spatial heterogeneity and flux over varying timescales.

At any moment in time, the physicochemical and functional properties of a cell membrane are dictated by two categories of drivers: (i) composition and structure, i.e., the component lipids and proteins and their organizational relationships, and (ii) influences extrinsic to the membrane proper, both intracellular (such as cytoskeleton and soluble cytoplasmic factors) and extracellular (such as matrix, signals arriving from a distance, and adjacent cells, viewed individually or as part of functionally integrated cellular networks). The process shaping a membrane's emergent properties is inherently iterative, as membrane changes differentially affect the very drivers dictating its current state, for example, via alterations in cell–cell interactions that modify adjoining cells or in membrane trafficking phenomena that modify membrane flows to and from the cell interior.

Cell membrane emergence can be approached from two distinct angles—drivers of emergent membrane properties and the emergent membrane properties in-and-of themselves. Membrane proteins offer a tractable handle for latching onto membrane emergence phenomena. The emergent structural and functional properties of the membrane protein repertoire are dictated by multiple factors: (i) membrane protein composition (fashioned by endogenous protein production and membrane trafficking), (ii) membrane lipid composition, (iii) membrane protein dynamics (changes in protein structure, in turn affecting cytoplasmic, transmembrane, and extracellular portions of the respective proteins), (iv) membrane protein distribution (higher order oligomers, both homo and hetero, multiprotein complexes, and membrane microdomains), and (v) protein and nonprotein molecular factors, both intramembranous and extrinsic to the membrane (intra- and extracellular, e.g., extracellular matrix and juxtacrine, cell-based signals) that impact protein functionality within a given spatiotemporal context.

A promising future path for membrane biology can be intuited from two contemporary lines of investigation that each showcase just how elegant the intersection can be between membrane protein structural dynamics and cell membrane biology, with the two informing and playing off of each other. These two examples illustrate routes whereby cell membrane emergence manifests itself.

One line of investigation focuses on membrane fusion, and in particular, the intracellular machinery for fusing membranous organelles in the context of precisely timed release of neurotransmitters for synaptic transmission. A fundamental insight, emanating from the studies of Sudhof and Rothman, is that a thermodynamically spontaneous process of protein folding drives perturbation of the membrane bilayer and membrane fusion (reviewed in Ref. [43]). The beauty of this evolving story stems from its molecular detail, with SNARE proteins zipping up, and SM proteins organizing trans-SNARE complexes (i.e., SNAREpins) spatially and temporally. Beyond the informative conformational changes of these pivotal proteins, we are also learning about additional proteins that can actually control this process. Intriguingly, "grapple" proteins function contextually, and by their nature, can inhibit the process, activate it, or both depending upon the conditions [44]. This one mechanism has, in-and-of-itself, sweeping implications, as the orderly execution of membrane fusion goes well beyond neurocrine communication, impacting the exquisite compartmental organization of all cells, mechanisms of hormone release, and various other membrane fusion-based phenomena.

The interplay between protein dynamics and cell membrane biology has been showcased in another realm of investigation, this one centered around G protein-coupled receptors (GPCRs) (reviewed in Ref. [45]). GPCRs are highly dynamic and exist in a multitude of functionally distinct conformations. Insights into these conformations, and how ligands affect their energetics and rates of interconversion, have provided key clues into how GPCRs actually work and make it clear that their functional versatility is attributable, in no small part, to their structural plasticity. Thus, GPCRs exist as ensembles of discrete conformations with energetics that can be influenced by a range of factors (lipids, cytosolic signaling and regulatory proteins, ligands, pH, ions, oligomerization, and possibly transmembrane voltage gradients). Whereas in the membrane fusion setting, protein structural dynamics generate force, in the case of GPCRs, they drive conformational changes at the cytoplasmic ends of transmembrane segments, that in turn provide interaction interfaces for cytosolic proteins, including heterotrimeric G proteins, GPCR kinases and arrestins, as well as localization to specific signaling compartments.

The experimental enabler for connecting protein dynamics to emergent cellular architectures and functions has been the growing array of biophysical methods that can be deployed to study protein structure and dynamics. These include crystallography, nuclear magnetic resonance (NMR) spectroscopy, fluorescence spectroscopy, electron microscopy, and electron paramagnetic resonance (EPR) spectroscopy. Though ideally, protein kinetics should be studied in native cell membranes, methodological challenges remain, resulting in reliance upon simplified membrane model systems. Ultimately, a more complete picture of protein dynamics and function will demand an ability to monitor conformational changes in single receptor molecules as they function in a native cell membrane, with receptor dynamic measurements at timescales of milliseconds to seconds [45].

The field of membrane biology is now replete with hot topics, many of which could be framed within a cell membrane emergence paradigm. As a sampling:

- Lipid bilayer substructure and the process of membrane protein insertion [46,47]
- Bidirectional functional interplay between intrinsic lipids and membrane proteins, including transient lipid–protein complexes and inducible co-clustering [48–52]

- Membrane thickness, shape, curvature, pressure, tension, and fluidity as determinants of membrane function, including implications for fusion and fission of the lipid bilayer and positioning of membrane proteins [53–57]
- Role of cytoskeletal and other cytoplasmic and intramembranous proteins in molding membrane substructure and specialization [58–62]
- Supramolecular membrane protein cluster formation [63]
- Coordination between transmembrane and GPI-anchored proteins [64]
- Single molecule membrane biology [65]
- Quantitative cell surface proteomics, both global and membrane substructure specific [66–68]
- Artificial membrane design
- Membrane alterations as drivers and reporters of pathogenesis.

Things will get more complicated before they get simpler. Lipidomics has started to hint at the vast complexity of cellular lipids, and the same holds for glycomics and the surface glycoprotein and glycolipid repertoires. Furthermore, variability in membrane composition and structure among different cell types, along with associated microdomain architectures, have hardly been explored.

Back in 1962, Kauzman and Tanford theorized about the effect of hydrophobicity on protein folding and stability, which was a first step to frame thinking around the emergence of membrane properties. Clearly, things have come a long away since then. The stage is now set for designing membrane engineering strategies with complex, emergent membrane states as a conceptual framework. One can think of surface engineering driven emergence as an overlay on existing, physiologically driven emergent membrane states. From a functional perspective, two pivotal emergent membrane states are the potentials for receptor (inward signal flow) and juxtacrine (outward signal flow) signaling. Attention will now be turned to the latter, in detailing a line of investigation where the early dividends have been flexible cell surface engineering tools and novel therapeutics, and with the ultimate goal of modulating cellular networks now coming into sight.

1.5 JUXTACRINE SIGNALING AND REWIRING CELLULAR NETWORKS

Juxtacrine signaling is a form of intercellular communication that depends on direct interaction of adjoining cells [69–72]. This cell contact-dependent mode of signaling, typically mediated by paired interactive proteins on apposed cells, is at the heart of diverse biological processes, in particular those requiring tight regulation and spatial control of stimulation of one cell by another (reviewed in Ref. [73]). The simplest juxtacrine links involve just a pair of interacting cells, for example, an APC interfacing with a T cell of the immune system. The triggering of T-cell receptors by APC-anchored major histocompatibility complex (MHC) antigenic peptide complexes represents a classic two-party juxtacrine signal encounter. Yet, even in this seemingly straightforward situation, greater complexity quickly arises as third-party cells enter the picture, along with an array of paracrine (soluble factor mediated) and more distant endocrine immunomodulatory signals. Thus, juxtacrine intercellular signaling is best thought of in a more encompassing way, as operating within cellular networks of varying complexity that frequently evolve in multistep fashion over varying timescales. Morphogenesis and developmental patterning exemplify how juxtacrine signals can cascade and effect phenotypic changes across a field of cells, with juxtacrine Delta-to-Notch serving as the classical paradigm [74–78].

Over the years, our laboratory has focused on strategies for engineering cell surfaces, with the express aim of modulating juxtacrine signaling potential. The ultimate goal is to alter, and even rewire, cellular networks by altering juxtacrine cues among the component cells. Interestingly, the concept of emergence also has relevance to such cellular network engineering, in a sense creating what amounts to a hierarchical series of emergent states that can continuously evolve. That is, cell surface engineering elicits emergent juxtacrine signaling arrays; the receiver cells manifest emergent phenotypes that influence other cells in the network; and the very network itself displays its own emergent properties. The "cellular choir" acquires its new voice. The emergence within engineered cells is a function of not only the particular molecules in play but also the particulars of the engineering method itself.

Emergent heterogeneity arising via juxtacrine signaling can be mathematically modeled [79]. It is not just a matter of how already networked cells are communicating among themselves but also how they embrace other cells outside of the immediate network. Advances in cytomics, systems analytics, and mathematical modeling should begin to enable the decoding of new cellular configurations and functionalities within such emergent cellular networks.

Much of juxtacrine signal engineering to date has revolved around tailoring the cell's membrane protein repertoire, by either adding, deleting, or otherwise modifying selected proteins with defined intercellular (*trans*) signaling properties. This membrane protein-focused engineering can be accomplished by a variety of means [80–84]. Genetic approaches may deal with individual genes, gene groupings, or synthetic gene circuits [80] (reviewed in Ref. [69]). Inherent limitations of gene transfer for certain applications have prompted the addition of nongenetic strategies to the cell surface engineer's toolbox, primarily ones that are protein transfer based. Mirroring synthetic cells, synthetic scaffold-based platforms, featuring juxtacrine signaling molecules detached from their endogenous cell backgrounds and immobilized on various matrices, are being developed as juxtacrine signaling surrogates [69,85].

Of note, juxtacrine signals are not restricted to surface glycoproteins. Platelet-activating factor (PAF), a biologically active phospholipid expressed on the surface of endothelial cells at inflammatory sites, mediates juxtacrine stimulation of leukocytes [86]. Another variant is molecular intersection between juxtacrine and paracrine signaling, with the membrane-anchored ligand virtually the same as, and often a precursor of, the secreted one [71]. The juxtacrine signaling molecule can even be an interloper, as an extraneous soluble factor that is immobilized via cell surface proteoglycan [87,88].

The design and deployment of synthetic cells with altered juxtacrine signaling potential is not without its complexities. Changes in juxtacrine signal arrays are being layered onto a backdrop of other regulatory inputs, encompassing bystander cells, extracellular matrix, and various paracrine, endocrine, and even autocrine signals. Additional complexity emanates from the interactive membrane proteins themselves, for instance, from their frequent capacity to signal bidirectionally. Another intriguing phenomenon that modulates juxtacrine signaling is "*cis* inhibition," wherein ligand and receptor are also neighbors on the same cell, and one interferes with the juxtacrine *trans* activating potential of the other. This serves to modulate signaling dynamics in significant ways [89,90]. There are also peculiar phenomena that influence the signal array, such as the internalization of membrane-associated juxtacrine signals through transcytosis by an adjacent cell [91,92].

1.6 PROTEIN PAINTING, ARTIFICIAL VETO CELL ENGINEERING

Our laboratory's path toward cell surface engineering and juxtacrine signal modulation started back in the 1980s with the development of what we termed *protein painting* strategies. The application was immune cell engineering, and more specifically, the engineering of APC. At the time, we were building on a discovery made several years earlier in the field of immunoregulation—that APC, typically immune activating, can also function as immune-inhibiting "veto" cells. This was a twist on one of cellular immunology's central dogmas, namely, that APC, such as dendritic cells, drive T-cell activation. This is mediated by surface-associated MHC • peptide antigen complexes and costimulators which provide APC-anchored first and second T-cell-activating juxtacrine signals, respectively. Thus, while the focus had traditionally been on the APC role as a T-cell activator, the early immunology literature was hinting at another APC functional dimension, casting it as a potential T-cell inhibitor. Thus emerged the concept of an immunoregulatory APC Trojan horse, a cell inhibiting a second cell that recognizes it [93,94].

Our seminal first step was to demonstrate that veto cells can be generated artificially through cell surface engineering. For expediency, we first invoked gene transfer as the means to neo-express surface proteins that can send inhibitory or pro-apoptotic juxtacrine signals to target T cells. Our group coined the term *coinhibitor* to characterize such surface-associated, *trans*-signaling inhibitory proteins, drawing a parallel between them and their well-established counterparts, the costimulators [95]. As defined, costimulators and coinhibitors alike mediate APC-to-T cell contact-dependent signaling, and operate alongside and coordinately with MHC • peptide antigen complexes to modulate antigen-specific T cells. In one case, there is T-cell activation, in the other inhibition. Having established that *artificial veto cells* can indeed be produced [82,95], we next sought to replicate this genetic feat via protein transfer. This led us to a series of technical advances in the protein transfer field, as we designed new classes of *protein paints* for cell surface engineering.

The first protein we invoked for juxtacrine coinhibition was the lymphoid surface molecule CD8. As it turned out, this classical *cis*-acting cytotoxic T-cell costimulator receptor could be co-opted to serve as a coinhibitor ligand, simply by artificially enforcing its expression on APC [96,97], thereby drawing upon its capacity to send inhibitory reverse signals through T-cell resident MHC class I, its cognate counter receptor [98]. This unusual choice of CD8 for coinhibition was prompted by our earlier demonstration, applying antisense RNA technology to T cells for the first time, that deletion of CD8 from alloantigen-presenting cloned $CD8^+$ T cells abrogates their veto function [99]. Our description of CD8 as coinhibitor, along with our formulation of the artificial veto cell paradigm [82,96,99–101], was amply confirmed by other investigators [102–104], albeit occasionally recast under new names [105–108].

From CD8, we turned to a more classic inhibitory surface protein, this one with the capacity to trigger T-cell apoptosis. The Fas ligand (FasL, CD95L, CD178): Fas (CD95, TNFRSF6) juxtacrine signaling axis plays a key role in shaping T- and B-cell repertoires and maintaining peripheral tolerance [109]. Naturally expressed FasL expresses veto-like function [110]. Chen et al. [111] and others proceeded to show that enforced FasL expression could induce alloantigen-specific T-cell tolerance and clonal deletion *in vivo* [105,110,112–115].

In principle, protein transfer offers a number of advantages over gene transfer, including finer control of surface protein levels, simplification of surface protein co-expression, shorter half-lives of exogenous protein on modified cells, and ultimate translatability into therapeutic applications in the clinic. We therefore shifted from gene to protein transfer for APC surface engineering. The concept is simply to exogenously coat cell surfaces by painting proteins on them. Over the years, the palette of *fusion protein paints* has grown, evolving from *ex vivo* to *in vivo* paints.

The first fusion protein paints were GPI-modified derivatives of transmembrane proteins, produced via chimeric gene expression. GPI proteins are "paintable" by virtue of their amphiphilicity, since they (i) remain soluble as pseudo-micelles after depletion of solubilizing detergents, and can thus be added back to cells without lysing them; and (ii) spontaneously incorporate into cell membranes. Tykocinski et al. [116] and Caras et al. [117] originally identified the GPI modification signal sequence within decay-accelerating factor (DAF; CD55), a natural GPI-anchored protein. The next step was to convey GPI anchors to proteins of interest by appending the DAF signal sequence to their carboxyl termini. The juxtacrine *trans*-signaling functionality of artificial GPI proteins, once appended to cell surfaces, was amply validated [118–123], creating the means to generate protein paints at will.

GPI protein paints, however, were hard to scale up, given the need to purify them from complex membrane lysates of transfectants. We therefore developed an alternative class of protein paints that can be produced as soluble recombinant proteins and are thus more scalable. These consisted of two-component protein complexes in which protein • $Fc_{\gamma1}$ derivatives are conjugated to chemically palmitated protein A, with both components produced efficiently as soluble recombinant proteins.

A series of immunobiological studies validated the functionality of both GPI protein paints and protein • $Fc_{\gamma1}$:palmitated-protein A conjugate paints. Painting of peptide antigen-loaded MHC class I • GPI complexes sensitized cellular targets to cytotoxic T-cell-mediated destruction, with the ability to titrate optimal antigenic density [119]. Moving from first to second APC-to-T juxtacrine cell signal, costimulator • GPI derivatives could be used to engineer tumor cell surfaces and thereby enhance their immunogenicity [120,124,125]. Significantly, the ability to finely titrate costimulator • $Fc_{\gamma1}$:palmitated-protein A conjugates onto APC surfaces via protein painting allowed us to uncover costimulator receptor activation thresholds [126]. Moreover, it enabled the generation of effective anti-cancer immunity by painting tumor cells *in situ* with a combination of four costimulator • $Fc_{\gamma1}$:palmitated-protein A conjugates [127]. Coinhibitors, such as FasL, could also be painted, which allowed for studies of how responding T cells quantitatively integrate opposing (costimulator and coinhibitor) juxtacrine signals concurrently delivered by APC [111].

1.7 TRANS SIGNAL CONVERSION

This early *ex vivo* cell surface engineering using membrane-incorporable protein paints beckoned the development of next-generation fusion protein paints that could be injected systemically and would home to their cellular targets *in vivo*. We referred to these generically as *signal converter proteins*, or *SCP*, since, once arriving at the target cell surface, they convert juxtacrine signals. We proceeded to develop a series of distinct types of multifunction signal converters, featuring a spectrum of signal-converting functionalities. Each new SCP seems to offer the unexpected, whether by its unusual potency, multiplicity of cellular targets, or mechanisms of action not anticipated from the known functions of the chimeric protein's components. Thus, SCP function can be framed in terms of emergence, beginning with the emergent states of the SCP-painted cell membrane and extending to the cellular targets of the converted juxtacrine signal and the cell network in which they are together embedded. Coupling a diverse and inherently multifunctional SCP repertoire to multicellular networks, one gets a tapestry of emergent cell states, matrices of cellular interaction, and ultimately, therapeutic outputs.

The following sections showcase a series of signal converter paradigms, which together highlight the robustness of this cell surface engineering modality. As the library of SCP has continued to grow, the original simple paradigm—conversion of juxtacrine signals between two interacting cells—has given way to more intricate ones. Converted juxtacrine signals can be redirected to third-party cells, and even to receptors on the very same cell, creating auto-signaling loops at the cell surface. A given SCP can in fact operate in more than one mode, perhaps converting signals between two cells in one context and redirecting them in another. In some instances, this functional multidimensionality allows for markedly different therapeutic applications for the same SCP.

Trans signal converter proteins (TSCP) were developed first (Figure 1.1A). The concept was to combine within a single fusion protein homing and effector elements, together geared toward converting a juxtacrine signal between two cells. Judicious selection of the homing element can elicit yet additional functions that reinforce the effector element. Our paradigmatic artificial veto cell-generating TSCP, CTLA-4 • FasL [128], illustrated this point well. The CTLA-4 (CD152) component serves a dual purpose, targeting the protein to APC surfaces bearing its counter receptors [B7-1 (CD80) and B7-2 (CD86)], and in the process, passively blocking their costimulator function. The FasL (CD95L) component sends a pro-apoptotic *trans* signal to its cognate Fas receptor (CD95), upregulated on activated T cells. Thus, this TSCP, with a costimulator receptor (CoSR) • coinhibitor (CoI) configuration, exchanges an active inhibitory (apoptosis-inducing) signal for a T-cell-activating costimulator one which is being passively blocked.

CTLA-4 • FasL's functional potential may go even further. First, given that increased B7 triggering of the CD28 costimulator receptor on T cells interferes with Fas-mediated apoptosis [129], CTLA-4 • FasL, in mimicking the B7 costimulator blockade of CTLA-4 • Ig [114,130], can potentiate its own pro-apoptotic, FasL-mediated activity. Second, by anchoring FasL to cell surfaces, CTLA-4 • FasL leverages the higher efficacy of surface-associated FasL [131,132]. Third, as a two-faced, Janus-like entity, CTLA-4 • FasL can be viewed from the opposite perspective, as anchoring CTLA-4 to activated T-cell surfaces. This creates the potential for two interesting CTLA-4-driven phenomena: reverse juxtacrine signaling, back through B7 ligand [133] and *trans*-endocytosis of B7 ligands away from APC surfaces [134,135].

CTLA-4 • FasL exhibits striking potency *in vitro*, when compared to CTLA-4 • Ig or soluble FasL, alone or in combination [128,136–138]. Interestingly, CTLA-4 • FasL not only deletes activated T cells but also induces anergic proliferative hyporesponsiveness in them [136]. CTLA-4 • FasL-mediated inhibition of allogeneic responses was documented in the *in vivo* context as well, with evidence for antigen-induced cell death of alloreactive cells [137]. The demonstration of CTLA-4 • FasL *in vivo* efficacy has also been extended to animal models of autoimmune and alloimmune pathogenesis, including autoimmune diabetes

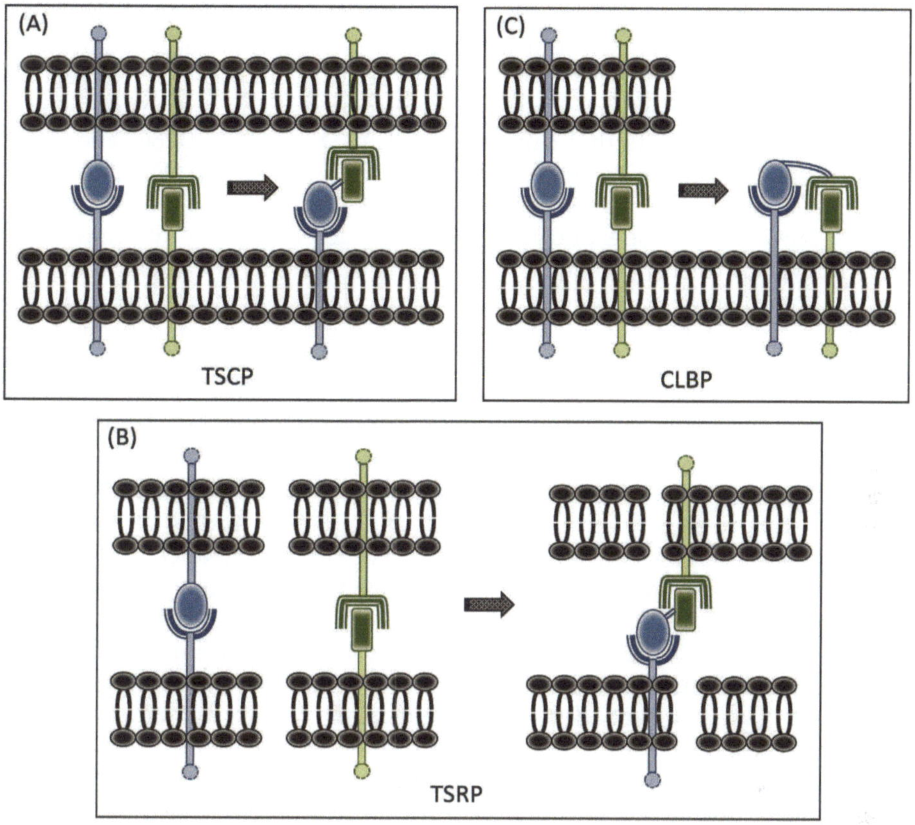

FIGURE 1.1 Three signal conversion paradigms. Each schematic displays two juxtacrine signaling transmembrane protein pairs to the left of the arrow and the respective SCP mode of action to its right. (A) *Trans* signal conversion, with the TSCP bridging two membrane protein ligand–receptor pairs that juxtacrine signal in parallel between the same two interactive cells; the *trans* signal converter protein (TSCP) blocks the outgoing signal of one signaling axis and simultaneously sends a "converted" juxtacrine signal through the receptor of the second signaling axis on the opposing cell. (B) *Trans* signal redirection, where the two membrane protein ligand–receptor pairs reside on different interacting cells; the *trans* signal redirecting protein (TSRP) blocks the outgoing signal of one signaling axis and simultaneously redirects a converted juxtacrine signal through the receptor of the second signaling axis on a "third-party" cell. (C) *Cis* loop-back autocrine stimulation, with the *cis* loop-back protein (CLBP) bridging two membrane protein receptors on the same cell; the CLBP blocks the outgoing signal of one signaling axis and simultaneously redirects a converted juxtacrine signal, in autocrine fashion, through a neighboring receptor on the same cell.

[139] and cardiac and corneal transplantation [140,141].

An important mechanistic insight into CTLA-4 • FasL function was the recognition of its ability to prevent the upregulation of anti-apoptotic c-FLIP which typically accompanies T-cell activation [142], serving to coordinately potentiate the fusion protein's own FasL pro-apoptotic signal [138]. Interestingly, this upregulation of an anti-apoptotic cytoplasmic protein, along with early post-activation apoptosis, does not occur when the T cells are simply treated with the two fusion protein components, alone or in combination. Thus, this functional activity represents an emergent property that is dependent on the chimeric structure of the perturbing agent and illustrates yet another nexus between SCP and the concept of emergence.

For all the SCP described herein, chimerization provides functional dividends that cannot be recapitulated by simply delivering the fused components as a combination of separate molecular entities. New mechanisms and functions are emergent within fusion proteins. The growing list of chimerization dividends for SCP include abilities to: (i) elicit new, sometimes unexpected, functionalities (as exemplified by CTLA-4 • FasL effect on c-FLIP); (ii) co-localize targeted molecules on cell membranes, sometimes generating macromolecular assemblies associated with enhanced function; (iii) deliver ligands to pre-selected cell types and organ sites *in vivo*, thereby concentrating therapeutic agent at the right location (as exemplified by Fn14 (fibroblast growth factor-inducible 14 kDa protein) • TRAIL (TNF-related apoptosis-inducing ligand) targeting TRAIL to inflamed TWEAK$^+$ (TNF-related weak inducer of apoptosis) endothelium, *vide infra*); (iv) tether ligands to membranes, thereby eliciting higher order ligand function (as exemplified by the various FasL-containing SCP that yield membrane-anchored FasL); and (v) enhance functional avidity by enforcing multimerization (discussed later in this chapter).

1.8 REDIRECTING JUXTACRINE SIGNALS

TSRP were a logical next step, as we created a new class of signal converters with the capacity to redirect converted signals to third-party cells, outside of the original cellular dyad (Figure 1.1B). In contrast to the situation with TSCP, where the cell being modulated bears receptors for both the blocked and converted signals, for TSRP, these two receptors reside on different cells.

We developed Fn14 • TRAIL as a paradigmatic TSRP, with inhibition of inflammation in mind. This fusion protein bridges two well-studied TNF superfamily signaling axes, TWEAK-to-Fn14 (CD266) and TRAIL (CD253)-to-TRAIL-R (TRAIL-R1, CD261; TRAIL-R2, CD262; TRAIL-R3, CD263; TRAIL-R4, CD264), which are linked to inflammation and immunoregulation, respectively [143,144]. TWEAK, a type II transmembrane protein, is expressed on a range of immune and nonimmune cell types [145] and is upregulated on activated monocyte/macrophages. TWEAK's counter receptor, Fn14 (fibroblast growth factor-inducible 14 kDa protein), is also widely expressed on many cell types, with the notable exceptions of T and B cells, and is overexpressed in tissue injury. Membrane-anchored TWEAK, functioning as a juxtacrine signal [146], contributes to autoimmune inflammatory disorders not only by promoting pro-inflammatory cytokine production but also through its pro-angiogenic effect and ability to increase the permeability of the neurovascular unit in the context of autoimmune encephalomyelitis [147,148]. Indeed, antibody-mediated blocking of this juxtacrine signaling axis ameliorates autoimmune encephalomyelitis manifestations in rodent models [149]. TRAIL (CD253) too has been linked to autoimmunity, with the documented capacity of this pro-apoptotic type II juxtacrine signaling protein to inhibit disease manifestations in autoimmune animal models [112,150–154].

By connecting the TWEAK:Fn14 and TRAIL:TRAIL-R signaling axes via chimeric Fn14 • TRAIL, we were introducing a protein that could at once bind to surface TWEAK at inflammatory sites (via its Fn14 component) and trigger inhibitory, pro-apoptotic TRAIL-R receptors on third-party (Fn14-negative) activated T cells in these locations (via its TRAIL ligand component) [155]. Thus, as a juxtacrine signal converter, Fn14 • TRAIL replaces a pro-inflammatory signal with one that knocks out bystander-activated T cells. There are other potential dividends of this fusion protein: (i) Fn14 • TRAIL in effect tethers TRAIL to the cell surface, leveraging the fact that surface-associated TRAIL is more functional than its soluble counterpart [156]; and (ii) as a TWEAK blocker, Fn14 • TRAIL likely interferes with both innate and adaptive immunity and TWEAK-driven pro-inflammatory activities.

This constellation of pleiotropic Fn14 • TRAIL functions all point to autoimmunity as an ideal therapeutic application. This proved to be the case, as we demonstrated that Fn14 • TRAIL prevents auto-antigen-induced autoimmune encephalomyelitis in mice, with an efficacy that is substantially greater than that for either of its component parts (as Fn14 • Ig and soluble TRAIL derivatives) when used in isolation or in combination [155]. This seminal study further suggested that Fn14 • TRAIL, through TWEAK blockade, can reverse the enhanced permeability of the blood–brain barrier in diseased animals, providing yet another explanation for the high therapeutic efficacy of this fusion protein in the encephalomyelitis model.

Fn14 • TRAIL, as well as the other SCP described in this chapter, are type I • II fusions, that is, a type I membrane protein at the amino terminus is linked to a type II membrane protein at the carboxyl terminus. This type I • II structural configuration results in a fusion protein that retains the extra-cytoplasmic ends of the respective constituent proteins, without constraining them. The added advantage is that one can draw upon members of both the immunoglobulin (predominantly type I) and TNF (predominantly type II) superfamilies, which are broadly represented in the cell membrane protein repertoire.

Fn14 • TRAIL also features the interesting kind of stoichiometries that can fortuitously emerge when one couples trimeric TNF superfamily members to nontrimeric type I membrane proteins, which favor higher order multimer formation. Our three-dimensional modeling suggested that trimerization, enforced by trimeric TWEAK on the otherwise monomeric Fn14 component, yields a neo-Fn14 trimer that can in turn stabilize the active, trimeric state of the fused TRAIL component [155]. Similarly, our CTLA-4 • FasL modeling suggested that interplay of dimeric CTLA-4 with trimeric FasL allows for a hexameric configuration, representing at once a dimer of FasL trimers and a trimer of CTLA-4 dimers [128]. This is functionally significant, since hexamerization empowers FasL and perhaps other TNF family members [131,157].

1.9 CREATING AUTO-SIGNALING LOOPS

For TSCP and TSRP, the receptor for the converted signal resides on second- and third-party cells, respectively. However, the receptor for the converted signal can just as well reside on the same cell that bears the surface signaling ligand that is being converted. This idea led us to yet another SCP category, designed to redirect converted signals back to the same cell, thereby generating auto-signaling loops. By judiciously choosing the component parts of such *CLBP* (Figure 1.1C), one can take advantage of forward and reverse signaling features of the fusion protein's surface molecular targets to amplify the converted signal. Thus, the CLBP is being configured to double as both signal converter and dual co-signaling agent.

Autocrine signaling has traditionally been framed in terms of secreted soluble factors modulating the very cell that produced them. This type of signaling calls for clustering of like cells in order to achieve higher local concentrations of the secreted signal, and engenders a community effect that encourages identical cells to respond coordinately as a group. This contrasts with auto-signaling mediated by surface-associated molecules, where the demand for cellular congregation is mitigated, as each cell carries its own triggers in immediate proximity to the cognate receptors. Thus, CLBP, by co-opting juxtacrine signals and converting them into autocrine ones, create the opportunity for community independence. The cell can become a lone ranger.

Our first paradigmatic CLBP, CD40 • FasL, was geared toward engendering autoinhibition of activated effector T cells [158]. The CD40 component binds to CD40L (CD154) neo-expressed on activated T cells, while the FasL (CD95L, CD178) component triggers the pro-apoptotic Fas receptor on these cells. The FasL: Fas signaling axis was an attractive choice for this fusion protein mediated, auto-apoptosis induction strategy, since data had already pointed to physiologic auto-apoptotic "suicide" signaling [159–162] between the Fas and FasL that are concurrently expressed side by side on activated T cells during an immune response [159,163,164]. CD40 • FasL artificially recapitulates this scenario, by converting the outbound immune-activating signal emanating from CD40L into a pro-apoptotic FasL signal redirected back to a neighboring Fas receptor on the same cell.

The logic for pairing these particular proteins was several fold. First, cell selectivity was achieved by choosing CLBP components whose counter receptors (CD40L and Fas) are neo-expressed, and acquire full functionality, following T-cell activation [165,166]. Second, the two chosen signaling pathways had a fortuitous connection, in that reverse signaling through CD40L sensitizes activated T cells to Fas-mediated apoptosis [167]. Thus, CD40 • FasL can function as a dual-triggering agent, driving a CD40L reverse signal that potentiates the Fas-mediated pro-apoptotic signal. Third, CD40 • FasL tethers FasL to cell surfaces, thereby invoking surface-associated FasL's higher efficacy [131,132]. This collection of mechanistic features exemplifies the kind of elegant design that is possible in pairing SCP components, where multiple mechanisms can be leveraged to achieve functional synergies and signal reinforcement.

We experimentally validated CD40 • FasL's *cis* autoinhibitory "loop-back" mechanism and immuomodulatory activity [158]. Evidence for the autocrine mechanism came from both soft agar clone formation and cell dilution analysis, showing that CD40 • FasL inhibition does not require intercellular contact. Of note, the idea that *cis* loop-back signaling can be artificially elicited using fusion proteins dove-tailed with our prior finding that painting of CoS • $Fc_{\gamma 1}$: palmitated-protein A conjugates can yield auto-activating T cells [168].

1.10 SCP THERAPEUTIC FLEXIBILITY

While the paradigmatic TSCP, TSRP, and CLBP fusion proteins were designed as distinct SCP categories, it turns out that the dividing lines are in reality blurred, with considerable functional overlap. That is, a fusion protein developed for redirecting juxtacrine signals to third-party cells can in a different cellular context trigger auto-signaling. In turn, this has served to extend the therapeutic reach of individual SCP.

A case example is CTLA-4 • FasL. It was first designed to be an immunoregulatory TSCP for the treatment of autoimmunity and transplantation diseases. In this context, it introduces an APC-to-T cell coinhibition signal (FasL/CD95L) in place of a costimulation one (B7-1/CD80; B7-2/CD86). However, we demonstrated that this very same fusion protein can function as a CLBP in instances where its cognate counter receptors, CD80/86 and CD95L, are co-expressed on the same cells, e.g., on malignant B cells. Here, CTLA-4 • FasL elicits auto-apoptosis [169]. A variation on this theme is seen with our other FasL-containing

SCP, CD40 • FasL. Developed originally for creating auto-apoptotic loops in activated autoimmune and alloimmune T cells, CD40 • FasL was subsequently shown to have utility for eradicating transformed T cells expressing the cognate counter receptors, CD40L (CD154) and Fas (CD95). In this case, we showed that it draws mechanistically on the same CLBP mode of activity [169]. Of note, despite the focus on its auto-apoptotic activity, there is no reason why CD40 • FasL can not in parallel bridge and drive Fas receptor-dependent apoptosis in neighboring cells.

Fn14 • TRAIL is also multifaceted. Originally conceived of as a TSRP for interfering with a pro-inflammatory intercellular signal (TWEAK) and redirecting a converted inhibitory signal (TRAIL) toward autoimmune T cells, Fn14 • TRAIL has since emerged as a cancer therapeutic. Neo-expression of this fusion protein's counter receptors (TWEAK and TRAIL-R) on various solid tumor types, including those of breast, liver, pancreas, brain, and colon [170–174], has made it possible to reinvent Fn14 • TRAIL as a CLBP to elicit tumor cell auto-apoptosis, as we have reported for hepatocellular carcinoma [175].

These examples illustrate how cancer cells, like activated T cells, can be prime targets for individual multifaceted SCP, and how therapeutic agents for autoimmunity can be reinvented as anticancer therapeutics. While we have keyed in on the auto-apoptotic mechanism in each instance, other supportive antitumor mechanisms come into play. For instance, as a TWEAK blocker, Fn14 • TRAIL can interfere with various TWEAK functions linked to tumor progression, such as promotion of proliferation, invasion, angiogenesis, and inflammation. Fn14 • TRAIL can also be cast as a next-generation soluble TRAIL derivative, more effective than the soluble TRAIL which has been under active development for cancer therapy [176]. The fusion protein's higher efficacy stems from its cell-associating and multimerization properties. This cancer therapeutic paradigm—a fusion protein inducing tumor cell auto-apoptosis—is now being extended through the use of tumor cell-directed antibody TRAIL fusion proteins [177].

We introduced the idea of an "SCP function bar code" to capture the multiple functions of individual signal-converting fusion proteins [84]. The mechanistic profile of each SCP, coded in this fashion, evolves over time, as new functional capabilities surface. SCP functional richness stems in part from that of membrane proteins themselves, an example being Escher-like surface receptor–ligand functional duality. That is, many surface proteins, regardless of whether they were initially labeled as "receptors" or "ligands," turn out to function as both, sending signals both inward and outward in signal-receiving and signal-sending modes, respectively. Molecular pairs associated with bidirectional signaling in the immune system include CD40L (CD154):CD40 (TNFRSF5), OX40L:OX40 (CD134), GITRL (TNFSF18):GITR (CD357; TNFRSF18), CD80/86:CTLA-4/CD28, and CD70:CD27 (TNFRSF7). The continuously growing list of "receiver" proteins that can send (forward signaling) and "sender" proteins that can receive (reverse signaling) seems to make bidirectional signaling a rule rather than an exception. The implication for SCP is that any given one can simultaneously function as an intercellular signal blocker, a reverse signaling ligand, and a competitive inhibitor of the receptor function.

1.11 THE CELL MEMBRANE FRONTIER

Introducing multifunctional SCP into multicellular networks *in vivo* creates a rich tableau of possibilities, some unexpected. Complexity in this setting emerges from a variety of factors, one of which is the promiscuous expression of many cognate counter receptors targeted by SCP. The dual binding capacity inhering in SCP contributes to target selectivity, but this is not absolute, with each SCP often able to touch more than one cellular network. This means that individual SCP can in principle impact more than one critical network node (for instance, via CLBP-mediated auto-apoptotic cell deletion) or intercellular link (for instance, via TSCP- or TSRP-mediated juxtacrine signal conversion) within one or more cellular networks. Further complexity arises from the constant flux in the *in vivo* cellular landscape, at the individual cell and cell population levels. Cell surface protein displays, along with the conformations and topologies of their component proteins, are dynamic, and SCP-elicited cell membrane emergence is layered onto this background flux.

Future SCP design will be guided by an ever-expanding knowledge base pertaining to cell surface protein repertoires, membrane protein dynamics, and functional cellular webs. Interesting leads will come from the growing compendium of single function antibody therapeutics [177]. A fully annotated *cell surface protein interactome (CSPI)*, mapping surface protein interaction networks and elaborating on those of particular relevance to juxtacrine communication, would provide a useful framework for charting SCP possibilities. Such a database could readily build on ongoing initiatives to catalog the full human interactome and disease-associated protein–protein interactions [178,179], as well as more focused exercises, such as keying in on the independently folding modular domains of proteins and mapping their interactive potential in a "fragmentome" [180]. The very definition of protein–protein interaction will need to be revisited, as confounding concepts such as "protein stickiness" [181] and functional "promiscuity and evolvability" enabled by protein conformational dynamism [182] gain traction. The SCP designer will clearly have to navigate a large matrix of juxtacrine-associated protein pairs in selecting those that are most targetable. The SCP library of the future will not only more fully embrace this large set of protein pairings but will likely contain SCP classes that go well beyond those featured in this chapter.

Fundamental advances in membrane biology will translate into entirely new classes of membrane-targeted therapeutics. Membrane protein dynamics and membrane microdomain physiology will provide the substratum, going beyond cell surface protein expression profiling *per se*. In the case of SCP, one can envision targeting selected active conformations of membrane proteins or modulating the trafficking of selected proteins into and out of membrane microdomains. The concept of induced "spatial mutation," purposefully changing the arrangement of membrane proteins on a cell surface, sets the stage for such therapeutic membrane modulation [183]. The tools for protein transfer will evolve in parallel. The possibilities include protein painting strategies that can introduce proteins into selected membrane subcompartments, or ones that can efficiently paint cells on the move within the blood circulation or other tissue compartments. The use of targeted liposomes for *in vivo* painting of circulating cells points to this type of next-generation cell surface engineering [184].

Ultimately, the cell membrane emergence paradigm may itself inform membrane-directed therapeutics design. Emergent membrane protein fingerprints, juxtacrine interactions, and cellular network states could eventually all become druggable. Deciphering emergent phenomena within the cell membrane could actually pave the way for simpler, smartly targeted interventions. By grappling with complexity, one can sometimes arrive at simplicity. There is now the prospect of bringing systems membrane biology together with therapeutics design, SCP among them, in compelling new ways.

References

[1] Gorter E, Grendel F. On bimolecular layers of lipoids on the chromocytes of the blood. J Exp Med 1925;41 (4):439–43.

[2] Tanford C. Ben Franklin stilled the waves—an informal history of pouring oil on water with reflections on the ups and downs of scientific life in general. New York, NY: Oxford University Press; 2004.

[3] Danielli JF. Some properties of lipoid films in relation to the structure of the plasma membrane. J Cell Comp Physiol 1936;7(3):393–408.

[4] Singer SJ, Nicolson GL. The fluid mosaic model of the structure of cell membranes. Science (New York, NY) 1972;175(4023):720–31.

[5] Palade GE. A study of fixation for electron microscopy. J Exp Med 1952;95(3):285–98.

[6] Robertson JD. The molecular structure and contact relationships of cell membranes. Prog Biophys Mol Biol 1960;10:343–418.

[7] Robertson JD. The ultrastructure of cell membranes and their derivatives. Biochem Soc Symp 1959;16:3–43.

[8] Frye LD, Edidin M. The rapid intermixing of cell surface antigens after formation of mouse–human heterokaryons. J Cell Sci 1970;7(2):319–35.

[9] Hladky SB, Haydon DA. Ion transfer across lipid membranes in the presence of gramicidin A. I: Studies of the unit conductance channel. Biochim Biophys Acta 1972;274(2):294–312.

[10] Mueller P, Rudin DO, Tien HT, Wescott WC. Reconstitution of cell membrane structure *in vitro* and its transformation into an excitable system. Nature 1962;194:979–80.

[11] Huang C. Studies on phosphatidylcholine vesicles. Formation and physical characteristics. Biochemistry 1969;8(1):344–52.

[12] Montal M, Mueller P. Formation of bimolecular membranes from lipid monolayers and a study of their electrical properties. Proc Natl Acad Sci USA 1972;69 (12):3561–6.

[13] Schindler H. Exchange and interactions between lipid layers at the surface of a liposome solution. Biochim Biophys Acta 1979;555(2):316–36.

[14] Chan YH, Boxer SG. Model membrane systems and their applications. Curr Opin Chem Biol 2007;11 (6):581–7.

[15] Percec V, Wilson DA, Leowanawat P, Wilson CJ, Hughes AD, Kaucher MS, et al. Self-assembly of Janus dendrimers into uniform dendrimersomes and other complex architectures. Science 2010;328(5981): 1009–14.

[16] Tomita M, Marchesi VT. Amino-acid sequence and oligosaccharide attachment sites of human erythrocyte glycophorin. Proc Natl Acad Sci USA 1975;72(8): 2964–8.

[17] Henderson R, Unwin PN. Three-dimensional model of purple membrane obtained by electron microscopy. Nature 1975;257(5521):28–32.

[18] Deisenhofer J, Epp O, Miki K, Huber R, Michel H. Structure of the protein subunits in the photosynthetic reaction centre of *Rhodopseudomonas viridis* at 3A resolution. Nature 1985;318(6047):618–24.

[19] Roux B, MacKinnon R. The cavity and pore helices in the KcsA K^+ channel: electrostatic stabilization of monovalent cations. Science (New York, NY) 1999;285 (5424):100–2.

[20] Overton E. On the osmotic properties of living plants and animals. Vierteljahrsschrift der Naturforschenden Gesellschaft in Zürich 1895;40:159–201.

[21] Brunton LL, Hayes JS, Mayer SE. Functional compartmentation of cyclic AMP and protein kinase in heart. Adv Cyclic Nucleotide Res 1981;14:391–7.

[22] Billcliff PG, Rollason R, Prior I, Owen DM, Gaus K, Banting G. CD317/tetherin is an organiser of membrane microdomains. J Cell Sci 2013;126(Pt 7):1553–64.

[23] Lingwood D, Simons K. Lipid rafts as a membrane-organizing principle. Science (New York, NY) 2010;327(5961):46–50.

[24] Simons K, Sampaio JL. Membrane organization and lipid rafts. Cold Spring Harb Perspect Biol 2011;3(10): a004697.

[25] Zuidscherwoude M, de Winde CM, Cambi A, van Spriel AB. Microdomains in the membrane landscape shape antigen-presenting cell function. J Leukoc Biol 2014;95(2):251–63.

[26] Simons K, Ikonen E. Functional rafts in cell membranes. Nature 1997;387(6633):569–72.

[27] Jacobson K, Sheets ED, Simson R. Revisiting the fluid mosaic model of membranes. Science (New York, NY) 1995;268(5216):1441–2.

[28] Boucheix C, Rubinstein E. Tetraspanins. Cell Mol Life Sci 2001;58(9):1189–205.

[29] Hemler ME. Tetraspanin functions and associated microdomains. Nat Rev Mol Cell Biol 2005;6(10): 801–11.

[30] Yanez-Mo M, Barreiro O, Gordon-Alonso M, Sala-Valdes M, Sanchez-Madrid F. Tetraspanin-enriched microdomains: a functional unit in cell plasma membranes. Trends Cell Biol 2009;19(9):434–46.

[31] Yang X, Claas C, Kraeft SK, Chen LB, Wang Z, Kreidberg JA, et al. Palmitoylation of tetraspanin proteins: modulation of CD151 lateral interactions, subcellular distribution, and integrin-dependent cell morphology. Mol Biol Cell 2002;13(3):767–81.

[32] Espenel C, Margeat E, Dosset P, Arduise C, Le Grimellec C, Royer CA, et al. Single-molecule analysis

of CD9 dynamics and partitioning reveals multiple modes of interaction in the tetraspanin web. J Cell Biol 2008;182(4):765–76.
[33] Claas C, Stipp CS, Hemler ME. Evaluation of prototype transmembrane 4 superfamily protein complexes and their relation to lipid rafts. J Biol Chem 2001;276 (11):7974–84.
[34] Pralle A, Keller P, Florin EL, Simons K, Horber JK. Sphingolipid-cholesterol rafts diffuse as small entities in the plasma membrane of mammalian cells. J Cell Biol 2000;148(5):997–1008.
[35] Zheng YZ, Foster LJ. Contributions of quantitative proteomics to understanding membrane microdomains. J Lipid Res 2009;50(10):1976–85.
[36] Neves SR, Iyengar R. Models of spatially restricted biochemical reaction systems. J Biol Chem 2009;284 (9):5445–9.
[37] Toomre D, Bewersdorf J. A new wave of cellular imaging. Annu Rev Cell Dev Biol 2010;26:285–314.
[38] Lidke DS, Lidke KA. Advances in high-resolution imaging techniques for three-dimensional imaging of cellular structures. J Cell Sci 2012;125(Pt 11):2571–80.
[39] Lidke DS, Wilson BS. Caught in the act: quantifying protein behaviour in living cells. Trends Cell Biol 2009;19(11):566–74.
[40] Kusumi A, Fujiwara TK, Chadda R, Xie M, Tsunoyama TA, Kalay Z, et al. Dynamic organizing principles of the plasma membrane that regulate signal transduction: commemorating the fortieth anniversary of Singer and Nicolson's fluid-mosaic model. Annu Rev Cell Dev Biol 2012;28:215–50.
[41] Sharma P, Varma R, Sarasij RC, Ira, Gousset K, Krishnamoorthy G, et al. Nanoscale organization of multiple GPI-anchored proteins in living cell membranes. Cell 2004;116(4):577–89.
[42] Groves JT, Kuriyan J. Molecular mechanisms in signal transduction at the membrane. Nat Struct Mol Biol 2010;17(6):659–65.
[43] Sudhof TC, Rothman JE. Membrane fusion: grappling with SNARE and SM proteins. Science (New York, NY) 2009;323(5913):474–7.
[44] Rizo J, Sudhof TC. The membrane fusion enigma: SNAREs, Sec1/Munc18 proteins, and their accomplices—guilty as charged? Annu Rev Cell Dev Biol 2012;28:279–308.
[45] Manglik A, Kobilka B. The role of protein dynamics in GPCR function: insights from the betaAR and rhodopsin. Curr Opin Cell Biol 2014;27C:136–43.
[46] White SH. Membrane protein insertion: the biology-physics nexus. J Gen Physiol 2007;129(5):363–9.
[47] Gofman Y, Haliloglu T, Ben-Tal N. Monte Carlo simulations of peptide–membrane interactions with the MCPep web server. Nucleic Acids Res 2012;40(Web Server issue):W358–63.
[48] Mueller V, Ringemann C, Honigmann A, Schwarzmann G, Medda R, Leutenegger M, et al. STED nanoscopy reveals molecular details of cholesterol- and cytoskeleton-modulated lipid interactions in living cells. Biophys J 2011;101(7):1651–60.
[49] Javanainen M, Hammaren H, Monticelli L, Jeon JH, Miettinen MS, Martinez-Seara H, et al. Anomalous and normal diffusion of proteins and lipids in crowded lipid membranes. Faraday Discuss 2013;161: 397–417 [discussion 419–59].
[50] Veatch SL, Chiang EN, Sengupta P, Holowka DA, Baird BA. Quantitative nanoscale analysis of IgE-FcepsilonRI clustering and coupling to early signaling proteins. J Phys Chem B 2012;116(23):6923–35.
[51] Aponte-Santamaria C, Briones R, Schenk AD, Walz T, de Groot BL. Molecular driving forces defining lipid positions around aquaporin-0. Proc Natl Acad Sci USA 2012;109(25):9887–92.
[52] Gustot A, Smriti, Ruysschaert JM, McHaourab H, Govaerts C. Lipid composition regulates the orientation of transmembrane helices in HorA, an ABC multidrug transporter. J Biol Chem 2010;285(19): 14144–51.
[53] Cybulski LE, Martin M, Mansilla MC, Fernandez A, de Mendoza D. Membrane thickness cue for cold sensing in a bacterium. Curr Biol 2010;20(17): 1539–44.
[54] Lomize AL, Pogozheva ID, Lomize MA, Mosberg HI. Positioning of proteins in membranes: a computational approach. Protein Sci 2006;15(6):1318–33.
[55] Herzig Y, Sharpe HJ, Elbaz Y, Munro S, Schuldiner M. A systematic approach to pair secretory cargo receptors with their cargo suggests a mechanism for cargo selection by Erv14. PLoS Biol 2012;10(5):e1001329.
[56] Parton DL, Klingelhoefer JW, Sansom MS. Aggregation of model membrane proteins, modulated by hydrophobic mismatch, membrane curvature, and protein class. Biophys J 2011;101(3):691–9.
[57] Lee HJ, Peterson EL, Phillips R, Klug WS, Wiggins PA. Membrane shape as a reporter for applied forces. Proc Natl Acad Sci USA 2008;105(49):19253–7.
[58] Ford MG, Mills IG, Peter BJ, Vallis Y, Praefcke GJ, Evans PR, et al. Curvature of clathrin-coated pits driven by epsin. Nature 2002;419(6905):361–6.
[59] Chen CT, Hehnly H, Doxsey SJ. Orchestrating vesicle transport, ESCRTs and kinase surveillance during abscission. Nat Rev Mol Cell Biol 2012;13(8):483–8.
[60] Neto H, Gould GW. The regulation of abscission by multi-protein complexes. J Cell Sci 2011;124(Pt 19): 3199–207.

[61] Parton RG, del Pozo MA. Caveolae as plasma membrane sensors, protectors and organizers. Nat Rev Mol Cell Biol 2013;14(2):98–112.
[62] Wollert T, Wunder C, Lippincott-Schwartz J, Hurley JH. Membrane scission by the ESCRT-III complex. Nature 2009;458(7235):172–7.
[63] Sieber JJ, Willig KI, Kutzner C, Gerding-Reimers C, Harke B, Donnert G, et al. Anatomy and dynamics of a supramolecular membrane protein cluster. Science (New York, NY) 2007;317(5841):1072–6.
[64] Chen Y, Veracini L, Benistant C, Jacobson K. The transmembrane protein CBP plays a role in transiently anchoring small clusters of Thy-1, a GPI-anchored protein, to the cytoskeleton. J Cell Sci 2009;122(Pt 21): 3966–72.
[65] Lenn T, Leake MC. Experimental approaches for addressing fundamental biological questions in living, functioning cells with single molecule precision. Open Biol 2012;2(6):120090.
[66] Schiess R, Mueller LN, Schmidt A, Mueller M, Wollscheid B, Aebersold R. Analysis of cell surface proteome changes via label-free, quantitative mass spectrometry. Mol Cell Proteomics 2009;8(4):624–38.
[67] Sun Z, Chen R, Cheng K, Liu H, Qin H, Ye M, et al. A new method for quantitative analysis of cell surface glycoproteome. Proteomics 2012;12(22):3328–37.
[68] Roncagalli R, Hauri S, Fiore F, Liang Y, Chen Z, Sansoni A, et al. Quantitative proteomics analysis of signalosome dynamics in primary T cells identifies the surface receptor CD6 as a Lat adaptor-independent TCR signaling hub. Nat Immunol 2014.
[69] Blagovic K, Gong ES, Milano DF, Natividad RJ, Asthagiri AR. Engineering cell–cell signaling. Curr Opin Biotechnol 2013;24(5):940–7.
[70] Anklesaria P, Teixido J, Laiho M, Pierce JH, Greenberger JS, Massague J. Cell–cell adhesion mediated by binding of membrane-anchored transforming growth factor alpha to epidermal growth factor receptors promotes cell proliferation. Proc Natl Acad Sci USA 1990;87(9):3289–93.
[71] Massague J. Transforming growth factor-alpha. A model for membrane-anchored growth factors. J Biol Chem 1990;265(35):21393–6.
[72] Artavanis-Tsakonas S, Muskavitch MA. Notch: the past, the present, and the future. Curr Top Dev Biol 2010;92:1–29.
[73] Zimmerman GA, Lorant DE, McIntyre TM, Prescott SM. Juxtacrine intercellular signaling: another way to do it. Am J Respir Cell Mol Biol 1993;9(6):573–7.
[74] Owen MR, Sherratt JA, Wearing HJ. Lateral induction by juxtacrine signaling is a new mechanism for pattern formation. Dev Biol 2000;217(1):54–61.
[75] Owen MR, Sherratt JA, Myers SR. How far can a juxtacrine signal travel? Proc Biol Sci 1999;266 (1419):579–85.
[76] O'Dea RD, King JR. The isolation of spatial patterning modes in a mathematical model of juxtacrine cell signaling. Math Med Biol 2013;30(2):95–113.
[77] Shaya O, Sprinzak D. From Notch signaling to fine-grained patterning: modeling meets experiments. Curr Opin Genet Dev 2011;21(6):732–9.
[78] Collier JR, Monk NA, Maini PK, Lewis JH. Pattern formation by lateral inhibition with feedback: a mathematical model of delta-notch intercellular signaling. J Theor Biol 1996;183(4):429–46.
[79] Walker DC, Georgopoulos NT, Southgate J. From pathway to population—a multiscale model of juxtacrine EGFR-MAPK signalling. BMC Syst Biol 2008;2: 102 0509-2-102.
[80] Auslander S, Fussenegger M. From gene switches to mammalian designer cells: present and future prospects. Trends Biotechnol 2013;31(3):155–68.
[81] Swartz MA, Hirosue S, Hubbell JA. Engineering approaches to immunotherapy. Sci Transl Med 2012;4 (148):148rv9.
[82] Tykocinski ML, Kaplan DR, Medof ME. Antigen-presenting cell engineering. The molecular toolbox. Am J Pathol 1996;148(1):1–16.
[83] Tykocinski ML, Chen A, Huang JH, Weber MC, Zheng G. New designs for cancer vaccine and artificial veto cells: an emerging palette of protein paints. Immunol Res 2003;27(2–3):565–74.
[84] Tykocinski ML. Signal converter proteins. Fusion protein technologies for biopharmaceuticals. Hoboken, New Jersey: John Wiley & Sons, Inc. 2013. pp. 453–473.
[85] Liu HW, Chen CH, Tsai CL, Hsiue GH. Targeted delivery system for juxtacrine signaling growth factor based on rhBMP-2-mediated carrier-protein conjugation. Bone 2006;39(4):825–36.
[86] Zimmerman GA, McIntyre TM, Mehra M, Prescott SM. Endothelial cell-associated platelet-activating factor: a novel mechanism for signaling intercellular adhesion. J Cell Biol 1990;110(2):529–40.
[87] Rot A. Endothelial cell binding of NAP-1/IL-8: role in neutrophil emigration. Immunol Today 1992;13 (8):291–4.
[88] Tanaka Y, Adams DH, Hubscher S, Hirano H, Siebenlist U, Shaw S. T-cell adhesion induced by proteoglycan-immobilized cytokine MIP-1 beta. Nature 1993;361(6407):79–82.
[89] Sprinzak D, Lakhanpal A, Lebon L, Santat LA, Fontes ME, Anderson GA, et al. Cis-interactions between Notch and Delta generate mutually exclusive signalling states. Nature 2010;465(7294):86–90.

[90] Yaron A, Sprinzak D. The cis side of juxtacrine signaling: a new role in the development of the nervous system. Trends Neurosci 2012;35(4):230–9.
[91] Marston DJ, Dickinson S, Nobes CD. Rac-dependent trans-endocytosis of ephrinBs regulates Eph-ephrin contact repulsion. Nat Cell Biol 2003;5(10):879–88.
[92] Kramer H. Patrilocal* cell–cell interactions: sevenless captures its bride. Trends Cell Biol 1993;3(4):103–5.
[93] Rammensee H-. Veto function *in vitro* and *in vivo*. Intern Rev Immunol 1989;4:175–91.
[94] Fink PJ, Shimenkovitz RP, Bevan MJ. Veto cells. Annu Rev Immunol 1988;6:115–37.
[95] Tykocinski ML, Kaplan DR. Prospects for anti-rejection therapies based upon CD8-dependent immunoregulation. Kidney Int Suppl 1993;39: S120–3.
[96] Hambor JE, Kaplan DR, Tykocinski ML. CD8 functions as an inhibitory ligand in mediating the immunoregulatory activity of $CD8^+$ cells. J Immunol 1990;145:1646–52.
[97] Hambor JE, Weber MC, Tykocinski ML, Kaplan DR. Regulation of allogeneic responses by expression of CD8 alpha chain on stimulator cells. Int Immunol 1990;2(9):879–83.
[98] Smith DM, Bluestone JA, Jeyarajah DR, Newberg MH, Engelhard VH, Thistlethwaite JR, et al. Inhibition of T cell activation by a monoclonal antibody reactive against the alpha3 domain of human MHC class I molecules. J Immunol 1994;153:1054–67.
[99] Kaplan DR, Hambor JE, Tykocinski ML. An immunoregulatory function for the CD8 molecule. Proc Natl Acad Sci USA 1989;86(21):8512–5.
[100] Hambor JE, Tykocinski ML, Kaplan DR. Functional consequences of anti-sense RNA-mediated inhibition of CD8 surface expression in a human T cell clone. J Exp Med 1988;168(4):1237–45.
[101] Tykocinski ML, Kaplan DR. A multifunctional perspective of the CD8 molecule. In: Herberman R, Lotzova E, editors. NK Cell mediated cytotoxicity. Receptors, signaling and mechanisms. Ann Arbor, MI: CRC Press; 1992. p. 393–408.
[102] Qi Y, Berg R, Singleton MA, Debrick JE, Staerz UD. Hybrid antibody mediated veto of cytotoxic T lymphocyte responses. J Exp Med 1996;183:1973–80.
[103] Sambhara SR, Miller RG. Programmed cell death of T cells signaled by the T cell receptor and the alpha3 domain of class I MHC. Science 1991;252:1424–7.
[104] Sambhara SR, Miller RG. Reduction of CTL antipeptide response mediated by $CD8^+$ cells whose class I MHC can bind the peptide. J Immunol 1994;152:1103–9.
[105] Matsue H, Matsue K, Walters M, Okumura K, Yagita H, Takashima A. Induction of antigen-specific immunosuppression by CD95L cDNA-transfected "killer" dendritic cells. Nat Med 1999;5:930–7.
[106] Min WP, Gorczynski R, Huang XY, Kushida M, Kim P, Obataki M, et al. Dendritic cells genetically engineered to express Fas ligand induce donor-specific hyporesponsiveness and prolong allograft survival. J Immunol 2000;164(1):161–7.
[107] Reich-Zeliger S, Zhao Y, Krauthgamer R, Bachar-Lustig E, Reisner Y. Anti-third party $CD8^+$ CTLs as potent veto cells: coexpression of CD8 and FasL is a prerequisite. Immunity 2000;13:507–15.
[108] Wu B, Wu J-M, Miagkov A, Adams RN, Levitsky HI, Drachman DB. Specific immunotherapy by genetically engineered APCs: the "guided missile" strategy. J Immunol 2001;166:4773–9.
[109] Krueger A, Fas SC, Baumann S, Krammer PH. The role of CD95 in the regulation of peripheral T-cell apoptosis. Immunol Rev 2003;193:58–69.
[110] Suss G, Shortman K. A subclass of dendritic cells kills CD4 T cells via Fas/Fas-ligand-induced apoptosis. J Exp Med 1996;183:1789–96.
[111] Chen A, Zheng G, Tykocinski ML. Quantitative interplay between activating and pro-apoptotic signals dictates T cell responses. Cell Immunol 2003;221:128–37.
[112] Mi QS, Ly D, Lamhamedi-Cherradi SE, Salojin KV, Zhou L, Grattan M, et al. Blockade of tumor necrosis factor-related apoptosis-inducing ligand exacerbates type 1 diabetes in NOD mice. Diabetes 2003;52 (8):1967–75.
[113] Zhang H-, Su X, Liu D, Liu W, Yang P, Wang Z, et al. Induction of specific T cell tolerance by Fas ligand-expressing antigen-presenting cells. J Immunol 1999;162:1423–30.
[114] Lu L, Qian S, Hershberger PA, Rudert WA, Lynch DH, Thomson AW. Fas ligand (CD95L) and B7 expression on dendritic cells provide counter-regulatory signals for T cell survival and proliferation. J Immunol 1997;158:5676–84.
[115] George JF, Sweeney SD, Kirklin JK, Simpson EM, Goldstein DR, Thomas JM. An essential role for Fas ligand in transplantation tolerance induced by donor bone marrow. Nat Med 1998;4:333.
[116] Tykocinski ML, Shu HK, Ayers DJ, Walter EI, Getty RR, Groger RK, et al. Glycolipid reanchoring of T-lymphocyte surface antigen CD8 using the 3′ end sequence of decay-accelerating factor's mRNA. Proc Natl Acad Sci USA 1988;85:3555–9.
[117] Caras IW, Weddell GN, Davitz MA, Nussenzweig V, Martin JDW. Signal for attachment of a phospholipid membrane anchor in decay accelerating factor. Science 1987;238:1280–3.

[118] Weber MC, Groger RK, Tykocinski ML. A glycosylphosphatidylinositol-anchored cytokine can function as an artificial cellular adhesion. Exp Cell Res 1994;145:1646–52.

[119] Huang JH, Greenspan NS, Tykocinski ML. Alloantigenic recognition of artificial glycosyl phosphatidylinositol-anchored HLA-A2.1. Mol Immunol 1994;31(13):1017–28.

[120] Brunschwig EB, Fayen JD, Medof ME, Tykocinski ML. Protein transfer of glycosyl-phosphatidylinositol (GPI)-modified murine B7-1 and B7-2 costimulators. J Immunother 1999;22:390–400.

[121] Geho DH, Fayen JD, Jackman RM, Moody BD, Porcelli SA, Tykocinski ML. Glycosyl-phosphatidylinositol reanchoring unmasks distinct antigen-presenting pathways for CD1b and CD1c. J Immunol 2000;65:1272–7.

[122] Medof ME, Nagarajan S, Tykocinski ML. Cell surface engineering with GPI-anchored proteins. FASEB J 1996;10:574–86.

[123] Fayen JD, Tykocinski ML, Medof ME. Glycerolphosphoinositide anchors for membrane-tethering proteins. Methods Enzymol 2000;327:351–68.

[124] Brunschwig EB, Levine E, Trefzer U, Tykocinski ML. Glycosylphosphatidylinositol-modified murine B7-1 and B7-2 retain costimulator function. J Immunol 1995;155:5498–505.

[125] McHugh RS, Ahmed SN, Wang YC, Sell KW, Selvaraj P. Construction, purification, and functional incorporation on tumor cells of glycolipid-anchored human B7-1 (CD80). Proc Natl Acad Sci USA 1995;92 (17):8059–63.

[126] Chen A, Zheng G, Tykocinski ML. Hierarchical costimulator thresholds for distinct immune responses: application of a novel two-step Fc fusion protein transfer method. J Immunol 2000;164(2):705–11.

[127] Zheng G, Chen A, Sterner RE, Zhang PJ, Pan T, Kiyatkin N, et al. Induction of antitumor immunity via intratumoral tetra-costimulator protein transfer. Cancer Res 2001;61:8127–34.

[128] Huang J-H, Tykocinski ML. CTLA-4-Fas ligand functions as a trans signal converter protein in bridging antigen-presenting cells and T cells. Int Immunol 2001;13:529–39.

[129] Noel PJ, Boise LH, Green JM, Thompson CB. CD28 costimulation prevents cell death during primary T cell activation. J Immunol 1996;157:636–42.

[130] Li Y, Li XC, Zheng XX, Wells AD, Turka LA, Strom TB. Blocking both signal 1 and signal 2 of T-cell activation prevents apoptosis of alloreactive T cells and induction of peripheral allograft tolerance. Nat Med 1999;5(11):1298–302.

[131] Holler N, Tardivel A, Kovacsovics-Bankowski M, Hertig S, Gaide O, Martinon F, et al. Two adjacent trimeric Fas ligands are required for Fas signaling and formation of a death-inducing signaling complex. Mol Cell Biol 2003;23(4):1428–40.

[132] Schneider P, Tschopp J. Apoptosis induced by death receptors. Pharm Acta Helv 2000;74(2–3):281–6.

[133] Baban B, Hansen AM, Chandler PR, Manlapat A, Bingaman A, Kahler DJ, et al. A minor population of splenic dendritic cells expressing CD19 mediates IDO-dependent T cell suppression via type I IFN signaling following B7 ligation. Int Immunol 2005;17 (7):909–19.

[134] Friedline RH, Brown DS, Nguyen H, Kornfeld H, Lee J, Zhang Y, et al. $CD4^+$ regulatory T cells require CTLA-4 for the maintenance of systemic tolerance. J Exp Med 2009;206(2):421–34.

[135] Schmidt EM, Wang CJ, Ryan GA, Clough LE, Qureshi OS, Goodall M, et al. Ctla-4 controls regulatory T cell peripheral homeostasis and is required for suppression of pancreatic islet autoimmunity. J Immunol (Baltimore, MD: 1950) 2009;182(1):274–82.

[136] Elhalel MD, Huang JH, Schmidt W, Rachmilewitz J, Tykocinski ML. CTLA-4-FasL induces alloantigen-specific hyporesponsiveness. J Immunol 2003;170 (12):5842–50.

[137] Dranitzki-Elhalel M, Huang JH, Rachmilewitz J, Pappo O, Parnas M, Schmidt W, et al. CTLA-4-FasL inhibits allogeneic responses *in vivo*. Cell Immunol 2006;239(2):129–35.

[138] Orbach A, Rachmilewitz J, Parnas M, Huang JH, Tykocinski ML, Dranitzki-Elhalel M. CTLA-4-FasL induces early apoptosis of activated T cells by interfering with anti-apoptotic signals. J Immunol 2007;179(11):7287–94.

[139] Jin Y, Qu A, Wang GM, Hao J, Gao X, Xie S. Simultaneous stimulation of Fas-mediated apoptosis and blockade of costimulation prevent autoimmune diabetes in mice induced by multiple low-dose streptozotocin. Gene Ther 2004;11(12):982–91.

[140] Feng YG, Jin YZ, Zhang QY, Hao J, Wang GM, Xie SS. CTLA4-Fas ligand gene transfer mediated by adenovirus induces long-time survival of murine cardiac allografts. Transplant Proc 2005;37(5):2379–81.

[141] Shi W, Chen M, Xie L. Prolongation of corneal allograft survival by CTLA4-FasL in a murine model. Graefes Arch Clin Exp Ophthalmol 2007;245(11):1691–7.

[142] Lavrik IN, Krammer PH. Regulation of CD95/Fas signaling at the DISC. Cell Death Differ 2012;19 (1):36–41.

[143] Winkles JA. The TWEAK-Fn14 cytokine-receptor axis: discovery, biology and therapeutic targeting. Nat Rev Drug Discov 2008;7(5):411–25.

[144] Falschlehner C, Schaefer U, Walczak H. Following TRAIL's path in the immune system. Immunology 2009;127(2):145–54.
[145] Maecker H, Varfolomeev E, Kischkel F, Lawrence D, LeBlanc H, Lee W, et al. TWEAK attenuates the transition from innate to adaptive immunity. Cell 2005;123(5):931–44.
[146] Galli SJ. Complexity and redundancy in the pathogenesis of asthma: reassessing the roles of mast cells and T cells. J Exp Med 1997;186(3):343–7.
[147] Desplat-Jego S, Varriale S, Creidy R, Terra R, Bernard D, Khrestchatisky M, et al. TWEAK is expressed by glial cells, induces astrocyte proliferation and increases EAE severity. J Neuroimmunol 2002;133(1–2):116–23.
[148] Potrovita I, Zhang W, Burkly L, Hahm K, Lincecum J, Wang MZ, et al. Tumor necrosis factor-like weak inducer of apoptosis-induced neurodegeneration. J Neurosci 2004;24(38):8237–44.
[149] Mueller AM, Pedre X, Kleiter I, Hornberg M, Steinbrecher A, Giegerich G. Targeting fibroblast growth factor-inducible-14 signaling protects from chronic relapsing experimental autoimmune encephalomyelitis. J Neuroimmunol 2005;159(1–2):55–65.
[150] Song K, Chen Y, Goke R, Wilmen A, Seidel C, Goke A, et al. Tumor necrosis factor-related apoptosis-inducing ligand (TRAIL) is an inhibitor of autoimmune inflammation and cell cycle progression. J Exp Med 2000;191:1095.
[151] Hilliard B, Wilmen A, Seidel C, Liu TS, Goke R, Chen Y. Roles of TNF-related apoptosis-inducing ligand in experimental autoimmune encephalomyelitis. J Immunol 2001;166(2):1314–9.
[152] Hirata S, Senju S, Matsuyoshi H, Fukuma D, Uemura Y, Nishimura Y. Prevention of experimental autoimmune encephalomyelitis by transfer of embryonic stem cell-derived dendritic cells expressing myelin oligodendrocyte glycoprotein peptide along with TRAIL or programmed death-1 ligand. J Immunol 2005;174(4):1888–97.
[153] Cretney E, McQualter JL, Kayagaki N, Yagita H, Bernard CC, Grewal IS, et al. TNF-related apoptosis-inducing ligand (TRAIL)/Apo2L suppresses experimental autoimmune encephalomyelitis in mice. Immunol Cell Biol 2005;83(5):511–9.
[154] Lamhamedi-Cherradi SE, Zheng S, Tisch RM, Chen YH. Critical roles of tumor necrosis factor-related apoptosis-inducing ligand in type 1 diabetes. Diabetes 2003;52(9):2274–8.
[155] Razmara M, Hilliard B, Ziarani AK, Murali R, Yellayi S, Ghazanfar M, et al. Fn14-TRAIL, a chimeric intercellular signal exchanger, attenuates experimental autoimmune encephalomyelitis. Am J Pathol 2009; 174(2):460–74.
[156] Wajant H, Moosmayer D, Wuest T, Bartke T, Gerlach E, Schonherr U, et al. Differential activation of TRAIL-R1 and -2 by soluble and membrane TRAIL allows selective surface antigen-directed activation of TRAIL-R2 by a soluble TRAIL derivative. Oncogene 2001;20(30):4101–6.
[157] Shirley S, Morizot A, Micheau O. Regulating TRAIL receptor-induced cell death at the membrane: a deadly discussion. Recent Patents Anticancer Drug Discov 2011;6(3):311–23.
[158] Dranitzki-Elhalel M, Huang JH, Sasson M, Rachmilewitz J, Parnas M, Tykocinski ML. CD40-FasL inhibits human T cells: evidence for an auto-inhibitory loop-back mechanism. Int Immunol 2007;19(4):355–63.
[159] Dhein J, Walczak H, Baumler C, Debatin K-, Krammer PH. Autocrine T-cell suicide mediated by APO-1 (Fas/CD95). Nature 1995;373:438–41.
[160] Brunner T, Mogil RM, LaFace D, Yoo NJ, Mahboubi A, Echeverri F, et al. Cell-autonomous Fas (CD95)/Fas-ligand interaction mediates activation-induced apoptosis in T-cell hybridomas. Nature 1995;373: 441–4.
[161] Ju ST, Panka DJ, Cui H, Ettinger R, el Khatib M, Sherr DH, et al. Fas (CD95)/FasL interactions required for programmed cell death after T-cell activation. Nature 1995;373:444–8.
[162] Russell JH, Rush B, Weaver C, Wang R. Mature T cells of autoimmune lpr/lpr mice have a defect in antigen-stimulated suicide. Proc Natl Acad Sci USA 1993;90:4409–13.
[163] Alderson MR, Tough TW, Davis-Smith T, Braddy S, Falk B, Schooley KA, et al. Fas ligand mediates activation-induced cell death in human T lymphocytes. J Exp Med 1995;181:71–7.
[164] Suda T, Okazaki T, Naito Y, Yokota T, Arai N, Ozaki S, et al. Expression of the Fas ligand in cells of T cell lineage. J Immunol 1995;154:3806–13.
[165] Elgueta R, Benson MJ, de Vries VC, Wasiuk A, Guo Y, Noelle RJ. Molecular mechanism and function of CD40/CD40L engagement in the immune system. Immunol Rev 2009;229(1):152–72.
[166] Ehrenschwender M, Wajant H. The role of FasL and Fas in health and disease. Adv Exp Med Biol 2009;647:64–93.
[167] Blair PJ, Riley JL, Harlan DM, Abe R, Tadaki DK, Hoffmann SC, et al. CD40 ligand (CD154) triggers a short-term $CD4^+$ T cell activation response that results in secretion of immunomodulatory cytokines and apoptosis. J Exp Med 2000;191(4):651–60.

[168] Zheng G, Chen A, Weber M, Tykocinski M. T cells painted with B7-1 and 4-1BB ligand auto-costimulate themselves. FASEB J 2003; vol. Abstr. 414.3.

[169] Orbach A, Rachmilewitz J, Shani N, Isenberg Y, Parnas M, Huang JH, et al. CD40.FasL and CTLA-4.FasL fusion proteins induce apoptosis in malignant cell lines by dual signaling. Am J Pathol 2010;177(6):3159–68.

[170] Tran NL, McDonough WS, Donohue PJ, Winkles JA, Berens TJ, Ross KR, et al. The human Fn14 receptor gene is up-regulated in migrating glioma cells *in vitro* and overexpressed in advanced glial tumors. Am J Pathol 2003;162(4):1313–21.

[171] Zhao H, Langerod A, Ji Y, Nowels KW, Nesland JM, Tibshirani R, et al. Different gene expression patterns in invasive lobular and ductal carcinomas of the breast. Mol Biol Cell 2004;15(6):2523–36.

[172] Ozawa F, Friess H, Kleeff J, Xu ZW, Zimmermann A, Sheikh MS, et al. Effects and expression of TRAIL and its apoptosis-promoting receptors in human pancreatic cancer. Cancer Lett 2001;163(1):71–81.

[173] Kawakita T, Shiraki K, Yamanaka Y, Yamaguchi Y, Saitou Y, Enokimura N, et al. Functional expression of TWEAK in human colonic adenocarcinoma cells. Int J Oncol 2005;26(1):87–93.

[174] Kawakita T, Shiraki K, Yamanaka Y, Yamaguchi Y, Saitou Y, Enokimura N, et al. Functional expression of TWEAK in human hepatocellular carcinoma: possible implication in cell proliferation and tumor angiogenesis. Biochem Biophys Res Commun 2004;318(3):726–33.

[175] Aronin A, Amsili S, Prigozhina TB, Tzdaka K, Rachmilewitz J, Shani N, et al. Fn14*TRAIL effectively inhibits hepatocellular carcinoma growth. PLoS ONE 2013;8(10):e77050.

[176] Wu GS. TRAIL as a target in anti-cancer therapy. Cancer Lett 2009;285(1):1–5.

[177] Melero I, Grimaldi AM, Perez-Gracia JL, Ascierto PA. Clinical development of immunostimulatory monoclonal antibodies and opportunities for combination. Clin Cancer Res 2013;19(5):997–1008.

[178] Elefsinioti A, Sarac OS, Hegele A, Plake C, Hubner NC, Poser I, et al. Large-scale *de novo* prediction of physical protein–protein association. Mol Cell Proteomics 2011;10(11) M111.010629.

[179] Sanchez C, Lachaize C, Janody F, Bellon B, Roder L, Euzenat J, et al. Grasping at molecular interactions and genetic networks in *Drosophila melanogaster* using FlyNets, an Internet database. Nucleic Acids Res 1999;27(1):89–94.

[180] Boxem M, Maliga Z, Klitgord N, Li N, Lemmens I, Mana M, et al. A protein domain-based interactome network for *C. elegans* early embryogenesis. Cell 2008;134(3):534–45.

[181] Brettner LM, Masel J. Protein stickiness, rather than number of functional protein-protein interactions, predicts expression noise and plasticity in yeast. BMC Syst Biol 2012;6:128 0509-6-128.

[182] Tokuriki N, Tawfik DS. Protein dynamism and evolvability. Science (New York, NY) 2009;324(5924): 203–7.

[183] Manz BN, Groves JT. Spatial organization and signal transduction at intercellular junctions. Nat Rev Mol Cell Biol 2010;11(5):342–52.

[184] Mitchell MJ, Wayne E, Rana K, Schaffer CB, King MR. TRAIL-coated leukocytes that kill cancer cells in the circulation. Proc Natl Acad Sci USA 2014;111 (3):930–5.

CHAPTER

2

Cell Surface Engineering by Chemical Reaction and Remodeling

Wade M. Fox[1] *and Debanjan Sarkar*[1,2]

[1]Department of Chemical and Biological Engineering, University at Buffalo, The State University of New York, Buffalo, NY, USA, [2]Department of Biomedical Engineering, University at Buffalo, The State University of New York, Buffalo, NY, USA

OUTLINE

2.1 INTRODUCTION

Molecular engineering of the cell membrane is a powerful tool to manipulate surface composition and for controlling the interactions between the cell and its surrounding environment. Since cells interact with the extracellular environment through the molecular receptors and ligands present on the membrane, it is important to control the molecular composition to regulate these interactions. Enzymatic strategies and genetic manipulation of cellular machineries are widely used to achieve this goal, particularly in molecular biology [1–3]. But recent advancements in material science and engineering have provided novel technologies to alter cell membranes through chemical methods [4,5]. Since cell membranes are molecularly composed of wide-ranging chemical functionalities, it provides a versatile ground for modification with characteristic molecules through defined pathways [5,6]. This enhances

J.M. Karp & W. Zhao (Eds): Micro- and Nanoengineering of the Cell Surface.
DOI: http://dx.doi.org/10.1016/B978-1-45-573146-6.00002-7

our ability to augment functioning of cells and manipulate their fate during different biological processes with precise control and defined mechanisms. As a result, chemical engineering of the cell membrane has been a powerful tool to interrogate fundamental cellular process and has emerged as a technology to develop cellular therapeutics and diagnostics. Particularly, chemical engineering of cell membranes has been used to investigate cell–cell and cell–matrix interactions [7–10] and to design targeted delivery [10], biosensing, and imaging techniques [11].

Cell membranes are chemically heterogeneous and dynamic in nature. In generic terms, cell membranes are characterized by the presence of different proteins, lipids, and carbohydrates. However, molecular structure and three-dimensional organization of these molecules on the cell surface are highly specific and overly complex and are dependent on the cell type and external environment. Importantly, cell membranes are continuously remodeled and reorganized in response to external conditions and stimuli to mediate cell–cell and cell–niche communication and intracellular signaling [12]. This indicates the dynamic nature of cell membranes where chemical signatures are continuously changing. Opposed to this process, which is recognized as a natural cellular mechanism derived from internal cellular machineries, exogenous modification of cell membrane is considered as a synthetic mechanism to alter cell surface composition.

Chemical modification of cell membranes by synthetic methodologies can be categorized either as a biological or physicochemical process. Genetic engineering and enzymatic transformations of cells are common biological mechanisms to alter the cell surface molecules. While the genetic approach involves introduction of exogenous genetic materials into the cell genome to express or regress specific cell surface molecules, the enzymatic approach transforms specific molecules into distinct types with defined functionality. Both of these approaches are widely used to study the fundamental biology of cell functioning, but these strategies have also found acceptance in developing cellular therapeutics and diagnostics. However, there are several limitations which potentially restrict the use of these approaches as generic methods. Particularly, genetic engineering of cells alters genomics of the host cell which can permanently alter genotype and can potentially induce malignant character in the long term with safety concerns. Alternately, enzymatic transformations are highly complex and require the presence of specific molecules on the cell surface for modifications. In contrast, cellular engineering using physicochemical principles can be broader and can provide platform technologies for developing molecular tools to modify the cell surface [4,10]. Manipulating living cells by physical and chemical techniques offers exciting avenues to decorate the cell membranes with molecules and specialized structures, for example, particles and patches. Physical processes utilize noncovalent molecular interactions, for example, hydrogen bonding, hydrophobic interactions, and electrostatic interactions to attach molecules and structures to cell membranes. Since these interactions are physical in nature, this technique does not involve any permanent chemical change on cell membranes to engineer cell surface. As a result, cell surface engineering by physical process is often transient in nature and probably not useful where permanent and long-term cell modification is required. Furthermore, physically immobilized ligands cannot sustain in a mechanical environment where a large force is exerted on the ligands. These limitations can be addressed by performing chemical reactions on cell membranes to attach molecules through covalent bond formation.

As cell membranes present a rich repertoire of wide-ranging functionalities, different chemical reaction methodologies can be used to react with cell membranes for covalent cell surface modifications. Contrary to biological transformation, chemical reaction engineering of cell membranes is not dependent on a specific surface molecule or receptor. Rather chemical reaction depends on having specific functional groups, independent of the constituent molecule, which can be modified through covalent reactions. This provides enhanced versatility of chemical reaction based cell engineering where different molecules can be attached to cell membranes through a diverse range of chemical reactions. Thus, chemical reaction based cell surface engineering is technologically a more adaptable method and is currently being explored as a molecular engineering tool to manipulate cell function and phenotype. In this chapter, we will primarily focus on the methods used to covalently engineer cell membranes and the resultant modifications. Use and effectiveness of these cell surface modifications will be briefly mentioned but not in detail.

2.2 METHODS AND TECHNOLOGY OF COVALENT CELL SURFACE REACTION

Chemical reactions on cell surfaces are dependent on the availability of specific functional groups on the surface molecules which can chemically react under acceptable conditions without impairing native cell functions. Since cell membranes display a diverse range of molecules, researchers have covalently modified the functional groups present in these molecules to attach specific molecules and structures. Furthermore, chemical reactions on the cell surface should be selective without any nonspecific modifications under guided cell-compatible conditions which are feasible in an aqueous environment.

2.2.1 Direct chemical modification of cell membrane

Selection of chemical reactions to covalently modify a cell membrane is dependent on the availability of a specific functional group which can react under defined conditions. The presence of proteins, carbohydrates, and lipids on cell membranes offers multiple sites to external reagents which can react to form a covalent linkage. However, based on feasibility and practical methods of modification, only a few selectable functional groups on cell membranes have been used for chemical reactions and conjugation. Most popular chemical targets include amines, sulfhydryl, carboxyl, and carbonyl groups present in proteins and other cell surface molecules. Proteins, with an unique composition and sequence of amino acids, on cell surfaces provide a rich environment for these functional groups. Although many amino acids do not have functional groups available for external chemical modifications, several amino acids possess these functional moieties which have been chemically modified on cell surfaces.

Amine groups, predominantly primary amines, are widely used for chemical modification of cell membranes due to well-established protocols, easily available conjugation linkers (which can conjugate cell membrane groups to the external molecules), and cell-compatible reaction conditions. Figure 2.1 schematically shows the chemical reactions which are primarily used for covalent transformation of primary amine groups on the cell surface. The most common reaction to modify primary amines involves reaction with *N*-hydroxysuccinimide (NHS) ester. NHS-ester-activated compounds react with primary amines in slightly alkaline conditions (pH 7.2–8.5) to form amide bonds with release of NHS which can be removed easily by dialysis or desalting. Since NHS-ester cross-linking reactions are usually performed in phosphate buffer at pH

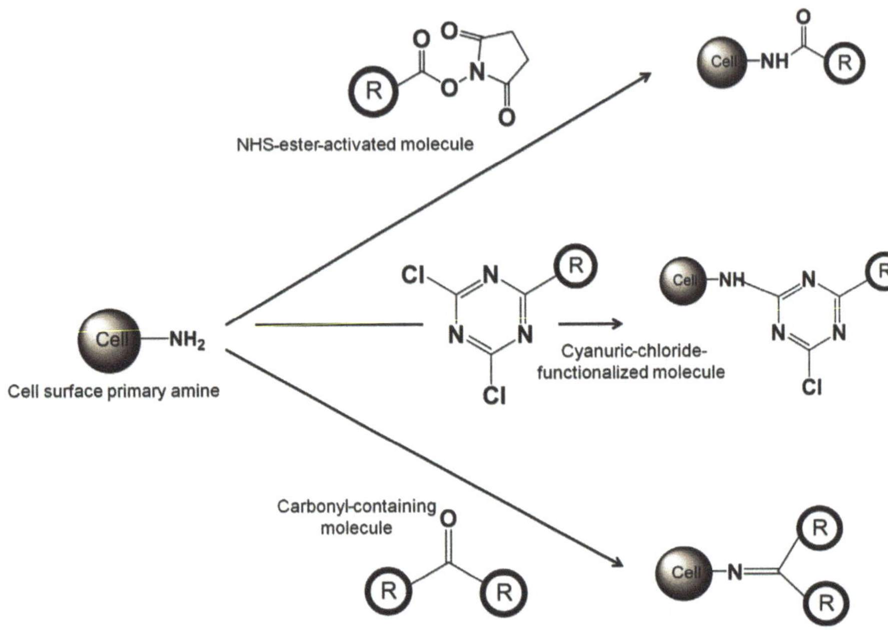

FIGURE 2.1 Covalent reaction of cell surface primary amine with amine-reactive reagents to form covalent adduct on cell membrane.

7.2–8.0 for 0.5–4 h at room temperature or 4°C, this reaction has been used extensively to covalently modify primary amines on the cell surface. To improve the solubility of NHS-ester molecules in aqueous medium, sulfo-NHS-esters are used which are identical to NHS-esters except that they contain a sulfonate ($-SO_3$) group on the NHS ring. This charged group has no effect on the reaction chemistry, but it increases the water solubility of the reaction reagent. In particular, the charged group prevents sulfo-NHS compounds from permeating cell membranes, enabling them for cell surface chemical modifications.

This chemical reaction has been used to covalently conjugate biotin on the cell membranes by using NHS-ester-activated biotin. Commercial availability of NHS-biotin and sulfo-NHS-biotin has enabled extensive use of this approach to covalently biotinylate the cell surface which can subsequently be functionalized through (strept) avidin. Biotin–streptavidin bridges on the cell surface allow immobilization of different functional molecules and enables this approach as a platform technology for covalent modification of cell membranes [13,14]. Since the biotin–streptavidin linkage is exceptionally strong and stable, this is an effective way to attach a multitude of molecules for adding new functionalities to cells. For example, this approach has been used to functionalize mesenchymal stem cells (MSCs) where cells were covalently biotinylated with NHS-activated biotin followed by addition of streptavidin and subsequent addition of biotinylated functional molecules. Using this approach, Sarkar et al. [15] immobilized Sialyl Lewis X (SLeX), a functional moiety expressed by leukocytes for their rolling response on activated endothelium, on MSCs and demonstrated that SLeX modified MSCs exhibited the rolling response on activated endothelium both *in vitro* and *in vivo*. Most importantly, this covalent modification was stable and did not induce any adverse effect on MSC functioning including their ability to adhere, differentiate, or secrete paracrine factors. The generic nature of this approach was further substantiated when similar covalent modifications of MSCs were used to immobilize platelet-derived growth factor (PDGF) sensing aptamer (single-stranded oligonucleotides) and modified MSCs were able to sense PDGF in a spatiotemporal manner [16]. Use of the NHS-biotin approach has also been applied to conjugate antibodies, quantum dots, and nanoparticles to the cell surfaces [17,18]. Cell surface functionalization, particularly covalent labeling of primary amines in cell surface proteins, by NHS-biotin has also been used as an analytical tool for enrichment and protein profiling studies [19,20]. This shows the wide-ranging utility for covalent functionalization of cell membranes with NHS-biotin.

Although use of NHS-biotin has gained acceptance due to well-established biotin–(strept)avidin conjugation technology, several other NHS-ester-activated molecules have been used for covalent modification of cells. NHS-activated small molecule, for example, 4-pentynoic acid, has been used to functionalize cells with alkyne groups which enable performing click chemistry on the cell membrane [21]. Similarly, NHS-activated macromolecules are developed for covalent functionalization of cells. In particular, NHS-activated polyethylene glycol (PEG) has been used to link PEG for several applications [22]. This approach has been used to functionalize islet cells, and no change in cell morphology and function was observed after PEG was covalently immobilized. Furthermore, NHS-PEG-modified islets were transplanted into the liver which showed transient normalization of blood glucose [23]. However, use of PEG is limited in its ability to link to therapeutically relevant molecules due to lack of functionalities which reduces the effectiveness in a wide range of applications [24]. This limitation can be addressed by using bifunctional PEG where one group specifically interacts with functional groups on the cell membrane while the other group is available for functionalization with an external reagent. To achieve this, amine groups on the cell surface were reacted with the linker molecule NHS-PEG-maleimide where NHS reacts with amines on the cell surface, leaves maleimide to react with a thiol group, the target molecule. This process allowed conjugating target peptides on the cell surface which induced interaction with E-selectin on endothelial cells for enhanced cell rolling [25]. However, a potential concern in this approach remains because cell surface thiols can also react to maleimide groups which can be within the same cell or in between multiple cells (leading to cross-linking). If this is the case, availability of second functional groups of bifunctional PEG is limited for further conjugation. This potential problem can be addressed by using multifunctional PEG where hyperbranched PEG with multiple functionalities can be conjugated to the cell membrane. Although it was not hyperbranched PEG, Chapanian et al. [24] have proposed a method of chemical conjugation to the cell surface using hyperbranched polyglycerol (HPG) molecules. The rationale for this approach was that while HPG and PEG have similar biocompatibility, HPG eliminates some of the shortcomings of PEG due to its functionality. In this study, red blood cells (RBCs) were decorated with HPG. The carboxyl groups were activated by addition of NHS, and the molecule was subsequently attached to the cell surface through reaction with the exposed amines of proteins present on the surface [24]. In addition to conjugation of small molecules and synthetic macromolecules, NHS-ester-based reactions have been used to conjugate biomolecules. For example, single-stranded DNA (ssDNA) was conjugated to cells by preparing NHS-DNA conjugates through reaction of NHS-PEG-maleimide and thiolated DNA [26]. Reaction of cells with NHS-DNA enabled conjugation of ssDNA to the cell surface through a reaction between NHS and cell surface amine groups. Using this approach, cellular assembly and cell–cell interaction were examined. This covalent immobilization is generic as it can be performed on any cells under cell-compatible conditions without eliciting any adverse reactions.

Although NHS-ester-based modification of primary amines is widely used for covalent modification of cell membranes, there are other reaction mechanisms which have been explored to modify cell surface amine groups. In this category, reaction of cell surface amines with cyanuric chloride is most prominent. Cyanuric-chloride-activated molecules have been used to covalently modify cell membranes due to their malleable chemistry, the efficacy in membrane protein modification, the chemical stability of the modified proteins, and the feasible reaction conditions [27]. Using this strategy, blood cells, mainly RBCs, are covalently modified with PEG [28]. Conditions for this covalent modification have been optimized with respect to pH, temperature, and reagent concentration. Using covalently modified RBC with PEG provides a tool to create "universal blood" by camouflaging the immune system as PEG provides a protective shell around the RBC that potentially excludes large molecules, such as antibodies, but does not appear to inhibit the passage of small molecules, such as glucose and oxygen [29,30]. Another strategy to modify cell surface amine group is through reaction with carbonyl functionality resulting in formation of Schiff's base. This approach has been primarily investigated by reacting a carbonyl-containing small molecule drug, tucaresol, with primary amine groups on T-cell surfaces to understand the immune response of T cells [31]. Reaction between cell surface amine and carbonyl functionality resulted in imine bond (Schiff's base) which mimicked the formation of Schiff bases between constitutive carbonyl and amino groups expressed by macromolecules at the surface of the T cell. However, this approach has not been explored as a platform technique for cell surface modification.

The second most important functionality on cell surfaces used for covalent modification is the sulfhydryl groups, that is, thiols. Cell surface thiols are present either in oxidized disulfide bridges or in reduced thiol groups. Common covalent reactions for cell surface thiol modification are schematically shown in Figure 2.2. Most widely used covalent reactions involve maleimide-activated molecules which specifically react with thiol groups (–SH) at near neutral conditions (pH 6.5–7.5) to form stable thioether linkages. These reactions are typically carried out at ambient temperature or at 37°C for 90–120 min and in serum-free media, typically phosphate buffer saline, to avoid any cross-reaction with thiol-containing compounds in the media. Maleimide-based covalent modification has been used for covalent immobilization of molecules and different structures, such as nanoparticles. For example, maleimide-functionalized dye, Alexa488, has been covalently linked to the surface of KG1a cells through covalent reaction which shows specific immobilization of the dye on the cell membrane [32]. In contrast, control dye with inactivated maleimide did not produce any signal indicating the specificity of this reaction. This approach has also been used to covalently conjugate macromolecules. For example, maleimide-functionalized PEG is used for covalent functionalization of RBCs to develop universal blood [33]. Maleimide-based covalent mechanisms have also been employed for covalent immobilization of nanoparticles on cell surfaces. Stephan et al. [34] have developed a method of nanoparticle conjugation in which the exposed free thiols of surface proteins on T cells were utilized. The nanoparticles are 100–300 nm in diameter and have a phospholipid surface layer which features maleimide headgroups available for reaction with the thiols on cell surface. This chemical conjugation is completed through a simple two-step process beginning with incubation of the nanoparticles with the cells to allow for

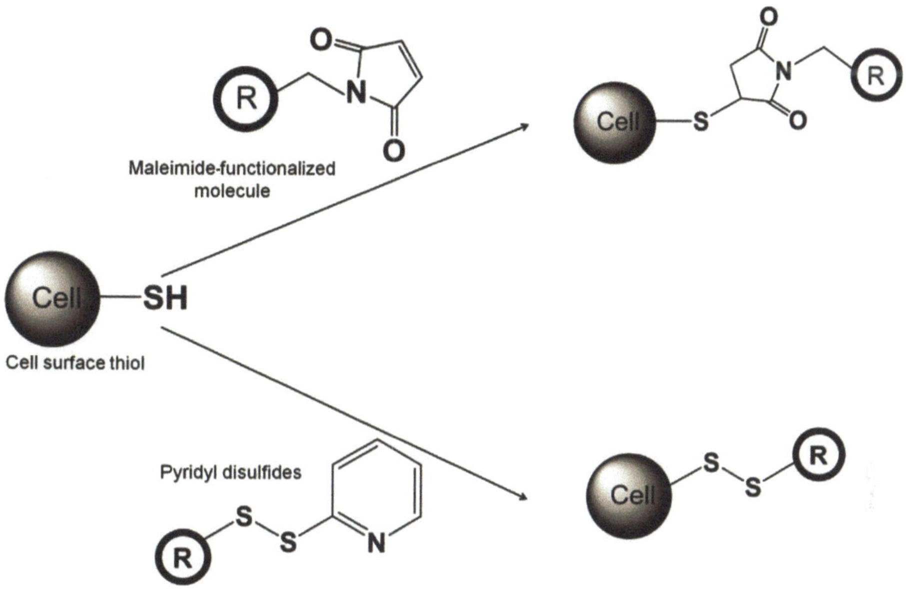

FIGURE 2.2 Common covalent reactions of cell surface thiols for covalent transformation.

maleimide–thiol bonding, and subsequent exposure of the nanoparticles to PEG molecules containing a thiol end to eliminate any remaining functional groups on the nanoparticle surface. Through this method, the nanoparticles can be attached to the cell at a count of approximately 140 ± 30 per cell without affecting the functionality of the cell [34]. This approach enabled sustained delivery of therapeutic molecules to control the immune function. In addition to maleimide-based covalent linking, other functional groups reactive to surface thiols are bromoacetamide and dithiopyridyl functionalities which can covalently bind to cell surface thiols [35]. This study also indicated that the reactivity of these reagents and maleimide-based compounds, particularly in neutral pH, was in the order of maleimide → dithiopyridyl → bromoacetamide (based on their reaction with mercaptoethanol) which can probably be extrapolated to analyze the relative potency of these reagents to cell surface thiols. Furthermore, it is important to note a delicate balance exists between the oxidized and reduced form of thiols depending on the redox environment [36,37]. Based on this microenvironmental condition, exchange between these two forms can lead to the formation of covalent linkages.

Although amine and thiol-based cell surface modification has been used extensively for cell surface modifications, cell surface carbonyls from aldehydes and ketones can also be covalently functionalized. Carbonyl groups can react with hydrazides at pH 5–7, typically carried out in phosphate buffer saline and at ambient temperature or 37°C, resulting in formation of hydrazone bonds. Similarly, under the same reaction conditions, carbonyl groups can react with alkoxy amines to form oxime bonds. However, the limited presence of carbonyl groups on cell surfaces requires an additional step through which carbonyls groups can be enhanced in enough surface concentration for covalent modification [38,39]. This process usually involves oxidation of diols in polysaccharides and glycoproteins which is either chemically mediated through periodate or enzymatically mediated through galactose oxidase. Carbonyl groups induced to the cell surface through the oxidation can react with hydrazide-activated molecules to form covalent adducts through hydrazine bond formation or with alkoxy-amine-activated molecules to form covalent adducts through oxime bond formation. This process was utilized to fluorescently label cells where hydrazide-modified dyes were allowed to react with the cell after oxidation which allowed covalent immobilization of dyes on the cell membrane [40]. Similar approaches have also been used to biotinylate cell surfaces by using hydrazide-activated biotin [41]. Apart from small molecule conjugation, researchers have used this strategy to covalently link nanoparticles to cell surfaces. Particularly, sialic acid residues of macrophage surfaces were oxidized with periodate to introduce aldehydes which were covalently conjugated to amine-functionalized polyamidoamine (PAMAM) and quantum dots [42]. With this approach, a targeted cell-based drug delivery system was developed. In addition to hydrazides, alkoxyamine-activated molecules can also react to the carbonyls to covalently label the cell membrane through formation of oxime bonds, for example, biotin was immobilized on the cell surface via oxime bond following oxidation of glycoprotein [43]. Most importantly, this covalent conjugation method represents a feasible chemical approach to detect cell surface glycoproteins for biomolecular tools in glycoproteomics [44]. Thus, covalent modification of the cell surface through aldehyde and ketone-based reactions is a practical approach for conjugation of molecules to the cell surface. Figure 2.3 schematically

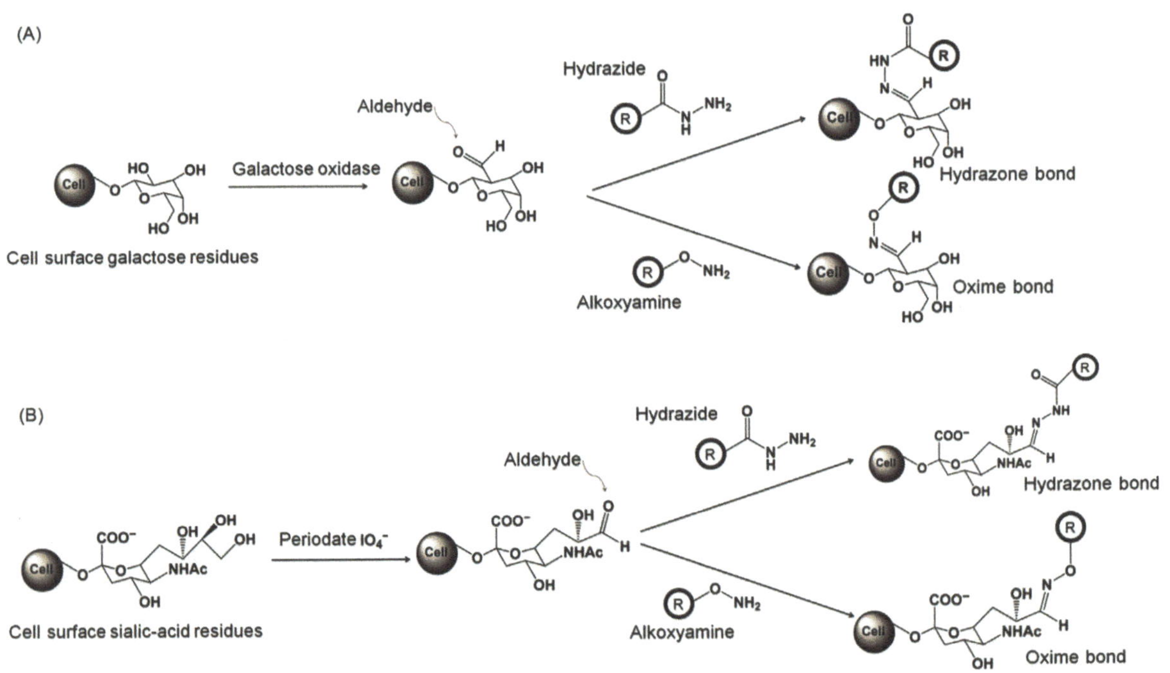

FIGURE 2.3 Enzymatic and chemical methods for introducing aldehyde groups into cell surface glycoconjugates unique functional groups for chemoselective ligation. (A) Treatment of cells with galactose oxidase converts exposed galactose residues to the corresponding aldehydes. (B) Treatment of cells with periodate under mild conditions results in the selective oxidation of sialic acid residues to produce aldehyde groups. Oxidation-induced aldehydes covalently react with hydrazide to form hydrazone bonds and with alkoxy amine to form oxime bonds.

summarizes these covalent reactions on the cell surface. Since carbonyl groups are not readily available on cell surfaces and it requires an additional oxidation step to introduce the carbonyl groups, this approach is not extensively used for cell surface modification. In particular, oxidation of the cell even under mild condition can induce an adverse effect on cells [45].

Direct chemical reaction of cell surface molecules is primarily centered on these functional groups: amine, thiol, and oxidation-induced carbonyls. Relatively simple reaction techniques performed under cell-compatible conditions, high selectivity, and minimal reaction steps are the primary reasons for the use of these approaches for direct covalent modification of cell membranes.

2.2.2 Indirect chemical modification of cell membrane

Although cell membranes represent a rich repertoire of chemical groups, only a few functionalities are available for direct covalent reactions under feasible conditions. This has limited the use of the covalent modification technique as a platform technology to immobilize wide ranges of molecules and functionalities. Direct covalent modification is dependent on the development of functionalized targets which can specifically react to cell surface functionalities. Thus to enhance the capability of covalent modification on cell surfaces, approaches are adopted where functionalities are introduced on cell surfaces through a non-covalent mechanism followed by a covalent reaction. Although, in a strict sense, this approach is not to be considered as a covalent modification of the cell membrane, the post-modification step is a covalent reaction and often enables performing a wide range of covalent chemistries on cell surfaces.

This indirect chemical modification strategy has been studied and explored in detail by the Bertozzi group which is schematically presented in Figure 2.4 [46]. In this approach, chemical functionality is introduced on the cell surface through a biosynthetic or a metabolic pathway. First, a chemically reactive moiety is incorporated into target glycans by metabolic labeling with an unnatural monosaccharide substrate. This chemically reactive group is known as the "chemical reporter." In the second step, the reporter group is covalently reacted with a target molecule through bioorthogonal ligation. Functionalities like ketones and azide groups are introduced through this strategy by biosynthesis of unnatural sialic acid (Figure 2.5) [47,48]. The ketone group was

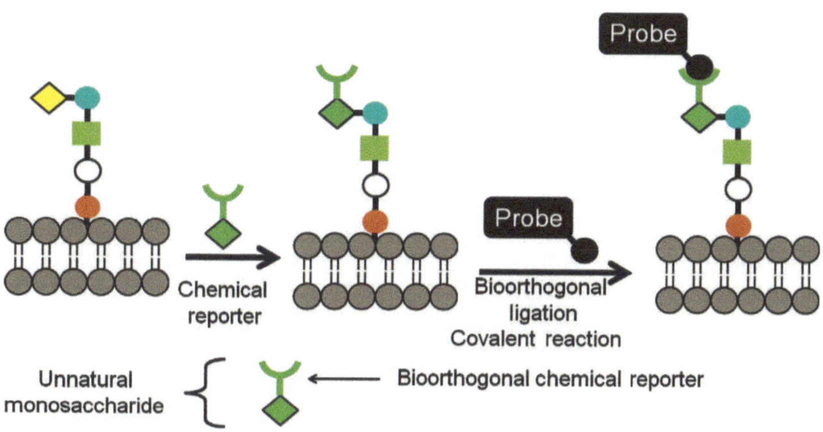

FIGURE 2.4 Bioorthogonal chemical reporter strategy to functionalize cell membranes.

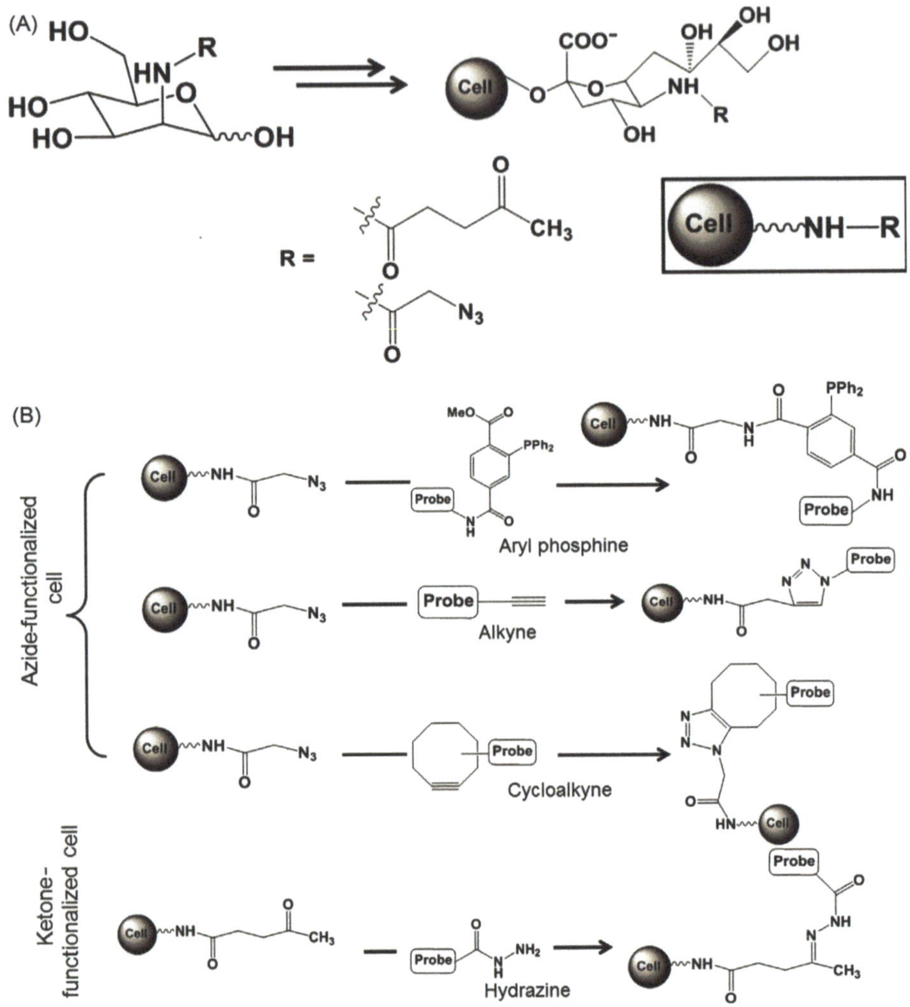

FIGURE 2.5 (A) Metabolic delivery of chemically orthogonal groups, ketones, and azides to cell surface by unnatural sialic acid biosynthesis, and (B) chemical ligation of azide with aryl phosphine, alkyne, and cycloalkyne and reaction of ketone with hydrazide. Probe represents the target molecule which can be immobilized to cell surface by this approach.

introduced by using a mannosamine analog derivatized with a levulinoyl side chain (ManLev) which transformed into the corresponding cell surface sialoside in human cells. This ketone group can undergo selective reaction with aminooxy, hydrazide, or related functional groups to form stable covalent adducts (Figure 2.5). As a result, cell surfaces can be covalently modified with an epitope of interest generated by chemical synthesis. At least 10^7 ketones on a single cell surface were achieved, offering significant potential for remodeling cells for specific cell–surface interactions [49]. This approach was further extended by introducing azide groups to cell surfaces using *N*-azidoacetylmannosamine. The presence of azide groups was verified through reaction of the azide group with a phosphine group and alkynes (Figure 2.5). Azides were ligated with triaryl phosphine using Staudinger reaction through formation of a stable amide bond. This approach was tested through attaching biotin to the cell surface by using biotinylated phosphine. Additionally, azides were covalently modified with linear alkynes via Cu-catalyzed azide–alkyne cycloaddition and with a variety of cyclooctynes via the strain-promoted azide–alkyne cycloaddition (also termed "Cu-free click chemistry") [50–52]. Since this cycloaddition avoids the usage of toxic metal catalysts, Cu-free click chemistry is more acceptable as a cell-compatible reaction. This biosynthetic metabolic labeling of cells is a versatile route to present chemical functionalities which are otherwise not present on the cell surface in appreciable surface concentrations for covalent modification. Additionally, this approach is highly chemoselective and can be performed in cell-compatible conditions without inducing any changes to the surrounding environment. Similar to this strategy, another approach involves enzymatic transformation of the cell surface proteins to introduce functional molecules for covalent transformation of the cell surface, for example, posttranslational modification of acyl carrier protein (ACP) on the cell surface by phosphopantetheine transferase (PPTase) transfer 4′-phosphopantetheine from co-enzyme A (CoA) to a serine residue of ACP [53,54]. By using several CoA derivatives, it was possible to functionalize cells with chemically diverse compounds such as fluorophores, affinity ligands, or quantum dots [54]. Immobilizing functional groups on cell surfaces through these strategies can provide tools for sensing and imaging, understanding cellular interactions, and enabling new molecular engineering tools for synthetic biology.

In addition to the introduction of functional groups by biosynthetic mechanism, functional groups can be introduced to cell membranes through physical methods for subsequent covalent cross-linking. By using this approach, ketone functionalities were physically introduced to the cell surface through vesicle-based delivery by spontaneous fusion of lipid to the cell membrane. Physically immobilized ketone was subsequently reacted to oxyamine for covalent modification, for example, covalent reaction of oxyamine-functionalized dye with ketone-functionalized cells resulted in dye conjugated cells [55]. This strategy is also useful for promoting cellular assembly where covalent reactions between ketone-functionalized cells and oxyamine-functionalized cells promoted three-dimensional cellular aggregations [8].

Thus, indirect covalent modification to the cell membrane via noncovalent approaches has significant importance. This approach is particularly relevant when appropriate functional groups, which can participate in direct covalent modification, are not available on the cell surface.

2.3 RELEVANCE OF COVALENT CELL SURFACE MODIFICATION

Modification of cell membranes using covalent strategies is extensively used to decorate cell surfaces with target molecules, both synthetic and biological, nanoparticle probes, and different structures. These target molecules can be small molecules, for example, biotin, pharmaceutical drugs, fluorescent probes, or they can be macromolecules including synthetic polymers, such as PEG or polyvinyl alcohol. Biological macromolecules like proteins, antibodies, aptamers, are also attached to cell surfaces to impart unique biofunctionalities to cells [25,56]. Structural moieties such as nanoparticles and quantum dots are immobilized on the cell for delivery of therapeutic molecules to the cell and noninvasive tracking of cells, respectively [16,34,57]. Choice of reaction depends on the type of target and the feasibility of a particular reaction mechanism to achieve the conjugation under cell-compatible conditions with high selectivity. Furthermore, site density, distribution, mobility, and stability of the covalently attached molecules are dependent on the reaction mechanism. Most importantly, these modifications should be able to fulfill the functional requirements of the cell for a given application with precise control.

In the previous sections, the specific use of covalent cell modifications is mentioned in the context of particular approaches. For example, human MSCs covalently modified by a Sialyl Lewis X through biotin–streptavidin were able to target the selectins in activated endothelium [15]; conjugation of aptamer on human MSC membranes were able to sense the PDGF [16]; and nanoparticles conjugated to T-cell surfaces influenced cellular function through controlled delivery of drugs [34]. These examples demonstrate that the covalent remodeling of cell membranes has significant therapeutic applicability and is potentially useful for interrogating fundamental properties of cells both *in vitro* and *in vivo*. Specific applications of cell surface modifications and detailed analysis of these technologies are mentioned in other chapters. In general, covalent cell surface engineering has been used for a significant period of time for various applications and purposes. Molecular biology and immunology of cells are studied by selectively conjugating molecules to the cell surface. Pharmaceutical and toxicological effects of synthetic molecules and their mechanistic pathways of action are also elucidated by using covalent cell surface modification approaches. From an analytical perspective, cell surface conjugation is traditionally used to label cells for imaging and sensing purposes. Thus, covalent cell surface modification is an established technology with multifaceted goals. However, the use of these approaches has seen a significant surge in recent years for developing novel tools and technologies which can be used for therapeutic and diagnostic purposes.

In particular, covalent modification of cells allows controlling their interactions with the extracellular matrix and neighboring cells for regenerative applications in tissue engineering. Furthermore, this approach has enabled targeted delivery of cells to appropriate tissue microenvironments and also the ability to sense the cellular niche in a spatiotemporal manner. This approach has significant impact on developing cell-based therapies where cells are systemically delivered to a particular site. Most importantly, chemically functionalized cells, following their localization, can exert controlled cellular responses that are observed therapeutically and diagnostically. Controlling cell fate and functioning by redesigning the cell membrane with defined functional molecules and structures will continue to impact on fundamental science and will provide improved biomedical technologies for several applications.

2.4 FUTURE PERSPECTIVES

With increased use of this covalent cell surface modification technique, it is important to understand the impact and usefulness of this strategy for a given purpose. Most studies indicate that covalent modification has minimal impact on cells, but it is important to examine its long-term effect in terms of toxicity and its potential to induce malignant features. Furthermore, localization and stability of covalently immobilized molecules and their fate should be critically investigated. These features vary significantly depending on the modification technique and the nature of the immobilized molecule. For example, cells modified with PEG-NHS are only stable for 48 h [22], whereas cells modified with streptavidin through covalent conjugation of biotin to cell surface showed enhanced stability up to 7 days [13]. Although long-term effects in terms of cell viability and functions were not detected, it is critical to monitor the cells for a longer period to maximize the efficiency of covalent cell surface modification. Thus, selection of proper modification techniques and optimization of strategies are crucial to ensure the availability and accessibility of the immobilized molecule for any given application. In the long term, it is important to characterize if the ligand is internalized or shredded by the cells and how cell division can change the density of functionalized ligand. Achieving proper spatial and temporal presentation of the ligands on the cell membrane is crucial for any application, particularly for therapeutic applications. Therefore, every covalent modification strategy needs careful examination to understand these features which are critical to ensure the effectiveness of this approach. Particularly, use of covalent cell engineering for *in vivo* and clinical applications requires proper study to ensure safety and higher efficacy. Finally, using cell surface modification *in situ* and in response to environmental stimulus can provide unique avenues to control cell functioning, both in physiological and disease conditions.

2.5 CONCLUSIONS

Cell surface modification is a powerful tool to modify the molecular composition of cell surfaces. Development of novel chemistries and material science tools has enabled the alteration of the cell surface through covalent modification techniques for a variety of applications. In particular, cell surfaces can be directly modified through covalent reactions if specific functional groups are present and can be accessed by external reagents under cell-compatible conditions. Alternately, covalent modifications can be performed following the noncovalent introduction of functional groups. Depending on the application, each reaction strategy needs optimization and complete investigation to ensure effective functionalization. In the future, development of novel molecules and conjugation strategies should be able to combine chemical mechanisms with biomolecular approaches to provide new avenues for controlling cell function and fate.

References

[1] Penn MS, Mangi AA. Genetic enhancement of stem cell engraftment, survival, and efficacy. Circ Res 2008;102(12):1471–82.
[2] Sackstein R, Merzaban JS, Cain DW, Dagia NM, Spencer JA, Lin CP, et al. Ex vivo glycan engineering of CD44 programs human multipotent mesenchymal stromal cell trafficking to bone. Nat Med 2008; 14(2):181–7.
[3] Verma IM, Weitzman MD. Gene therapy: twenty-first century medicine. Annu Rev Biochem 2005;74(1): 711–38.
[4] Zhao W, Teo GSL, Kumar N, Karp JM. Chemistry and material science at the cell surface. Mater Today 2010;13(4):14–21.
[5] Kellam B, De Bank PA, Shakesheff KM. Chemical modification of mammalian cell surfaces. Chem Soc Rev 2003;32(6):327–37.
[6] Mahal LK, Bertozzi CR. Engineered cell surfaces: fertile ground for molecular landscaping. Chem Biol 1997;4(6):415–22.
[7] Ko IK, Kean TJ, Dennis JE. Targeting mesenchymal stem cells to activated endothelial cells. Biomaterials 2009;30(22):3702–10.
[8] Dutta D, Pulsipher A, Luo W, Yousaf MN. Synthetic chemoselective rewiring of cell surfaces: generation of three-dimensional tissue structures. J Am Chem Soc 2011;133(22):8704–13.
[9] Stevens MM, George JH. Exploring and engineering the cell surface interface. Science 2005;310(5751): 1135–8.
[10] Stephan MT, Irvine DJ. Enhancing cell therapies from the outside in: Cell surface engineering using synthetic nanomaterials. Nano Today 2011;6(3):309–25.
[11] Giepmans BNG, Adams SR, Ellisman MH, Tsien RY. The fluorescent toolbox for assessing protein location and function. Science 2006;312(5771):217–24.
[12] McMahon HT, Gallop JL. Membrane curvature and mechanisms of dynamic cell membrane remodelling. Nature 2005;438(7068):590–6.
[13] Sarkar D, Vemula PK, Teo GSL, Spelke D, Karnik R, Wee LY, et al. Chemical engineering of mesenchymal stem cells to induce a cell rolling response. Bioconjug Chem 2008;19(11):2105–9.

[14] Sarkar D, Zhao W, Gupta A, Loh W, Karnik R, Karp J. Cell surface engineering of mesenchymal stem cells. In: Vemuri M, Chase LG, Rao MS, editors. Mesenchymal stem cell assays and applications. New York: Humana Press; 2011. p. 505–23.
[15] Sarkar D, Spencer JA, Phillips JA, Zhao W, Schafer S, Spelke DP, et al. Engineered cell homing. Blood 2011;118(25):e184–91.
[16] Zhao W, Schafer S, Choi J, Yamanaka YJ, Lombardi ML, Bose S, et al. Cell-surface sensors for real-time probing of cellular environments. Nat Nanotechnol 2011;6(8):524–31.
[17] Mason J, Tomlinson I, Rosenthal S, Blakely R. Labeling cell-surface proteins via antibody quantum dot streptavidin conjugates. In: Rosenthal S, Wright D, editors. Nano Biotechnology Protocols. New York: Humana Press; 2005. p. 35–50.
[18] Cheng H, Kastrup CJ, Ramanathan R, Siegwart DJ, Ma M, Bogatyrev SR, et al. Nanoparticulate cellular patches for cell-mediated tumoritropic delivery. ACS Nano 2010;4(2):625–31.
[19] deBlaquiere J, Burgess AW. Affinity purification of plasma membranes. J Biomol Tech 1999;10(2):64–71.
[20] Rybak JN, Ettorre A, Kaissling B, Giavazzi R, Neri D, Elia G. In vivo protein biotinylation for identification of organ-specific antigens accessible from the vasculature. Nat Methods 2005;2(4):291–8.
[21] Lim CW, Le HT, Han JH, Kim DH, Jang JG, Kim TW. New Water-soluble Alkynylating Agent for Cell Surface Protein: Sulfosuccinimidyl 4-Pentynoate. Bull Korean Chem Soc 2013;34(6):1895–8.
[22] Teramura Y, Kaneda Y, Totani T, Iwata H. Behavior of synthetic polymers immobilized on a cell membrane. Biomaterials 2008;29(10):1345–55.
[23] Contreras JL, Xie D, Mays J, Smyth CA, Eckstein C, Rahemtulla FG, et al. A novel approach to xenotransplantation combining surface engineering and genetic modification of isolated adult porcine islets. Surgery 2004;136(3):537–47.
[24] Chapanian R, Constantinescu I, Medvedev N, Scott MD, Brooks DE, Kizhakkedathu JN. Therapeutic cells via functional modification: influence of molecular properties of polymer grafts on in vivo circulation, clearance, immunogenicity, and antigen protection. Biomacromolecules 2013;14(6):2052–62.
[25] Cheng H, Byrska-Bishop M, Zhang CT, Kastrup CJ, Hwang NS, Tai AK, et al. Stem cell membrane engineering for cell rolling using peptide conjugation and tuning of cell–selectin interaction kinetics. Biomaterials 2012;33(20):5004–12.
[26] Hsiao SC, Shum BJ, Onoe H, Douglas ES, Gartner ZJ, Mathies RA, et al. Direct cell surface modification with DNA for the capture of primary cells and the investigation of myotube formation on defined patterns. Langmuir 2009;25(12):6985–91.
[27] Hashemi-Najafabadi S, Vasheghani-Farahani E, Shojaosadati SA, Rasaee MJ, Armstrong JK, Moin M, et al. A method To optimize PEG-coating of red blood cells. Bioconjug Chem 2006;17(5):1288–93.
[28] Blackall DP, Armstrong JK, Meiselman HJ, Fisher TC. Polyethylene glycol–coated red blood cells fail to bind glycophorin A–specific antibodies and are impervious to invasion by the Plasmodium falciparum malaria parasite. Blood 2001;97(2):551–6.
[29] Fisher TC. PEG-coated red blood cells-simplifying blood transfusion in the new millennium? Immunohematology 2000;16(1):37–48.
[30] Lublin DM. Universal RBCs. Transfusion 2000;40 (11):1285–9.
[31] Chen H, Hall S, Heffernan B, Thompson NT, Rogers MV, Rhodes J. Convergence of schiff base costimulatory signaling and TCR signaling at the level of mitogen-activated protein kinase ERK2. J Immunol 1997;159(5):2274–81.
[32] Fretz MM, Penning NA, Al-Taei S, Futaki S, Takeuchi T, Nakase I, et al. Temperature-, concentration- and cholesterol-dependent translocation of L- and D-octa-arginine across the plasma and nuclear membrane of CD34+ leukaemia cells. Biochem J 2007;403(2): 335–42.
[33] Nacharaju P, Boctor FN, Manjula BN, Acharya SA. Surface decoration of red blood cells with maleimidophenyl-polyethylene glycol facilitated by thiolation with iminothiolane: an approach to mask A, B, and D antigens to generate universal red blood cells. Transfusion 2005;45(3):374–83.
[34] Stephan MT, Moon JJ, Um SH, Bershteyn A, Irvine DJ. Therapeutic cell engineering with surface-conjugated synthetic nanoparticles. Nat Med 2010;16(9):1035–41.
[35] Kichler A, Remy JS, Boussif O, Frisch B, Boeckler C, Behr JP, et al. Efficient gene delivery with neutral complexes of lipospermine and thiol-reactive phospholipids. Biochem Biophys Res Commun 1995;209 (2):444–50.
[36] Torres AG, Gait MJ. Exploiting cell surface thiols to enhance cellular uptake. Trends Biotechnol 2012;30 (4):185–90.
[37] Sahaf B, Heydari K, Herzenberg LA, Herzenberg LA. The extracellular microenvironment plays a key role in regulating the redox status of cell surface proteins in HIV-infected subjects. Arch Biochem Biophys 2005;434(1):26–32.
[38] Gahmberg CG, Tolvanen M. Nonmetabolic radiolabeling and tagging of glycoconjugates. In: Lennarz WJ, Hart GW, editors. Methods in enzymology. San Diego: Academic Press; 1994. p. 32–44.

[39] Norgard KE, Han H, Powell L, Kriegler M, Varki A, Varki NM. Enhanced interaction of L-selectin with the high endothelial venule ligand via selectively oxidized sialic acids. Proc Natl Acad Sci USA 1993;90(3):1068–72.

[40] Wilchek M, Spiegel S, Spiegel Y. Fluorescent reagents for the labeling of glycoconjugates in solution and on cell surfaces. Biochem Biophys Res Commun 1980;92 (4):1215–22.

[41] Bayer EA, Ben-Hur H, Wilchek M. Biocytin hydrazide—A selective label for sialic acids, galactose, and other sugars in glycoconjugates using avidin-biotin technology. Anal Biochem 1988;170(2):271–81.

[42] Holden CA, Yuan Q, Yeudall WA, Lebman DA, Yang H. Surface engineering of macrophages with nanoparticles to generate a cell-nanoparticle hybrid vehicle for hypoxia-targeted drug delivery. Int J Nanomedicine 2010;5:25–36.

[43] Zeng Y, Ramya TN, Dirksen A, Dawson PE, Paulson JC. High-efficiency labeling of sialylated glycoproteins on living cells. Nat Methods 2009;6(3):207–9.

[44] Ramya TNC, Weerapana E, Cravatt BF, Paulson JC. Glycoproteomics enabled by tagging sialic acid- or galactose-terminated glycans. Glycobiology 2013;23(2): 211–21.

[45] Srinivas L, Colburn NH. Preferential oxidation of cell surface sialic acid by periodate leads to promotion of transformation in JB6 cells. Carcinogenesis 1984;5(4): 515–9.

[46] Laughlin ST, Bertozzi CR. Imaging the glycome. Proc Natl Acad Sci 2009;106(1):12–7.

[47] Mahal LK, Yarema KJ, Bertozzi CR. Engineering chemical reactivity on cell surfaces through oligosaccharide biosynthesis. Science 1997;276(5315):1125–8.

[48] Saxon E, Bertozzi CR. Cell surface engineering by a modified staudinger reaction. Science 2000;287(5460): 2007–10.

[49] Yarema KJ, Mahal LK, Bruehl RE, Rodriguez EC, Bertozzi CR. Metabolic delivery of ketone groups to sialic acid residues: application to cell surface glycoform engineering. J Biol Chem 1998; 273(47):31168–79.

[50] Hong V, Steinmetz NF, Manchester M, Finn MG. Labeling live cells by copper-catalyzed alkyne–azide click chemistry. Bioconjug Chem 2010;21(10):1912–6.

[51] Agard NJ, Prescher JA, Bertozzi CR. A strain-promoted [3 + 2] azide–alkyne cycloaddition for covalent modification of biomolecules in living systems. J Am Chem Soc 2004;126(46):15046–7.

[52] Baskin JM, Prescher JA, Laughlin ST, Agard NJ, Chang PV, Miller IA, et al. Copper-free click chemistry for dynamic in vivo imaging. Proc Natl Acad Sci 2007;104(43):16793–7.

[53] Mootz HD, Finking R, Marahiel MA. 4′-phosphopantetheine transfer in primary and secondary metabolism of bacillus subtilis. J Biol Chem 2001;276(40): 37289–98.

[54] George N, Pick H, Vogel H, Johnsson N, Johnsson K. Specific labeling of cell surface proteins with chemically diverse compounds. J Am Chem Soc 2004;126(29): 8896–7.

[55] Dutta D, Pulsipher A, Luo W, Mak H, Yousaf MN. Engineering cell surfaces via liposome fusion. Bioconjug Chem 2011;22(12):2423–33.

[56] Zhao W, Loh W, Droujinine IA, Teo W, Kumar N, Schafer S, et al. Mimicking the inflammatory cell adhesion cascade by nucleic acid aptamer programmed cell-cell interactions. FASEB J 2011;25(9): 3045–56.

[57] Jaiswal JK, Mattoussi H, Mauro JM, Simon SM. Long-term multiple color imaging of live cells using quantum dot bioconjugates. Nat Biotechnol 2003;21(1): 47–51.

CHAPTER

3

Bioconjugation Reactions in Living Cells: Development, Advances, and Applications of Glycan-Specific Technologies

Laura Alberch and Kevin J. Yarema

Department of Biomedical Engineering and the Translational Tissue Engineering Center, The Johns Hopkins University, Baltimore, MD, USA

OUTLINE

J.M. Karp & W. Zhao (Eds): *Micro- and Nanoengineering of the Cell Surface.*
DOI: http://dx.doi.org/10.1016/B978-1-45-573146-6.00003-9

3.1 INTRODUCTION

The surfaces of living cells are covered by a layer of complex carbohydrates, collectively known as the glycocalyx (Figure 3.1A), that play critical roles in their interactions with the microenvironment, influence development from conception to adulthood, and are implicated in many diseases. Over the past half century, the study of these sugars has lagged behind other biopolymers (e.g., nucleic acids and proteins) due in part to the complexity of these molecules [1] and in part due to a lack of experimental methods capable of addressing unique challenges that they pose (e.g., their lack of template-driven synthesis precludes methods such as polymer chain reaction that allows the detection and characterization of vanishingly small amounts of nucleic acids) [2]. In recent years, the increased sensitivity of methods such as mass spectrometry [3] coupled with bioinformatics [4,5] and computational tools [6] as well as high-throughput approaches such as carbohydrate [7] and lectin arrays [8,9] has led to major advances in the characterization of complex sugars and their roles in biological systems.

A major limitation of the majority of tools currently available to investigate complex carbohydrates, including the just-mentioned mass spectrometry and array technologies, is that the samples must be isolated from a biological sample for analysis, thus precluding *in vivo* analysis. The major exception to this limitation is provided by the combined application of bioorthogonal ligation chemistries and metabolic oligosaccharide engineering (MOE) methods, which are well suited even for demanding tasks, such as the *in vivo* study of the dynamic glycome that has already led to the visualization of glycans in living *Caenorhabditis elegans* [10] and zebra fish [11] with minimal disturbance to native biological processes. This chapter provides a detailed description of both bioorthogonal ligation reactions (in Section 2) and their utilization in MOE experiments (in Section 3) but first briefly introduces each concept here in the introduction.

A bioorthogonal ligation is defined as a conjugation that occurs in a living system without disturbing native cellular processes. Ideally, they should be highly chemoselective, nontoxic, and form stable products under physiological conditions [12]. The high selectivity of these reactions was initially exploited for drug delivery against Ehrlich ascites carcinoma cells through bioorthogonal aldehyde-based chemistry [13]. Since then, the ability of certain monosaccharides (primarily sialic acid) to be selectively oxidized (Figure 3.1B) has allowed this approach to be extended toward the labeling of cell surface glycans and the imaging of living L6 myoblasts in cell culture [14]. Although this method has long offered a fast and convenient way of measuring sialic acid [15] and more recently additional sugars such as galactose [16] *in vitro*, it is not applicable for *in vivo* experiments due to the cytotoxicity of the oxidizing reagents or other chemicals required. Moreover, the detection of naturally occurring, oxidized monosaccharides suffers from the limited repertoire of chemical functional groups available, namely aminoxy [17], hydrazide [14], and thiosemicarbazide [18] groups.

The limited chemical repertoire available naturally for exploitation through bioorthogonal

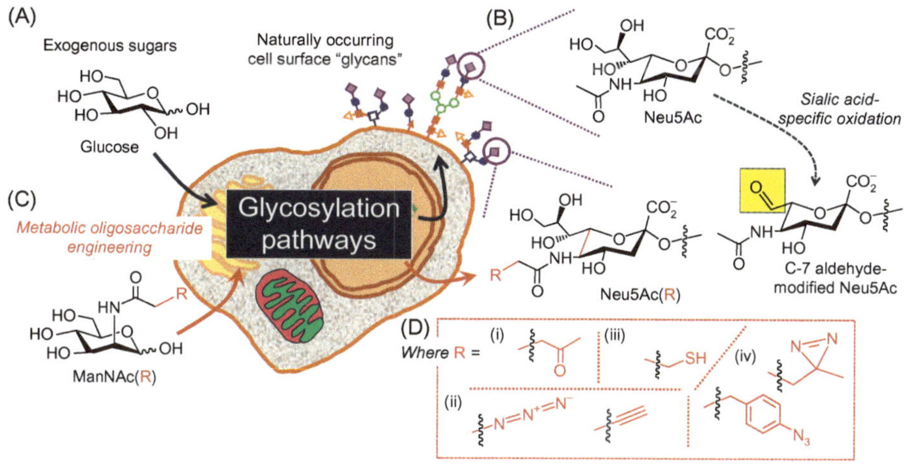

FIGURE 3.1 Overview of cell surface glycosylation and the installation of chemical functional groups suitable for bioorthogonal conjugation reactions. (A) Natural cell surface glycans are ultimately derived from the intake of exogenous sugars (primarily dietary glucose, shown) and their subsequent processing by the glycosylation pathways of a cell. (B) Once on the cell surface, certain sugars—primarily sialic acid (the most common human form of this sugar, Neu5Ac, is shown)—can be selectively oxidized to install aldehyde groups that can be exploited for bioorthogonal ligation reactions as discussed in the main text. (C) MOE is a strategy where nonnatural monosaccharide analogs (such as "R"-modified *N*-acetylmannosamine (ManNAc), shown) are introduced into the glycosylation pathways of a cell and metabolically installed on the cell surface in place of the natural biosynthetic products (in this example, as "R"-modified forms of Neu5Ac). (D) By using this approach, numerous chemical groups appropriate for bioorthogonal ligation reactions can be installed into glycans including (i) ketones, which have chemical reactivity similar to the aldehydes shown in (B); (ii) azides and alkynes, the functional groups required for the increasingly ubiquitous click chemistries; (iii) thiol groups, which can adhere to gold or chemical functional groups such as maleimide; and (iv) diazirines and arylazides used for photoactivated ligation reactions.

ligation conjugation reactions was overcome through MOE. MOE is a two-step technique where biological macromolecules can be modified by the addition of a synthetic precursor carrying a chemical reporter group (Figure 3.1C). These modified sugars are taken up by the cell, intercept the metabolic pathway without being perceived as being unnatural by the endogenous enzymes, and are finally incorporated in the glycocalyx. In this manner, MOE allows for the noninvasive manipulation of cell surface glycans. This technology has dramatically increased the chemical functional groups available in cell surface glycans for conjugation via bioorthogonal ligation; these groups include ketones, azides, alkynes, thiols, and phenylaryl azides as shown in Figure 3.1D and discussed in detail in review articles [19] as well as in the next section of this chapter in the context of the ability of these functional groups to participate in bioorthogonal ligation reactions.

3.2 BIOORTHOGONAL CHEMICAL LIGATION REACTIONS FOR GLYCAN LABELING

This section provides an overview of the main bioorthogonal ligation reactions that have been applied toward the study of surface carbohydrates on living cells starting with a brief mention of aldehydes that can be installed in glycans via chemical oxidation (Figure 3.1B) and then covering the expanded repertoire of functional groups now available through MOE (Figure 3.1D) in greater depth. The advantages and limitations of each existing method are discussed to show how a "chemical biology" strategy has already overcome significant hurdles in developing methods to investigate complex carbohydrates and also to illustrate how opportunities remain to further improve this promising approach.

3.2.1 Aldehyde- and Ketone-Based Bioorthogonal Reactions

3.2.1.1 *Aldehyde-Based Reactions*

Aldehyde-based ligation has been widely applied toward the quantification of cell surface sialic acids that do not *a priori* carry a bioorthogonal group for labeling (Figure 3.2). In this method, the vicinal diols in this sugar are selectively oxidized by sodium periodate to provide an aldehyde that—in early colorimetric experiments such as the periodate-resorcinol assay—was reacted with resorcinol and copper and the resulting light absorbance was measured [15,20]. Subsequently, conjugation strategies have been devised where the sugar bearing the aldehyde is subjected to aniline-catalyzed oxime ligation (known as **p**eriodate oxidation and **a**niline-catalyzed oxime **l**igation (PAL)) to label sialic acid; the aniline catalyst allows the use of low concentrations of aminoxy-biotin at physiological pH [17,21].

Although used for decades, new developments in a general approach to chemically modify cell surface glycans and then label them (or quantify via colorimetric methods) continue to this day as evidenced by recent reports of new methods to detect sialic acids, for example by using a thiosemicarbazide method [18]. In addition, these techniques are being expanded to include additional monosaccharides, such as galactose [17]. The use of bioorthogonal ligation reactions for glycan labeling, however, has increased much more dramatically through the metabolic incorporation of unnatural sugars that carry reactive functional groups. The rich chemical diversity of these groups (outlined in Figure 3.1D and reviewed extensively elsewhere [19,22–24]) has encouraged the development of a wide range of bioorthogonal chemical ligations. Thus, because of the incorporation of a wide range of functional groups into cell surface glycans via MOE (Figure 3.1B), we are no longer limited to using aldehydes as reactive handles; perhaps ironically however, the first demonstration of the ability of MOE to install bioorthogonal functional groups into glycans was accomplished with a ketone-modified analog [25,26] that had very similar chemical features as nonmetabolically produced aldehydes.

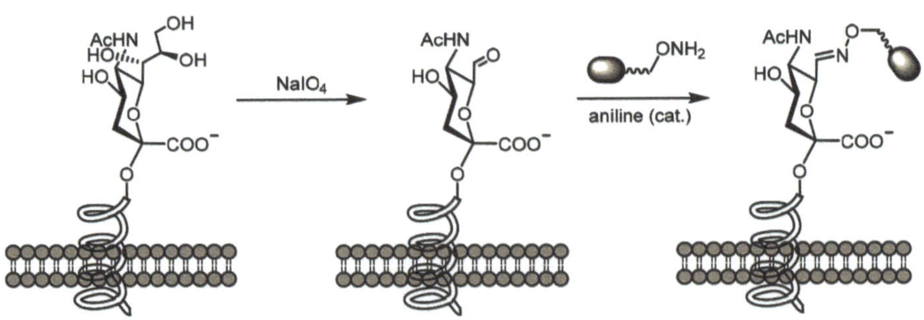

FIGURE 3.2 Aniline-catalyzed oxime ligation for the labeling of sialic acids that do not carry a bioorthogonal chemical reporter.

3.2.1.2 *Metabolic Incorporation Expands Bioorthogonal Ligation to Include Ketone-Based Reactions*

The transition of MOE into the realm of bioorthogonal ligation reactions occurred when the Bertozzi group showed that alkyl "R" groups originally demonstrated by the Reutter group in experiments that pioneered this technology in the early 1990s [27] could be replaced with ketone-modified moieties [25]. Because ketones have chemical reactivity similar to aldehydes, their metabolic incorporation into carbohydrates on the cell surface suffers from many of the limitations of aldehyde-oxidized glycans mentioned above, but as a "proof-of-principle" model the similar chemical reactivity of the two groups allowed the wealth of bioorthogonal ligation established with aldehydes to be applied to MOE rapidly. In particular, cells expressing metabolically ketone-modified sialic acids could be readily conjugated with aminoxy- or hydrazide-conjugated probes (Figure 3.3) [26].

Each reagent used for bioorthogonal ligation of aldehydes and ketones has benefits and drawbacks. For example, hydrazide conjugation reagents are less expensive but the formation of the hydrazone product only occurs readily at subphysiological pH (e.g., pH 5.0) and is reversible (i.e., 100-fold less stable than a reaction between an oxime and a ketone [28,29]) resulting in fewer conjugated molecules. This drawback can be partially alleviated by using excess reagents to shift the equilibrium toward the product. In some cases, the unique features of hydrazide- and aminoxy-conjugated probes can be used in tandem to gain new insights into biological systems; one example is their comparative use to measure kinetic parameters for small-molecule drug delivery by covalent cell surface targeting [29]. Regardless of whether oximes or hydrazides are used for conjugation with ketones or aldehydes, their use *in vivo* is limited by competition with endogenous keto-metabolites [12]. Moreover, the metabolic incorporation of sugar analogs functionalized with a ketone group, exemplified by Ac_4ManLev, is severely restricted in many nonhuman cell lines [19,26] and is rarely if ever as efficient as more recently developed azido-analogs [30], which will be discussed next.

3.2.2 Azide- and Alkyne-Based Bioorthogonal Reactions

A major advance in the use of bioorthogonal ligation reactions to investigate glycans came when the Bertozzi group showed that azide groups could be metabolically incorporated into cell surface carbohydrates [30]. As discussed in more detail below, the cell surface azides were first subject to a modified Staudinger ligation reaction and later to the well-known copper-catalyzed "click" reaction and recently also to

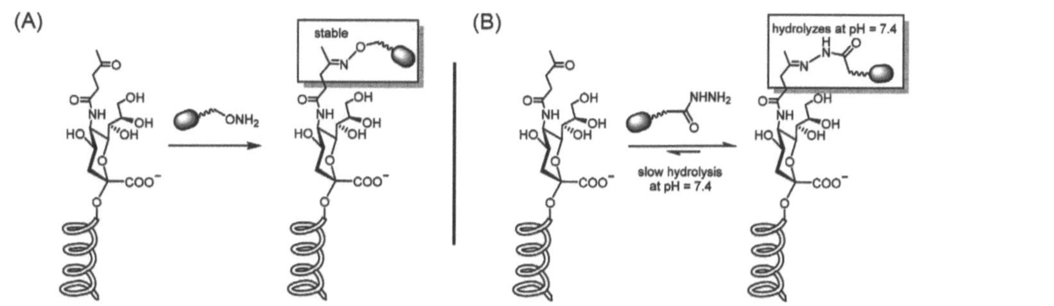

FIGURE 3.3 Strategies for labeling cell-surface keto-sialic acids using (A) oxime- and (B) hydrazide-conjugated probes.

strain-promoted cycloaddition reactions. It should be noted that the discussion below is focused on using MOE to install the azide group onto the cell surface, but in many cases the complementary coupling partner for click chemistries (e.g., alkynes [31,32]) can also be metabolically incorporated into glycans thereby providing this methodology with great flexibility and versatility.

3.2.2.1 *Classical and Modified Staudinger Ligation Reactions*

A benign chemoselective ligation was developed by the Bertozzi group based on the Staudinger ligation where an azide reacts with a phosphine (Figure 3.4) [30,33]. While the classical Staudinger reaction reduces azides to amines, the intermediate aza-ylide readily hydrolyzes to the phosphine oxide and an amide in the presence of water making it unsuitable for use in physiological conditions. The Bertozzi group envisioned a modification of this reaction as a potential method for bioconjugation by installing a methyl ester to the aryl phosphine, which acts as an electrophilic trap. In this manner, the aza-ylide reacts with the ester moiety faster than hydrolysis of the aza-ylide, resulting in a stable amide bond for the ligation of bioorthogonal probes [12]. The carboxylic acid group allows for facile attachment of any biological probe. This ligation strategy was first used for proteins [34] and protein-associated lipids [35] and subsequently has been used extensively for the chemoselective targeting of cell surface glycans [36–38]. Another step forward was in the development of the "traceless" Staudinger ligation, where the ester [39] or thioester [40] attached to the aza-ylide intermediate is displaced and hydrolysis of the resulting intermediate affords an amide bond and liberates the aryl phosphine oxide.

In contrast to the hydrazone ligation used with ketone-modified MOE analogs, the efficiency of the modified Staudinger ligation does not vary with pH. Moreover, the azide tag is abiotic and its small size allows the use of azido sugars to undergo metabolic incorporation more efficiently than their keto counterparts. However, the use of this reaction has been limited by the oxygen sensitivity of phosphines [41]. The slow reaction kinetics of phosphine

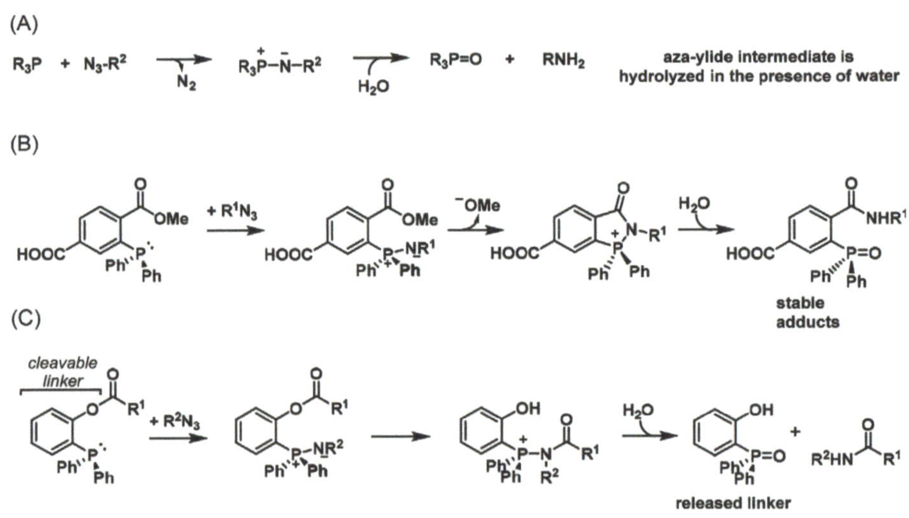

FIGURE 3.4 Staudinger ligations showing the (A) classical, (B) Bertozzi-modified, and (C) traceless variations of this reaction.

reagents require high concentrations for *in vivo* experiments, which results in an elevated background signal when using fluorogenic probes. This problem can be alleviated by the increased sensitivity observed when a luciferin-conjugated phosphine designed to release luciferin upon Staudinger ligation is used. Once luciferin diffuses into the cell, it reacts with luciferase and is converted to oxyluciferin with the release of a photon of visible light [36]. This is an example of bioluminescence imaging, where the visible light generated upon reaction with luciferase can be detected by a highly sensitive CCD camera [42].

3.2.2.2 Click Reactions: the Copper-Catalyzed [3+2] Azide–Alkyne Cycloaddition

The most common and simplest procedure for the conjugation of cells to biomolecules via an azide functional group is through a click reaction. The term "click chemistry" was coined by Barry Sharpless to describe any reaction which would follow a set of criteria such as high yields, high stereospecificity, physiologically stable, does not require chromatographic separation, forms only benign by-products, and can be carried out in water [43]. The most widely used click reaction is the copper-catalyzed [3 + 2] azide–alkyne cycloaddition, often abbreviated as CuAAC reaction (Figure 3.5). Its great success is rooted in the fact that the azide and the triazole product are essentially inert to most components in a biological environment. In addition, the azide is small in size and usually is ignored by endogenous enzymes. The CuAAC-reaction mechanistic details have been thoroughly studied and reviewed [44,45]. Its applications are countless, as recently reviewed in depth [46], and so we will focus on its recent advances in bioconjugation of living cells.

A major limitation to the use of the CuAAC reaction in living systems is the cytotoxicity imparted by the copper catalyst, which is required to speed up the rate of reaction at levels useful for biological systems. The high concentrations of copper catalyst can be reduced by the addition of Cu(I) stabilizing ligands (Figure 3.6). These ligands mainly function by protecting the copper catalyst from oxidation by oxygen reactive species and by preventing the formation of copper clusters, thus increasing the effective concentration of copper. Although *tris*-(benzyltriazolylmethyl)amine (TBTA) has been the most widely used ligand [47,48], others have been designed with modifications at the triazole side chain to improve their physical properties and kinetics. For example, the acetic acid group in (2-[4-{(bis[(1-tert-butyl-1H-1,2,3-triazol-4-yl)methyl]-amino)methyl}-1H-1,2,3-triazol-1-yl]acetic acid) (BTTAA) improves its water solubility as it becomes acetate at physiological pH, while the two *tert*-butyl groups prevent the polymerization of copper acetylides [49]. Introducing a highly polar sulfate group as in

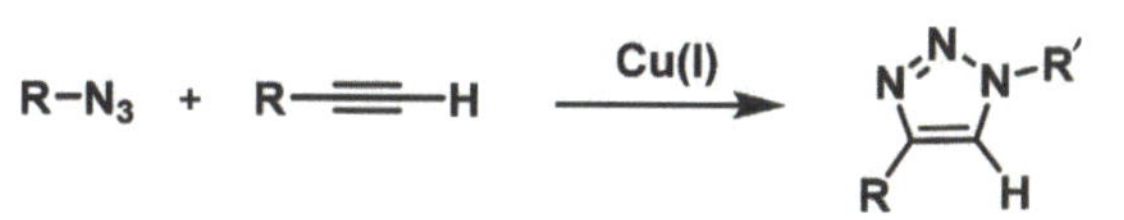

FIGURE 3.5 The copper-catalyzed azide–alkyne cycloaddition reaction.

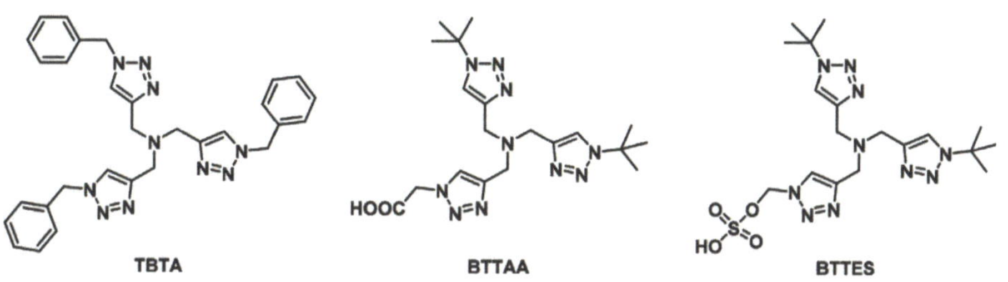

FIGURE 3.6 Examples of ligands for the CuAAC.

the case of 2-[4-{(bis[(1-tert-butyl-1H-1,2,3-triazol-4-yl)methyl]amino)-methyl}-1H-1,2,3-triazol-1-yl]-ethyl hydrogen sulfate (BTTES) minimizes the cell membrane permeability of the coordinated copper [50]. The development of these ligands has encouraged the revisiting of the CuAAC for use in cell surface labeling, as it is still the fastest bioorthogonal reaction and it is compatible with both azide- and alkyne-tagged carbohydrates.

3.2.2.3 *Strain-Promoted Azide–Alkyne Cycloaddition Reactions*

An ingenious alternative to the use of copper in the azide–alkyne cycloaddition is to increase the reactivity of the alkyne (lowering LUMO energy) by constraining it in a ring. This strategy, known as a strain-promoted azide–alkyne cycloaddition (SPAAC), can be used under physiological conditions without the need of a catalyst, and consequently it is often referred to as the copper-free click reaction (Figure 3.7).

Novel cyclooctynes are constantly being developed (Figure 3.8). Their design has been assisted by computational models and theoretical studies that explain and predict the enhanced reactivity of novel cyclooctynes [51,52]. An important design strategy is centered in decreasing the electron density around the triple bond and thus lowering the LUMO energy, either by incorporation of propargylic electron withdrawing fluorine groups (DIFO) [53] or by installing two adjacent aromatic rings which allow the delocalization of the alkyne π-system (DIBO) [54]. The fast reaction kinetics of these highly strained biarylazacylooctynones with azides allow for their use in visualizing metabolically labeled glycans [55,56].

Other modifications to the cyclooctynes focus on increasing their solubility in water, a common problem encountered with the first generation of cyclooctyne derivatives. Their poor hydrophilicity resulted in high nonspecific binding to serum proteins and promoted sequestration by

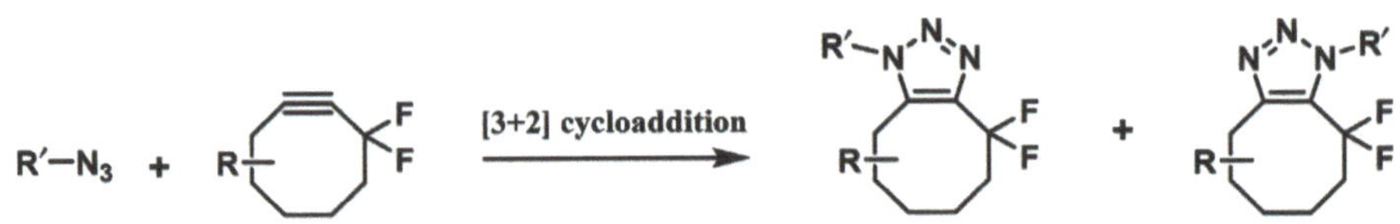

FIGURE 3.7 Strain-promoted azide–alkyne cycloaddition.

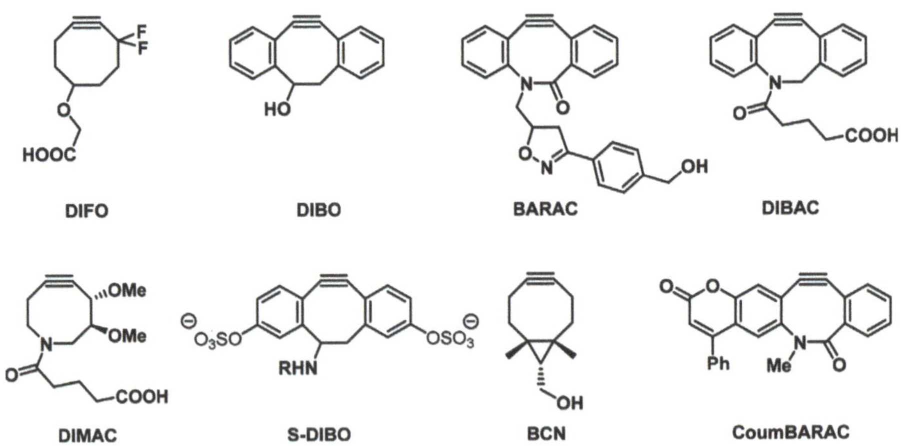

FIGURE 3.8 Examples of cyclooctyne reagents used for labeling with azide-bearing chemical reporters.

membranes thereby reducing bioavailable concentrations [57]. BARAC, DIBAC, and DIMAC possess a heteroatom in the eight-membered ring, which greatly improves aqueous solubility and pharmacokinetics [57]. Incorporation of sulfated groups such as in S-DIBO renders the compounds completely water soluble, while retaining excellent reaction kinetics. Moreover, unlike other biarylcyclooctynes, S-DIBO does not pass through the cell membrane and so exclusively labels extracellular glycoconjugates further decreasing the background signal [58]. Novel bicyclo−[6.1.0] nonyne, which has been applied for imaging of glycans on the cell surface of melanoma MV3 cells, offers the advantage that it can be prepared in only two synthetic steps [59].

While the use of benign cyclooctynes has provided an advance in the field of biological imaging *in vivo*, challenges with non-specificity and sensitivity remain. For example, nonspecific biotinylation, presumably by reaction of the alkyne moiety with cysteine residues, reduces the sensitivity of detection [60,61]. Protection of the peptidylcysteines by treatment with iodoacetamide prior to labeling results in a strong reduction of nonspecific labeling [62]. In an alternative approach, a cyclooctyne probe that becomes fluorogenic upon reaction with an azide, CoumBARAC, was developed by the Bertozzi laboratory [63,64]. Although this compound paved the way toward the development of other such probes, the low-fluorescence quantum yields of CoumBARAC have so far limited its use for biological imaging applications. The design of fluorogenic probes that are activated upon triazole formation would prove useful in the dynamic imaging of the glycome in situations (e.g., in living animals) where it is not possible to wash away unbound probes.

3.2.3 Thiol-based Chemistry

Although bioconjugation with thiol reagents is the most well known for labeling cysteine residues on proteins, this approach is also becoming useful for labeling unnatural sugars synthetically modified with a thiol group on the cell surface. Because most native surface thiols are complexed with metals, form disulfide bonds, or are in membrane proteins shielded by the glycocalyx, the metabolic incorporation of thiol-containing sugar analogs (Ac_5ManNTGc) allows for chemoselective labeling at the outermost periphery of the cell surface [65]. A major advantage of thiol-based ligation strategies is the plethora of commercially available reagents for labeling these functional groups [66]. One option, maleimide-functionalized probes (Figure 3.9), offer the advantage that they are highly efficient at physiological pH, unreactive with tyrosine and histidine side chains, and are not light sensitive. Alternatively, the strong coordinate−covalent interaction between thiols and gold allows for cell surface thiols to be directly attached to gold surfaces, which has a strong impact on stem cells [67] and can even influence cancer cells to dramatically alter their behavior [68,69].

The oxidative nature surrounding the cell surface under *in vitro* culture conditions, which can also occur *in vivo*, results in approximately 90% of thiols from modified sialic acid forming *cis*-disulfides [65], and these must be reduced to free thiols to maximize their labeling in bioorthogonal conjugation reactions. For this purpose, treatment with tris-(carboxyethyl)-phosphine (TCEP) is preferred over dithiothreitol, as the former does not contain any thiols and so does not require washing before reacting with the maleimide reagent [70].

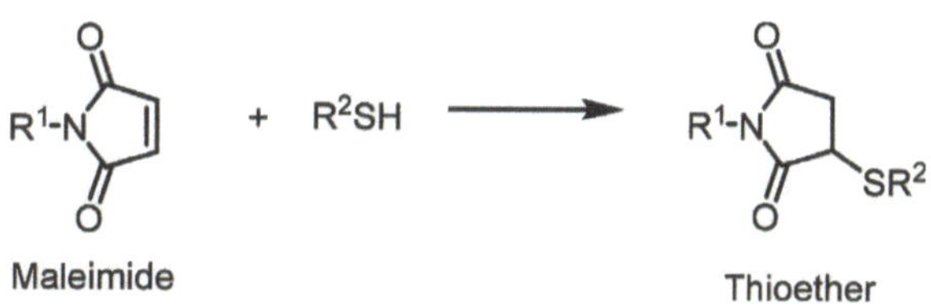

FIGURE 3.9 Reaction of a thiol with a maleimide.

The use of TCEP in sufficiently high concentrations nevertheless remains toxic to cells limiting the use of this reagent in biological systems.

3.2.4 Photoactivated Ligation Reactions

The demonstration that functional groups suited for photoaffinity labeling (PAL) such as phenylaryl azides [71] and diazirines [72,73] could be metabolically incorporated into cell surface glycans further expanded the potential of bioorthogonal ligation reactions for the investigation and understanding of complex biological carbohydrates. In PAL, UV light converts the inert photoreactive crosslinker to a highly reactive carbene or nitrene, which forms a covalent bond with its target molecule at predetermined points during the experiment, for example, after binding of the photoreactive ligand to a receptor as exemplified by the Paulson's group's efforts to identify the *in situ* binding partners for CD22 [71]. When using this strategy, it is important that the photoreactive group of choice causes minimal perturbation to the binding affinity of the ligand to its receptor and its functionality. When this criterion is met, the ability of PAL strategies to form robust covalent bonds to capture noncovalent interactions [74] comprises a powerful tool to capture biologically relevant binding events in their cellular context.

Two types of carbenes can be generated upon irradiation: a singlet carbene, where both electrons are in a single orbital and a triplet carbene, where each electron occupies a separate orbital. Understanding the ground-state carbene multiplicity is important as it dictates the reactivity of the unpaired electron and thus the coupling efficiency of the carbene. The carbene's electronic distribution is determined by the steric and inductive effects exerted by each substituent. Although a full explanation is beyond the scope of this book (but is reviewed elsewhere [75–77]), in brief it is well established that singlet carbenes are stabilized by σ-electron withdrawing groups and π-electron donating groups that are delocalized into the empty p-orbital (Figure 3.10). For example, fluorine is often incorporated due to its high electronegativity (σ-direction) and its ability to donate its lone pair of electrons into the p-orbital, stabilizing the singlet carbene.

An ideal photoactivatable chemical group for bioconjugation has several requirements including that: (i) it is activated at a long UV wavelength to minimize damage to other biomolecules, (ii) it is similar in structure to the parent molecule that its biological activity is not significantly altered, (iii) it forms a stable covalent bond with the receptor upon activation, and (iv) it generates short-lived intermediates to minimize nonspecific binding. Largely based on these requirements, the three most popular photoactivatable groups used in bioconjugate labeling are benzophenones, arylazides, and diazirines [78]. Each group has advantages and limitations and these will be highlighted in further sections.

3.2.4.1 *Benzophenones*

Benzophenones generate triplet carbenes upon irradiation (Figure 3.11). Although they require prolonged exposure, which often leads to nonspecific labeling, the long UV range for activation and their inertness to solvent have made them

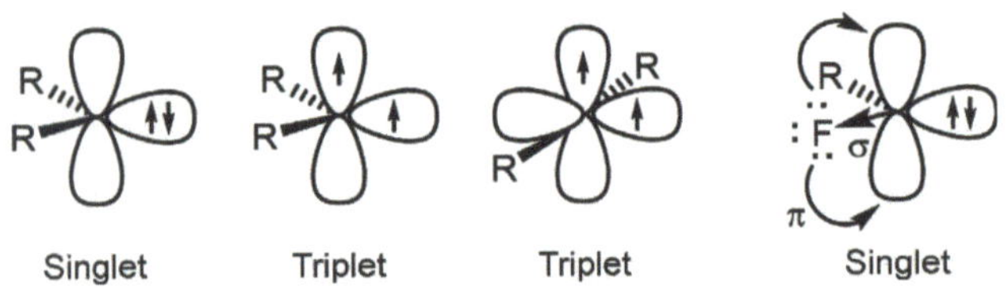

FIGURE 3.10 Electron distribution in singlet and triplet carbene ground states.

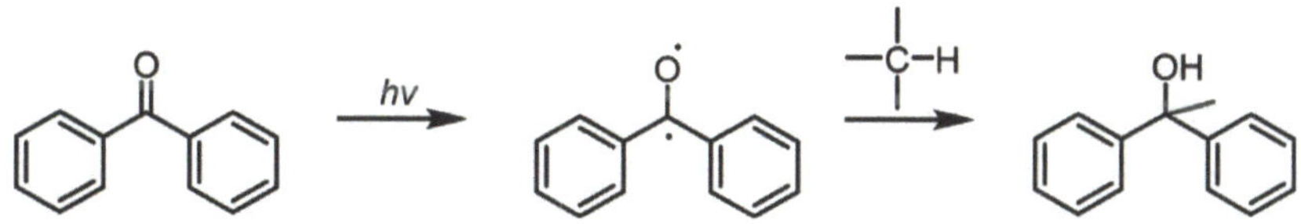

FIGURE 3.11 C–H insertion into carbenes generated from benzophenones.

FIGURE 3.12 Rearrangement of unsubstituted arylazides to dehydroazepenes.

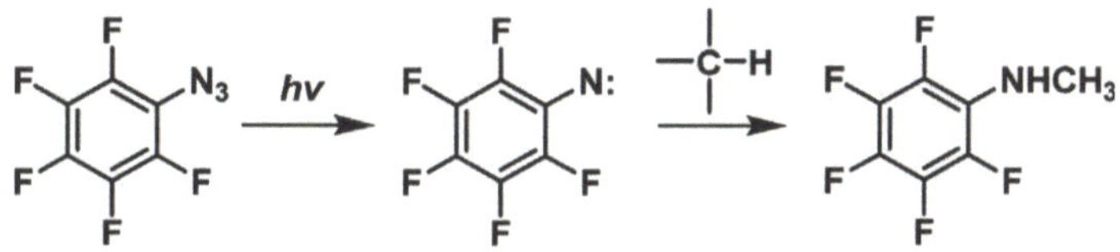

FIGURES 3.13 C–H insertion of perfluorinated arylazides.

attractive photoactivatible crosslinkers [79,80]. However, their bulkiness may prohibit their use for certain biological imaging applications [81].

3.2.4.2 *Arylazides*

Arylazide-modified sugars can be metabolically incorporated into glycans of interest, and upon exposure to UV light, the arylazide generates a nitrene that can insert into C–H and N–H sites of a neighboring protein. While an advantage of using arylazide-modified sugars is their ease of preparation, a drawback is that they are activated at short UV wavelengths (254 nm) that have the potential to damage other biomolecules, such as nucleic acids that have a strong absorbance at this wavelength. Moreover, simple arylazides are rarely used as photoreactive crosslinkers because they rapidly undergo ring expansion to dehydroazepenes, which tend to react with nucleophiles such as amines and thiols, rather than C–H insertion (Figure 3.12) [82,83].

Substitution of the phenyl ring with electron withdrawing groups prevents rearrangement and allows these nitrenes to undergo C–H insertion with high efficiency (Figure 3.13) [84].

A limitation of using arylazides for photoaffinity labeling is that the relatively large phenyl group can interfere with the biological function of the original compound. Moreover, in the cell, the arylazides can be reduced to anilines by endogenous thiols; interestingly, Chang's group has taken this side reaction as an opportunity to detect intracellular hydrogen sulfide levels [85,86].

3.2.4.3 *Diazirines*

While traditionally, arylazides were the most widely used photoreactive group, diazirines are growing in popularity, and increasingly the photoreactive group of choice. Like benzophenones, diazirines have the advantage that they are activated at longer UV range (330–370 nm), which minimizes damage to other biomolecules present in a cell. In addition, the small size of the diazirine makes it particularly suited for PAL because it causes the least structure perturbation compared to the original substrate. Finally, the small size of this group has allowed sugars bearing a diazirine at the *N*-acyl position to be metabolically incorporated on the cell surface to capture carbohydrate—protein interactions [72].

3.3 BIOORTHOGONAL LIGATION REACTIONS: EXPLOITATION IN MOE-BASED APPLICATIONS

The previous section of this chapter outlined emerging strategies to exploit chemical functional groups that now can be installed on the cell surface via MOE by using bioorthogonal reaction chemistries. For these spectacular "chemical biology" advances ultimately to be meaningful, however, the hope is that they will find use in solving real-world biological or medical problems. This section addresses this issue, first by a general discussion of nuances of analog incorporation into various glycoforms that ultimately determine biological or medical significance, and then by giving several specific examples of biological problems (e.g., imaging) or medical conditions (e.g., treatment of degenerative disease or cancer) where a combination of MOE and bioorthogonal ligation chemistries appear to be particularly promising.

3.3.1 MOE Extends Beyond N-Acyl-Modified Sialic Acid

The examples illustrated in Figure 3.1, which are largely used to illuminate the bioorthogonal chemistries discussed in Section 2 of this chapter, where chemical modifications focused on the *N*-acyl position of sialic acid, which comprised the original *in vivo* demonstration of MOE [27] and which remains the best studied example of this technology to date [19,24]. In recent years, however, additional sites of modification on sialic acid, for example, the C-7 position accessed through modifications to the C-4 position of ManNAc [87], have been successfully functionalized with chemical reporter groups. In other cases, modifications at the C-6 position of ManNAc to express C-9-modified sialic acid are not tolerated because one of the steps during its biosynthetic conversion to sialic acid involves phosphorylation at the C-6 position. Instead, the C-9-modified sialic acid can be administered directly to the cells [88].

The metabolic incorporation of sialic acid and its precursor ManNAc has been the most extensively studied, as sialic acid is conveniently located at the outermost periphery where it is readily accessible for binding with exogenous substances, such as labeling reagents, lectins, and toxins [89]. Moreover, sialyltransferases tolerate a wide range of modified substrates [22] ranging from compounds with reactive chemical handles for imaging the glycome [10] to alkyl-modified analogs that regulate cellular processes during embryonic neural development [90] or the neurite outgrowth of PC12 cells [91]. Despite the focus on sialic acid in this report, we wish to emphasize that over the past decade, there has been an increasing focus on expanding MOE technologies to additional monosaccharide vehicles, which have been successfully demonstrated to date for GlcNAc, GalNAc, and fucose analogs [19]. By exploiting these additional monosaccharides, we believe that a vast realm of currently unexplored glycan-based biological activities can be fruitfully investigated in the future. In the remainder of this report, however, we offer only a small sampling of how a combination of MOE and bioorthogonal ligation chemistries have already contributed to biology and medicine.

3.3.2 Imaging of Living Cells

Cell surface glycans comprise a fingerprint of a cell and are specifically recognized by carbohydrate-binding proteins generically known as lectins. The dynamic display of cellular glycans is constantly changing throughout the cell cycle and abnormal changes can be indicative of disease. Imaging these glycans through bioorthogonal chemical ligations provides information on the cell's state. The biosynthesis of glycans is not directly encoded by the genome and so the imaging techniques that rely on genetic reporters for proteins and DNA are not applicable. The imaging of the cell surface is further complicated by the high heterogeneity of glycans structures, differing in linkage, composition, branching, and anomericity. While the use of lectins and antibodies has been used extensively to image glycoconjugates, they require prior isolation of the glycan from the cells. Moreover, they are not suitable for *in vivo* dynamic imaging due to their low-binding affinity and tissue impermeability. Toward this end, bioorthogonal chemical reporters have been developed to image glycans on the cell surface and the advantages and limitations of each existing method have been described above.

The labeling of cell surface glycans is dominated by the SPAAC due to the benign nature and faster reaction kinetics of this reaction compared to the modified Staudinger ligation. However, the development of copper ligands which reduce catalyst loading has encouraged the revisiting of the CuAAC for use in biological reactions, as it is still the fastest bioorthogonal reaction and is compatible with both azide- and alkyne-tagged biomolecules [49].

The small size and inertness of an azide are ideal for MOE experiments, and once expressed on the cell surface it can be reacted with other biomolecules via the Staudinger ligation, the SPAAC, or the CuAAC. The promiscuity of the enzymes involved in the hexosamine pathway allows for the efficient enzymatic conversion of azide-modified precursors to SialNAz, GlcNAz, and GalNAz. However, this is not the case for fucose-modified analogs since 90% of GDP fucose utilized by the cell originates from GDP mannose through the *de novo* biosynthetic pathway. Intercepting the fucose salvage pathway using enzymatically modified azide or alkyne GDP-fucose analogs significantly improves its metabolic uptake [50].

While metabolic incorporation of Ac_4ManNAz exclusively yields the azide-modified sialic acid, it is more difficult to achieve precise pathway targeting with analogs such as Ac_4GlcNAz because a portion of this compound, or its metabolites, is reversibly converted by GlcNAc-2-epimerase into Ac_4GalNAz. This problem can be avoided by treating the cells with the alkyne-functionalized GlcNAc analog (Ac_4GlcNAlk), which is not recognized by the enzyme and thus is a more specific reporter of *O*-GlcNAc modification. Moreover, the metabolic efficiency of Ac_4GlcNAlk is comparable to that of Ac_4GlcNAz [92], and the use of azide-detection tags has improved sensitivity over the reverse copper-catalyzed click chemistry orientation due to decreased background signal [93,94].

Photocrosslinking has been used successfully for the identification of *cis*-ligands of CD22, a glycoprotein that regulates B-cell signaling. Metabolic incorporation of the 9-arylazide derivative of sialic acid (9AAzNeuAc) in B cells followed by irradiation established that CD22 mostly binds to sialic acids on *cis*-ligands of neighboring CD22 resulting in homomultimeric complexes [71,95]. This carbohydrate-mediated interaction has also been demonstrated by installing a diazirine at the *N*-acyl side chain of mannosamine (Ac_4-ManDAz) and sialic acid (Ac_5-5-SiaDAz). *N*-acyl modification is preferred over the 9-position of sialic acid, as the latter can interfere with biological processes [96]. Metabolic incorporation of an arylazide at the *N*-acyl position of sialic acid was extremely poor [97], while substitution with a diazirine led to high levels of expressed Ac_5-5-SiaDAz [72]. Ideally, the sialic acid possesses both a diazirine moiety for the capture of carbohydrate–protein interactions and another functional group for conjugation with a chemical probe. Toward this end, bifunctional sialic acids have been developed (9AzSiaDAz) and incorporated into CHO, HeLa, Daudi, and K20 cells [98].

3.3.3 *In vivo* Labeling/Imaging

With the continuing success of *in vitro* imaging and the wide variety of bioorthogonal ligation strategies, the *in vivo* labeling of living organism has been made possible. Zebra fish and *C. elegans* are ideal models for imaging the glycome as they are transparent, their development is well studied, and exogenous substances are easily incorporated. Opaque organisms such as mice can be visualized using far-red fluorescence imaging and single-photon emission computed tomography. The MOE approach has been successfully extended towards the *in vivo* imaging of cell surface sialylated [99], fucosylated [100], and polylactosamine [101] glycans in zebra fish by metabolic incorporation of the appropriate azide-modified sugar precursor.

Using a multicolored labeling strategy, the mucin-type *O*-glycans in zebra fish [11] and *C. elegans* [10] were imaged from embryo to adult stage by inoculating GalNAz, followed by copper-free click chemistry of the azide with a DIFO reagent conjugated to red and green fluorescent probes. In this manner, it was shown that the biosynthesis of mucin-type *O*-glycans for these two models is highly dynamic and tissue dependent. This methodology was extended toward the imaging of sialic acids in zebra fish by incubation with Ac_4ManNAz [99].

A limitation of MOE is that the metabolic incorporation of Ac_4ManNAz results in Sia5Az into both N- and O-linked glycans. However, it was found that Ac_3-4-azido-ManNAc is selectively expressed as Sia7Az in *O*-glycosylated proteins, and this analog was used for the *in vivo* imaging of the midbrain and hindbrain regions in zebra fish, which contain high levels of *O*-glycosylated proteins [87].

Another challenge in labeling glycans *in vivo* is that the typical biotin–streptavidin system used for detection in living cells has limited vessel permeability due to the large size of streptavidin. As a result, this approach cannot be applied for the imaging of normal glycans and is largely restricted to the imaging of tumor glycans. Usually, glycans in optically opaque organisms such as mice are labeled *in vivo* but are analyzed *ex vivo*. Toward this end, Brindle's group has developed fluorescently labeled tetrazine reagents which are small enough to access normal tissue and have lower levels of nonspecific retention. The tetrazene moiety undergoes a Diels–Alder reaction with an alkene appended to the cyclooctyne probe and its fluorescence measured [102].

3.3.4 Tissue Engineering, Stem Cell Research, and Regenerative Medicine

The oxidative nature around the glycocalyx allows for thioglycolyl sugars to readily form disulfide bonds with adjacent thiol-containing cells, which results in cell–cell clustering. These adhesive properties imparted by the thiol group can be extended to attachment of the cell to matrix materials, which is particularly challenging [103]. The strong gold–thiol interaction and the small spacing of the sulfur anchors on gold (4.97 Å) make thiol-bearing sialic acids ideal for attachment to gold substrates. Alternatively, the analog-treated cells can be anchored to maleimide-functionalized glass surfaces. The physical properties of the biomaterial, such as its elasticity and topography, can modulate the morphology of stem cells, which in turn directs the stem cell differentiation into specific lineages [104]. Yarema's group demonstrated that treating hEBD stem cells with Ac_5ManNTGc induced differentiation to a neuronal cell linage [67]. When Jurkat cells were grown on gold-coated nanofibers, they developed a distinctive "spreading" morphology which was exclusive to sialosides of N-linked glycoproteins [69]. Extending this methodology from 2D- to 3D-scaffolds (gold- or maleimide-coated nanofibers) significantly increased cell adhesion (66–81%) compared to the 2D films (23–35%) [68].

The early experiments with thiol-modified sialic acids highlight the role of these sugars on adhesion processes involved in stem cell fate. In follow-up studies, a more sophisticated understanding of the underlying mechanism for these compounds, for example, through modulation of potassium channel activity [105], is being gained. From a more practical point of view, the thiol analogs illustrate an important aspect of nonnatural monosaccharide residues installed in the cell surface via MOE. Specifically, the thioglycolyl sialic acids, similar to other sialosides installed via MOE such as ketone groups [26], are slowly removed from the cell surface over time thus allowing thiol-dependent responses to be reversed in some cases. In other cases, such as stem cell differentiation, however, it appears that thiol-analog-driven biological responses are irreversible. Nevertheless, the levels of surface display can be maintained by continuous addition of the analog precursor Ac_5ManNTGc while the removal of the synthetic sugars can be accelerated by the addition of the natural sialic acid precursor ManNAc [26,65]. The removal of nonnatural sugars from the cell surface is potentially important for tissue engineering applications of MOE to avoid possible deleterious immune responses triggered by the nonnatural glycan.

3.3.5 Cancer Therapy

It is well established that the composition of cell surface glycans in diseased cells, especially cancer cells [106,107], is different than those of healthy cells. For example, elevated levels of sialic acid [108] and tri- and tetra-antennary *N*-glycans [109–111] are an indication of cancer. Some of the earliest MOE experiments showed that a tagged sialic acid can be introduced to the cancer cell by MOE and quantified for diagnostic purposes [112] or used as a handle for delivery of toxins for cancer therapy [113]. An ongoing challenge in using MOE for cancer therapy, however, is the selective targeting of cancer cells over healthy cells. This problem is partially alleviated because sialic acid is overexpressed in metastatic cells and so they are most affected by any modifications in sialic acid. A strategy to increase the tissue-specific delivery of Ac_3ManNAz was to install a peptide recognized by the prostate-specific antigen (PSA) protease at the O-6 position of the sugar. In this manner, the caged sugar could only be released upon reaction with the PSA secreted by prostate tumor tissue [114]. Another approach to selectively deliver azido sugars to cancer cells is by encapsulating them in ligand-targeted liposomes [115]. Controlled release of cancer prodrug Bu_4ManNAc over a 10-plus day period was achieved by encapsulating the sugar in polyethylene glycol-sebacic acid [116] resulting in a substantial improvement compared to the otherwise twice daily administration required for *in vivo* studies [117].

3.4 CONCLUDING COMMENTS

This chapter provides an overview of bioorthogonal ligation reactions that—when applied to the study of glycans—were first used to label selectively oxidized sialic acids. The development of MOE, which allows a profusion of nonnatural chemical functional groups suitable for various bioorthogonal ligation approaches to be incorporated into glycans in living cells and animals, provided a major impetus to this field. Although the development of underlying chemistry-based technologies continues, some of the most exciting recent developments in the bioorthogonal ligation field include the burgeoning use of this approach to confront pressing biomedical research and clinical challenges. The introduction to these issues provided by this chapter is designed to provide a foundation for the interested reader to independently investigate the application of bioorthogonal ligation to glycobiology in more depth and hopefully begin making experimental contributions to this exciting field.

Acknowledgments

Funding for this contribution was provided by the National Institutes of Health (R01CA112314 and P01HL107153).

References

[1] Werz DB, Ranzinger R, Herget S, Adibekian A, Von der Lieth CW, Seeberger PH. Exploring the structural diversity of mammalian carbohydrates ("glycospace") by statistical databank analysis. ACS Chem Biol 2007;2:685–91.

[2] Campbell CT, Yarema KJ. Large-scale approaches for glycobiology. Gen Biol 2005;6:236 (Article).

[3] Zaia J. Mass spectrometry and glycomics. OMICS 2010;14:401–8.

[4] Zhang H, Li X-j, Martin DB, Aebersold R. Identification and quantification of N-linked glycoproteins using hydrazide chemistry, stable isotope labeling and mass spectrometry. Nat Biotechnol 2003; 21:660–6.

[5] Zhang H, Aebersold R. Isolation of glycoproteins and identification of their N-linked glycosylation sites. Methods Mol Biol 2006;328:177–85.

[6] Krambeck FJ, Bennun SV, Narang S, Choi S, Yarema KJ, Betenbaugh MJ. A mathematical model to derive *N*-glycan structures and cellular enzyme activities from mass spectrometric data. Glycobiology 2009; 19:1163–75.

[7] Park S, Gildersleeve JC, Blixt O, Shin I. Carbohydrate microarrays. Chem Soc Rev 2013;42:4310–26.

[8] Pilobello KT, Krishnamoorthy L, Slawek D, Mahal LK. Development of a lectin microarray for the rapid analysis of protein glycopatterns. ChemBioChem 2005;6: 985–9.

[9] Gupta G, Surolia A, Sampathkumar S-G. Lectin microarrays for glycomic analysis. OMICS 2010; 14:419–36.

[10] Laughlin ST, Bertozzi CR. *In vivo* imaging of *Caenorhabditis elegans* glycans. ACS Chem Biol 2009;4:1068–72.

[11] Laughlin ST, Baskin JM, Amacher SL, Bertozzi CR. *In vivo* imaging of membrane-associated glycans in developing zebrafish. Science 2008;320:664–7.

[12] Saxon E, Bertozzi CR. Cell surface engineering by a modified Staudinger reaction. Science 2000;287:2007–10.

[13] Rideout D, Calogeropoulou T, Jaworski J, McCarthy M. Synergism through direct covalent bonding between agents: a strategy for rational design of chemotherapeutic combinations. Biopolymers 1990;29: 247–62.

[14] De Bank PA, Kellam B, Kendall DA, Shakesheff KM. Surface engineering of living myoblasts via selective periodate oxidation. Biotechnol Bioeng 2003;81:800–8.

[15] Jourdian GW, Dean L, Roseman S. The sialic acids. XI. A periodate-resorcinol method for the quantitative estimation of free sialic acids and their glycosides. J Biol Chem 1971;246:430–5.

[16] Ramya TNC, Weerapana E, Cravatt BF, Paulson JC. Glycoproteomics enabled by tagging sialic acid- or galactose-terminated glycans. Glycobiology 2013; 23:211–21.

[17] Zeng Y, Ramya TNC, Dirksen A, Dawson PE, Paulson JC. High-efficiency labeling of sialylated glycoproteins on living cells. Nat Meth 2009;6:207–9.

[18] Zhang Y, Yuan J, Song J, Wang Z, Huang L. An efficient method for selectively imaging and quantifying *in situ* the expression of sialylated glycoproteins on living cells. Glycobiology 2013;23:643–53.

[19] Du J, Meledeo MA, Wang Z, Khanna HS, Paruchuri VDP, Yarema KJ. Metabolic glycoengineering: sialic acid and beyond. Glycobiology 2009;19:1382–401.

[20] Peters BP, Aronson Jr NN. Reactivity of the sialic acid derivative 5-acetamido-3,5-dideoxy-L-arabino-heptulosonic acid in the resorcinol and thiobarbituric acid assays. Carbohydr Res 1976;47:345–53.

[21] Kohler JJ. Aniline: a catalyst for sialic acid detection. ChemBioChem 2009;10:2147–50.

[22] Campbell CT, Sampathkumar S-G, Weier C, Yarema KJ. Metabolic oligosaccharide engineering: perspectives, applications, and future directions. Mol Biosyst 2007;3:187–94.

[23] Aich U, Yarema KJ. Non-natural sugar analogues: Chemical probes for metabolic oligosaccharide engineering. In: Fraser-Reid BO, Tatsuta K, Thiem J, editors. Glycoscience. Springer-Verlag (Berlin, Heidelberg). 2008. p. 2133–2190.

[24] Almaraz RT, Mathew MP, Tan E, Yarema KJ. Metabolic oligosaccharide engineering: Implications for selectin-mediated adhesion and leukocyte extravasation. Ann Biomed Eng 2012;40:806–15.

[25] Mahal LK, Yarema KJ, Bertozzi CR. Engineering chemical reactivity on cell surfaces through oligosaccharide biosynthesis. Science 1997;276:1125–8.

[26] Yarema KJ, Mahal LK, Bruehl RE, Rodriguez EC, Bertozzi CR. Metabolic delivery of ketone groups to sialic acid residues. Application to cell surface glycoform engineering. J Biol Chem 1998;273: 31168–79.

[27] Kayser H, Zeitler R, Kannicht C, Grunow D, Nuck R, Reutter W. Biosynthesis of a nonphysiological sialic acid in different rat organs, using *N*-propanoyl-D-hexosamines as precursors. J Biol Chem 1992;267: 16934–8.

[28] Jencks WP. Studies on the mechanism of oxime and semicarbazone formation. J Am Chem Soc 1959;81: 475–81.

[29] Nauman DA, Bertozzi CR. Kinetic parameters for small-molecule drug delivery by covalent cell surface targeting. Biochim Biophys Acta 2001;1568:147–54.

[30] Saxon E, Luchansky SJ, Hang HC, Yu C, Lee SC, Bertozzi CR. Investigating cellular metabolism of synthetic azidosugars with the Staudinger ligation. J Am Chem Soc 2002;124:14893–902.

[31] Hsu T-L, Hanson SR, Kishikawa K, Wang S-K, Sawa M, Wong C-H. Alkynyl sugar analogs for the labeling and visualization of glycoconjugates in cells. Proc Natl Acad Sci USA 2007;104:2614–9.

[32] Chang PV, Chen X, Smyrniotis C, Xenakis A, Hu T, Bertozzi CR, Wu P. Metabolic labeling of sialic acids in living animals with alkynyl sugars. Angew Chem Int Ed Engl 2009;48:4030–3.

[33] Schilling CI, Jung N, Biskup M, Schepers U, Brase S. Bioconjugation via azide-Staudinger ligation: an overview. Chem Soc Rev 2011;40:4840–71.

[34] Dawson PE, Muir TW, Clark-Lewis I, Kent SBH. Synthesis of proteins by native chemical ligation. Science 1994;266:776–9.

[35] Hang HC, Gentjes EJ, Grotenbreg G, Pollington AM, Bijlmakers MJ, Ploegh HL. Chemical probes for the rapid detection of fatty-acylated proteins in mammalian cells. JACS 129:2744–2745.

[36] Cohen AS, Dubikovskaya EA, Rush JS, Bertozzi CR. Real-time bioluminescence imaging of glycans on live cells. J Am Chem Soc 2010;132:8563–5.

[37] Prescher JA, Dube DH, Bertozzi CR. Chemical remodelling of cell surfaces in living animals. Nature 2004;430:873–7.

[38] Kaewsapsak P, Esonu O, Dube DH. Recruiting the host's immune system to target *Helicobacter pylori's* surface glycans. ChemBioChem 2013;14:721–6.

[39] Saxon E, Armstrong JI, Bertozzi CR. A "traceless" Staudinger ligation for the chemoselective synthesis of amide bonds. Org Lett 2000;2:2141–3.

[40] Nilsson BL, Kiessling LL, Raines RT. High-yielding Staudinger ligation of a phosphinothioester and azide to form a peptide. Org Lett 2000;3:9–12.

[41] Bernardes GJL, Linderoth L, Doores KJ, Boutureira O, Davis BG. Site-selective traceless Staudinger ligation for glycoprotein synthesis reveals scope and limitations. ChemBioChem 2011;12:1383–6.

[42] Li C, Wang X-h, Zheng L, Huang J-f. Automated protein subcellular localization based on local invariant features. Protein J 2013;32:230–7.

[43] Kolb HC, Finn MG, Sharpless KB. Click chemistry: diverse chemical function from a few good reactions. Angew Chem Int Ed 2001;40:2004–21.

[44] Hein JE, Fokin VV. Copper-catalyzed azide-alkyne cycloaddition (CuAAC) and beyond: new reactivity of copper(I) acetylides. Chem Soc Rev 2010;39:1302–15.

[45] Himo F, Lovell T, Hilgraf R, Rostovtsev VV, Noodleman L, Sharpless KB, Fokin VV. Copper(I)-catalyzed synthesis of azoles. DFT study predicts unprecedented reactivity and intermediates. J Am Chem Soc 2004;127:210–6.

[46] Thirumurugan P, Matosiuk D, Jozwiak K. Click chemistry for drug development and diverse chemical–biology applications. Chem Rev 2013;113: 4905–79.

[47] Chan TR, Hilgraf R, Sharpless KB, Fokin VV. Polytriazoles as copper(I)-stabilizing ligands in catalysis. Org Lett 2004;6:2853–5.

[48] Wang Q, Chan TR, Hilgraf R, Fokin VV, Sharpless KB, Finn MG. Bioconjugation by copper(I)-catalyzed azide-alkyne [3 + 2] cycloaddition. J Am Chem Soc 2003;125:3192–3.

[49] Besanceney-Webler C, Jiang H, Zheng T, Feng L, Soriano del Amo D, Wang W, Klivansky LM, Marlow FL, Liu Y, Wu P. Increasing the efficacy of bioorthogonal click reactions for bioconjugation: a comparative study. Angew Chem Int Ed Engl 2011;2001:35.

[50] Soriano del Amo D, Wang W, Jiang H, Besanceney C, Yan AC, Levy M, Liu Y, Marlow FL, Wu P. Biocompatible copper(I) catalysts for *in vivo* imaging of glycans. J Am Chem Soc 2010;132:16893–9.

[51] Ess DH, Jones GO, Houk KN. Transition states of strain-promoted metal-free click chemistry: 1,3-dipolar cycloadditions of phenyl azide and cyclooctynes. Org Lett 2008;10:1633–6.

[52] Gordon CG, Mackey JL, Jewett JC, Sletten EM, Houk KN, Bertozzi CR. Reactivity of biarylazacyclooctynones in copper-free click chemistry. J Am Chem Soc 2012;134:9199–208.

[53] Baskin JM, Prescher JA, Laughlin ST, Agard NJ, Chang PV, Miller IA, Lo A, Codelli JA, Bertozzi CR. Copper-free click chemistry for dynamic *in vivo* imaging. Proc Natl Acad Sci USA 2007;104:16793–7.

[54] Jewett JC, Sletten EM, Bertozzi CR. Rapid Cu-free click chemistry with readily synthesized biarylazacyclooctynones. J Am Chem Soc 2010;132:3688–90.
[55] Ning X, Guo J, Wolfert MA, Boons G-J. Visualizing metabolically labeled glycoconjugates of living cells by copper-free and fast Huisgen cycloadditions. Angew Chem Int Ed 2008;120:2285–7.
[56] Mbua NE, Guo J, Wolfert MA, Steet R, Boons G-J. Strain-promoted alkyne-azide cycloadditions (SPAAC) reveal new features of glycoconjugate biosynthesis. ChemBioChem 2011;12:1912–21.
[57] Sletten EM, Bertozzi CR. A hydrophilic azacyclooctyne for Cu-free click chemistry. Org Lett 2008;10: 3097–9.
[58] Friscourt F, Ledin PA, Mbua NE, Flanagan-Steet HR, Wolfert MA, Steet R, Boons G-J. Polar dibenzocyclooctynes for selective labeling of extracellular glycoconjugates of living cells. J Am Chem Soc 2012;134:5381–9.
[59] Dommerholt J, Schmidt S, Temming R, Hendriks LJA, Rutjes FPJT, van Hest JCM, Lefeber DJ, Friedl P, van Delft FL. Readily accessible bicyclononynes for bioorthogonal labeling and three-dimensional imaging of living cells. Angew Chem Int Ed Engl 2010;49: 9422–5.
[60] Chang PV, Prescher JA, Sletten EM, Baskin JM, Miller IA, Agard NJ, Lo A, Bertozzi CR. Copper-free click chemistry in living animals. Proc Natl Acad Sci USA 2010;107:1821–6.
[61] Beatty KE, Fisk JD, Smart BP, Lu YY, Szychowski J, Hangauer MJ, Baskin JM, Bertozzi CR, Tirrell DA. Live-cell imaging of cellular proteins by a strain-promoted azide-alkyne cycloaddition. ChemBioChem 2010;11:2092–5.
[62] van Geel R, Pruijn GJM, van Delft FL, Boelens WC. Preventing thiol-yne addition improves the specificity of strain-promoted azide-alkyne cycloaddition. Bioconjug Chem 2012;23:392–8.
[63] Jewett JC, Bertozzi CR. Synthesis of a fluorogenic cyclooctyne activated by Cu-free click chemistry. Org Lett 2011;13:5937–9.
[64] Shieh P, Hangauer MJ, Bertozzi CR. Fluorogenic azidofluoresceins for biological imaging. J Am Chem Soc 2012;132:17428–31.
[65] Sampathkumar S-G, Jones MB, Yarema KJ. Metabolic expression of thiol-derivatized sialic acids on the cell surface and their quantitative estimation by flow cytometry. Nat Protoc 2006;1:1840–51.
[66] Tyagarajan K, Pretzer E, Wiktorowicz JE. Thiol-reactive dyes for fluorescence labeling of proteomic samples. Electrophoresis 2003;24:2348–58.
[67] Sampathkumar S-G, Li AV, Jones MB, Sun Z, Yarema KJ. Metabolic installation of thiols into sialic acid modulates adhesion and stem cell biology. Nat Chem Biol 2006;2:149–52.
[68] Du J, Che P-L, Wang Z-Y, Aich U, Yarema KJ. Designing a binding interface for control of cancer cell adhesion via 3D topography and metabolic oligosaccharide engineering. Biomaterials 2011;32:5427–37.
[69] Du J, Che P-L, Aich U, Tan E, Kim HJ, Sampathkumar S-G, Yarema KJ. Deciphering glycan linkages involved in Jurkat cell interactions with gold-coated nanofibers via sugar-displayed thiols. Bioorg Med Chem Lett 2011;21:4980–4.
[70] Getz EB, Xiao M, Chakrabarty T, Cooke R, Selvin PR. A comparison between the sulfhydryl reductants tris (2-carboxyethy)phosphine and dithiothreitol for use in protein biochemistry. Anal Biochem 1999;273: 73–80.
[71] Han S, Collins BE, Bengtson P, Paulson JC. Homomultimeric complexes of CD22 revealed by *in situ* photoaffinity protein-glycan crosslinking. Nat Chem Biol 2005;1:93–7.
[72] Tanaka Y, Kohler JJ. Photoactivatable crosslinking sugars for capturing glycoprotein interactions. J Am Chem Soc 2008;130:3278–9.
[73] Bond MR, Zhang H, Vu PD, Kohler JJ. Photocrosslinking of glycoconjugates using metabolically incorporated diazirine-containing sugars. Nat Protoc 2009;4:1044–63.
[74] Yu S-H, Boyce M, Wands AM, Bond MR, Bertozzi CR, Kohler JJ. Metabolic labeling enables selective photocrosslinking of *O*-GlcNac-modified proteins to their binding partners. Proc Natl Acad Sci USA 2012; 109:4834–9.
[75] Brahms DLS, Dailey WP. Fluorinated carbenes. Chem Rev 1996;96:1585–632.
[76] Bourissou D, Guerret O, Gabbaï FP, Bertrand G. Stable carbenes. Chem Rev 1999;100:39–92.
[77] Dormán G, Prestwich GD. Using photolabile ligands in drug discovery and development. Trends Biotechnol 2000;18:64–77.
[78] Dubinsky L, Krom BP, Meijler MM. Diazirine based photoaffinity labeling. Bioorg Med Chem 2012;20:554–70.
[79] Hilbold B, Perrault M, Ehret C, Niu S-L, Frisch B, Pécheur E-I, Bourel-Bonnet L. Benzophenone-containing fatty acids and their related photosensitive fluorescent new probes: design, physico-chemical properties and preliminary functional investigations. Bioorg Med Chem 2011;19:7464–73.
[80] Torkar A, Bregant S, Devel L, Novinec M, Lenarčič B, Lah T, Dive V. A novel photoaffinity-based probe for selective detection of cathepsin L active form. ChemBioChem 2012;13:2616–21.

[81] Lapinsky DJ. Tandem photoaffinity labeling-bioorthogonal conjugation in medicinal chemistry. Bioorg Med Chem Lett 2012;20:6237–47.
[82] Shields CJ, Chrisope DR, Schuster GB, Dixon AJ, Poliakoff M, Turner JJ. Photochemistry of aryl azides: detection and characterization of a dehydroazepine by time-resolved infrared spectroscopy and flash photolysis at room temperature. J Am Chem Soc 1987;109:4723–6.
[83] Li YZ, Kirby JP, George MW, Poliakoff M, Schuster GB. 1,2-Didehydroazepines from the photolysis of substituted aryl azides: analysis of their chemical and physical properties by time-resolved spectroscopic methods. J Am Chem Soc 1988;110:8092–8.
[84] Keana JFW, Cai SX. New reagents for photoaffinity labeling: synthesis and photolysis of functionalized perfluorophenyl azides. J Org Chem 1990;55:3640–7.
[85] Chen S, Chen Z-J, Ren W, Ai H-W. Reaction-based genetically encoded fluorescent hydrogen sulfide sensors. J Am Chem Soc 2012;134:9589–92.
[86] Lippert AR, New EJ, Chang CJ. Reaction-based fluorescent probes for selective imaging of hydrogen sulfide in living cells. J Am Chem Soc 2011; 133:10078–80.
[87] Moeller H, Boehrsch V, Bentrop J, Bender J, Hinderlich S, Hackenberger CPR. Glycan-specific metabolic oligosaccharide engineering of C7-substituted sialic acids. Angew Chem Int Ed Engl 2012;51:5986–90.
[88] Oetke C, Brossmer R, Mantey LR, Hinderlich S, Isecke R, Reutter W, Keppler OT, Pawlita M. Versatile biosynthetic engineering of sialic acid in living cells using synthetic sialic acid analogues. J Biol Chem 2002;277:6688–95.
[89] Varki A, Chen X. Advances in the biology and chemistry of sialic acids. ACS Chem Biol 2010;5:163–76.
[90] Schmidt C, Stehling P, Schnitzer J, Reutter W, Horstkorte R. Biochemical engineering of neural cell surfaces by the synthetic *N*-propanoyl-substituted neuraminic acid precursor. J Biol Chem 1998;273:19146–52.
[91] Kontou M, Bauer C, Reutter W, Horstkorte R. Sialic acid metabolism is involved in the regulation of gene expression during neuronal differentiation of PC12 cells. Glycoconjug J 2008;25:237–44.
[92] Zaro BW, Yang Y-Y, Hang HC, Pratt MR. Chemical reporters for fluorescent detection and identification of *O*-GlcNAc-modified proteins reveal glycosylation of the ubiquitin ligase NEDD4-1. Proc Natl Acad Sci USA 2011;108:8146–51.
[93] Agard NJ, Baskin JM, Prescher JA, Lo A, Bertozzi CR. A comparative study of bioorthogonal reactions with azides. ACS Chem Biol 2006;1:644–8.
[94] Speers AE, Cravatt BF. Profiling enzyme activities *in vivo* using click chemistry methods. Chem Biol 2004;11:535–46.
[95] Yarema KJ, Sun Z. A photochemical snapshot of CD22 binding. Nat Chem Biol 2005;1:69–70.
[96] Yu H, Chen X. Carbohydrate post-glycosylational modifications. Org Biomol Chem 2007;5:865–72.
[97] Luchansky SJ, Goon S, Bertozzi CR. Expanding the diversity of unnatural cell-surface sialic acids. ChemBioChem 2004;5:371–4.
[98] Feng L, Hong S, Rong J, You Q, Dai P, Huang R, Tan Y, Hong W, Xie C, Zhao J, Chen X. Bifunctional unnatural sialic acids for dual metabolic labeling of cell-surface sialylated glycans. J Am Chem Soc 2013;135:9244–7.
[99] Dehnert KW, Baskin JM, Laughlin ST, Beahm BJ, Naidu NN, Amacher SL, Bertozzi CR. Imaging the sialome during zebrafish development with copper-free click chemistry. ChemBioChem 2012;13:353–7.
[100] Dehnert KW, Beahm BJ, Huynh TT, Baskin JM, Laughlin ST, Wang W, Wu P, Amacher SL, Bertozzi CR. Metabolic labeling of fucosylated glycans in developing zebrafish. ACS Chem Biol 2011;6:547–55.
[101] Zheng T, Jiang H, Gros M, Soriano dAD, Sundaram S, Lauvau G, Marlow F, Liu Y, Stanley P, Wu P. Tracking *N*-acetyllactosamine on cell-surface glycans. Angew Chem Int Ed Engl 2011;50:4113–8.
[102] Neves AA, Stöckmann H, Wainman YA, Kuo JCH, Fawcett S, Leeper FJ, Brindle KM. Imaging cell surface glycosylation *in vivo* using "double click" chemistry. Bioconjug Chem 2013;24:934–41.
[103] Langer R, Tirrell DA. Designing materials for biology and medicine. Nature 2004;428:487–92.
[104] Higuchi A, Ling Q-D, Chang Y, Hsu S-T, Umezawa A. Physical cues of biomaterials guide stem cell differentiation fate. Chem Rev 2013;113:3297–328.
[105] Hua Z, Lvov A, Morin TJ, Kobertz WR. Chemical control of metabolically-engineered voltage-gated K^+ channels. Bioorg Med Chem Lett 2011;21:5021–4.
[106] Sell S. Cancer-associated carbohydrates identified by monoclonal antibodies. Hum Pathol 1990;21:1003–19.
[107] Fuster MM, Esko JD. The sweet and sour of cancer: glycans as novel therapeutic targets. Nat Rev Cancer 2005;5:526–42.
[108] Malykh YN, Schauer R, Shaw L. *N*-Glycolylneuraminic acid in human tumours. Biochimie 2001;83:623–34.
[109] Pierce M, Arango J. Rous sarcoma virus-transformed baby hamster kidney cells express higher levels of asparagine-linked tri- and tetraantennary glycopeptides containing [GlcNAc-β(1,6)Man-α(1,6)Man] and poly-*N*-acetyllactosamine sequences than baby hamster kidney cells. J Biol Chem 1986;261:10772–7.

[110] Dennis JW, Laferte S, Waghorne C, Breitman ML, Kerbel RS. β1-6 branching of Asn-linked oligosaccharides is directly associated with metastasis. Science 1987;236:236–9.

[111] Miura Y, Hato M, Shinohara Y, Kuramoto H, Furukawa J, Kurogochi M, Shimaoka H, Tada M, Nakanishi K, Ozaki M, Todo S, Nishimura S-I. BlotGlycoABC™, an integrated glycoblotting technique for rapid and large scale clinical glycomics. Mol Cell Proteomics 2008;7:370–7.

[112] Lemieux GA, Yarema KJ, Jacobs CL, Bertozzi CR. Exploiting differences in sialoside expression for selective targeting of MRI contrast reagents. J Am Chem Soc 1999;121:4278–9.

[113] Mahal LK, Yarema KJ, Lemieux GA, Bertozzi CR. In: Inoue Y, Lee YC, Troy II FA, editors. Sialobiology and Other Novel Forms of Glycosylation. Osaka, Japan: Gakushin Publishing Company; 1999. p. 273–80.

[114] Chang PV, Dube DH, Sletten EM, Bertozzi CR. A strategy for the selective imaging of glycans using caged metabolic precursors. J Am Chem Soc 2010;132:9516–8.

[115] Xie R, Hong S, Feng L, Rong J, Chen X. Cell-selective metabolic glycan labeling based on ligand-targeted liposomes. J Am Chem Soc 2012; 134:9914–7.

[116] Aich U, Meledeo MA, Sampathkumar S-G, Fu J, Jones MB, Weier CA, Chung SY, Tang BC, Yang M, Hanes J, Yarema KJ. Development of delivery methods for carbohydrate-based drugs: controlled release of biologically-active short chain fatty acid-hexosamine analogs. Glycoconjug J 2010;27:445–59.

[117] Gagiannis D, Gossrau R, Reutter W, Zimmermann-Kordmann M, Horstkorte R. Engineering the sialic acid in organs of mice using *N*-propanoylmannosamine. Biochim Biophys Acta 2007;1770:297–306.

CHAPTER

4

Pushing the Bacterial Envelope: Strategies for Re-Engineering Bacterial Surfaces with Heterologous Proteins and Sugars

Samir Gautam[1] *and David A. Spiegel*[2,3]

[1]Department of Cell Biology, Yale School of Medicine, New Haven, CT, USA, [2]Department of Pharmacology, Yale School of Medicine, New Haven, CT, USA, [3]Department of Chemistry, Yale University, New Haven, CT, USA

OUTLINE

J.M. Karp & W. Zhao (Eds): Micro- and Nanoengineering of the Cell Surface.
DOI: http://dx.doi.org/10.1016/B978-1-45-573146-6.00004-0

4.1 BACTERIAL SURFACE DISPLAY

The bacterial envelope is a multilayered organelle that surrounds the cell, providing morphological structure and protection from environmental stressors. In gram-negative organisms, the envelope comprises an inner membrane (IM), an outer membrane (OM), and a periplasmic space in between that contains the peptidoglycan cell wall—a highly cross-linked mesh-like macromolecule that confers mechanical strength to the cell (Figure 4.1A). Gram-positive bacteria, in contrast, lack an OM and have a thicker, surface-exposed cell wall (Figure 4.1B). These structures have been studied intensely for over 60 years by bacterial cell biologists and antibiotic developers, among others, due to their critical roles in bacterial growth and survival [1,2].

As our understanding of the bacterial envelope has grown, however, so has its utility; we now possess a powerful set of techniques for expressing proteins on the cell surface that has transformed the envelope from a target of basic and pharmaceutical research into a versatile platform for biotechnological innovation. Since the first examples in 1986 [3,4], protein display in bacteria has matured into a robust, streamlined methodology with an expansive range of applications, including environmental remediation [5], biofuel production [6], biocatalysis [6,7], biosensing [8], protein library screening [9], cancer therapy [10], and vaccinology [7,11–14].

More recently, bacterial surface engineering has extended beyond the polypeptide paradigm to include sugars as well. This technology, referred to as "glycoengineering," [15–17] involves heterologous expression of genes that mediate saccharide biosynthesis, enabling the display of a variety of nonnative proteoglycans, glycolipids, and polysaccharides on the surface of recombinant bacteria. Glycoengineered organisms, in turn, have been utilized as whole cell vaccines [18–21], anti-infective probiotics [22], and "living factories" for glycoconjugate synthesis [16].

In this chapter, we describe the repertoire of techniques that have been established for displaying heterologous proteins and sugars on the surface of bacteria. We then discuss how re-engineered bacteria have been exploited in basic and applied biology thus far. Throughout, we attempt to highlight recently developed technologies across disciplines that may prove useful in the context of bacterial surface display, improving existing applications and perhaps enabling new ones. Finally, we address an important caveat: despite the tantalizing clinical and biotechnological prospects for surface-engineered bacteria, their implementation outside the laboratory has been greatly hindered due to concerns over the environmental dangers of genetically modified organisms (GMOs). We therefore include a discussion of these theoretical risks and present a set of recombinant techniques for circumventing them (see Box 4.1).

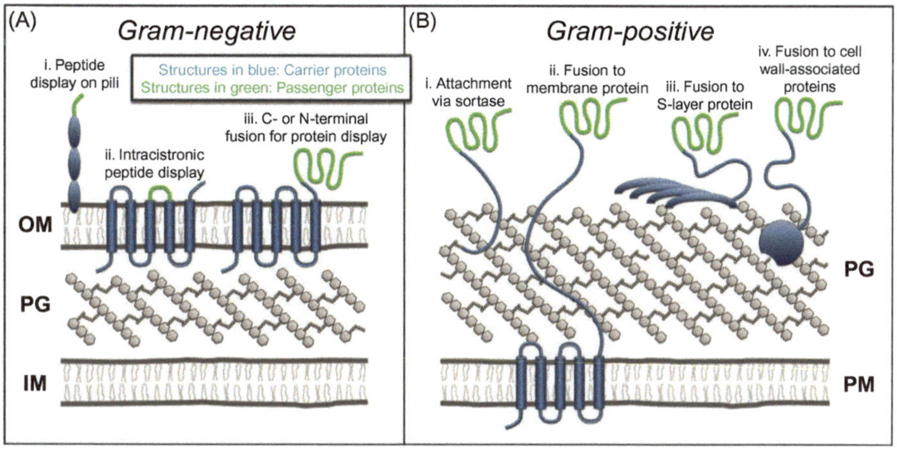

FIGURE 4.1 Techniques for displaying heterologous proteins on bacteria. Passenger proteins (colored green) are fused to carrier proteins (colored blue) and expressed at the cell surface. (A) Techniques relevant to gram-negative bacteria include: (i) genetic fusion of a peptide to a fimbrial protein, (ii) insertion of a peptide into an extracellular loop of a surface protein, and (iii) N- or C-terminal fusion of a large passenger protein to a carrier. (B) Techniques relevant to gram-positive bacteria include: (i) fusion to a sortase-incorporated surface protein, (ii) fusion to the surface-exposed domain of a transmembrane protein, (iii) fusion to S-layer (surface layer) proteins, and (iv) fusion to cell wall-associated, non-covalently attached proteins (e.g., autolysins). IM, inner membrane; INP, ice nucleation protein; OM, outer membrane; PG, peptidoglycan; PM, plasma membrane.

BOX 4.1

RISKS ASSOCIATED WITH GMOs AND RECOMBINANT GENETIC SOLUTIONS

A number of potentially transformative applications have been devised for surface-engineered bacteria, but their practical use has been relatively limited due to four major concerns about the potential hazards of releasing GMOs into the environment [23–25].

First is the risk of parallel transfer of foreign DNA to other organisms, a particular issue in the case of virulence gene expression in vaccine vectors. This risk can be minimized by increasing the stability of the genetic modification, both through chromosomal integration of the foreign DNA, and by ensuring that the chromosomal insertion site is distant from especially mobile genetic elements such as transposons.

Second is the possibility of releasing an organism with significant fitness advantage, leading to unrestrained propagation of the GMO and damaging environmental consequences. Fears of this outcome may be allayed through the use of auxotrophic strains unable to survive in the environment (i.e., "biologically contained" organisms) or through use of non-living vehicles such as OMVs, BGs, and inactivated spores.

Third is a risk unique to vaccine delivery: the possibility of an attenuated vaccine strain reverting to a virulent genotype, thus placing the vaccinee and/or close contacts in danger of contracting disease. Attenuation is usually accomplished by disrupting a gene involved with either virulence or central metabolic pathways. The former is associated with greater risk as virulence factors are generally located on mobilizable genetic elements that may be acquired from wild-type organisms. Metabolic genes, meanwhile, are generally located on the chromosome, so attenuations at these sites are more stable [23].

The *fourth* concern—introduction of drug resistance genes into the environment—is perhaps the most worrisome. In the laboratory setting, maintaining heterologous genes episomally in bacteria using antibiotic resistance selection markers (the simplest of recombinant techniques) is perfectly acceptable practice. However, it is imperative for genetically engineered bacteria released into the environment—such as vaccines and bioremediation devices—to be free of such markers, so as to prevent spread of drug resistance. Two genetic methods that satisfy this requirement have been established: the balanced lethal system (more widely used) [26,27] and recombineering [28]—both of which are effective in gram-negative and gram-positive [29] organisms.

The *balanced lethal approach* uses an episomal plasmid to express the gene of interest, but importantly, the plasmid is maintained through complementation rather than antibiotic resistance [26,27]. Generation of a markerless carrier strain using the balanced lethal system proceeds as follows. A key metabolic gene such as one involved in amino acid synthesis is deleted, rendering the strain auxotrophic. This genetic deletion is achieved through (i) heterologous recombination between the host chromosome and a suicide vector encoding a null allele and a resistance marker flanked by recombinase target sites, (ii) antibiotic-mediated selection for transformants, and (iii) subsequent excision of the resistance marker through transient expression of recombinase. (*Note*: markerless strains may be identified using a negative selection marker such as sacB, which confers sensitivity to sucrose.) The mutant strain is then transformed with a plasmid containing two elements: the gene of interest and the metabolic gene deleted from the host. Thus, the episomal element is stably maintained in the recombinant strain, as its loss would remove complementation with the essential gene.

In *recombineering*, stabilization of the gene of interest is guaranteed by direct integration into the bacterial chromosome [28]. This is achieved through the use of a modified strain expressing the lamba-red bacteriophage system—a set of enzymes that greatly increase the efficiency of homologous recombination. After induction of the recombination enzymes, the strain is electroporated with a linear DNA construct that contains (i) the heterologous gene, (ii) flanks consisting of sequences complementary to the chromosomal insertion site, and (iii) a selection marker which is itself flanked by recombinase target sites. After selection for integrants, the marker is removed using recombinase to produce the desired strain.

4.2 STRATEGIES FOR RE-ENGINEERING BACTERIAL SURFACES WITH HETEROLOGOUS PROTEINS

The majority of bacterial protein display methods employ a common strategy in which a heterologous protein of interest (passenger) is genetically fused to an endogenous surface protein (carrier). The structure of the cell envelope, in turn, determines where these extracellular passenger–carrier chimeras reside; in gram-negative organisms they associate with the OM (Figure 4.1A), while in gram-positive bacteria chimeras are usually attached to the peptidoglycan cell wall (Figure 4.1B) [30].

4.2.1 Gram-Negative Organisms

A number of recombinant techniques have been developed for expressing passenger–carrier fusion proteins on the surface of gram-negative species (Figure 4.1A). For peptides and small proteins up to approximately 60 amino acids [31], exogenous sequences may be inserted into (i) proteins that comprise surface appendages such as pili and flagella, (ii) extracellular loops of outer membrane proteins such as LamB and OmpA, or (iii) lipoproteins and virulence factors anchored to the surface [13,32]. However, for the display of larger proteins such as enzymes, passengers must be fused to the N- or C-terminus of carriers rather than intracistronically (Figure 4.1C,iii) [13].

Autodisplay is one such system commonly used for the display of large proteins. The strategy involves fusion of passenger proteins to autotransporters, a class of virulence factors that includes IgA1 protease from *Neisseria gonorrhoeae*, MisL from *Salmonella enterica* serovar Typhimurium, and AIDA-I (adhesin involved in diffuse adherence) from *Escherichia coli* [7]. During normal expression, autotransporters insert into the OM, forming a porin-like β-barrel structure with their C-terminal domain. The N-terminal domain, connected to the β-barrel via a long, flexible linker is subsequently translocated through the pore to access the extracellular milieu. Therefore, autotransporters may be 'hijacked' to allow surface display by replacing the native N-terminal domain with a heterologous passenger [33]. A notable advantage of this approach is the lateral mobility of carriers within the OM, which enables multimerization of passenger proteins, as has been demonstrated with surface-displayed streptavidin [34], nitrilase from *Klebsiella pneumoniae* [35], and sorbitol dehydrogenase [36].

The other major system for expressing large proteins at the bacterial surface involves fusion to ice nucleation protein (INP), an abundant lipoprotein present in the OM of many plant pathogens [13,37]. A unique property of INP is its mechanism of surface expression; rather than threading through the IM in an unfolded state like the majority of bacterial proteins, INP is translocated fully folded. Thus, INP-mediated surface display can accommodate passenger proteins that associate with bulky cofactors, such as heme- and diflavin-containing cytochrome P450 enzymes [13,38].

Since their introduction in the 1990s [39,40], autodisplay and INP-based expression systems have developed into highly optimized, well-characterized biotechnological platforms. Passenger–carrier chimeras expressed using these methods (i) efficiently traverse the IM and periplasm to reach the cell surface in great abundance (more than 100,000 molecules per cell in the case of autodisplay [41]), (ii) reside stably at the surface without substantial detachment and loss of the recombinant protein, (iii) resist degradation by proteases within the periplasmic space, and (iv) do not disrupt the integrity of the cell envelope or induce growth defects [42]. Given these properties, along with the well-defined genetic techniques required for autodisplay- and INP-based expression, surface display in gram-negative bacteria can be considered a straightforward and robust class of techniques.

4.2.2 Gram-Positive Organisms

Whereas surface display in gram-negative bacteria is achieved through fusion to carrier proteins expressed on the OM, a different strategy must be undertaken in gram-positive organisms, which lack this outer vestment (Figure 4.1B). The most common method involves attaching proteins directly to the surface-exposed peptidoglycan cell wall, using a class of enzymes called sortases (Figure 4.1B,i) [43,44]. These membrane-associated enzymes recognize specific pentapeptide sorting motifs in secreted proteins, cleave within the motif, and covalently attach the N-terminal portion to the cell wall [45]. Thus, fusion of a passenger protein to the N-terminus of a sortase substrate protein enables efficient expression on the gram-positive cell surface. Sortase-mediated display of heterologous proteins has proven especially robust both because of the covalent nature of attachment, and because of the rigid peptidoglycan substrate, which ensures stability of surface expression [11].

Alternatives to sortase-based surface display have been explored as well. For example, passengers fused to transmembrane proteins that extend from the plasma membrane through the cell wall and into the extracellular space are effectively displayed at the cell surface (Figure 4.1B,ii) [46]. Surface expression can also be achieved using chimeras between passengers and proteins that comprise the bacterial cell surface layer (S-layer), a self-assembling crystalline lattice that coats several species of bacteria (Figure 4.1B,iii) [47–52]. Notably, the absence of an OM on gram-positive organisms removes a significant obstacle to protein translocation, facilitating the display of large proteins when using sortase-mediated, transmembrane fusion, and S-layer fusion protein methodologies [11].

A final, somewhat unique approach involves fusion to autolysins—cell wall degradative enzymes that bind *non-covalently* to components of the gram-positive cell wall (Figure 4.1B,iv) [53–57]. The key advantage of this methodology is that autolysin-passenger chimeras may be expressed in a model organism (e.g., *E. coli*), purified, and adsorbed to the cell wall of a wild-type gram-positive strain. Thus, using this strategy, gram-positive bacteria can be re-engineered with heterologous surface proteins, but remain genetically unchanged, thereby circumventing the potential environmental risks of GMOs (see Box 4.1) [55].

4.2.3 Spores

Certain bacterial species, such as those within the *Clostridium* and *Bacillus* genera, display a unique behavior known as sporulation, in which bacteria transform from metabolically active (vegetative) cells into dormant structures (spores) in response to nutrient depleted conditions [12]. During this process, bacterial DNA becomes cocooned within a multilayered wall of peptidoglycan and concentric protein shells, rendering the DNA impervious to toxic chemicals, radiation, and extremes of pH, temperature, and dehydration [12]. Spores are thus able to survive in inhospitable environments for extended periods—in one report, 25 million years [12,58]. Despite the spore's extraordinary inertness, its surface proteins remain functional (in part due to their attachment to a stable matrix, see Section 4.3.4), ready to sense nutrient repletion, and germinate into a fully active state [59]. Numerous methods have been established for displaying heterologous proteins on these remarkable structures, all involving fusion to carrier proteins expressed at the spore surface [12]. These carriers include CotB, CotC, and CotG in *Bacillus subtilis*; [60]. BclA and BclB in *Bacillus anthracis* [61]; and Cry1Ac in *Bacillus thuringiensis* [62]. Spores re-engineered by these means have proven useful as biosensors [8,63–69], biocatalysts [70], and vaccines [71,72].

While the applications of spores are discussed in more detail below, it is worth describing here the unique advantages spores offer over conventional bacteria in three biotechnological settings. First, similar to vegetative cell systems, spores are well suited to biosensing—the detection and reporting of analytes in solution (see Section 4.3.5.1)—because of their sensitivity, low cost, ease of use, and small size. The exceptional stability of spores, however, allows them to be used outside of the laboratory in harsh natural environments, where conventional cellular systems cannot survive [70]. Second, spores are unique in that proteins need not traverse a membrane in order to be expressed on the cell surface; instead, proteins are attached to the outermost layer of the spore, which is assembled within the cytoplasm of the mother cell during sporulation [73]. This property allows expression of multimers such as streptavidin [74], proteins with problematic hydrophobic domains that preclude membrane translocation [60], enzymes requiring cofactors [60], and very large proteins [31]. Finally, the spore's ability to germinate confers an advantage in the context of vaccine delivery, as it remains stable during storage at room temperature, but then transforms once delivered to the nutrient-rich environments of the tissue into vegetative cells with potent immunostimulatory properties (see Section 4.3.1) [75,76].

4.2.4 Outer Membrane Vesicles and Bacterial Ghosts

Outer membrane vesicles (OMVs) and bacterial ghosts (BGs) are acellular derivatives of the gram-negative envelope that contain the parent cell's surface-expressed proteins [77–79]. OMVs consist of outer membrane-derived bilamellar vesicles, 50–250 nm in diameter, which are continuously elaborated by growing gram-negative bacteria [77]. Efficient surface display on OMVs has been achieved by fusing passenger proteins to the *E. coli* toxin ClyA, which becomes concentrated on these structures during vesiculation [80]. OMVs have particular utility as vaccine delivery vehicles because they are stable during prolonged storage [81], and strongly immunogenic (by virtue of the many immunostimulatory components of the OM these vesicles retain), but do not present the same environmental safety hazards as intact cells because they are unable to propagate (see Box 4.1) [82–84].

BGs, in contrast, are complete gram-negative bacterial envelopes produced through heterologous expression of the bacteriophage-derived lysis gene *E* (Figure 4.2). Upon expression, Protein E inserts into the IM via its N-terminus, while the C-terminal domain translocates across the IM (Figure 4.2A) and integrates into the OM (Figure 4.2B), inducing

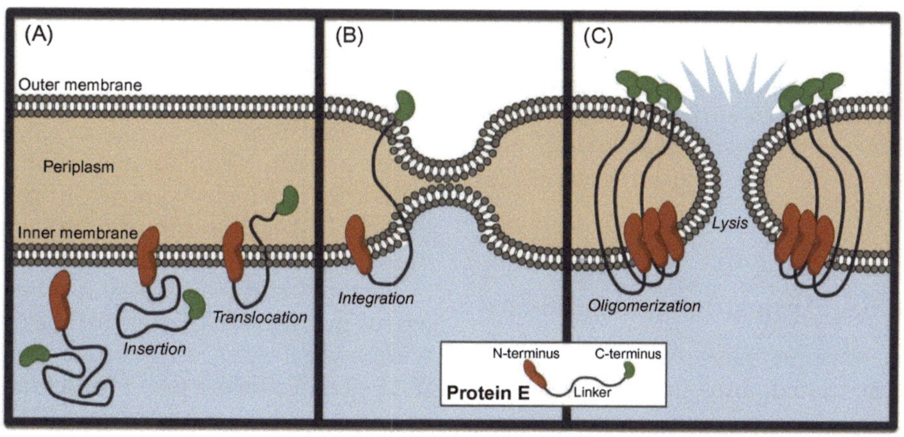

FIGURE 4.2 Schematic representation of Protein E-induced lysis and BG generation. (A) The N-terminal domain of Protein E inserts into the IM and the C-terminus translocates into the periplasm. (B) The C-terminus then integrates into the OM to induce fusion with the IM. (C) Finally, Protein E oligomerizes to form the lysis tunnel. Source: *Adapted from Langemann et al. [78].*

fusion of the two membranes. Protein E oligomerization leads to the formation of a "lysis tunnel" (Figure 4.2C), through which the cytoplasm is released extracellularly. However, due to the controlled nature of the lytic event and the effective seal between IM and OM, cell envelope morphology and periplasmic content are maintained [78]. Most commonly, anchoring of foreign proteins on BGs is achieved through fusion of passengers with the outer membrane protein, OmpA. Finally, like OMVs, BGs are excellent vehicles for vaccine delivery because of their potent immunostimulatory properties and inability to replicate [85,86].

4.3 APPLICATIONS OF BACTERIA EXPRESSING HETEROLOGOUS SURFACE PROTEINS

Bacterial surface display has lent itself to a host of biotechnological and biomedical applications. These can be organized into four categories: (i) display of antigens in vaccine design, (ii) expression of antitumor proteins for cancer therapy, (iii) immobilization of enzymes for biocatalysis, and (iv) a set of related applications that involve binding between surface proteins and molecules in solution (biosensing, biosorption, and library screening). These topics will be discussed here only in brief; please see focused reviews for greater detail [6,9,11–13].

4.3.1 Vaccines Against Infectious Disease

In general, vaccines have two components: a pathogen-associated antigen that serves as the target for adaptive immune responses and an adjuvant—a material that potentiates and tunes immune responses to the co-delivered antigen [87]. Uptake of these components by antigen presenting cells (APCs) such as dendritic cells represents the crucial initiating step in the generation of immune responses that protect against future infection.

Surface-engineered bacterial vaccines are constructed by expressing a pathogen-associated antigen on an inactivated bacterium, live-attenuated pathogen, or food grade (nonpathogenic) bacterial species. The bacterial vector itself is able to serve as an adjuvant due to the numerous immunostimulatory pathogen-associated molecular patterns (PAMPs) present within the cytoplasm and cell envelope. These include bacterial DNA, peptidoglycan, gram-positive lipoproteins, gram-negative lipopolysaccharide (LPS), and a set of PAMPs associated with viable cells (vita-PAMPs) which includes mRNA and cyclic dinucleotides [88].

Localization of antigen at the bacterial surface *per se* appears to be important for the immunogenicity of these recombinant vaccine vectors. In almost all direct comparisons between surface, cytoplasmic, and secreted expression systems, surface-engineered vaccines have proven most efficacious (even when normalized to magnitude of expression) [89–100]. One reason for the improved immunogenicity of surface-anchored antigen is that it is efficiently co-delivered with adjuvant (bacterial PAMPs) to APCs, which is essential for generating effective immune responses [101]. Furthermore, surface expression produces a multivalent display of antigen on the cell surface, which is known to promote cross-linking of B-cell receptors and stimulation of humoral immune responses [102].

Bacterial surface display has been especially useful in the development of mucosal vaccines with several advancing as far as clinical trials [7,11–14]. Table 4.1 provides a summary of the bacterial vectors used in mucosal vaccinology, and a catalog of the pathogens these vaccines have targeted.

Several arguments have been proposed for why surface-engineered bacteria have proven such effective mucosal immunogens. The first

TABLE 4.1 Examples of Surface-Engineered Bacteria in Mucosal Vaccinology

Bacterial Vector	Pathogenic Target
Salmonella spp. [103,104]	Dengue virus [105], Hepatitis B virus [106], Japanese encephalitis virus [107], Rotavirus [108], Sendai Virus [109], Transmissible gastroenteritis coronavirus [100,110,111], *Bacillus anthracis* [112,113], Enterotoxigenic *E. coli* [100,114,115], *Helicobacter pylori* [116], *Listeria monocytogenes* [117], *Streptococcus pneumoniae* [118,119], *Yersinia enterocolitica* [120], *Yersinia pestis* [121,122], *Plasmodium falciparum* [123–125], *Porphyromonas gingivalis* [126], *Trichinella spiralis* [123,127]
Bordetella pertussis [128]	Enterovirus 71 [129], Influenza virus [130], *Haemophilus influenzae* [131], *Neisseria meningitidis* [132], *Schistosoma mansoni* [133]
Escherichia coli	*Borrelia burgdorferi* [134], Enterohemorrhagic *E. coli* [134,135], Enterotoxigenic *E. coli* [136,137], *L. monocytogenes* [134], *Salmonella* spp. [138], *Y. enterocolitica* [120], *P. falciparum* [135]
Shigella spp.	Enterotoxigenic *E. coli* [139–143]
Vibrio cholerae [144]	Attaching and effacing *E. coli* [145], Enterotoxigenic *E. coli* [146]
Lactic acid bacteria [147–150]: *Lactobacillus* spp., *Leuconostoc* spp., *Pediococcus* spp., and *Streptococcus* spp.	Human immunodeficiency virus [151,152], Human papillomavirus [153–158], Influenza virus [159], Rotavirus [160–163], SARS Coronavirus [164], *B. burgdorferi* [165], *Clostridium tetani* [166], Enterotoxigenic *E. coli* [167–170], *H. pylori* [171], *N. meningitidis* [172], *Proteus mirabilis* [173], *Salmonella enteritidis* [174–176], *Streptococcus agalactiae* [177], *Streptococcus mutans* [178], *S. pneumoniae* [179–184], *Streptococcus pyogenes* [185], *Yersinia pseudotuberculosis* [186], *Giardia lamblia* [187], *P. falciparum* [188–190]
Staphylococcus spp. [191]	Respiratory syncytial virus [192–194], *Salmonella* spp. [138]
B. subtilis spores [12]	*B. anthracis* [195,196], *Clostridium perfringens* [197], *C. tetani* [198,199], Enterotoxigenic *E. coli* [199], *Clonorchis sinensis* [200,201], *Schistosoma japonicum* [202]
OMVs [82]	*S. pneumoniae* [203]
BGs [85,86]	Foot-and-mouth disease virus [204], Hepatitis B virus [205], Human immunodeficiency virus [206], *B. anthracis* [112], *Chlamydia trachomatis* [207,208], *N. meningitidis* [209–211], *P. falciparum* [99]

Lactic acid bacteria include *Lactobacillus, Leuconostoc, Pediococcus,* and *Streptococcus* spp.

derives simply from the size of the bacterial vector, which at >500 nm in diameter is considered "particulate." Delivery of antigen by such particulate vehicles is known to greatly improve antigen presentation by APCs, leading to stronger elicitation of immune responses at the mucosa than the same antigen in soluble form [88]. Second, the intrinsic adjuvanticity of bacteria provides a potent immunostimulatory signal to overcome the bias toward tolerance that exists in mucosal immune tissue [212]. Third, the transient colonization and/or mild infection induced by live bacterial vaccines leads to a sustained, local generation of antigen at the mucosa that greatly improves delivery to APCs [213]. Finally, bacterial vaccines are uniquely efficient at gaining access to APCs within mucosal immune tissue, by virtue of surface properties that prevent trapping within the mucus layer [214], and surface adhesion proteins and carbohydrates that promote attachment to mucosal epithelium and translocation into immune inductive sites [215,216]. With respect to bacterial surface display, it is worth noting that surface expression of targeting proteins, such as antibodies that bind to proteins on epithelial cells, has been shown to further increase adhesion and translocation of bacterial vaccine vectors into immune tissue [193,217].

4.3.2 Anticancer Therapeutics

The use of live bacteria in cancer therapy has fascinated researchers since the early nineteenth century, when tumor regression was observed in a set of cancer patients after they contracted gas gangrene (an aggressive soft tissue infection caused by *C. perfringens*) [10,218–220]. The anticancer effect of bacteria has since been traced to a number of intrinsic properties of these organisms, such as their ability to home to tumor tissue with up to 1000-fold selectivity over healthy tissues [221]. In the case of *C. perfringens*, this ability arises from the fact that spores selectively germinate within the core of solid tumors, which is hypoxic due to poor vascularization [222]. Facultative anaerobes like *Salmonella* species and *L. monocytogenes*, meanwhile, accumulate in cancer tissue due to entrapment in the tortuous tumor vasculature [223], chemotaxis toward nutrients released from dying cancer cells [224], and preferential growth in the tumor's unique metabolic milieu [224], where they are also protected from attack by white blood cells due to local immunosuppression [225,226]. Upon accumulation in the tumor, bacteria promote cell death both by competing for nutrients and stimulating normally suppressed intratumoral immune cells to attack cancer cells [225].

With the advent of genomics and recombinant technology, antitumor bacteria—which had largely remained curiosities in the field of cancer treatment—have now been transformed into finely tuned therapeutic devices through surface expression of proteins with antitumoral properties. For instance, the invasin adhesion protein from *Y. pseudotuberculosis*, which mediates entry of the bacterium into nonphagocytic mammalian cells, has been expressed on the surface of nonpathogenic *E. coli*-based therapeutic strains to promote invasion into cancer cells [227]. This strategy was utilized to deliver "trans-kingdom" RNAi in a mouse model of colon cancer to silence expression of the epithelial oncoprotein, β-catenin [228]. In another example, *S. enterica* serovar Typhimurium was engineered to express the apoptosis-promoting protein, Fas ligand, on its cell surface. The modified bacterium was shown to inhibit growth of primary breast and colon cancers as well as pulmonary metastases in mice [229]. Finally, bacteria have been engineered to express cytotoxic prokaryotic proteins such as cytolysin A, a surface-localized pore-forming toxin, producing strains that significantly inhibit tumor growth and metastasis in murine models of colon and lung cancer [230,231].

A number of other non-surface related genetic modifications have been shown to potentiate the anticancer effects of bacterial vectors as well, many of which could conceivably be combined with surface display to create more effective, multimodal therapeutics. For instance, *Salmonella* spp. have been designed to secrete pro-inflammatory cytokines such as IL-2 that promote adaptive immune responses [232], as well as cytotoxic agents such as tumor necrosis factor-related apoptosis-inducing ligand (TRAIL), which directly induce tumor lysis [233]. A second effective modification is expression of genetic circuits that selectively activate gene expression in response to cues in the tumor microenvironment [234]. Such circuits have been used to couple cytotoxin production to arrival at tumors, which reduces the off-target toxicity of these tissue-damaging proteins. Third, through expression of prodrug converting enzymes, tumor-localized bacteria can be used to activate chemotherapeutic prodrugs *in situ*, increasing drug concentrations in the poorly accessible tumor core, and reducing collateral damage to healthy tissue. The efficacy of this approach has been confirmed not only in murine models but also in patients as well, as phase I trials have shown that an attenuated *Salmonella* strain, VNP20009, expressing the *E. coli* cytosine deaminase gene—which converts the prodrug 5-fluorocytosine into the active chemotherapeutic 5-fluorouracil (5-FU)—homed to tumors and generated significantly elevated intratumoral 5-FU levels with respect to serum [235].

Of course, intravenous administration of live bacteria is not without risk of side effects. For instance, preliminary tolerability studies with VNP20009 showed dose-dependent elicitation of cytokines, such as IL-1β, IL-6, and TNF-α, which were associated with hypotension, fever, thrombocytopenia, anemia, diarrhea, nausea, and vomiting [236]. However, through detoxification of pro-inflammatory bacterial components such as LPS (see Section 4.3.1), attenuating mutations to virulence factors and genes in central metabolic pathways (see Box 4.1), titration of tolerable doses, and careful clinical monitoring post-administration, the immunotoxicity of live bacteria may be minimized, allowing these potentially valuable therapeutics to be tested further in clinical settings.

4.3.3 Vaccines Against Cancer

In addition to serving as devices designed to directly kill cancer cells, genetically engineered bacteria have also proven a highly effective platform for development of anticancer vaccines [237–241]. In actuality, these two approaches overlap in many ways; for example, vaccine vectors such as *L. monocytogenes* colonize tumors and induce direct cytotoxicity, while conversely, oncotherapeutic bacteria lead not only to tumor damage but also to tumor-specific adaptive responses due to release of cancer cell antigens during bacterially induced lysis [240,242]. Nevertheless, in cancer vaccinology, elicitation of immune responses is the primary objective, the ultimate goal being immune-mediated clearance of both primary tumors and metastases as well. In this context, the immunostimulatory properties of the live bacterial vector are leveraged to break the inherent tolerance to cancer-associated antigens that exists because of their similarity to self proteins and their residence in an immunosuppressive tumor milieu [243]. Notably, translation of bacteria-based anticancer vaccines to the clinic has recently begun, with phase I trials demonstrating the safety of a promising listerial vaccine against cervical carcinoma [244].

Bacterial surface display has been used in the design of anticancer vaccines extensively as a means of antigen expression. For instance, T-cell epitopes from NY-ESO-1, a commonly expressed tumor antigen, were inserted into the fimbrial proteins of *S. enterica* serovar Typhimurium to create a vaccine capable of generating robust antigen-specific T-cell responses in mice [245]. Similarly, the E7 oncoprotein (which is involved with the development of cervical cancer) and the 37 kDa oncofetoprotein (which is found in many human cancers) were surface expressed on lactic acid bacteria to generate effective mucosal vaccines against human papillomavirus-induced tumors [246–249]. Further, it was shown in a mouse model of cervical cancer that the immunogenicity of live E7-expressing bacterial vaccines was greatly potentiated when the vector simultaneously secreted the pro-inflammatory cytokine, IL-12 [155]. This important finding illustrates the potential of combining multiple recombinant modifications in the same therapeutic. It also suggests that vaccine-generated cytokine secretion is a strategy that could, in principle, boost the immunogenicity of any live vaccine.

In addition to antigen expression, bacterial surface display has been used to improve targeting of bacterial vectors to tumors, in an analogous manner to the targeting of vaccines to mucosal immune tissue (described in Section 4.3.1). On an attenuated *Salmonella* vaccine strain, for example, surface display of single chain antibody fragments specific for carcinoembryonic antigen—a glycoprotein overexpressed on many epithelial-derived cancers—was found to significantly improve intratumoral accumulation of the recombinant bacteria *in vivo* [250].

4.3.4 Biocatalysis

The use of microbially expressed enzymes for (bio)catalysis of chemical reactions has a long history that includes the ancient practices of bread and cheese making, beer and wine fermentation, and, more recently, antibiotic production. The value of enzymes in chemical synthesis lies in their potential to catalyze reactions with remarkable rapidity in a highly chemo-, regio-, and enantioselective manner under mild conditions, while minimizing the toxic waste generation, protection/deprotection steps, and energy expenditure involved in conventional synthetic chemistry [251,252]. Furthermore, through directed evolution and other protein engineering techniques (described in Section 4.3.5.3), modified enzymes may be designed to (i) remain stable at elevated temperatures in solution with organic solvents, (ii) accept nonnative substrates, and (iii) catalyze novel reactions [251]. These properties have led to a rapidly growing interest in the use of biocatalysts in industrial production of detergents, food, cosmetics, textiles, fine chemicals, biofuels, and pharmaceuticals [253–255]. In fact, over 500 commercial products are produced using enzymes, generating an industrial enzyme market valued at US$5.1 billion as of 2009 [255].

Immobilization of enzymes on the bacterial cell surface offers key advantages over the use of free enzyme in biocatalysis. First, attachment of enzymes to a solid matrix, such as the bacterial envelope, vastly increases their stability [7,256]; in one example, a surface-expressed organophosphorus hydrolase retained 100% catalytic efficiency for an astounding 45 days in culture [257]. Second, the production of enzyme on the cell surface obviates the need for protein purification, which can be laborious and costly. Third, once genetic engineering of a biocatalyst strain is complete, there is access to a nearly infinite supply of enzyme, requiring only culture in inexpensive growth medium—again greatly reducing the price of production. Enzyme display on the cell surface also has advantages over cytoplasmic expression, in that it allows use of substrates and generation of products that are impermeable to lipid membranes.

A wide array of enzymes have been expressed on the surface of bacteria, including oxidoreductases (e.g., cytochrome P450 enzymes [38,258]) and various hydrolases (e.g., glycosyltransferases [259], phosphatases [260], lipases [261], other esterases [262]) [13,33]. Organisms re-engineered with such enzymes have been used to degrade organic pollutants, generate fuels from renewable energy sources (e.g., plant biomass), and manufacture chemical compounds for pharmaceutical and industrial purposes [6].

4.3.5 Interactions Between Surface Proteins and Solutes

In this section, we discuss applications of bacterial surface display that are based on binding between a surface-bound protein and a molecule in solution, which may be a protein, a small molecule, or an element. A myriad of inventive uses for this binding interaction have been established, falling under the headings of biosensing, biosorption, and screening (Figure 4.3).

4.3.5.1 Biosensing: Binding for Purposes of Detection

Over the last two decades, researchers have endeavored to genetically engineered bacteria

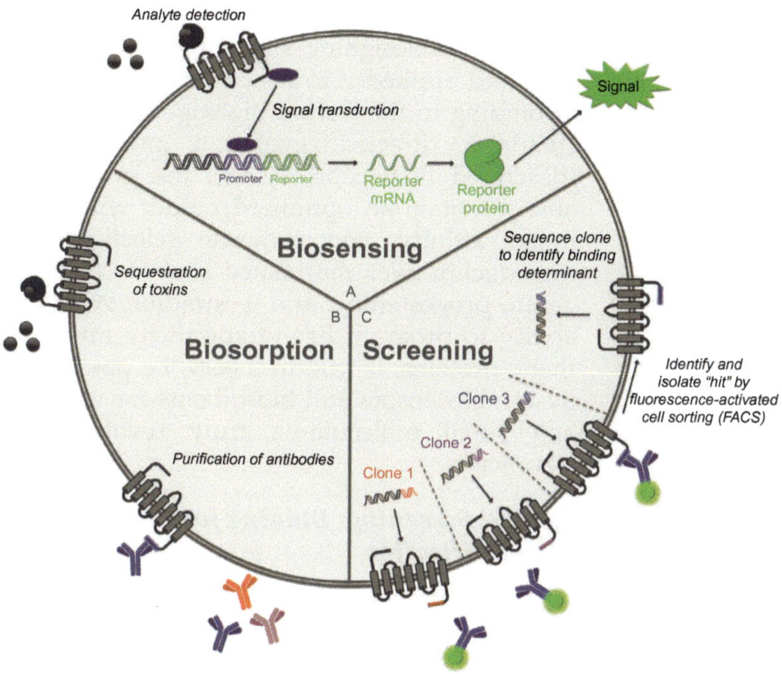

FIGURE 4.3 Applications based on interactions between surface proteins and solutes. (A) Biosensing involves detection of an analyte and dose-dependent transduction of the binding signal into a change in expression of a reporter gene. (B) In biosorption, surface-engineered bacteria are used to purify materials from solutions, e.g., toxins from water supplies or epitope-specific antibodies from polyclonal serum samples. (C) Surface display has been used in a variety of screening applications, such as antibody epitope mapping, as depicted here.

that serve as devices for detecting and quantifying molecules in solution. To this end, an entire class of "two-component biosensors" (Figure 4.3A) has been devised. These bacteria express (i) a genetic sensor that binds to an analyte and transduces the binding event into a dose-dependent change in activity of a promoter and (ii) a reporter gene driven by said promoter that produces a measurable signal (e.g., luciferase, green fluorescent protein, or β-galactosidase) [263]. In one notable example, *E. coli* was converted into a biosensor for arsenite-containing compounds through transformation with a plasmid containing the *ars* operon, which mediates bacterial resistance to such toxins. This regulatory circuit works as follows: in the absence of arsenite, the protein arsR represses the *ars* promoter, but upon binding the metalloid, it unbinds the promoter to permit expression of several genes that detoxify and extrude arsenite-containing compounds. Thus, by expressing a promoterless bacterial luciferase (luxAB) gene [264] downstream of the *ars* promoter in *E. coli*, a simple biological sensor for arsenite could be constructed. Remarkably, due to the sensitivity of arsR for its substrate and the signal amplification inherent in such operons, this re-engineered bacterial strain was capable of detecting arsenite at *subattomolar* concentrations—1000 times below the detection limits of conventional techniques [263,265]. Similarly designed biosensors have been used to measure concentrations of (i) various heavy metals and organic toxins in the setting of environmental monitoring [266], (ii) nutrients in the study of soil microbe metabolism [267], and (iii) analytes with potential clinical relevance such as quorum-sensing molecules [268] and antimicrobial drugs [12,269].

To date, two-component biosensors have been constructed only with intracellular sensing mechanisms. However, surface-engineered bacteria have been used as the basis for a novel

ELISA-like quantitative immunoassay. In this approach, single chain variable fragments (scFvs) of antibody proteins are expressed on the surface of bacteria, analogous to the immobilization of antibodies on ELISA plates. Samples are incubated with these recombinant bacteria, and antibody-bound analyte is then separated from unbound analyte via centrifugation. Quantification of bound analyte is performed using horseradish peroxidase–antibody conjugates or other detection mechanisms. Along these lines, Chen et al. [270] have constructed scFv-expressing bacteria capable of measuring concentrations of the cardiac glycoside drug, digoxin, in the nanomolar range. The success of this approach indicates the enticing possibility that such diagnostic devices may complement or even replace traditional immunoassays like ELISA due to their cost efficiency, ability to self-regenerate, and relatively high sensitivity.

4.3.5.2 *Biosorption: Binding for Purposes of Purification*

Bacteria expressing peptide or protein ligands for soluble binding partners have been used as effective chromatographic media for purification of a range of molecules (Figure 4.3B). For example, biosorption of heavy metal toxins to metal-binding peptides on the surface of *E. coli* and *Staphylococcus* species has been used for environmental remediation of polluted water [5]. Meanwhile, expression of peptide epitopes has enabled isolation of single antibody clones from polyclonal serum samples (produced in immunized rabbits)—an approach that may serve as an alternative to hybridoma development for accessing monoclonal antibodies [271].

The scope of this technology, in fact, is likely to broaden further in light of a recent landmark study in protein engineering, which showed that novel proteins capable of binding to ligands of interest can be rationally designed *in silico* [272]. The authors of this study described a general computational method for designing small molecule-binding sites and applied it to create a protein capable of binding to the steroid digoxigenin. Through standard directed evolution methodology (described in Section 4.3.5.3), the group was able to obtain an optimized binder with picomolar affinity, and exquisite selectivity for digoxigenin over the related steroids digitoxigenin, progesterone, and β-estradiol. With this ability to program ligand specificity into synthetic proteins, it will, in theory, be possible to design biosensors and biosorbents for virtually any small molecule—a truly revolutionary advance.

4.3.5.3 *Screening: Binding for Purposes of Identification*

Similar to the more commonly used phage display technique [273], bacterial surface display enables the identification of proteins and peptides with desirable binding attributes by linking the identity of displayed proteins with their genetic sequence. To achieve this, large sets of heterogeneous bacterial clones are generated (by transforming bacteria with surface display vector libraries containing either cDNA or randomly mutated sequences), subjected to iterative rounds of screening and amplification, and sequenced to identify the surface protein(s) responsible for mediating binding interactions (Figure 4.3C). The use of bacteria rather than phage in such screens is advantageous because of the density of surface protein expression, which, at >100,000 per cell, dwarfs the few molecules displayed per phage [274], and greatly enhances the sensitivity of screens. Additional advantages are that bacteria, in contrast to phages, replicate autonomously and are of sufficient size to be detected by fluorescence-activated cell sorting (FACS) instruments, which allow quantitation of fluorescence of individual cells at rates upwards of 30,000 per second. This technology enables highly efficient identification of peptides and proteins with desired properties. For instance, through surface display of peptide libraries, the linear epitopes of antibodies have been rapidly determined through incubation of fluorescently labeled antibody with individual clones and FACS-mediated selection of antibody-bound bacteria (Figure 4.3C) [275,276]. Similar approaches have been taken to discover peptides with a range of targets including tumor cells [277,278], human signaling proteins [279,280], neural stem cells [281], viral proteins [282], etc. [9,283–287]. In addition, novel inhibitors for enzymes such as cathepsin G and trypsin have been culled from surface-displayed peptide libraries based on inhibitor enzyme binding affinities [288,289].

One particularly powerful application of combinatorial bacterial display is directed evolution, a process through which proteins are subjected to rounds of random mutagenesis and screening (often through FACS sorting) to identify "evolved" proteins with improved or unique function [290]. The use of bacteria is essential for directed evolution experiments because of their short generation time, amenability to genetic manipulation, and compatibility with sorting instruments.

Surface display-based directed evolution has found several applications thus far. In one example, antibodies with improved affinity were generated through anchored periplasmic expression (APEx) of antibody libraries in *E. coli*, permeabilization of the OM, incubation with fluorescent antigen, and FACS sorting of labeled cells [291–294]. Enzymes with novel activity have also been evolved using bacterial surface display [295–301]. Importantly, identification of hits in such screens differs from previously described methods, in that clones are not identified through interaction with fluorescent binding partners, but through attachment of fluorescent reaction products to the cell surface. For instance, in the development of an endopeptidase with novel substrate specificity, a library of bacterial clones displaying enzyme variants was incubated with a peptide substrate containing a BODIPY fluorophore on one side of the scissile bond and a quencher on the other. Clones expressing an enzyme capable of cleaving the bond would electrostatically capture an unquenched BODIPY-containing cleavage product on their membrane (by virtue of the product's +3 overall charge), allowing for rapid detection and isolation by FACS [298].

4.4 STRATEGIES FOR RE-ENGINEERING BACTERIAL SURFACES WITH HETEROLOGOUS SUGARS

As described above, surface display of exogenous proteins is accomplished through fusion of passengers to extracellular carriers. Heterologous expression of glycans, meanwhile, is more challenging because their biosynthesis is not template directed. Thus, installation of nonnative sugars on the bacterial cell surface must be achieved indirectly, via recombinant expression of enzymes that mediate synthesis, attachment, and remodeling of oligosaccharides, or deletion of endogenous genes involved with these processes (see Box 4.2 for a detailed description of endogenous bacterial glycans).

Three basic strategies have been used in bacterial glycoengineering. In the first, the operon encoding the entire metabolic pathway responsible for synthesis of a given sugar is cloned and transferred to a carrier strain (Figure 4.4A). This approach has been used to express *Pseudomonas aeruginosa* capsular polysaccharide [18,307–309] and *Shigella* LPS [310–313] in a number of heterologous species. The second strategy involves site-specific N-linked glycosylation of target proteins (Figure 4.4A). This is accomplished through expression of (i) the genes required for biosynthesis of the nonnative sugar; (ii) PglB, an oligosaccharyltransferase cloned from *Campylobacter jejuni* that catalyzes

BOX 4.2

GLYCANS EXPRESSED AT ON BACTERIAL SURFACES

The surface glycome in gram-positive organisms consists of glycoproteins (such as S-layer proteins) [302], secondary cell wall polysaccharides (which link the S-layer to peptidoglycan), teichoic acids (anionic glycopolymers which thread through the cell wall), and capsular polysaccharides (in those species expressing capsule) [303]. The biosynthesis of capsule, which represents one of the major targets of glycoengineering, takes place in the cytoplasm, where nucleotide phosphosugars are assembled into massive polymers of species-specific oligosaccharide repeat units by membrane-associated enzymes and translocated to the cell surface through dedicated transport pathways [304,305]. These capsular polysaccharides shield bacteria from the immune system by preventing opsonization with antibodies and complement, consequently reducing phagocytosis by host immune cells.

Gram-negative bacteria also possess glycoproteins (such as pili [15]) as well as capsular polysaccharides (in many species), but are unique in their expression of the complex glycolipid, LPS—another favorite target of glycoengineers (Figure 4.3). LPS, also referred to as endotoxin for its ability to induce toxic shock in mammals, constitutes the OM of most gram-negative bacteria, and has three basic structural components (Figure 4.3) [306]. Lipid A, which is highly conserved across genera and mediates endotoxicity, represents the hydrophobic moiety that anchors LPS to the membrane. The second component, core oligosaccharide, is synthesized in the cytoplasm as well and ligated to Lipid A. This glycolipid intermediate is then flipped to the periplasmic face of the IM, and the final component—O antigen—is attached before trafficking to the OM. Similar to capsule, O antigen is a polymer of oligosaccharide units that shows considerable diversity across species and may grow to very large sizes: up to 164 units of a branched tetrasaccharide in *E. coli* [306].

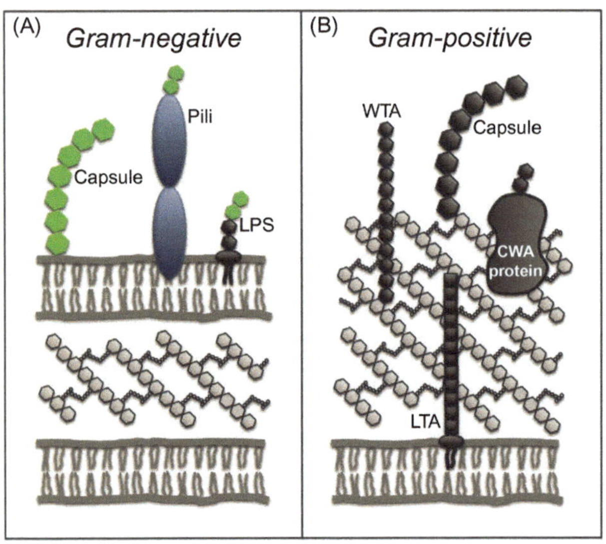

FIGURE 4.4 Glycoengineering strategies for gram-negative and gram-positive bacteria. Saccharides colored green have been modified using glycoengineering techniques. (A) Techniques relevant to gram-negative bacteria include heterologous expression of genes mediating capsule synthesis, site-specific N-linked glycosylation of pili, and tailoring of LPSs. (B) Glycans expressed on the gram-positive envelope are depicted here, although to our knowledge, there have not yet been any examples of glycoengineering in gram-positive bacteria. CWA protein, cell wall-associated protein; LPS, lipopolysaccharide; LTA, lipoteichoic acid; WTA, wall teichoic acid.

attachment of glycans to an asparagine residue within the glycosylation sequon, D/EYNXS/T; and (iii) a modified surface protein containing the heptapeptide glycosylation sequon [15–17]. A third strategy entails glycan remodeling through addition and/or deletion of enzymes that modify cell surface oligosaccharides—most commonly LPS, as described in the following section (Figure 4.4A) [314,315].

4.5 APPLICATIONS OF BACTERIA EXPRESSING HETEROLOGOUS SURFACE SUGARS

4.5.1 Vaccines Against Infectious Disease

Glycoengineering methods have found a number of useful applications—the most successful, perhaps, being vaccine design. One effective vaccine platform involves the heterologous expression of carbohydrate antigens (such as glycoproteins [19], capsular polysaccharides [18,307–309,316–319], and LPS [311–313,320]) on the surface of attenuated vaccine strains. Analogously to the recombinant protein-expressing vaccines described above, the bacterial vector serves as a potent adjuvant stimulating adaptive immune responses against the pathogen-associated glycoantigen. The second major approach involves attenuation of pathogens through genetic modification of their LPS, allowing direct vaccination with avirulent pathogenic strains. Common molecular alterations to LPS, diagrammed in Figure 4.5, include addition of 4′-phosphate groups (through deletion of the phosphatase, LpxF) or of fatty acids

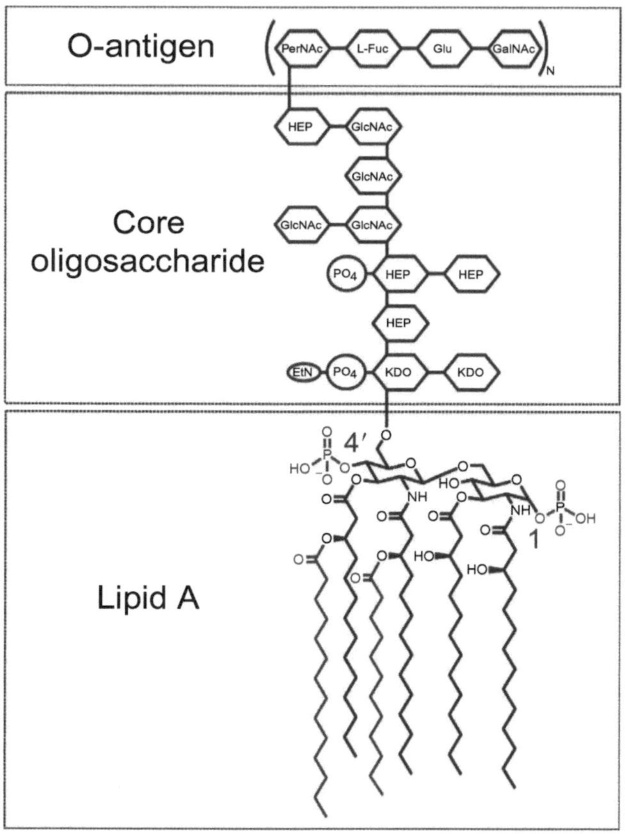

FIGURE 4.5 LPS structure. The predominant LPS glycoform expressed in *E. coli* O157:H7 is represented here [321,322]. Removal of the 1′-phosphate group indicated in blue (through LpxE expression) renders Lipid A nontoxic while maintaining its adjuvanticity [323]. Removal of the 4′-phosphate group (via LpxF expression) and/or acyl groups indicated in blue (via deletion of LpxL and LpxM) allows Lipid A to evade detection by TLR4. Abbreviations: EtN, ethanolamine; GalNAc, D-*N*-acetylgalactose; Glc, D-glucose; GlcNAc, D-*N*-acetylglucose; Hep, L-glycero-D-manno-heptose; KDO, 3-deoxy-D-manno-octulosonic acid (2-keto-3-deoxyoctulosonic acid); L-Fuc, L-fucose; PerNAc, D-*N*-acetylperosamine; PO_4, phosphate.

to produce hexa-acetylated Lipid A (through expression of acyltransferases, LpxL, and LpxM) [314]. In contrast to the immunoevasive native glycoforms, re-engineered LPS variants potently activate toll-like receptor 4 (TLR4), leading to the elicitation of pro-inflammatory immune responses that mediate rapid clearance of the bacteria, and generation of adaptive immunity against the bacterial vector [315]. This method has been applied to construct attenuated vaccines for *Shigella flexneri* [324], *Y. pestis* [21,325,326], *Salmonella* spp. [327,328], *Francisella tularensis* [20,329], *Leptospira interrogans* [330], *Neisseria meningitides* [331], *Moraxella catarrhalis* [332], and *K. pneumoniae* [333].

4.5.2 Anti-Infective Probiotic Therapy

Surface glycoengineering has also been used to develop a potentially revolutionary class of anti-infective probiotics: avirulent bacterial strains expressing oligosaccharide receptors for the bacterial exotoxins that cause cholera, shigellosis, and traveller's diarrhea, which together claim hundreds of thousands of lives per year worldwide [22,334–338]. The rationale for this approach is that exotoxins are sequestered by receptor decoys on probiotic strains in the gut lumen rather than binding to their targets on colonic epithelial cells, which induces diarrheal disease. Due to the density and multivalency of receptor expression on the surface of probiotic strains, exotoxins are intercepted *in vivo* with extraordinary efficiency. In one study, 1 mg of recombinant bacteria was shown to neutralize over 150 μg of shiga toxin, a potent exotoxin released by *Shigella* spp. that leads to dysentery. In mice, twice daily administration of the therapeutic strain provided 100% protection against otherwise fatal shigellosis, even in killed form. Thus, in addition to being cost effective (given that large-scale production can be achieved through simple fermentation), this therapeutic approach is highly effective and safe. Furthermore, these re-engineered probiotics apply no selective pressure for evolution of resistant pathogenic strains because they specifically intercept disease-causing toxins rather than killing the pathogen itself. Recently, similar principles have guided the development of a probiotic expressing the glycan receptor for adhesion proteins on uropathogenic strains of *E. coli*. Prophylactic treatment with the recombinant strain abrogated binding of the pathogen to the urinary epithelium, preventing development of urinary tract infection in mice [339]. Unfortunately, despite the great promise such glycoengineered probiotics hold for the treatment of infectious disease, they have been slowed in their clinical development due to apprehension surrounding the release of GMOs (see Box 4.1).

4.5.3 Glycoprotein Production

The final application of glycoengineering relates to production of recombinant glycoproteins. The importance of this process is highlighted by the fact that approximately 70% of therapeutically relevant proteins have saccharide modifications, including catabolic enzymes used for treating inborn errors of metabolism and erythropoietin, which is used in the treatment of chronic kidney disease [340]. These carbohydrate moieties affect several pharmacological properties of glycoproteins, including *in vivo* activity, half-life, proteolytic stability, and tissue targeting [341–346]. Glycoproteins are also used as vaccines; in fact, some of the most commonly used vaccines—including those against *H. influenzae*, *N. meningitidis*, and *S. pneumoniae*—consist of capsular polysaccharide–protein conjugates. Conventional methods for producing such compounds involve isolation of polysaccharide capsule from pathogens, purification of recombinant carrier proteins, chemical coupling of polysaccharides to carrier proteins, and purification of glycoconjugates—a typically low yielding and expensive process [15]. Alternatively, researchers have pursued total synthesis of glycans, but despite having led

BOX 4.3

CHEMICAL RE-ENGINEERING OF BACTERIAL SURFACE PROTEINS AND SUGARS

While the strategies presented in this chapter for expressing heterologous proteins and sugars at the bacterial cell surface have proven useful in a myriad of applications, they nevertheless fall short in one important regard: they offer no way of installing non-genetically encoded molecules such as fluorophores, affinity tags, and cytotoxic drugs on bacteria. For such compounds, an alternative set of approaches, pioneered by chemical biologists over the last decade, must be employed.

In general, there are two basic strategies for installing small molecules onto bacterial surfaces: direct and two-step labeling. In direct labeling, small molecule–monomer conjugates are incubated with actively growing cells so that cells take up the conjugate and incorporate it via endogenous biosynthetic pathways into polymeric structures on the cell surface (such as proteins or sugars). In contrast, two-step labeling involves (i) incorporation of a monomer equipped with a small, reactive chemical handle (such as an azide) into a cell surface structure, and (ii) subsequent bioorthogonal ligation of a small molecule functionalized with a complementary reactive group (an alkyne in this example) to the surface-exposed handle. This two-step approach is particularly useful when installing large compounds that are unable to pass the plasma membrane, or too bulky to allow direct incorporation by the enzymes involved with polymer biosynthesis. Besides the azide/terminal alkyne click reaction pair, commonly used bioorthogonal reaction pairs include: azides and phosphines (Staudinger ligation), aldehydes/ketones and hydrazines or hydroxylamines, and tetrazines and strained alkenes such as trans-cyclooctenes [349,350].

Two-step metabolic labeling of glycans, initially developed by Bertozzi et al. [351], represents the paradigmatic example of chemical surface engineering. In this methodology, monosaccharides modified with chemical handles are first assimilated into surface-associated sugars and then small molecules of interest are attached to the functionalized glycan at the cell surface. The outer membrane glycolipid, LPS, has served as the most common target for modification [352–354], although recent studies have shown that glycoproteins are amenable to chemical manipulation as well [355–357].

Polypeptides at the cell surface have been similarly re-engineered via two-step metabolic labeling techniques. In analogy to the functionalized monosaccharides used to label glycans, chemically tagged noncanonical amino acids (ncAAs) are incorporated into proteins and presented on the cell surface for subsequent ligation. To this end, ncAAs such as azidohomoalanine, azidonorvaline, and azidonorleucine have been shown to serve as methionine surrogates, replacing the endogenous amino acid during translation [358–362]. Alkyne-functionalized compounds, in turn, are attached to these protein-bound, surface-exposed azides. Through this approach, methionines are globally replaced with ncAAs—a particularly valuable feature in certain settings (e.g., proteomic analysis). However, for site-specific modification, an alternative technique—pioneered by Schultz et al.—must be used. First, a genetically engineered tRNA–tRNA synthetase pair, designed to incorporate a ketone-containing ncAA (m-acetyl-L-phenylalanine) into a recombinant surface protein in response to the amber nonsense stop codon, is introduced into a bacterial strain. Then, the strain is incubated with m-acetyl-L-phenylalanine, so that the ncAA is selectively incorporated into the recombinant protein. Finally, the surface-exposed ketone functionality is used as an attachment point for a small molecule containing a hydrazine or hydroxylamine group [363].

to elegant chemical developments [347,348], these approaches have been hampered by the extreme chemical complexity of oligosaccharides, deriving from the stereoisomeric diversity of monosaccharide subunits, numerous possible glycosidic linkages, and frequent branching.

Heterologous expression of N-linked glycosylation systems in *E. coli*, meanwhile, allows for scalable, inexpensive production of easily purifiable, homogeneous glycoconjugates, with site-specific installation of glycans on target proteins. While still in need of improvements (such as identification of bacterial oligosaccharyltransferases that can attach glycans with non-reducing sugars to acceptor proteins), bacterial glycotagging technology has the potential to revolutionize the production of therapeutic and immunomodulatory glycoproteins in the future [16,17].

4.6 CONCLUSION

Owing to the self-regenerating nature of the bacterial vector, its immunogenic properties, strong genotype–phenotype linkage, and amenability to genetic manipulation, bacterial surface display has proven an extraordinarily versatile and economical biotechnological system. Applications of these strategies in vaccinology have led to agents in clinical trials, while catalytic "biofactories" have played a key role in industrial and pharmaceutical synthesis, and bacteria-based screening technologies have driven forward basic research. However, we believe the surface engineering paradigm may be further improved by combining heterologous expression of protein and saccharides with the many complementary technologies elucidated above including computational protein engineering. Even further, genetic surface display can benefit from integration with an exciting set of methods emerging in chemical biology that allow installation of synthetic compounds on the cell surface (see Box 4.3) [364,365]. Together, these powerful techniques would impart incredible versatility to the surface engineering platform, transforming the cell envelope into a veritable tabula rasa, able to accommodate all manner of protein, sugar, and synthetic chemical in virtually any combination. With these tools, for instance, one may imagine constructing an anticancer therapeutic strain that expresses (i) a tumor antigen to trigger adaptive immune responses; (ii) an oligosaccharide targeting moiety to promote uptake by APCs and efficient antigen presentation (e.g., the DC-SIGN ligand, Sialyl-Lewis^{X}) [366–368]; and (iii) a chemotherapeutic payload that is released upon arrival in the tumor microenvironment [225]. Thus, through interdisciplinary efforts involving synthetic chemists, bacteriologists, and biomedical engineers, we imagine the development of a new class of "synthetic organisms"—chemically and genetically re-engineered bacteria that will serve as novel multimodal therapeutics, biotechnological devices, and tools to advance fundamental research.

Acknowledgment

We deeply appreciate the contributions of Dr. Thihan Padukkavidana toward the completion of this manuscript.

References

[1] Duguid JP. The sensitivity of bacteria to the action of penicillin. Edinb Med J 1946;53:401–12.

[2] Mudd S, Heinmets F, Anderson TF. The pneumococcal capsular swelling reaction, studied with the aid of the electron microscope. J Exp Med 1943;78(5):327–32.

[3] Freudl R, MacIntyre S, Degen M, Henning U. Cell surface exposure of the outer membrane protein OmpA of *Escherichia coli* K-12. J Mol Biol 1986;188(3):491–4.

[4] Charbit A, Boulain JC, Ryter A, Hofnung M. Probing the topology of a bacterial membrane protein by genetic insertion of a foreign epitope; expression at the cell surface. EMBO J 1986;5(11):3029–37.

[5] Kuroda K, Ueda M. Molecular design of the microbial cell surface toward the recovery of metal ions. Curr Opin Biotechnol 2011;22(3):427–33.

[6] Wu CH, Mulchandani A, Chen W. Versatile microbial surface-display for environmental remediation and biofuels production. Trends Microbiol 2008;16(4):181–8.

[7] Jose J, Meyer TF. The autodisplay story, from discovery to biotechnical and biomedical applications. Microbiol Mol Biol Rev 2007;71(4):600–19.

[8] Date A, Pasini P, Daunert S. Fluorescent and bioluminescent cell-based sensors: strategies for their preservation. Adv Biochem Eng Biotechnol 2010;117:57–75.

[9] Daugherty PS. Protein engineering with bacterial display. Curr Opin Struct Biol 2007;17(4):474–80.

[10] Paton AW, Morona R, Paton JC. Bioengineered microbes in disease therapy. Trends Mol Med 2012;18 (7):417–25.

[11] Wernerus H, Stahl S. Biotechnological applications for surface-engineered bacteria. Biotechnol Appl Biochem 2004;40(Pt 3):209–28.

[12] Knecht LD, Pasini P, Daunert S. Bacterial spores as platforms for bioanalytical and biomedical applications. Anal Bioanal Chem 2011;400(4):977–89.

[13] van Bloois E, Winter RT, Kolmar H, Fraaije MW. Decorating microbes: surface display of proteins on *Escherichia coli*. Trends Biotechnol 2011;29(2):79–86.

[14] Wells J. Mucosal vaccination and therapy with genetically modified lactic acid bacteria. Annu Rev Food Sci Technol 2011;2:423–45.

[15] Terra VS, Mills DC, Yates LE, Abouelhadid S, Cuccui J, Wren BW. Recent developments in bacterial protein glycan coupling technology and glycoconjugate vaccine design. J Med Microbiol 2012;61(Pt 7):919–26.

[16] Baker JL, Celik E, DeLisa MP. Expanding the glycoengineering toolbox: the rise of bacterial N-linked protein glycosylation. Trends Biotechnol 2013;31(5):313–23.

[17] Merritt JH, Ollis AA, Fisher AC, Delisa MP. Glycans-by-design: engineering bacteria for the biosynthesis of complex glycans and glycoconjugates. Biotechnol Bioeng 2013;110(6):1550–64.

[18] DiGiandomenico A, Rao J, Harcher K, Zaidi TS, Gardner J, Neely AN, et al. Intranasal immunization with heterologously expressed polysaccharide protects against multiple *Pseudomonas aeruginosa* infections. Proc Natl Acad Sci USA 2007;104(11):4624–9.

[19] Iwashkiw JA, Fentabil MA, Faridmoayer A, Mills DC, Peppler M, Czibener C, et al. Exploiting the *Campylobacter jejuni* protein glycosylation system for glycoengineering vaccines and diagnostic tools directed against brucellosis. Microb Cell Fact 2012;11:13.

[20] Wang X, Ribeiro AA, Guan Z, Abraham SN, Raetz CR. Attenuated virulence of a Francisella mutant lacking the lipid A 4′-phosphatase. Proc Natl Acad Sci USA 2007;104(10):4136–41.

[21] Montminy SW, Khan N, McGrath S, Walkowicz MJ, Sharp F, Conlon JE, et al. Virulence factors of *Yersinia pestis* are overcome by a strong lipopolysaccharide response. Nat Immunol 2006;7(10):1066–73.

[22] Paton AW, Morona R, Paton JC. Bioengineered bugs expressing oligosaccharide receptor mimics: toxin-binding probiotics for treatment and prevention of enteric infections. Bioeng Bugs 2010;1(3):172–7.

[23] Frey J. Biological safety concepts of genetically modified live bacterial vaccines. Vaccine 2007;25(30): 5598–605.

[24] Detmer A, Glenting J. Live bacterial vaccines—a review and identification of potential hazards. Microb Cell Fact 2006;5:23.

[25] Spreng S, Viret JF. Plasmid maintenance systems suitable for GMO-based bacterial vaccines. Vaccine 2005;23(17–18):2060–5.

[26] Galan JE, Nakayama K, Curtiss 3rd R. Cloning and characterization of the asd gene of *Salmonella typhimurium*: use in stable maintenance of recombinant plasmids in Salmonella vaccine strains. Gene 1990;94 (1):29–35.

[27] Curtiss 3rd R, Galan JE, Nakayama K, Kelly SM. Stabilization of recombinant avirulent vaccine strains *in vivo*. Res Microbiol 1990;141(7-8):797–805.

[28] Thomason L, Court DL, Bubunenko M, Costantino N, Wilson H, Datta S, et al. Recombineering: genetic engineering in bacteria using homologous recombination. Curr Protoc Mol Biol 2007; (Chapter 1: Unit 1.16).

[29] van Pijkeren JP, Britton RA. High efficiency recombineering in lactic acid bacteria. Nucleic Acids Res 2012;40(10):e76.

[30] Silhavy TJ, Kahne D, Walker S. The bacterial cell envelope. Cold Spring Harb Perspect Biol 2010;2(5): a000414.

[31] Charbit A, Molla A, Saurin W, Hofnung M. Versatility of a vector for expressing foreign polypeptides at the surface of gram-negative bacteria. Gene 1988;70 (1):181–9.

[32] Klemm P, Schembri MA. Fimbrial surface display systems in bacteria: from vaccines to random libraries. Microbiology 2000;146(**Pt 12**):3025–32.

[33] Jose J, Maas RM, Teese MG. Autodisplay of enzymes-molecular basis and perspectives. J Biotechnol 2012;161(2):92–103.

[34] Park M, Jose J, Thommes S, Kim JI, Kang MJ, Pyun JC. Autodisplay of streptavidin. Enzyme Microb Technol 2011;48(4–5):307–11.

[35] Detzel C, Maas R, Tubeleviciute A, Jose J. Autodisplay of nitrilase from *Klebsiella pneumoniae* and whole-cell degradation of oxynil herbicides and related compounds. Appl Microbiol Biotechnol 2012;97(11): 4887–96.

[36] Jose J, von Schwichow S. Autodisplay of active sorbitol dehydrogenase (SDH) yields a whole cell biocatalyst for the synthesis of rare sugars. Chembiochem 2004;5(4):491–9.

[37] Kawahara H. The structures and functions of ice crystal-controlling proteins from bacteria. J Biosci Bioeng 2002;94(6):492–6.

[38] Yim SK, Kim DH, Jung HC, Pan JG, Kang HS, Ahn T, et al. Surface display of heme-and diflavin-containing cytochrome P450 BM3 in *Escherichia coli*: a whole cell biocatalyst for oxidation. J Microbiol Biotechnol 2010;20(4):712–7.

[39] Klauser T, Pohlner J, Meyer TF. Extracellular transport of cholera toxin B subunit using Neisseria IgA protease beta-domain: conformation-dependent outer membrane translocation. EMBO J 1990;9(6): 1991–9.

[40] Jung HC, Lebeault JM, Pan JG. Surface display of *Zymomonas mobilis* levansucrase by using the ice-nucleation protein of *Pseudomonas syringae*. Nat Biotechnol 1998;16(6):576–80.

[41] Jose J, Bernhardt R, Hannemann F. Functional display of active bovine adrenodoxin on the surface of *E. coli* by chemical incorporation of the [2Fe–2S] cluster. Chembiochem 2001;2(9):695–701.

[42] Lee SY, Choi JH, Xu Z. Microbial cell-surface display. Trends Biotechnol 2003;21(1):45–52.

[43] Samuelson P, Gunneriusson E, Nygren PA, Stahl S. Display of proteins on bacteria. J Biotechnol 2002;96 (2):129–54.

[44] Leenhouts K, Buist G, Kok J. Anchoring of proteins to lactic acid bacteria. Antonie Van Leeuwenhoek 1999;76(1–4):367–76.

[45] Marraffini LA, Dedent AC, Schneewind O. Sortases and the art of anchoring proteins to the envelopes of gram-positive bacteria. Microbiol Mol Biol Rev 2006;70(1):192–221.

[46] Narita J, Okano K, Kitao T, Ishida S, Sewaki T, Sung MH, et al. Display of alpha-amylase on the surface of *Lactobacillus casei* cells by use of the PgsA anchor protein, and production of lactic acid from starch. Appl Environ Microbiol 2006;72(1):269–75.

[47] Avall-Jaaskelainen S, Kyla-Nikkila K, Kahala M, Miikkulainen-Lahti T, Palva A. Surface display of foreign epitopes on the *Lactobacillus brevis* S-layer. Appl Environ Microbiol 2002;68(12):5943–51.

[48] Vollenkle C, Weigert S, Ilk N, Egelseer E, Weber V, Loth F, et al. Construction of a functional S-layer fusion protein comprising an immunoglobulin G-binding domain for development of specific adsorbents for extracorporeal blood purification. Appl Environ Microbiol 2004;70(3):1514–21.

[49] Nomellini JF, Duncan G, Dorocicz IR, Smit J. S-layer-mediated display of the immunoglobulin G-binding domain of streptococcal protein G on the surface of *Caulobacter crescentus*: development of an immunoactive reagent. Appl Environ Microbiol 2007;73 (10):3245–53.

[50] Zarschler K, Janesch B, Kainz B, Ristl R, Messner P, Schaffer C. Cell surface display of chimeric glycoproteins via the S-layer of *Paenibacillus alvei*. Carbohydr Res 2010;345(10):1422–31.

[51] Ilk N, Egelseer EM, Sleytr UB. S-layer fusion proteins—construction principles and applications. Curr Opin Biotechnol 2011;22(6):824–31.

[52] Engelhardt H. Are S-layers exoskeletons? The basic function of protein surface layers revisited. J Struct Biol 2007;160(2):115–24.

[53] Vollmer W, Joris B, Charlier P, Foster S. Bacterial peptidoglycan (murein) hydrolases. FEMS Microbiol Rev 2008;32(2):259–86.

[54] Okano K, Zhang Q, Kimura S, Narita J, Tanaka T, Fukuda H, et al. System using tandem repeats of the cA peptidoglycan-binding domain from *Lactococcus lactis* for display of both N-and C-terminal fusions on cell surfaces of lactic acid bacteria. Appl Environ Microbiol 2008;74(4):1117–23.

[55] Bosma T, Kanninga R, Neef J, Audouy SA, van Roosmalen ML, Steen A, et al. Novel surface display system for proteins on non-genetically modified gram-positive bacteria. Appl Environ Microbiol 2006;72(1):880–9.

[56] Raha AR, Varma NR, Yusoff K, Ross E, Foo HL. Cell surface display system for *Lactococcus lactis*: a novel development for oral vaccine. Appl Microbiol Biotechnol 2005;68(1):75–81.

[57] Hu S, Kong J, Kong W, Guo T, Ji M. Characterization of a novel LysM domain from *Lactobacillus fermentum* bacteriophage endolysin and its use as an anchor to display heterologous proteins on the surfaces of lactic acid bacteria. Appl Environ Microbiol 2010;76 (8):2410–8.

[58] Cano RJ, Borucki MK. Revival and identification of bacterial spores in 25-to 40-million-year-old *Dominican amber*. Science 1995;268(5213):1060–4.

[59] Hudson KD, Corfe BM, Kemp EH, Feavers IM, Coote PJ, Moir A. Localization of GerAA and GerAC germination proteins in the *Bacillus subtilis* spore. J Bacteriol 2001;183(14):4317–22.

[60] Kim J, Schumann W. Display of proteins on *Bacillus subtilis* endospores. Cell Mol Life Sci 2009;66(19): 3127–36.

[61] Thompson BM, Stewart GC. Targeting of the BclA and BclB proteins to the *Bacillus anthracis* spore surface. Mol Microbiol 2008;70(2):421–34.

[62] Du C, Chan WC, McKeithan TW, Nickerson KW. Surface display of recombinant proteins on *Bacillus thuringiensis* spores. Appl Environ Microbiol 2005;71 (6):3337–41.

[63] Date A, Pasini P, Daunert S. Integration of spore-based genetically engineered whole-cell sensing systems into portable centrifugal microfluidic platforms. Anal Bioanal Chem 2010;398(1):349–56.

[64] Date A, Pasini P, Sangal A, Daunert S. Packaging sensing cells in spores for long-term preservation of sensors: a tool for biomedical and environmental analysis. Anal Chem 2010;82(14):6098–103.

[65] Date A, Pasini P, Daunert S. Construction of spores for portable bacterial whole-cell biosensing systems. Anal Chem 2007;79(24):9391–7.

[66] Fantino JR, Barras F, Denizot F. Sposensor: a whole-bacterial biosensor that uses immobilized *Bacillus subtilis* spores and a one-step incubation/ detection process. J Mol Microbiol Biotechnol 2009;17(2):90–5.

[67] Rotman B, Cote MA. Application of a real-time biosensor to detect bacteria in platelet concentrates. Biochem Biophys Res Commun 2003;300 (1):197–200.

[68] Dodatko T, Akoachere M, Jimenez N, Alvarez Z, Abel-Santos E. Dissecting interactions between nucleosides and germination receptors in *Bacillus cereus* 569 spores. Microbiology 2010;156(Pt 4):1244–55.

[69] Serp D, von Stockar U, Marison IW. Immobilized bacterial spores for use as bioindicators in the validation of thermal sterilization processes. J Food Prot 2002;65(7):1134–41.

[70] Kwon SJ, Jung HC, Pan JG. Transgalactosylation in a water-solvent biphasic reaction system with beta-galactosidase displayed on the surfaces of *Bacillus subtilis* spores. Appl Environ Microbiol 2007;73 (7):2251–6.

[71] Cutting SM, Hong HA, Baccigalupi L, Ricca E. Oral vaccine delivery by recombinant spore probiotics. Int Rev Immunol 2009;28(6):487–505.

[72] Oggioni MR, Ciabattini A, Cuppone AM, Pozzi G. Bacillus spores for vaccine delivery. Vaccine 2003;21 (Suppl. 2):S96–101.

[73] Henriques AO, Moran Jr. CP. Structure, assembly, and function of the spore surface layers. Annu Rev Microbiol 2007;61:555–88.

[74] Kim JH, Lee CS, Kim BG. Spore-displayed streptavidin: a live diagnostic tool in biotechnology. Biochem Biophys Res Commun 2005;331(1):210–4.

[75] Uyen NQ, Hong HA, Cutting SM. Enhanced immunisation and expression strategies using bacterial spores as heat-stable vaccine delivery vehicles. Vaccine 2007;25(2):356–65.

[76] Tam NK, Uyen NQ, Hong HA, Duc lH, Hoa TT, Serra CR, et al. The intestinal life cycle of *Bacillus subtilis* and close relatives. J Bacteriol 2006; 188(7):2692–700.

[77] Kulp A, Kuehn MJ. Biological functions and biogenesis of secreted bacterial outer membrane vesicles. Annu Rev Microbiol 2010;64:163–84.

[78] Langemann T, Koller VJ, Muhammad A, Kudela P, Mayr UB, Lubitz W. The Bacterial Ghost platform system: production and applications. Bioeng Bugs 2010;1(5):326–36.

[79] Ellis TN, Kuehn MJ. Virulence and immunomodulatory roles of bacterial outer membrane vesicles. Microbiol Mol Biol Rev 2010;74(1):81–94.

[80] Kim JY, Doody AM, Chen DJ, Cremona GH, Shuler ML, Putnam D, et al. Engineered bacterial outer membrane vesicles with enhanced functionality. J Mol Biol 2008; 380(1):51–66.

[81] Danzig L. Meningococcal vaccines. Pediatr Infect Dis J 2004;23(12 Suppl.):S285–92.

[82] Collins BS. Gram-negative outer membrane vesicles in vaccine development. Discov Med 2011;12(62): 7–15.

[83] Oster P, O'Hallahan J, Aaberge I, Tilman S, Ypma E, Martin D. Immunogenicity and safety of a strain-specific MenB OMV vaccine delivered to under 5-year olds in New Zealand. Vaccine 2007;25 (16):3075–9.

[84] Mashburn-Warren L, McLean RJ, Whiteley M. Gram-negative outer membrane vesicles: beyond the cell surface. Geobiology 2008;6(3):214–9.

[85] Kudela P, Koller VJ, Lubitz W. Bacterial ghosts (BGs)—advanced antigen and drug delivery system. Vaccine 2010;28(36):5760–7.

[86] Riedmann EM, Kyd JM, Cripps AW, Lubitz W. Bacterial ghosts as adjuvant particles. Expert Rev Vaccines 2007;6(2):241–53.

[87] Batista-Duharte A, Lindblad EB, Oviedo-Orta E. Progress in understanding adjuvant immunotoxicity mechanisms. Toxicol Lett 2011;203(2):97–105.

[88] Blander JM, Sander LE. Beyond pattern recognition: five immune checkpoints for scaling the microbial threat. Nat Rev Immunol 2012;12(3):215–25.

[89] Stover CK, Bansal GP, Hanson MS, Burlein JE, Palaszynski SR, Young JF, et al. Protective immunity elicited by recombinant bacille Calmette-Guerin (BCG) expressing outer surface protein A (OspA) lipoprotein: a candidate Lyme disease vaccine. J Exp Med 1993;178(1):197–209.

[90] Bastos RG, Dellagostin OA, Barletta RG, Doster AR, Nelson E, Osorio FA. Construction and immunogenicity of recombinant *Mycobacterium bovis* BCG expressing GP5 and M protein of porcine reproductive respiratory syndrome virus. Vaccine 2002;21 (1-2):21–9.

[91] Kang HY, Curtiss R. Immune responses dependent on antigen location in recombinant attenuated *Salmonella typhimurium* vaccines following oral immunization. FEMS Immunol Med Microbiol 2003;37(2–3):99–104.

[92] Kaufmann SH, Hess J. Impact of intracellular location of and antigen display by intracellular bacteria: implications for vaccine development. Immunol Lett 1999;65(1–2):81–4.

[93] Hess J, Gentschev I, Miko D, Welzel M, Ladel C, Goebel W, et al. Superior efficacy of secreted over somatic antigen display in recombinant Salmonella vaccine induced protection against listeriosis. Proc Natl Acad Sci USA 1996;93(4): 1458–63.

[94] Bermúdez-Humarán LG, Cortes-Perez NG, Le Loir Y, Alcocer-González JM, Tamez-Guerra RS, de Oca-Luna RM, et al. An inducible surface presentation system improves cellular immunity against human papillomavirus type 16 E7 antigen in mice after nasal administration with recombinant lactococci. J Med Microbiol 2004;53(Pt 5):427–33.

[95] Titball RW, Howells AM, Oyston PC, Williamson ED. Expression of the *Yersinia pestis* capsular antigen (F1 antigen) on the surface of an aroA mutant of *Salmonella typhimurium* induces high levels of protection against plague. Infect Immun 1997;65(5):1926–30.

[96] Norton PM, Brown HW, Wells JM, Macpherson AM, Wilson PW, Le Page RW. Factors affecting the immunogenicity of tetanus toxin fragment C expressed in *Lactococcus lactis*. FEMS Immunol Med Microbiol 1996;14(2–3):167–77.

[97] Reveneau N, Geoffroy MC, Locht C, Chagnaud P, Mercenier A. Comparison of the immune responses induced by local immunizations with recombinant *Lactobacillus plantarum* producing tetanus toxin fragment C in different cellular locations. Vaccine 2002;20(13–14):1769–77.

[98] Dieye Y, Hoekman AJ, Clier F, Juillard V, Boot HJ, Piard JC. Ability of *Lactococcus lactis* to export viral capsid antigens: a crucial step for development of live vaccines. Appl Environ Microbiol 2003;69 (12):7281–8.

[99] Haddad D, Liljeqvist S, Kumar S, Hansson M, Ståhl S, Perlmann H, et al. Surface display compared to periplasmic expression of a malarial antigen in *Salmonella typhimurium* and its implications for immunogenicity. FEMS Immunol Med Microbiol 1995;12(3–4):175–86.

[100] Chen H, Schifferli DM. Mucosal and systemic immune responses to chimeric fimbriae expressed by *Salmonella enterica* serovar Typhimurium vaccine strains. Infect Immun 2000;68(6):3129–39.

[101] Blander JM, Medzhitov R. Toll-dependent selection of microbial antigens for presentation by dendritic cells. Nature 2006;440(7085):808–12.

[102] Bachmann MF, Jennings GT. Vaccine delivery: a matter of size, geometry, kinetics and molecular patterns. Nat Rev Immunol 2010;10(11):787–96.

[103] Galen JE, Pasetti MF, Tennant S, Ruiz-Olvera P, Sztein MB, Levine MM. *Salmonella enterica* serovar Typhi live vector vaccines finally come of age. Immunol Cell Biol 2009;87(5):400–12.

[104] Cheminay C, Hensel M. Rational design of Salmonella recombinant vaccines. Int J Med Microbiol 2008;298(1–2):87–98.

[105] Luria-Perez R, Cedillo-Barron L, Santos-Argumedo L, Ortiz-Navarrete VF, Ocaña-Mondragon A, Gonzalez-Bonilla CR. A fusogenic peptide expressed on the surface of *Salmonella enterica* elicits CTL responses to a dengue virus epitope. Vaccine 2007;25(27):5071–85.

[106] Lee JS, Shin KS, Pan JG, Kim CJ. Surface-displayed viral antigens on Salmonella carrier vaccine. Nat Biotechnol 2000;18(6):645–8.

[107] Dou JL, Jing T, Fan JJ, Yuan ZM. Surface display of domain III of Japanese encephalitis virus E protein on *Salmonella typhimurium* by using an ice nucleation protein. Virol Sin 2011;26(6):409–17.

[108] Wang L, Huang JA, Nagesha HS, Smith SC, Phelps A, Holmes I, et al. Bacterial expression of the major antigenic regions of porcine rotavirus VP7 induces a neutralizing immune response in mice. Vaccine 1999;17 (20–21):2636–45.

[109] Huang H, Wang YJ, White AP, Meng JZ, Liu GR, Liu SL, et al. Salmonella expressing a T-cell epitope from Sendai virus are able to induce anti-infection immunity. J Med Microbiol 2009;58(Pt 9):1236–42.

[110] Chen H, Schifferli DM. Construction, characterization, and immunogenicity of an attenuated *Salmonella enterica* serovar Typhimurium pgtE vaccine expressing fimbriae with integrated viral

epitopes from the spiC promoter. Infect Immun 2003;71(8):4664–73.

[111] Chen H, Schifferli DM. Comparison of a fimbrial versus an autotransporter display system for viral epitopes on an attenuated Salmonella vaccine vector. Vaccine 2007;25(9):1626–33.

[112] Baillie LW, Rodriguez AL, Moore S, Atkins HS, Feng C, Nataro JP, et al. Towards a human oral vaccine for anthrax: the utility of a Salmonella Typhi Ty21a-based prime-boost immunization strategy. Vaccine 2008;26(48):6083–91.

[113] Stokes MG, Titball RW, Neeson BN, Galen JE, Walker NJ, Stagg AJ, et al. Oral administration of a *Salmonella enterica*-based vaccine expressing *Bacillus anthracis* protective antigen confers protection against aerosolized *B. anthracis*. Infect Immun 2007;75 (4):1827–34.

[114] Ascón MA, Hone DM, Walters N, Pascual DW. Oral immunization with a *Salmonella typhimurium* vaccine vector expressing recombinant enterotoxigenic *Escherichia coli* K99 fimbriae elicits elevated antibody titers for protective immunity. Infect Immun 1998;66 (11):5470–6.

[115] Pascual DW, Hone DM, Hall S, van Ginkel FW, Yamamoto M, Walters N, et al. Expression of recombinant enterotoxigenic *Escherichia coli* colonization factor antigen I by *Salmonella typhimurium* elicits a biphasic T helper cell response. Infect Immun 1999;67(12):6249–56.

[116] Rizos K, Lattemann CT, Bumann D, Meyer TF, Aebischer T. Autodisplay: efficacious surface exposure of antigenic UreA fragments from *Helicobacter pylori* in Salmonella vaccine strains. Infect Immun 2003;71(11):6320–8.

[117] Spreng S, Dietrich G, Goebel W, Gentschev I. Protection against murine listeriosis by oral vaccination with recombinant Salmonella expressing protective listerial epitopes within a surface-exposed loop of the TolC-protein. Vaccine 2003;21(7-8):746–52.

[118] Nayak AR, Tinge SA, Tart RC, McDaniel LS, Briles DE, Curtiss 3rd R. A live recombinant avirulent oral Salmonella vaccine expressing pneumococcal surface protein A induces protective responses against *Streptococcus pneumoniae*. Infect Immun 1998;66(8): 3744–51.

[119] Xin W, Li Y, Mo H, Roland KL, Curtiss 3rd R. PspA family fusion proteins delivered by attenuated *Salmonella enterica* serovar Typhimurium extend and enhance protection against *Streptococcus pneumoniae*. Infect Immun 2009;77(10):4518–28.

[120] Kramer U, Rizos K, Apfel H, Autenrieth IB, Lattemann CT. Autodisplay: development of an efficacious system for surface display of antigenic determinants in Salmonella vaccine strains. Infect Immun 2003;71(4):1944–52.

[121] Liu WT, Hsu HL, Liang CC, Chuang CC, Lin HC, Liu YT. A comparison of immunogenicity and protective immunity against experimental plague by intranasal and/or combined with oral immunization of mice with attenuated *Salmonella* serovar Typhimurium expressing secreted *Yersinia pestis* F1 and V antigen. FEMS Immunol Med Microbiol 2007;51(1):58–69.

[122] Morton M, Garmory HS, Perkins SD, O'Dowd AM, Griffin KF, Turner AK, et al. A *Salmonella enterica* serovar Typhi vaccine expressing *Yersinia pestis* F1 antigen on its surface provides protection against plague in mice. Vaccine 2004;22(20):2524–32.

[123] Chinchilla M, Pasetti MF, Medina-Moreno S, Wang JY, Gomez-Duarte OG, Stout R. Enhanced immunity to *Plasmodium falciparum* circumsporozoite protein (PfCSP) by using *Salmonella enterica* serovar Typhi expressing PfCSP and a PfCSP-encoding DNA vaccine in a heterologous prime-boost strategy. Infect Immun 2007;75(8):3769–79.

[124] Ruiz-Pérez F, León-Kempis R, Santiago-Machuca A, Ortega-Pierres G, Barry E, Levine M, et al. Expression of the *Plasmodium falciparum* immunodominant epitope (NANP)(4) on the surface of *Salmonella enterica* using the autotransporter MisL. Infect Immun 2002;70(7):3611–20.

[125] Ruiz-Olvera P, Ruiz-Pérez F, Sepulveda NV, Santiago-Machuca A, Maldonado-Rodríguez R, Garcia-Elorriaga G, et al. Display and release of the *Plasmodium falciparum* circumsporozoite protein using the autotransporter MisL of *Salmonella enterica*. Plasmid 2003;50(1):12–27.

[126] Isoda R, Simanski SP, Pathangey L, Stone AE, Brown TA. Expression of a *Porphyromonas gingivalis* hemagglutinin on the surface of a Salmonella vaccine vector. Vaccine 2007;25(1):117–26.

[127] Pompa-Mera EN, Yépez-Mulia L, Ocaña-Mondragón A, García-Zepeda EA, Ortega-Pierres G, González-Bonilla CR. *Trichinella spiralis*: intranasal immunization with attenuated *Salmonella enterica* carrying a gp43 antigen-derived 30mer epitope elicits protection in BALB/c mice. Exp Parasitol 2011;129(4):393–401.

[128] Li R, Lim A, Alonso S. Attenuated *Bordetella pertussis* BPZE1 as a live vehicle for heterologous vaccine antigens delivery through the nasal route. Bioeng Bugs 2011;2(6):315–9.

[129] Ho SY, Chua SQ, Foo DG, Locht C, Chow VT, Poh CL, et al. Highly attenuated *Bordetella pertussis* strain BPZE1 as a potential live vehicle for delivery of heterologous vaccine candidates. Infect Immun 2008;76(1):111–9.

[130] Li R, Lim A, Ow ST, Phoon MC, Locht C, Chow VT, et al. Development of live attenuated *Bordetella pertussis* strains expressing the universal influenza vaccine candidate M2e. Vaccine 2011;29(33):5502–11.

[131] Alonso S, Willery E, Renauld-Mongénie G, Locht C. Production of nontypeable *Haemophilus influenzae* HtrA by recombinant *Bordetella pertussis* with the use of filamentous hemagglutinin as a carrier. Infect Immun 2005;73(7):4295–301.

[132] Coppens I, Alonso S, Antoine R, Jacob-Dubuisson F, Renauld-Mongénie G, Jacobs E, et al. Production of *Neisseria meningitidis* transferrin-binding protein B by recombinant *Bordetella pertussis*. Infect Immun 2001;69(9):5440–6.

[133] Renauld-Mongénie G, Mielcarek N, Cornette J, Schacht AM, Capron A, Riveau G, et al. Induction of mucosal immune responses against a heterologous antigen fused to filamentous hemagglutinin after intranasal immunization with recombinant *Bordetella pertussis*. Proc Natl Acad Sci USA 1996;93 (15):7944–9.

[134] Buddenborg C, Daudel D, Liebrecht S, Greune L, Humberg V, Schmidt MA. Development of a tripartite vector system for live oral immunization using a gram-negative probiotic carrier. Int J Med Microbiol 2008;298(1–2):105–14.

[135] Zhu C, Ruiz-Perez F, Yang Z, Mao Y, Hackethal VL, Greco KM, et al. Delivery of heterologous protein antigens via hemolysin or autotransporter systems by an attenuated ler mutant of rabbit enteropathogenic *Escherichia coli*. Vaccine 2006;24(18):3821–31.

[136] Tobias J, Holmgren J, Hellman M, Nygren E, Lebens M, Svennerholm AM. Over-expression of major colonization factors of enterotoxigenic *Escherichia coli*, alone or together, on non-toxigenic *E. coli* bacteria. Vaccine 2010;28(43):6977–84.

[137] Remer KA, Bartrow M, Roeger B, Moll H, Sonnenborn U, Oelschlaeger TA. Split immune response after oral vaccination of mice with recombinant *Escherichia coli* Nissle 1917 expressing fimbrial adhesin K88. Int J Med Microbiol 2009;299(7):467–78.

[138] Nhan NT, Gonzalez de Valdivia E, Gustavsson M, Hai TN, Larsson G. Surface display of Salmonella epitopes in *Escherichia coli* and *Staphylococcus carnosus*. Microb Cell Fact 2011;10:22.

[139] Zheng JP, Zhang ZS, Li SQ, Liu XX, Yuan SL, Wang P, et al. Construction of a novel Shigella live-vector strain co-expressing CS3 and LTB/STm of enterotoxigenic *E. coli*. World J Gastroenterol 2005;11(22):3411–8.

[140] Altboum Z, Levine MM, Galen JE, Barry EM. Genetic characterization and immunogenicity of coli surface antigen 4 from enterotoxigenic *Escherichia coli* when it is expressed in a Shigella live-vector strain. Infect Immun 2003;71(3):1352–60.

[141] Barry EM, Wang J, Wu T, Davis T, Levine MM. Immunogenicity of multivalent Shigella-ETEC candidate vaccine strains in a guinea pig model. Vaccine 2006;24(18):3727–34.

[142] Ranallo RT, Fonseka CP, Cassels F, Srinivasan J, Venkatesan MM. Construction and characterization of bivalent *Shigella flexneri* 2a vaccine strains SC608 (pCFAI) and SC608(pCFAI/LTB) that express antigens from enterotoxigenic *Escherichia coli*. Infect Immun 2005;73(1):258–67.

[143] Koprowski H, Levine MM, Anderson RJ, Losonsky G, Pizza M, Barry EM. Attenuated *Shigella flexneri* 2a vaccine strain CVD 1204 expressing colonization factor antigen I and mutant heat-labile enterotoxin of enterotoxigenic *Escherichia coli*. Infect Immun 2000;68(9):4884–92.

[144] Silva AJ, Eko FO, Benitez JA. Exploiting cholera vaccines as a versatile antigen delivery platform. Biotechnol Lett 2008;30(4):571–9.

[145] Keller R, Hilton TD, Rios H, Boedeker EC, Kaper JB. Development of a live oral attaching and effacing *Escherichia coli* vaccine candidate using *Vibrio cholerae* CVD 103-HgR as antigen vector. Microb Pathog 2010;48(1):1–8.

[146] Tobias J, Lebens M, Bölin I, Wiklund G, Svennerholm AM. Construction of non-toxic *Escherichia coli* and *Vibrio cholerae* strains expressing high and immunogenic levels of enterotoxigenic *E. coli* colonization factor I fimbriae. Vaccine 2008;26 (6):743–52.

[147] Wells JM, Mercenier A. Mucosal delivery of therapeutic and prophylactic molecules using lactic acid bacteria. Nat Rev Microbiol 2008;6(5):349–62.

[148] Pontes DS, de Azevedo MS, Chatel JM, Langella P, Azevedo V, Miyoshi A. *Lactococcus lactis* as a live vector: heterologous protein production and DNA delivery systems. Protein Exp Purif 2011;79 (2):165–75.

[149] Tarahomjoo S. Development of vaccine delivery vehicles based on lactic acid bacteria. Mol Biotechnol 2012;51(2):183–99.

[150] Villatoro-Hernandez J, Montes-de-Oca-Luna R, Kuipers OP. Targeting diseases with genetically engineered *Lactococcus lactis* and its course towards medical translation. Expert Opin Biol Ther 2011;11 (3):261–7.

[151] Xin KQ, Hoshino Y, Toda Y, Igimi S, Kojima Y, Jounai N, et al. Immunogenicity and protective efficacy of orally administered recombinant *Lactococcus lactis* expressing surface-bound HIV Env. Blood 2003;102(1):223–8.

[152] Di Fabio S, Medaglini D, Rush CM, Corrias F, Panzini GL, Pace M, et al. Vaginal immunization of Cynomolgus monkeys with *Streptococcus gordonii* expressing HIV-1 and HPV 16 antigens. Vaccine 1998;16(5):485–92.

[153] Poo H, Pyo HM, Lee TY, Yoon SW, Lee JS, Kim CJ, et al. Oral administration of human papillomavirus type 16 E7 displayed on *Lactobacillus casei* induces E7-specific antitumor effects in C57/BL6 mice. Int J Cancer 2006;119(7):1702–9.

[154] Cortes-Perez NG, Lefèvre F, Corthier G, Adel-Patient K, Langella P, Bermúdez-Humarán LG. Influence of the route of immunization and the nature of the bacterial vector on immunogenicity of mucosal vaccines based on lactic acid bacteria. Vaccine 2007;25(36):6581–8.

[155] Bermúdez-Humarán LG, Cortes-Perez NG, Lefèvre F, Guimarães V, Rabot S, Alcocer-Gonzalez JM, et al. A novel mucosal vaccine based on live Lactococci expressing E7 antigen and IL-12 induces systemic and mucosal immune responses and protects mice against human papillomavirus type 16-induced tumors. J Immunol 2005;175(11):7297–302.

[156] Medaglini D, Rush CM, Sestini P, Pozzi G. Commensal bacteria as vectors for mucosal vaccines against sexually transmitted diseases: vaginal colonization with recombinant streptococci induces local and systemic antibodies in mice. Vaccine 1997;15(12–13):1330–7.

[157] Oggioni MR, Manganelli R, Contorni M, Tommasino M, Pozzi G. Immunization of mice by oral colonization with live recombinant commensal streptococci. Vaccine 1995;13(8):775–9.

[158] Chang TL, Chang CH, Simpson DA, Xu Q, Martin PK, Lagenaur LA, et al. Inhibition of HIV infectivity by a natural human isolate of *Lactobacillus jensenii* engineered to express functional two-domain CD4. Proc Natl Acad Sci USA 2003;100(20):11672–7.

[159] Lei H, Sheng Z, Ding Q, Chen J, Wei X, Lam DM, et al. Evaluation of oral immunization with recombinant avian influenza virus HA1 displayed on the *Lactococcus lactis* surface and combined with the mucosal adjuvant cholera toxin subunit B. Clin Vaccine Immunol 2011;18(7):1046–51.

[160] Marelli B, Perez AR, Banchio C, de Mendoza D, Magni C. Oral immunization with live *Lactococcus lactis* expressing rotavirus VP8 subunit induces specific immune response in mice. J Virol Methods 2011;175(1):28–37.

[161] Perez CA, Eichwald C, Burrone O, Mendoza D. Rotavirus vp7 antigen produced by *Lactococcus lactis* induces neutralizing antibodies in mice. J Appl Microbiol 2005;99(5):1158–64.

[162] Qiao X, Li G, Wang X, Li X, Liu M, Li Y. Recombinant porcine rotavirus VP4 and VP4-LTB expressed in *Lactobacillus casei* induced mucosal and systemic antibody responses in mice. BMC Microbiol 2009;9:249.

[163] Li YJ, Ma GP, Li GW, Qiao XY, Ge JW, Tang LJ, et al. Oral vaccination with the porcine rotavirus VP4 outer capsid protein expressed by *Lactococcus lactis* induces specific antibody production. J Biomed Biotechnol 2010;2010 708460.

[164] Lee JS, Poo H, Han DP, Hong SP, Kim K, Cho MW, et al. Mucosal immunization with surface-displayed severe acute respiratory syndrome coronavirus spike protein on *Lactobacillus casei* induces neutralizing antibodies in mice. J Virol 2006;80(8):4079–87.

[165] del Rio B, Dattwyler RJ, Aroso M, Neves V, Meirelles L, Seegers JF, et al. Oral immunization with recombinant *Lactobacillus plantarum* induces a protective immune response in mice with Lyme disease. Clin Vaccine Immunol 2008;15(9):1429–35.

[166] Robinson K, Chamberlain LM, Schofield KM, Wells JM, Le Page RW. Oral vaccination of mice against tetanus with recombinant *Lactococcus lactis*. Nat Biotechnol 1997;15(7):653–7.

[167] Liu JK, Hou XL, Wei CH, Yu LY, He XJ, Wang GH, et al. Induction of immune responses in mice after oral immunization with recombinant *Lactobacillus casei* strains expressing enterotoxigenic *Escherichia coli* F41 fimbrial protein. Appl Environ Microbiol 2009;75(13):4491–7.

[168] Wei CH, Liu JK, Hou XL, Yu LY, Lee JS, Kim CJ. Immunogenicity and protective efficacy of orally or intranasally administered recombinant *Lactobacillus casei* expressing ETEC K99. Vaccine 2010;28(24):4113–8.

[169] Hu CX, Xu ZR, Li WF, Niu D, Lu P, Fu LL. Secretory expression of K88 (F4) fimbrial adhesin FaeG by recombinant *Lactococcus lactis* for oral vaccination and its protective immune response in mice. Biotechnol Lett 2009;31(7):991–7.

[170] Chu H, Kang S, Ha S, Cho K, Park SM, Han KH, et al. *Lactobacillus acidophilus* expressing recombinant K99 adhesive fimbriae has an inhibitory effect on adhesion of enterotoxigenic *Escherichia coli*. Microbiol Immunol 2005;49(11):941–8.

[171] Gu Q, Song D, Zhu M. Oral vaccination of mice against *Helicobacter pylori* with recombinant *Lactococcus lactis* expressing urease subunit B. FEMS Immunol Med Microbiol 2009;56(3):197–203.

[172] Ciabattini A, Giomarelli B, Parigi R, Chiavolini D, Pettini E, Aricò B, et al. Intranasal immunization of mice with recombinant *Streptococcus gordonii* expressing NadA of *Neisseria meningitidis* induces systemic

bactericidal antibodies and local IgA. Vaccine 2008;26(33):4244–50.

[173] Scavone P, Miyoshi A, Rial A, Chabalgoity A, Langella P, Azevedo V, et al. Intranasal immunisation with recombinant *Lactococcus lactis* displaying either anchored or secreted forms of *Proteus mirabilis* MrpA fimbrial protein confers specific immune response and induces a significant reduction of kidney bacterial colonisation in mice. Microbes Infect 2007;9(7):821–8.

[174] Kajikawa A, Satoh E, Leer RJ, Yamamoto S, Igimi S. Intragastric immunization with recombinant *Lactobacillus casei* expressing flagellar antigen confers antibody-independent protective immunity against *Salmonella enterica* serovar Enteritidis. Vaccine 2007;25(18):3599–605.

[175] Kajikawa A, Igimi S. Innate and acquired immune responses induced by recombinant *Lactobacillus casei* displaying flagellin-fusion antigen on the cell surface. Vaccine 2010;28(19):3409–15.

[176] Kajikawa A, Nordone SK, Zhang L, Stoeker LL, LaVoy AS, Klaenhammer TR, et al. Dissimilar properties of two recombinant *Lactobacillus acidophilus* strains displaying Salmonella FliC with different anchoring motifs. Appl Environ Microbiol 2011;77(18):6587–96.

[177] Buccato S, Maione D, Rinaudo CD, Volpini G, Taddei AR, Rosini R, et al. Use of *Lactococcus lactis* expressing pili from group B Streptococcus as a broad-coverage vaccine against streptococcal disease. J Infect Dis 2006;194(3):331–40.

[178] Iwaki M, Okahashi N, Takahashi I, Kanamoto T, Sugita-Konishi Y, Aibara K, et al. Oral immunization with recombinant *Streptococcus lactis* carrying the *Streptococcus mutans* surface protein antigen gene. Infect Immun 1990;58(9):2929–34.

[179] Hanniffy SB, Carter AT, Hitchin E, Wells JM. Mucosal delivery of a pneumococcal vaccine using *Lactococcus lactis* affords protection against respiratory infection. J Infect Dis 2007;195(2):185–93.

[180] Villena J, Medina M, Racedo S, Alvarez S. Resistance of young mice to pneumococcal infection can be improved by oral vaccination with recombinant *Lactococcus lactis*. J Microbiol Immunol Infect 2010;43 (1):1–10.

[181] Medina M, Villena J, Vintiñi E, Hebert EM, Raya R, Alvarez S. Nasal immunization with *Lactococcus lactis* expressing the pneumococcal protective protein A induces protective immunity in mice. Infect Immun 2008;76(6):2696–705.

[182] Campos IB, Darrieux M, Ferreira DM, Miyaji EN, Silva DA, Arêas AP, et al. Nasal immunization of mice with *Lactobacillus casei* expressing the Pneumococcal surface protein A: induction of antibodies, complement deposition and partial protection against *Streptococcus pneumoniae* challenge. Microbes Infect 2008;10(5):481–8.

[183] Hernani MeL Ferreira PC, Ferreira DM, Miyaji EN, Ho PL, Oliveira ML. Nasal immunization of mice with *Lactobacillus casei* expressing the pneumococcal surface protein C primes the immune system and decreases pneumococcal nasopharyngeal colonization in mice. FEMS Immunol Med Microbiol 2011;62 (3):263–72.

[184] Villena J, Medina M, Raya R, Alvarez S. Oral immunization with recombinant *Lactococcus lactis* confers protection against respiratory pneumococcal infection. Can J Microbiol 2008;54(10):845–53.

[185] Mannam P, Jones KF, Geller BL. Mucosal vaccine made from live, recombinant *Lactococcus lactis* protects mice against pharyngeal infection with *Streptococcus pyogenes*. Infect Immun 2004;72(6):3444–50.

[186] Daniel C, Sebbane F, Poiret S, Goudercourt D, Dewulf J, Mullet C, et al. Protection against *Yersinia pseudotuberculosis* infection conferred by a *Lactococcus lactis* mucosal delivery vector secreting LcrV. Vaccine 2009;27(8):1141–4.

[187] Lee P, Faubert GM. Expression of the *Giardia lamblia* cyst wall protein 2 in *Lactococcus lactis*. Microbiology 2006;152(Pt 7):1981–90.

[188] Ramasamy R, Yasawardena S, Zomer A, Venema G, Kok J, Leenhouts K. Immunogenicity of a malaria parasite antigen displayed by *Lactococcus lactis* in oral immunisations. Vaccine 2006;24(18):3900–8.

[189] Moorthy G, Ramasamy R. Mucosal immunisation of mice with malaria protein on lactic acid bacterial cell walls. Vaccine 2007;25(18):3636–45.

[190] Moorthy SA, Yasawardena SG, Ramasamy R. Age-dependent systemic antibody responses and immunisation-associated changes in mice orally and nasally immunised with *Lactococcus lactis* expressing a malaria parasite protein. Vaccine 2009;27(36):4947–52.

[191] Wernerus H, Lehtio J, Samuelson P, Stahl S. Engineering of staphylococcal surfaces for biotechnological applications. J Biotechnol 2002;96(1):67–78.

[192] Cano F, Plotnicky-Gilquin H, Nguyen TN, Liljeqvist S, Samuelson P, Bonnefoy J, et al. Partial protection to respiratory syncytial virus (RSV) elicited in mice by intranasal immunization using live staphylococci with surface-displayed RSV-peptides. Vaccine 2000;18 (24):2743–52.

[193] Cano F, Liljeqvist S, Nguyen TN, Samuelson P, Bonnefoy JY, Stahl S, et al. A surface-displayed cholera toxin B peptide improves antibody responses using food-grade staphylococci for mucosal subunit vaccine delivery. FEMS Immunol Med Microbiol 1999;25(3):289–98.

[194] Nguyen TN, Hansson M, Stahl S, Bachi T, Robert A, Domzig W, et al. Cell-surface display of heterologous epitopes on *Staphylococcus xylosus* as a potential delivery system for oral vaccination. Gene 1993;128 (1):89–94.

[195] Duc LH, Hong HA, Atkins HS, Flick-Smith HC, Durrani Z, Rijpkema S, et al. Immunization against anthrax using *Bacillus subtilis* spores expressing the anthrax protective antigen. Vaccine 2007;25 (2):346–55.

[196] Cote CK, Rossi CA, Kang AS, Morrow PR, Lee JS, Welkos SL. The detection of protective antigen (PA) associated with spores of *Bacillus anthracis* and the effects of anti-PA antibodies on spore germination and macrophage interactions. Microb Pathog 2005;38 (5–6):209–25.

[197] Huang JM, Hong HA, Van Tong H, Hoang TH, Brisson A, Cutting SM. Mucosal delivery of antigens using adsorption to bacterial spores. Vaccine 2010;28 (4):1021–30.

[198] Isticato R, Cangiano G, Tran HT, Ciabattini A, Medaglini D, Oggioni MR, et al. Surface display of recombinant proteins on *Bacillus subtilis* spores. J Bacteriol 2001;183(21):6294–301.

[199] Mauriello EM, Duc LH, Isticato R, Cangiano G, Hong HA, De Felice M, et al. Display of heterologous antigens on the *Bacillus subtilis* spore coat using CotC as a fusion partner. Vaccine 2004;22(9–10):1177–87.

[200] Zhou Z, Xia H, Hu X, Huang Y, Ma C, Chen X, et al. Immunogenicity of recombinant *Bacillus subtilis* spores expressing *Clonorchis sinensis* tegumental protein. Parasitol Res 2008;102(2):293–7.

[201] Zhou Z, Xia H, Hu X, Huang Y, Li Y, Li L, et al. Oral administration of a *Bacillus subtilis* spore-based vaccine expressing *Clonorchis sinensis* tegumental protein 22.3 kDa confers protection against *Clonorchis sinensis.* Vaccine 2008;26(15):1817–25.

[202] Li L, Hu X, Wu Z, Xiong S, Zhou Z, Wang X, et al. Immunogenicity of self-adjuvanticity oral vaccine candidate based on use of *Bacillus subtilis* spore displaying Schistosoma japonicum 26 KDa GST protein. Parasitol Res 2009;105(6):1643–51.

[203] Muralinath M, Kuehn MJ, Roland KL, Curtiss R. Immunization with *Salmonella enterica* serovar Typhimurium-derived outer membrane vesicles delivering the pneumococcal protein PspA confers protection against challenge with *Streptococcus pneumoniae.* Infect Immun 2011;79(2):887–94.

[204] Ruppert A, Arnold N, Hobom G. OmpA-FMDV VP1 fusion proteins: production, cell surface exposure and immune responses to the major antigenic domain of foot-and-mouth disease virus. Vaccine 1994;12(6):492–8.

[205] Jechlinger W, Haller C, Resch S, Hofmann A, Szostak MP, Lubitz W. Comparative immunogenicity of the hepatitis B virus core 149 antigen displayed on the inner and outer membrane of bacterial ghosts. Vaccine 2005;23(27):3609–17.

[206] Szostak MP, Hensel A, Eko FO, Klein R, Auer T, Mader H, et al. Bacterial ghosts: non-living candidate vaccines. J Biotechnol 1996;44(1–3):161–70.

[207] Eko FO, Lubitz W, McMillan L, Ramey K, Moore TT, Ananaba GA, et al. Recombinant *Vibrio cholerae* ghosts as a delivery vehicle for vaccinating against *Chlamydia trachomatis*. Vaccine 2003;21(15):1694–703.

[208] Eko FO, He Q, Brown T, McMillan L, Ifere GO, Ananaba GA, et al. A novel recombinant multisubunit vaccine against Chlamydia. J Immunol 2004;173 (5):3375–82.

[209] Koeberling O, Giuntini S, Seubert A, Granoff DM. Meningococcal outer membrane vesicle vaccines derived from mutant strains engineered to express factor H binding proteins from antigenic variant groups 1 and 2. Clin Vaccine Immunol 2009;16(2):156–62.

[210] Hou VC, Koeberling O, Welsch JA, Granoff DM. Protective antibody responses elicited by a meningococcal outer membrane vesicle vaccine with overexpressed genome-derived neisserial antigen 1870. J Infect Dis 2005;192(4):580–90.

[211] Koeberling O, Welsch JA, Granoff DM. Improved immunogenicity of a H44/76 group B outer membrane vesicle vaccine with over-expressed genome-derived Neisserial antigen 1870. Vaccine 2007;25 (10):1912–20.

[212] Otczyk DC, Cripps AW. Mucosal immunization: a realistic alternative. Hum Vaccin 2010;6(12):978–1006.

[213] Bianchi ME. DAMPs, PAMPs and alarmins: all we need to know about danger. J Leukoc Biol 2007;81(1):1–5.

[214] Nelson AL, Roche AM, Gould JM, Chim K, Ratner AJ, Weiser JN. Capsule enhances pneumococcal colonization by limiting mucus-mediated clearance. Infect Immun 2007;75(1):83–90.

[215] Ohno H, Hase K. Glycoprotein 2 (GP2): grabbing the FimH bacteria into M cells for mucosal immunity. Gut Microbes 2010;1(6):407–10.

[216] Yamamoto M, Pascual DW, Kiyono H. M cell-targeted mucosal vaccine strategies. Curr Top Microbiol Immunol 2012;354:39–52.

[217] Maggi T, Spinosa M, Ricci S, Medaglini D, Pozzi G, Oggioni MR. Genetic engineering of *Streptococcus gordonii* for the simultaneous display of two heterologous proteins at the bacterial surface. FEMS Microbiol Lett 2002;210(1):135–41.

[218] Barbe S, Van Mellaert L, Anne J. The use of clostridial spores for cancer treatment. J Appl Microbiol 2006;101(3):571–8.

[219] Coley WB. The treatment of inoperable sarcoma by bacterial toxins (the mixed toxins of the streptococcus erysipelas and the bacillus prodigiosus). Proc R Soc Med 1910;3(Surg Sect):1–48.
[220] Morales A, Eidinger D, Bruce AW. Intracavitary Bacillus Calmette-Guerin in the treatment of superficial bladder tumors. J Urol 1976;116(2):180–3.
[221] Clairmont C, Lee KC, Pike J, Ittensohn M, Low KB, Pawelek J, et al. Biodistribution and genetic stability of the novel antitumor agent VNP20009, a genetically modified strain of *Salmonella typhimurium*. J Infect Dis 2000;181(6):1996–2002.
[222] Lambin P, Theys J, Landuyt W, Rijken P, van der Kogel A, van der Schueren E, et al. Colonisation of clostridium in the body is restricted to hypoxic and necrotic areas of tumours. Anaerobe 1998;4(4):183–8.
[223] Forbes NS, Munn LL, Fukumura D, Jain RK. Sparse initial entrapment of systemically injected *Salmonella typhimurium* leads to heterogeneous accumulation within tumors. Cancer Res 2003;63(17):5188–93.
[224] Kasinskas RW, Forbes NS. *Salmonella typhimurium* specifically chemotax and proliferate in heterogeneous tumor tissue *in vitro*. Biotechnol Bioeng 2006;94(4):710–21.
[225] Forbes NS. Engineering the perfect (bacterial) cancer therapy. Nat Rev Cancer 2010;10(11):785–94.
[226] Sznol M, Lin SL, Bermudes D, Zheng LM, King I. Use of preferentially replicating bacteria for the treatment of cancer. J Clin Invest 2000;105(8):1027–30.
[227] Anderson JC, Clarke EJ, Arkin AP, Voigt CA. Environmentally controlled invasion of cancer cells by engineered bacteria. J Mol Biol 2006;355 (4):619–27.
[228] Xiang S, Fruehauf J, Li CJ. Short hairpin RNA-expressing bacteria elicit RNA interference in mammals. Nat Biotechnol 2006;24(6):697–702.
[229] Loeffler M, Le'Negrate G, Krajewska M, Reed JC. Inhibition of tumor growth using salmonella expressing Fas ligand. J Natl Cancer Inst 2008;100 (15):1113–6.
[230] Jiang SN, Phan TX, Nam TK, Nguyen VH, Kim HS, Bom HS, et al. Inhibition of tumor growth and metastasis by a combination of *Escherichia coli*-mediated cytolytic therapy and radiotherapy. Mol Ther 2010;18(3):635–42.
[231] Nguyen VH, Kim HS, Ha JM, Hong Y, Choy HE, Min JJ. Genetically engineered *Salmonella typhimurium* as an imageable therapeutic probe for cancer. Cancer Res 2010;70(1):18–23.
[232] Al-Ramadi BK, Fernandez-Cabezudo MJ, El-Hasasna H, Al-Salam S, Bashir G, Chouaib S. Potent anti-tumor activity of systemically-administered IL2-expressing Salmonella correlates with decreased angiogenesis and enhanced tumor apoptosis. Clin Immunol 2009;130(1):89–97.
[233] Ganai S, Arenas RB, Forbes NS. Tumour-targeted delivery of TRAIL using *Salmonella typhimurium* enhances breast cancer survival in mice. Br J Cancer 2009;101(10):1683–91.
[234] Arrach N, Zhao M, Porwollik S, Hoffman RM, McClelland M. Salmonella promoters preferentially activated inside tumors. Cancer Res 2008;68 (12):4827–32.
[235] Nemunaitis J, Cunningham C, Senzer N, Kuhn J, Cramm J, Litz C, et al. Pilot trial of genetically modified, attenuated Salmonella expressing the *E. coli* cytosine deaminase gene in refractory cancer patients. Cancer Gene Ther 2003;10 (10):737–44.
[236] Toso JF, Gill VJ, Hwu P, Marincola FM, Restifo NP, Schwartzentruber DJ, et al. Phase I study of the intravenous administration of attenuated *Salmonella typhimurium* to patients with metastatic melanoma. J Clin Oncol 2002;20(1):142–52.
[237] Shahabi V, Maciag PC, Rivera S, Wallecha A. Live, attenuated strains of Listeria and Salmonella as vaccine vectors in cancer treatment. Bioeng Bugs 2010;1 (4):235–43.
[238] Bolhassani A, Zahedifard F. Therapeutic live vaccines as a potential anticancer strategy. Int J Cancer 2012;131(8):1733–43.
[239] Le Gouellec A, Chauchet X, Polack B, Buffat L, Toussaint B. Bacterial vectors for active immunotherapy reach clinical and industrial stages. Hum Vaccin Immunother 2012;8:10.
[240] Guirnalda P, Wood L, Paterson Y. *Listeria monocytogenes* and its products as agents for cancer immunotherapy. Adv Immunol 2012;113:81–118.
[241] Hegazy WA, Hensel M. *Salmonella enterica* as a vaccine carrier. Future Microbiol 2012;7(1):111–27.
[242] Avogadri F, Martinoli C, Petrovska L, Chiodoni C, Transidico P, Bronte V, et al. Cancer immunotherapy based on killing of Salmonella-infected tumor cells. Cancer Res 2005;65(9):3920–7.
[243] Buonaguro L, Petrizzo A, Tornesello ML, Buonaguro FM. Translating tumor antigens into cancer vaccines. Clin Vaccine Immunol 2011;18(1):23–34.
[244] Maciag PC, Radulovic S, Rothman J. The first clinical use of a live-attenuated *Listeria monocytogenes* vaccine: a Phase I safety study of Lm-LLO-E7 in patients with advanced carcinoma of the cervix. Vaccine 2009;27(30):3975–83.
[245] Meng JZ, Dong YJ, Huang H, Li S, Zhong Y, Liu SL, et al. Oral vaccination with attenuated *Salmonella enterica* strains encoding T-cell epitopes from tumor antigen NY-ESO-1 induces specific cytotoxic

T-lymphocyte responses. Clin Vaccine Immunol 2010;17(6):889–94.

[246] Cortes-Perez NG, Bermudez-Humaran LG, Le Loir Y, Rodriguez-Padilla C, Gruss A, Saucedo-Cardenas O, et al. Mice immunization with live lactococci displaying a surface anchored HPV-16 E7 oncoprotein. FEMS Microbiol Lett 2003;229(1):37–42.

[247] Bermudez-Humaran LG, Cortes-Perez NG, Lefevre F, Guimaraes V, Rabot S, Alcocer-Gonzalez JM, et al. A novel mucosal vaccine based on live Lactococci expressing E7 antigen and IL-12 induces systemic and mucosal immune responses and protects mice against human papillomavirus type 16-induced tumors. J Immunol 2005;175(11):7297–302.

[248] Bermudez-Humaran LG, Cortes-Perez NG, Lefevre F, Guimaraes V, Rabot S, Alcocer-Gonzalez JM, et al. An inducible surface presentation system improves cellular immunity against human papillomavirus type 16 E7 antigen in mice after nasal administration with recombinant lactococci. J Med Microbiol 2004;53(Pt 5):427–33.

[249] Fredriksen L, Mathiesen G, Sioud M, Eijsink VG. Cell wall anchoring of the 37-kilodalton oncofetal antigen by *Lactobacillus plantarum* for mucosal cancer vaccine delivery. Appl Environ Microbiol 2010;76 (21):7359–62.

[250] Bereta M, Hayhurst A, Gajda M, Chorobik P, Targosz M, Marcinkiewicz J, et al. Improving tumor targeting and therapeutic potential of Salmonella VNP20009 by displaying cell surface CEA-specific antibodies. Vaccine 2007;25(21):4183–92.

[251] Bornscheuer UT, Huisman GW, Kazlauskas RJ, Lutz S, Moore JC, Robins K. Engineering the third wave of biocatalysis. Nature 2012;485(7397):185–94.

[252] Woodley JM. New opportunities for biocatalysis: making pharmaceutical processes greener. Trends Biotechnol 2008;26(6):321–7.

[253] Ran N, Rui E, Liu J, Tao J. Chemoenzymatic synthesis of small molecule human therapeutics. Curr Pharm Des 2009;15(2):134–52.

[254] Illanes A, Cauerhff A, Wilson L, Castro GR. Recent trends in biocatalysis engineering. Bioresour Technol 2012;115:48–57.

[255] Sanchez S, Demain AL. Enzymes and bioconversions of industrial, pharmaceutical, and biotechnological significance. Org Process Res Dev 2011;15(1):224–30.

[256] Zhu G, Wang P. Polymer–enzyme conjugates can self-assemble at oil/water interfaces and effect interfacial biotransformations. J Am Chem Soc 2004;126 (36):11132–3.

[257] Wang AA, Mulchandani A, Chen W. Specific adhesion to cellulose and hydrolysis of organophosphate nerve agents by a genetically engineered *Escherichia coli* strain with a surface-expressed cellulose-binding domain and organophosphorus hydrolase. Appl Environ Microbiol 2002;68(4):1684–9.

[258] Schumacher SD, Hannemann F, Teese MG, Bernhardt R, Jose J. Autodisplay of functional CYP106A2 in *Escherichia coli*. J Biotechnol 2012;161 (2):104–12.

[259] Wu PH, Giridhar R, Wu WT. Surface display of transglucosidase on *Escherichia coli* by using the ice nucleation protein of *Xanthomonas campestris* and its application in glucosylation of hydroquinone. Biotechnol Bioeng 2006;95(6):1138–47.

[260] Suzuki T, Lett MC, Sasakawa C. Extracellular transport of VirG protein in Shigella. J Biol Chem 1995;270(52):30874–80.

[261] Yang TH, Kwon MA, Song JK, Pan JG, Rhee JS. Functional display of pseudomonas and Burkholderia lipases using a translocator domain of EstA autotransporter on the cell surface of *Escherichia coli*. J Biotechnol 2010;146(3):126–9.

[262] Schultheiss E, Weiss S, Winterer E, Maas R, Heinzle E, Jose J. Esterase autodisplay: enzyme engineering and whole-cell activity determination in microplates with pH sensors. Appl Environ Microbiol 2008;74(15):4782–91.

[263] Daunert S, Barrett G, Feliciano JS, Shetty RS, Shrestha S, Smith-Spencer W. Genetically engineered whole-cell sensing systems: coupling biological recognition with reporter genes. Chem Rev 2000;100 (7):2705–38.

[264] Kim T, Spiegel DA. Serendipitous discovery of two highly selective inhibitors of bacterial luciferase. Tetrahedron 2013;69(36):7692–8.

[265] Ramanathan S, Shi W, Rosen BP, Daunert S. Sensing antimonite and arsenite at the subattomole level with genetically engineered bioluminescent bacteria. Anal Chem 1997;69(16):3380–4.

[266] Belkin S. Microbial whole-cell sensing systems of environmental pollutants. Curr Opin Microbiol 2003;6(3):206–12.

[267] Standing D, Meharg AA, Killham K. A tripartite microbial reporter gene system for real-time assays of soil nutrient status. FEMS Microbiol Lett 2003;220 (1):35–9.

[268] Kumari A, Pasini P, Deo SK, Flomenhoft D, Shashidhar H, Daunert S. Biosensing systems for the detection of bacterial quorum signaling molecules. Anal Chem 2006;78(22):7603–9.

[269] Galluzzi L, Karp M. Whole cell strategies based on lux genes for high throughput applications toward new antimicrobials. Comb Chem High Throughput Screen 2006;9(7):501–14.

[270] Chen G, Cloud J, Georgiou G, Iverson BL. A quantitative immunoassay utilizing *Escherichia coli* cells

possessing surface-expressed single chain Fv molecules. Biotechnol Prog 1996;12(4):572–4.

[271] Christmann A, Wentzel A, Meyer C, Meyers G, Kolmar H. Epitope mapping and affinity purification of monospecific antibodies by *Escherichia coli* cell surface display of gene-derived random peptide libraries. J Immunol Methods 2001;257(1–2):163–73.

[272] Tinberg CE, Khare SD, Dou J, Doyle L, Nelson JW, Schena A, et al. Computational design of ligand-binding proteins with high affinity and selectivity. Nature 2013;501(7466):212–6.

[273] Pande J, Szewczyk MM, Grover AK. Phage display: concept, innovations, applications and future. Biotechnol Adv 2010;28(6):849–58.

[274] Lowman HB, Bass SH, Simpson N, Wells JA. Selecting high-affinity binding proteins by monovalent phage display. Biochemistry 1991;30 (45):10832–8.

[275] Rockberg J, Lofblom J, Hjelm B, Stahl S, Uhlen M. Epitope mapping using gram-positive surface display. Curr Protoc Immunol 2010; (Chapter 9: Unit 9.9).

[276] Rockberg J, Lofblom J, Hjelm B, Uhlen M, Stahl S. Epitope mapping of antibodies using bacterial surface display. Nat Methods 2008;5(12):1039–45.

[277] Zitzmann S, Kramer S, Mier W, Mahmut M, Fleig J, Altmann A, et al. Identification of a new prostate-specific cyclic peptide with the bacterial FliTrx system. J Nucl Med 2005;46(5):782–5.

[278] Li W, Lei P, Yu B, Wu S, Peng J, Zhao X, et al. Screening and identification of a novel target specific for hepatoma cell line HepG2 from the FliTrx bacterial peptide library. Acta Biochim Biophys Sin (Shanghai) 2008;40(5):443–51.

[279] Kronqvist N, Lofblom J, Jonsson A, Wernerus H, Stahl S. A novel affinity protein selection system based on staphylococcal cell surface display and flow cytometry. Protein Eng Des Sel 2008;21 (4):247–55.

[280] Kenrick SA, Daugherty PS. Bacterial display enables efficient and quantitative peptide affinity maturation. Protein Eng Des Sel 2010;23(1):9–17.

[281] Little LE, Dane KY, Daugherty PS, Healy KE, Schaffer DV. Exploiting bacterial peptide display technology to engineer biomaterials for neural stem cell culture. Biomaterials 2011;32(6):1484–94.

[282] Bessette PH, Rice JJ, Daugherty PS. Rapid isolation of high-affinity protein binding peptides using bacterial display. Protein Eng Des Sel 2004;17 (10):731–9.

[283] Taschner S, Meinke A, von Gabain A, Boyd AP. Selection of peptide entry motifs by bacterial surface display. Biochem J 2002;367(Pt 2):393–402.

[284] Nakajima H, Shimbara N, Shimonishi Y, Mimori T, Niwa S, Saya H. Expression of random peptide fused to invasin on bacterial cell surface for selection of cell-targeting peptides. Gene 2000;260(1–2):121–31.

[285] Lofblom J. Bacterial display in combinatorial protein engineering. Biotechnol J 2011;6(9):1115–29.

[286] Jostock T, Dubel S. Screening of molecular repertoires by microbial surface display. Comb Chem High Throughput Screen 2005;8(2):127–33.

[287] Getz JA, Schoep TD, Daugherty PS. Peptide discovery using bacterial display and flow cytometry. Methods Enzymol 2012;503:75–97.

[288] Jose J, Betscheider D, Zangen D. Bacterial surface display library screening by target enzyme labeling: identification of new human cathepsin G inhibitors. Anal Biochem 2005;346(2):258–67.

[289] Wentzel A, Christmann A, Kratzner R, Kolmar H. Sequence requirements of the GPNG beta-turn of the *Ecballium elaterium* trypsin inhibitor II explored by combinatorial library screening. J Biol Chem 1999;274 (30):21037–43.

[290] Yang G, Withers SG. Ultrahigh-throughput FACS-based screening for directed enzyme evolution. Chembiochem 2009;10(17):2704–15.

[291] Harvey BR, Shanafelt AB, Baburina I, Hui R, Vitone S, Iverson BL, et al. Engineering of recombinant antibody fragments to methamphetamine by anchored periplasmic expression. J Immunol Methods 2006;308(1–2):43–52.

[292] Harvey BR, Georgiou G, Hayhurst A, Jeong KJ, Iverson BL, Rogers GK. Anchored periplasmic expression, a versatile technology for the isolation of high-affinity antibodies from *Escherichia coli*-expressed libraries. Proc Natl Acad Sci USA 2004;101 (25):9193–8.

[293] Maynard JA, Maassen CB, Leppla SH, Brasky K, Patterson JL, Iverson BL, et al. Protection against anthrax toxin by recombinant antibody fragments correlates with antigen affinity. Nat Biotechnol 2002;20(6):597–601.

[294] Mazor Y, Van Blarcom T, Mabry R, Iverson BL, Georgiou G. Isolation of engineered, full-length antibodies from libraries expressed in *Escherichia coli.* Nat Biotechnol 2007;25(5):563–5.

[295] Olsen MJ, Stephens D, Griffiths D, Daugherty P, Georgiou G, Iverson BL. Function-based isolation of novel enzymes from a large library. Nat Biotechnol 2000;18(10):1071–4.

[296] Varadarajan N, Gam J, Olsen MJ, Georgiou G, Iverson BL. Engineering of protease variants exhibiting high catalytic activity and exquisite substrate selectivity. Proc Natl Acad Sci USA 2005;102 (19):6855–60.

[297] Varadarajan N, Georgiou G, Iverson BL. An engineered protease that cleaves specifically after sulfated tyrosine. Angew Chem Int Ed Engl 2008;47 (41):7861–3.

[298] Varadarajan N, Rodriguez S, Hwang BY, Georgiou G, Iverson BL. Highly active and selective endopeptidases with programmed substrate specificities. Nat Chem Biol 2008;4(5):290–4.

[299] Becker S, Hobenreich H, Vogel A, Knorr J, Wilhelm S, Rosenau F, et al. Single-cell high-throughput screening to identify enantioselective hydrolytic enzymes. Angew Chem Int Ed Engl 2008;47(27):5085–8.

[300] Aharoni A, Thieme K, Chiu CP, Buchini S, Lairson LL, Chen H, et al. High-throughput screening methodology for the directed evolution of glycosyltransferases. Nat Methods 2006;3(8):609–14.

[301] Yang G, Rich JR, Gilbert M, Wakarchuk WW, Feng Y, Withers SG. Fluorescence activated cell sorting as a general ultra-high-throughput screening method for directed evolution of glycosyltransferases. J Am Chem Soc 2010;132(30):10570–7.

[302] Dube DH, Champasa K, Wang B. Chemical tools to discover and target bacterial glycoproteins. Chem Commun (Camb) 2011;47(1):87–101.

[303] Schneewind O, Missiakas DM. Protein secretion and surface display in gram-positive bacteria. Philos Trans R Soc Lond B Biol Sci 2012;367(1592):1123–39.

[304] Yother J. Capsules of *Streptococcus pneumoniae* and other bacteria: paradigms for polysaccharide biosynthesis and regulation. Annu Rev Microbiol 2011;65:563–81.

[305] Whitfield C. Biosynthesis and assembly of capsular polysaccharides in *Escherichia coli*. Annu Rev Biochem 2006;75:39–68.

[306] Wang X, Quinn PJ. Lipopolysaccharide: biosynthetic pathway and structure modification. Prog Lipid Res 2010;49(2):97–107.

[307] DiGiandomenico A, Rao J, Goldberg JB. Oral vaccination of BALB/c mice with *Salmonella enterica* serovar Typhimurium expressing *Pseudomonas aeruginosa* O antigen promotes increased survival in an acute fatal pneumonia model. Infect Immun 2004;72(12):7012–21.

[308] Gilbert C, Robinson K, Le Page RW, Wells JM. Heterologous expression of an immunogenic pneumococcal type 3 capsular polysaccharide in *Lactococcus lactis*. Infect Immun 2000;68(6):3251–60.

[309] Nierop Groot MN, Godefrooij J, Kleerebezem M. Heterologous expression of the pneumococcal serotype 14 polysaccharide in *Lactococcus lactis* requires lactococcal epsABC regulatory genes. Appl Environ Microbiol 2008;74(3):912–5.

[310] Viret JF, Cryz Jr. SJ, Lang AB, Favre D. Molecular cloning and characterization of the genetic determinants that express the complete Shigella serotype D (*Shigella sonnei*) lipopolysaccharide in heterologous live attenuated vaccine strains. Mol Microbiol 1993;7 (2):239–52.

[311] Xu DQ, Cisar JO, Ambulos Jr N, Burr DH, Kopecko DJ. Molecular cloning and characterization of genes for *Shigella sonnei* form I O polysaccharide: proposed biosynthetic pathway and stable expression in a live salmonella vaccine vector. Infect Immun 2002;70 (8):4414–23.

[312] Falt IC, Schweda EK, Klee S, Singh M, Floderus E, Timmis KN, et al. Expression of *Shigella dysenteriae* serotype 1 O-antigenic polysaccharide by *Shigella flexneri* aroD vaccine candidates and different *S. flexneri* serotypes. J Bacteriol 1995;177(18):5310–5.

[313] Jennison AV, Roberts F, Verma NK. Construction of a multivalent vaccine strain of *Shigella flexneri* and evaluation of serotype-specific immunity. FEMS Immunol Med Microbiol 2006;46(3):444–51.

[314] Raetz CR, Reynolds CM, Trent MS, Bishop RE. Lipid A modification systems in gram-negative bacteria. Annu Rev Biochem 2007;76:295–329.

[315] Needham BD, Trent MS. Fortifying the barrier: the impact of lipid A remodelling on bacterial pathogenesis. Nat Rev Microbiol 2013;11(7):467–81.

[316] Arnold H, Bumann D, Felies M, Gewecke B, Sorensen M, Gessner JE, et al. Enhanced immunogenicity in the murine airway mucosa with an attenuated Salmonella live vaccine expressing OprF–OprI from *Pseudomonas aeruginosa*. Infect Immun 2004;72 (11):6546–53.

[317] Cartee RT, Forsee WT, Jensen JW, Yother J. Expression of the *Streptococcus pneumoniae* type 3 synthase in *Escherichia coli*. Assembly of type 3 polysaccharide on a lipid primer. J Biol Chem 2001;276 (52):48831–9.

[318] Ofek I, Kabha K, Athamna A, Frankel G, Wozniak DJ, Hasty DL, et al. Genetic exchange of determinants for capsular polysaccharide biosynthesis between *Klebsiella pneumoniae* strains expressing serotypes K2 and K21a. Infect Immun 1993;61 (10):4208–16.

[319] Sahly H, Keisari Y, Ofek I. Manno(rhamno)biose-containing capsular polysaccharides of *Klebsiella pneumoniae* enhance opsono-stimulation of human polymorphonuclear leukocytes. J Innate Immun 2009;1(2):136–44.

[320] dQ Xu, Cisar JO, Osorio M, Wai TT, Kopecko DJ. Core-linked LPS expression of *Shigella dysenteriae* serotype 1 O-antigen in live Salmonella Typhi vaccine vector Ty21a: preclinical evidence of immunogenicity and protection. Vaccine 2007;25(33):6167–75.

[321] Kim SH, Jia W, Parreira VR, Bishop RE, Gyles CL. Phosphoethanolamine substitution in the lipid A of

Escherichia coli O157:H7 and its association with PmrC. Microbiology 2006;152(Pt 3):657–66.

[322] Chng SS, Ruiz N, Chimalakonda G, Silhavy TJ, Kahne D. Characterization of the two-protein complex in *Escherichia coli* responsible for lipopolysaccharide assembly at the outer membrane. Proc Natl Acad Sci USA 2010;107(12):5363–8.

[323] Casella CR, Mitchell TC. Putting endotoxin to work for us: monophosphoryl lipid A as a safe and effective vaccine adjuvant. Cell Mol Life Sci 2008;65(20):3231–40.

[324] D'Hauteville H, Khan S, Maskell DJ, Kussak A, Weintraub A, Mathison J, et al. Two msbB genes encoding maximal acylation of lipid A are required for invasive *Shigella flexneri* to mediate inflammatory rupture and destruction of the intestinal epithelium. J Immunol 2002;168(10):5240–51.

[325] Sun W, Six D, Kuang X, Roland KL, Raetz CR, Curtiss 3rd R. A live attenuated strain of *Yersinia pestis* KIM as a vaccine against plague. Vaccine 2011;29(16):2986–98.

[326] Feodorova VA, Pan'kina LN, Savostina EP, Sayapina LV, Motin VL, Dentovskaya SV, et al. A *Yersinia pestis* lpxM-mutant live vaccine induces enhanced immunity against bubonic plague in mice and guinea pigs. Vaccine 2007;25(44):7620–8.

[327] Kong Q, Six DA, Roland KL, Liu Q, Gu L, Reynolds CM, et al. Salmonella synthesizing 1-dephosphorylated [corrected] lipopolysaccharide exhibits low endotoxic activity while retaining its immunogenicity. J Immunol 2011;187(1):412–23.

[328] Kong Q, et al. Phosphate groups of lipid A are essential for *Salmonella enterica* serovar Typhimurium virulence and affect innate and adaptive immunity. Infect Immun 2012;80(9):3215–24.

[329] Kim TH, Pinkham JT, Heninger SJ, Chalabaev S, Kasper DL. Genetic modification of the O-polysaccharide of *Francisella tularensis* results in an avirulent live attenuated vaccine. J Infect Dis 2012;205(7):1056–65.

[330] Srikram A, Zhang K, Bartpho T, Lo M, Hoke DE, Sermswan RW, et al. Cross-protective immunity against leptospirosis elicited by a live, attenuated lipopolysaccharide mutant. J Infect Dis 2011;203(6):870–9.

[331] van der Ley P, van den Dobbelsteen G. Next-generation outer membrane vesicle vaccines against *Neisseria meningitidis* based on nontoxic LPS mutants. Hum Vaccin 2011;7(8):886–90.

[332] Peng D, Hong W, Choudhury BP, Carlson RW, Gu XX. *Moraxella catarrhalis* bacterium without endotoxin, a potential vaccine candidate. Infect Immun 2005;73(11):7569–77.

[333] Clements A, Tull D, Jenney AW, Farn JL, Kim SH, Bishop RE, et al. Secondary acylation of *Klebsiella pneumoniae* lipopolysaccharide contributes to sensitivity to antibacterial peptides. J Biol Chem 2007;282(21):15569–77.

[334] Focareta A, Paton JC, Morona R, Cook J, Paton AW. A recombinant probiotic for treatment and prevention of cholera. Gastroenterology 2006;130(6):1688–95.

[335] Paton AW, Jennings MP, Morona R, Wang H, Focareta A, Roddam LF, et al. Recombinant probiotics for treatment and prevention of enterotoxigenic *Escherichia coli* diarrhea. Gastroenterology 2005;128(5):1219–28.

[336] Paton AW, Morona R, Paton JC. A new biological agent for treatment of Shiga toxigenic *Escherichia coli* infections and dysentery in humans. Nat Med 2000;6(3):265–70.

[337] Paton JC, Rogers TJ, Morona R, Paton AW. Oral administration of formaldehyde-killed recombinant bacteria expressing a mimic of the Shiga toxin receptor protects mice from fatal challenge with Shiga-toxigenic *Escherichia coli*. Infect Immun 2001;69(3):1389–93.

[338] Pinyon RA, Paton JC, Paton AW, Botten JA, Morona R. Refinement of a therapeutic Shiga toxin-binding probiotic for human trials. J Infect Dis 2004;189(9):1547–55.

[339] Watts RE, Tan CK, Ulett GC, Carey AJ, Totsika M, Idris A, et al. *Escherichia coli* 83972 expressing a *P. fimbriae* oligosaccharide receptor mimic impairs adhesion of uropathogenic *E. coli*. J Infect Dis 2012;206(8):1242–9.

[340] Sethuraman N, Stadheim TA. Challenges in therapeutic glycoprotein production. Curr Opin Biotechnol 2006;17(4):341–6.

[341] Sinclair AM, Elliott S. Glycoengineering: the effect of glycosylation on the properties of therapeutic proteins. J Pharm Sci 2005;94(8):1626–35.

[342] Sola RJ, Griebenow K. Effects of glycosylation on the stability of protein pharmaceuticals. J Pharm Sci 2009;98(4):1223–45.

[343] Sola RJ, Griebenow K. Glycosylation of therapeutic proteins: an effective strategy to optimize efficacy. BioDrugs 2010;24(1):9–21.

[344] Werner A, Horstkorte R, Glanz D, Biskup K, Blanchard V, Berger M, et al. Glycoengineering the N-acyl side chain of sialic acid of human erythropoietin affects its resistance to sialidase. Biol Chem 2012;393(8):777–83.

[345] Macdougall IC. Optimizing the use of erythropoietic agents–pharmacokinetic and pharmacodynamic considerations. Nephrol Dial Transplant 2002;17(Suppl. 5):66–70.

[346] Friedman B, Vaddi K, Preston C, Mahon E, Cataldo JR, McPherson JM. A comparison of the pharmacological properties of carbohydrate remodeled recombinant and placental-derived beta-glucocerebrosidase: implications for clinical efficacy in treatment of Gaucher disease. Blood 1999;93(9):2807–16.

[347] Boltje TJ, Buskas T, Boons GJ. Opportunities and challenges in synthetic oligosaccharide and glycoconjugate research. Nat Chem 2009;1(8):611–22.

[348] Werz DB. Chemical synthesis of carbohydrates and their surface immobilization: a brief introduction. Methods Mol Biol 2012;808:13–29.

[349] Prescher JA, Bertozzi CR. Chemistry in living systems. Nat Chem Biol 2005;1(1):13–21.

[350] Kolb HC, Finn MG, Sharpless KB. Click chemistry: diverse chemical function from a few good reactions. Angew Chem Int Ed Engl 2001;40(11):2004–21.

[351] Mahal LK, Yarema KJ, Bertozzi CR. Engineering chemical reactivity on cell surfaces through oligosaccharide biosynthesis. Science 1997;276 (5315):1125–8.

[352] Yi W, Liu X, Li Y, Li J, Xia C, Zhou G, et al. Remodeling bacterial polysaccharides by metabolic pathway engineering. Proc Natl Acad Sci USA 2009;106(11):4207–12.

[353] Dumont A, Malleron A, Awwad M, Dukan S, Vauzeilles B. Click-mediated labeling of bacterial membranes through metabolic modification of the lipopolysaccharide inner core. Angew Chem Int Ed Engl 2012;51(13):3143–6.

[354] Goon S, Schilling B, Tullius MV, Gibson BW, Bertozzi CR. Metabolic incorporation of unnatural sialic acids into *Haemophilus ducreyi* lipooligosaccharides. Proc Natl Acad Sci USA 2003;100(6):3089–94.

[355] Besanceney-Webler C, Jiang H, Wang W, Baughn AD, Wu P. Metabolic labeling of fucosylated glycoproteins in Bacteroidales species. Bioorg Med Chem Lett 2011;21(17):4989–92.

[356] Koenigs MB, Richardson EA, Dube DH. Metabolic profiling of *Helicobacter pylori* glycosylation. Mol Biosyst 2009;5(9):909–12.

[357] Liu F, Aubry AJ, Schoenhofen IC, Logan SM, Tanner ME. The engineering of bacteria bearing azido-pseudaminic acid-modified flagella. Chembiochem 2009;10(8):1317–20.

[358] Tanrikulu IC, Schmitt E, Mechulam Y, Goddard 3rd WA, Tirrell DA. Discovery of *Escherichia coli* methionyl-tRNA synthetase mutants for efficient labeling of proteins with azidonorleucine *in vivo*. Proc Natl Acad Sci USA 2009;106(36):15285–90.

[359] Ngo JT, Tirrell DA. Noncanonical amino acids in the interrogation of cellular protein synthesis. Acc Chem Res 2011;44(9):677–85.

[360] Link AJ, Tirrell DA. Cell surface labeling of *Escherichia coli* via copper(I)-catalyzed [3 + 2] cycloaddition. J Am Chem Soc 2003;125(37):11164–5.

[361] Link AJ, Vink MK, Agard NJ, Prescher JA, Bertozzi CR, Tirrell DA. Discovery of aminoacyl-tRNA synthetase activity through cell-surface display of noncanonical amino acids. Proc Natl Acad Sci USA 2006;103(27):10180–5.

[362] Link AJ, Vink MK, Tirrell DA. Presentation and detection of azide functionality in bacterial cell surface proteins. J Am Chem Soc 2004;126 (34):10598–602.

[363] Zhang Z, Smith BA, Wang L, Brock A, Cho C, Schultz PG. A new strategy for the site-specific modification of proteins *in vivo*. Biochemistry 2003;42 (22):6735–46.

[364] Gautam S, Gniadek TJ, Kim T, Spiegel DA. Exterior design: strategies for redecorating the bacterial surface with small molecules. Trends Biotechnol 2013;31 (4):258–67.

[365] Nelson JW, Chamessian AG, McEnaney PJ, Murelli RP, Kazmierczak BI, Spiegel DA. A biosynthetic strategy for re-engineering the *Staphylococcus aureus* cell wall with non-native small molecules. ACS Chem Biol 2010;5 (12):1147–55.

[366] Hug I, Zheng B, Reiz B, Whittal RM, Fentabil MA, Klassen JS, et al. Exploiting bacterial glycosylation machineries for the synthesis of a Lewis antigen-containing glycoprotein. J Biol Chem 2011;286 (43):37887–94.

[367] Lepenies B, Lee J, Sonkaria S. Targeting C-type lectin receptors with multivalent carbohydrate ligands. Adv Drug Deliv Rev 2013;.

[368] van Kooyk Y, Unger WW, Fehres CM, Kalay H, Garcia-Vallejo JJ. Glycan-based DC-SIGN targeting vaccines to enhance antigen cross-presentation. Mol Immunol 2013;55(2):143–5.

CHAPTER

5

Noncovalent Functionalization of Cell Surface

Debjit Dutta, Srujan K. Marepally and Praveen Kumar Vemula

Institute for Stem Cell Biology and Regenerative Medicine, National Centre for Biological Sciences-Campus, UAS-GKVK, Bangalore, India

OUTLINE

J.M. Karp & W. Zhao (Eds): Micro- and Nanoengineering of the Cell Surface.
DOI: http://dx.doi.org/10.1016/B978-1-45-573146-6.00005-2

5.1 INTRODUCTION

Surface of the cell plays a pivotal role in defining the cell fate. A wide range of functional biomolecules that anchored on the cell surface perform different roles. A cell communicates with other cells, environment, pathogens, and nutrients through its surface. In addition, cell migration is highly dependent on the cell membrane. Thus, cell fate could be modulated through engineering cell surface (Figure 5.1). Cell surface is a fertile ground to manipulate cell phenotypes and biological fates. While these manipulations are key tools to answer some of the basic biological questions [1], they could be instrumental in developing novel diagnostic and therapeutic systems [2] through engineering of cell surface. Applications of genetic tools for manipulating proteins and probing cellular functions have revolutionized biology and medicine. These tools were proven to be successful, but their applications are restricted due to their inability to introduce novel functionality into biomolecules or interrogate the distinct chemical entities that control cellular responses. To address these limitations, multiple research teams have developed a set of chemoselective reactions, known as bioorthogonal reactions, with minimal cross-reactivity and toxicity in biological systems.

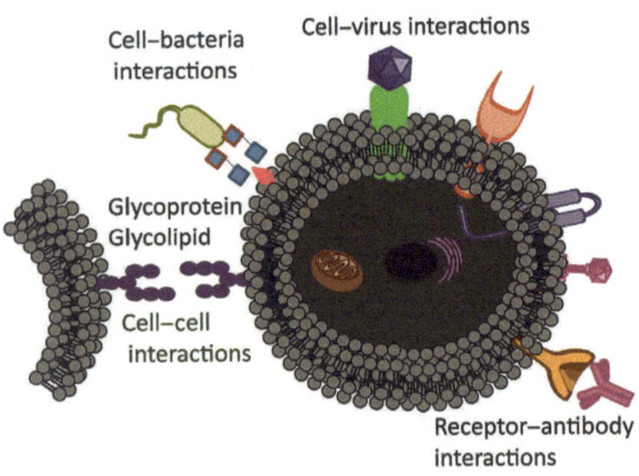

FIGURE 5.1 Schematic of cell surface that plays a vital role in numerous cell functions.

A number of synergistic approaches combining chemistry, material science, and biology have been applied to engineer and manipulate living cells. Desirable chemical functionalities and materials can be introduced onto the cell surface by covalent and noncovalent methodologies, as well as through specific biological recognition events, including antibody–antigen and ligand–receptor interactions. These strategies have been used in a wide range of applications that include designing the biomaterial scaffolds [3,4] to control cell fate, labeling of cells with molecular and nanoparticle probes to visualize cellular processes and molecular pathways [5–7], delivering diverse species into cells [8,9], and patterning cells for drug discovery [10,11]. To date, cell surface engineering has been primarily achieved through the use of molecular biology and genetic engineering [12]. However, perturbations of inherent cellular physiology with genetic engineering may significantly interfere with cellular functions. Thus, there is a need to develop tools that will provide simple alternatives to genetic and biosynthetic pathways.

Nongenetic cell surface modifications have been generally achieved using the following methods: (i) covalent conjugation of desired molecules to surface functional groups like amines [11,13–20], (ii) incorporation of amphiphilic polymers into the lipid bilayer membrane of cells by harnessing hydrophobic interactions [21], (iii) electrostatic interactions between cationic polymers and a negatively charged surface [15], and (iv) metabolic engineering of ketone- or azide-modified carbohydrates to display desired functional groups on the cell surface, followed by bioorthogonal click reactions to tether molecule of choice [22–24]. Schematic representation of some examples of engineering of cell surface is shown in Figures 5.2 and 5.3.

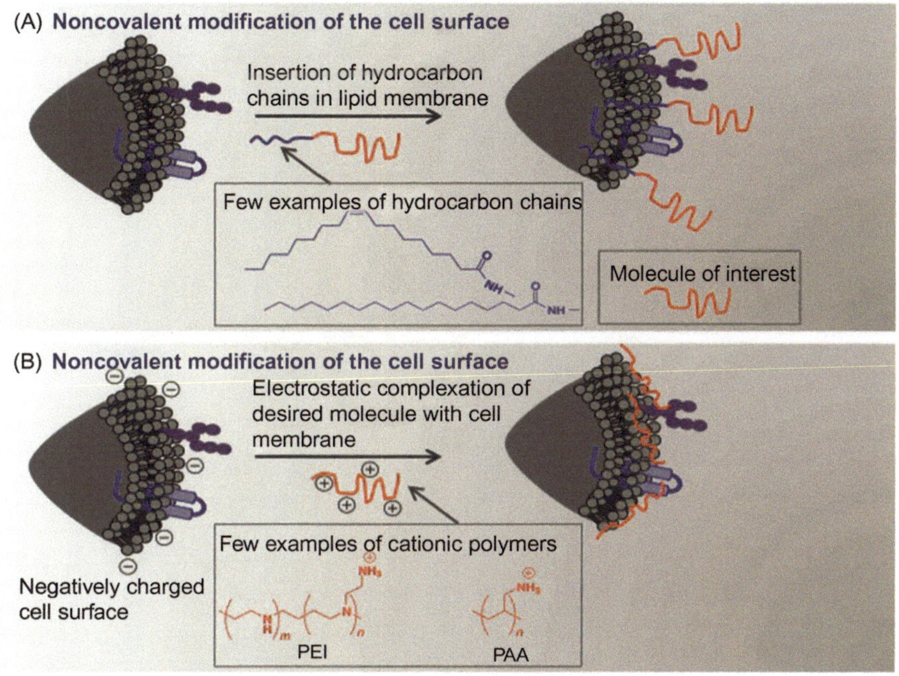

FIGURE 5.2 (A) and (B) Schematic of engineering the cell surface using noncovalent methods. Typically, hydrophobic and ionic interactions have been utilized to immobilize the molecules of interest on the surface of the cell.

This chapter focuses on nanotechnology approaches that could be applied for the application of cell surface engineering and their limitations.

5.2 METHODS OF CELL SURFACE ENGINEERING—APPLICATIONS AND RECENT DEVELOPMENTS

5.2.1 Targeting Ability of Surface Engineered Cells

Cell therapy involving systemic infusion and targeted delivery of progenitor/mesenchymal stem cells (MSCs) and other cell types to damage and diseased tissues hold tremendous promise for the treatment of a number of diseases such as cardiomyopathy, cancer, and fibrosis [25]. MSCs are capable of differentiating into multiple lineages and produce extracellular matrix (bone, cartilage, fat) and can also secrete immunomodulatory cytokines to reduce inflammation. Although they are currently being tested in more than 100 clinical trials for treatment of multiple diseases, including graft versus host disease, myocardial infarction, multiple sclerosis, and skeletal tissue repair, the major challenge in MSC therapy is the delivery of MSCs efficiently to targeted location. Systemically administered therapeutic MSCs have poor expression of homing ligands or lose them during the culture expansion and only less than 1% of culture-expanded MSCs can actually home to the target tissue [26]. Therefore, efforts have been made to introduce cell homing ligands onto cell membranes. Cell homing ligands (on the homing cell) and receptors (on the endothelium) allow homing cells to tether, roll, adhere, and then transmigrate on

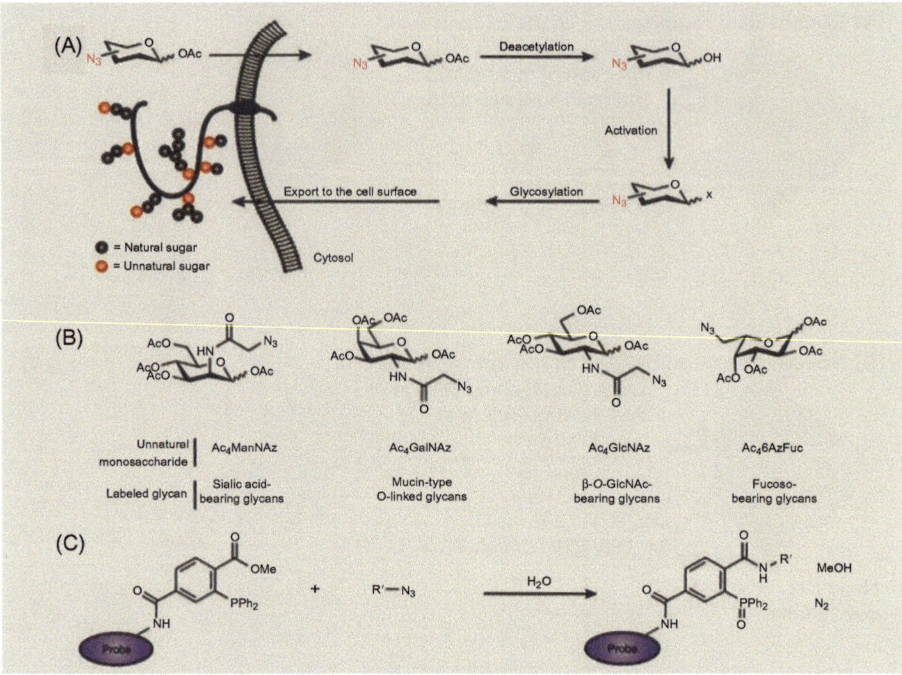

FIGURE 5.3 Schematics of metabolic approach of cell surface engineering. (A) The per-O-acetylated azido sugars passively diffuse across the cell membrane. In the cytosol, the acetyl groups are removed by intracellular esterases, the azido sugar is activated to a nucleotide sugar donor (X = uridine diphosphate or cytidine monophosphate) and exported to the cell surface through glycosylation. (B) Different chemical structures of azido sugars and the glycan subtypes labeled. (C) The Staudinger ligation: triaryl phosphines react selectively with alkyl azides to form an amide bond to the alkyl substituent, such as an azido sugar. *Source: Reprinted with permission from Ref. [24] © NPG 2007.*

endothelium as part of the cell homing cascade. A great effort has been made to achieve targeted delivery of cells, and a wide range of methods such as genetic and enzymatic engineering [27–29] treatment with cytokines [30] and chemical approaches have been developed [31,32].

5.2.2 Noncovalent Immobilization of Agents on the Surface of Cells

Although a covalent approach offered a method to modify the cell surface for longer duration, often, covalent functionalization might interfere with the function of native proteins or other molecules that are present on the cell surface. Thus, to overcome these potential limitations, a robust noncovalent approach to transiently engineer the cell surface has been developed [33]. In this method, unilamellar vesicles composed of biotinylated lipid were fused with the cell membrane to anchor biotin on the cell membrane of MSCs (Figure 5.4). Intercalation of the lipid vesicles with lipid bilayer of the cell membrane is a robust phenomenon; it offers numerous opportunities to immobilize any desired molecules on the surface of the cell membrane. The biotinylated surface of the cell was further hybridized with multivalent streptavidin and followed by biotinylated SLeX (Figure 5.4). As cell membrane undergoes a flip–flop process, lipids in the membrane are rearranged and recycled.

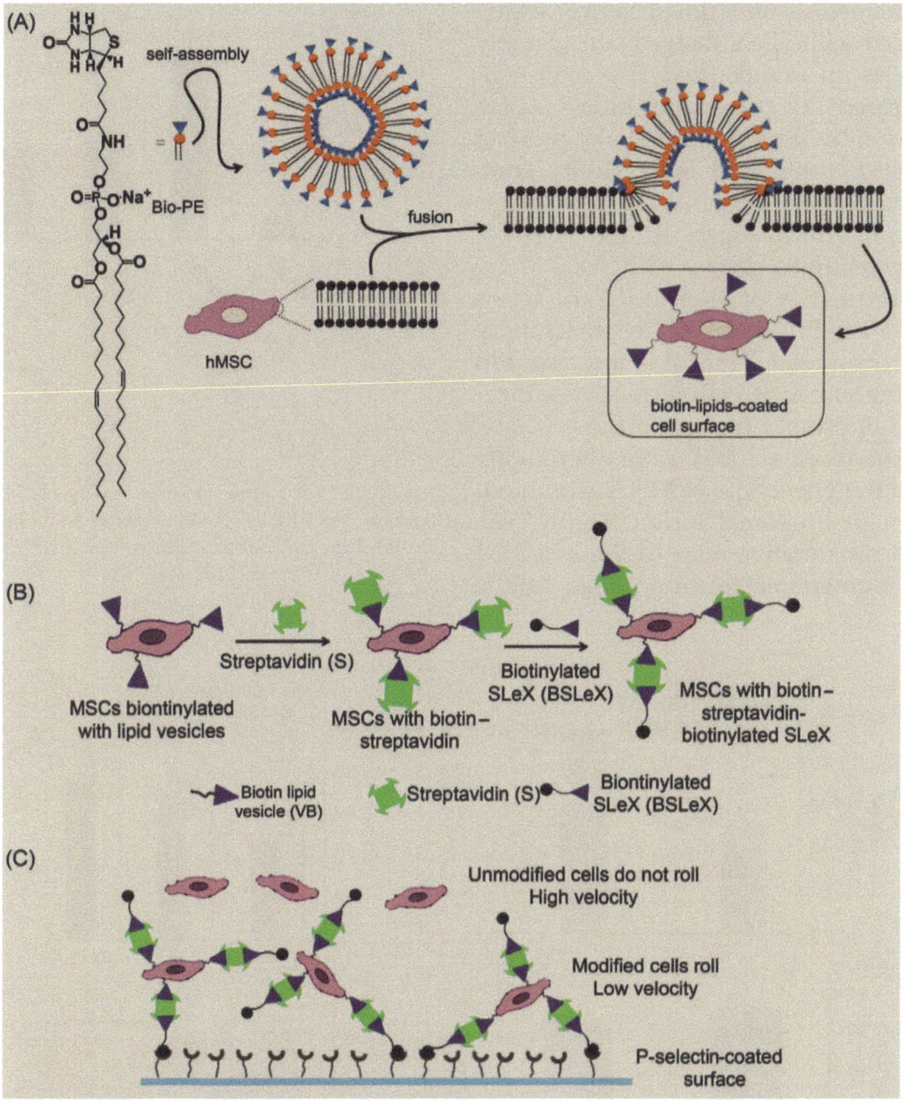

FIGURE 5.4 (A) Schematic showing formation of vesicles from biotinylated lipid and modification of MSCs with biotinylated lipid vesicles. (B) Biotinylated cell surface was complexed with streptavidin followed by biotinylated SLeX. (C) Induced rolling response of SLeX-modified MSCs on a P-selectin-coated surface. hMSC, human mesenchymal stem cell. *Source: Reprinted with permission from Ref. [33] © Elsevier 2010.*

Thus, modification on the cell surface through noncovalent liposomal hybridization offers a transient cell surface medication and avoids nonspecific modification of cell surface molecules. In addition, transient immobilization of adhesion ligands offered a great advantage through allowing for initial adhesion and cytokine activation, not impeding subsequent extravasation.

Furthermore, the rolling response of the SLeX-modified MSCs with P-selectin-coated substrates has been examined using a flow chamber assay. MSCs modified with SLeX showed considerably lower velocities on immobilized P-selectin substrates compared to phosphate buffer saline (PBS) treated cells (Figure 5.5). SLeX-modified MSCs exhibited ~eightfold less velocity compared to unmodified MSCs at a shear stress of 0.5 dyn/cm^2 which suggested the effect of noncovalently modified surface with SLeX.

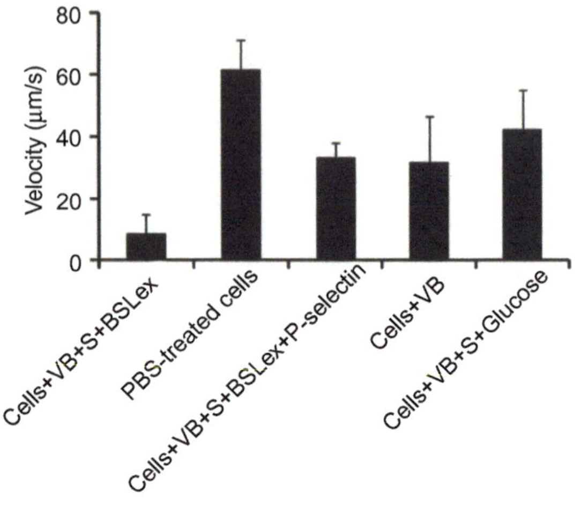

FIGURE 5.5 Flow chamber assay on P-selectin-coated surface. Velocity of SLeX-modified and PBS-treated MSCs at 0.5 dyn/cm^2. *Source: Reprinted with permission from Ref. [33] © Elsevier 2010.*

Inherent properties of stem cells are sensitive to minor modifications of cell machinery and microenvironment. Thus, it is important to investigate the effect of cell surface modification through liposomal fusion on the cell fate. Therefore, Sarkar et al. [33] have thoroughly characterized the phenotype of MSCs after modification through liposomal fusion (Figure 5.6). Viability and proliferative rates of the modified MSCs were comparable to unmodified MSCs.

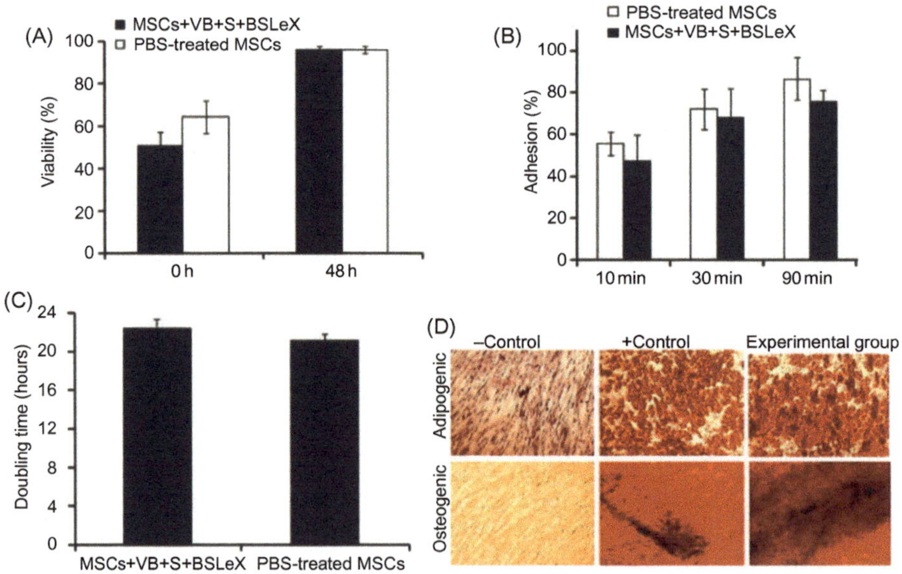

FIGURE 5.6 (A) Viability of the SLeX-modified MSCs at 0 h and after 48 h. (B) Adherence of SLeX-modified cells compared to the PBS-treated cells. (C) Doubling time of the SLeX-modified cells compared to PBS-treated cells. (D) Alkaline phosphatase and Oil Red O staining, 23 days after addition of osteogenic and adipogenic differentiation media to SLeX-modified MSCs. VB, biotinylated vesicles; BSLeX, biotinylated SLeX; S, streptavidine (− Control: SLeX-modified MSCs in cell expansion media; + Control: MSCs in differentiation media; Experimental group: SLeX-modified MSCs in differentiation media). *Source: Reprinted with permission from Ref. [33] © Elsevier 2010.*

In addition, the adhesion kinetics of modified MSCs and untreated MSCs are similar on tissue culture polystyrene plates, which indicated that inherent properties of the cell surface did not compromise due to noncovalent liposomal fusion. Another important characteristic of MSCs is the ability to differentiate into multiple lineages that did not compromise modification of the cell surface with biotinylated lipids and subsequent complexation of streptavidin and biotinylated SLeX. Modified MSCs were efficiently differentiated into osteogenic and adipogenic lineages as shown by alkaline phosphatase activity and Oil Red O staining, respectively (Figure 5.6). Those results clearly suggested that modifying the cells with rolling ligands through a noncovalent lipid vesicle approach did not affect the cell phenotype and thus cells may maintain their normal functional characteristics that will be important for cell therapy. This is a compelling example for the broad scope of cell surface modification through noncovalent approaches.

5.2.3 Noncovalent Cell Surface Modification Followed by Covalent Chemistry to Enhance the Cell Interactions

As described in previous sections, where cell surface molecules were covalently reacted with biotinylated-*N*-hydroxyl-succinimidyl (NHS) followed by immobilization of a molecule of interest through noncovalent complexation, often it may lead to nonspecific functionalization of cell surface molecules. To overcome this limitation, Dutta et al. [34] have developed an alternative method where low-molecular-weight amphiphile with a desired functional group was noncovalently fused with the cell membrane; subsequently, they have covalently conjugated those immobilized functional groups selectively with complementary reactive groups (Figure 5.7). In this approach, they have eliminated the nonspecific functionalization of biomolecules that are present on the cell surface.

Dutta et al. [34] have presented a method to tether chemoselective ketone and oxyamine groups from mouse fibroblast surfaces by liposome fusion toward the goal of rewiring the cell surface. The modified cells were patterned on solid surfaces using bioorthogonal oxime bond formation and without the use of any cell-adhesive ligand (Figure 5.7) and could be quantified using flow cytometry (Figure 5.8). The ketone and oxyamine groups act as synthetic cell surface receptors and can immobilize a wide range of ligands and probes for subsequent use in cell targeting and tracking. Thus, rewiring cell adhesion was achieved using chemoselective oxime bond formation and without using any cell-adhesive ligands such as Arg-Gly-Asp (RGD) and fibronectin. Through avoiding the use of any additional biomolecules in this strategy, one can eliminate the potential long-term stability and degradation limitation in complex cell culture media or cell targeting *in vivo*. This methodology also bypasses any metabolic or genetic approach to modify the cell surface, which holds the risk of perturbing any cellular processes by altering cellular physiology.

5.2.4 Programmed Cell–Substrate and Cell–Cell Assembly

Cellular communication in a three-dimensional (3D) environment often requires physical contact between neighboring cells that is mediated by the surrounding extracellular matrix [35]. Efforts to mimic the 3D cellular environment and assembly have been mostly limited to using artificial solid scaffolds as platforms to seed cells and manipulating them by external forces and tools [36–39]. Recently, chemists and material scientists have begun to develop minimal invasive ways to engineer

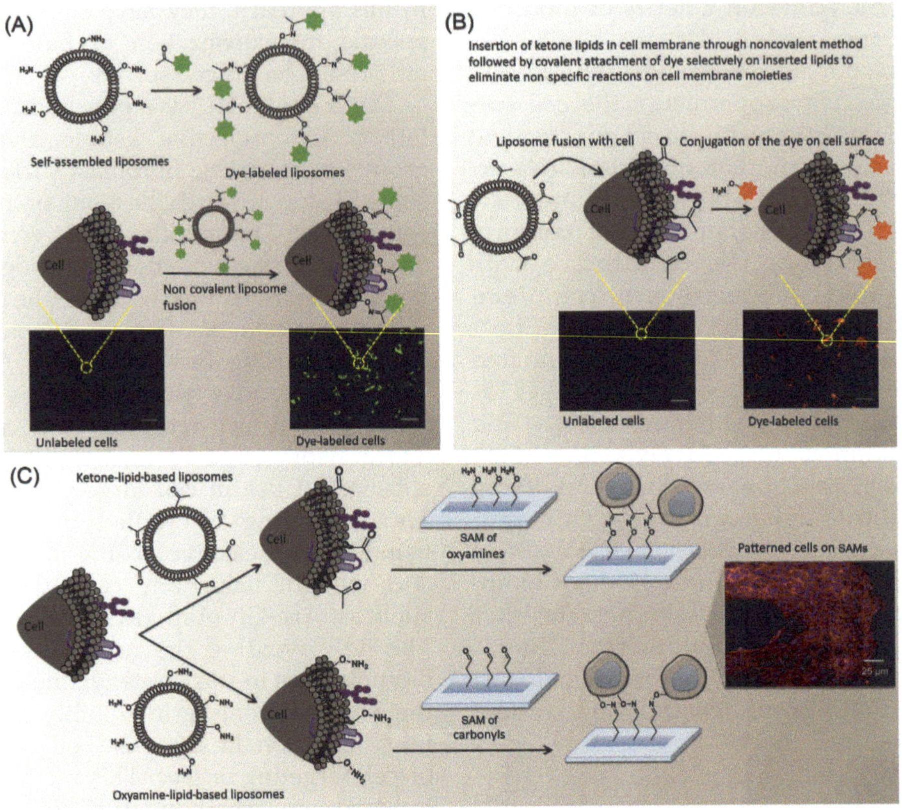

FIGURE 5.7 (A) Formation of oxyamine-based liposomes followed by dye functionalization. Subsequently, liposomes were noncovalently fused with cells to immobilize dye-based lipids on the cell surface. (B) Ketone-based lipids were immobilized on the cell surface followed by covalent attachment of dyes on ketones selectively. (C) Schematics and fluorescent micrographs of rewired cells adhered to pattern on self-assembled monolayers (SAMs) of alkanethiolates on gold substrates. *Source: Reprinted with permission from Ref. [34] © ACS 2011.*

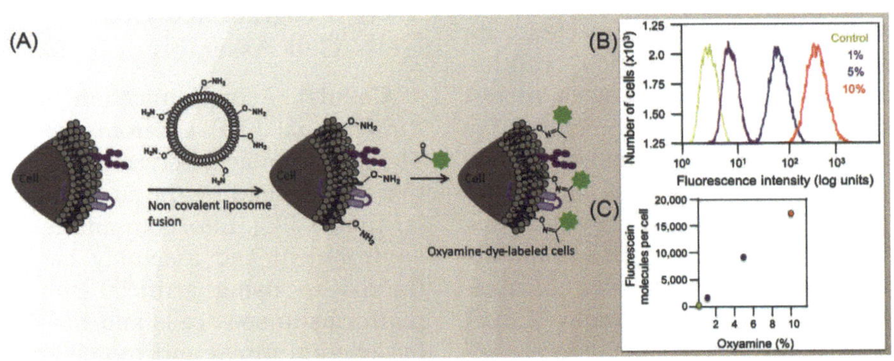

FIGURE 5.8 (A) Immobilization of dye through liposome fusion with cell membrane. (B) and (C) Cell surface molecule quantification using flow cytometry. *Source: Reprinted with permission from Ref. [34] © ACS 2011.*

cell surfaces by biorecognition moieties that act as synthetic cell surface receptors to control cell–cell and cell–substrate interactions. Several approaches, including the use of dielectric forces [40–42], laser-guided writing [43,44], surface manipulation [45], and a number of lithographic printing techniques [11,46,47], have been integrated with 3D scaffold designs to produce multitype cellular arrays or 3D cell clusters or spheroids.

A DNA-based approach has been used for the programmable assembly of 3D cellular structure, which may find applications in synthesizing artificial tissues [38]. Recently, Gartner and Bertozzi [38] have demonstrated the assembly of DNA-modified cells into microtissues in a well-defined manner (i.e., structures were controlled by varying the DNA site density on the cell surface, cell concentration, and DNA sequence) as demonstrated in Figures 5.9 and 5.10. Briefly, by adjusting the stoichiometry (i.e., 1:50) of two Jurkat cell populations that were modified with complementary DNA strands, single cell clusters were produced where the limiting cell type was surrounded by cells added in excess. DNA strands were introduced to the cell surface by Staudinger ligation between phosphine-modified single-stranded DNA (ssDNA) and azide-modified cell surfaces where the azide functionality was introduced by metabolic engineering of peracetylated *N*-α-azidoacetylmannosamine ($Ac_4ManNAz$) and subsequent display of the corresponding sialic acid in the cell membrane. Interestingly, using fluorescently labeled DNA, they showed that DNA strands are clustered at the cell–cell interface, confirming their specific role in cell–cell assembly. Significantly, such assembled cell clusters could be isolated and purified using flow cytometry and subsequently used as the core for further iterative cell assembly (i.e., adding another layer of DNA-modified cells).

More importantly, they have demonstrated that cells within the assembled microtissues communicate via a paracrine mechanism, which is a prerequisite for the development of therapeutically functional tissues. Specifically, in a proof-of-concept model, a CHO cell line engineered to express growth factor interleukin-3 (IL-3) was used as a first building block and the second was an untransformed hematopoietic progenitor cell line (FL5.12) whose survival and replication depend on the presence of IL-3. Cell composites were assembled using DNA hybridization at the interface and then embedded within a 3D agarose matrix. After 16 h, the cell structures underwent accelerated growth due to the presence of IL-3. By contrast, when CHO cells lacking the gene that codes IL-3 were used in the microtissue, FL5.12 cells showed no growth and instead underwent a morphological change corresponding to the absence of IL-3-induced apoptosis. Additional advantages of using DNA assembly in such systems include their reversibility and versatility with respect to engineering multiple orthogonal cell–cell interactions. In addition to ssDNA, self-assembled cargo-carrying DNA arrays can also be attached to the surface of cells. Reiche and colleagues attached hexagonal DNA arrays on cells using streptavidin and antibody as bridges [48]. In their study, it was observed that one or two DNA array patches often attached to one cell and that single DNA arrays often to bridged cells. Occasionally, micron-sized DNA arrays seemed capable of binding to and enveloping multiple cells into larger cellular structures. These larger structures may have the potential to form the foundation for the construction of tissues or organs for transplantation in tissue engineering.

Engineered cell–cell interactions in a bottom-up tissue engineering approach can also be achieved by presenting bioorthogonal functional groups on the cell surface. Dutta et al. [49] have developed a simple liposome fusion method to display ketone or oxyamine functional groups from cell surfaces to

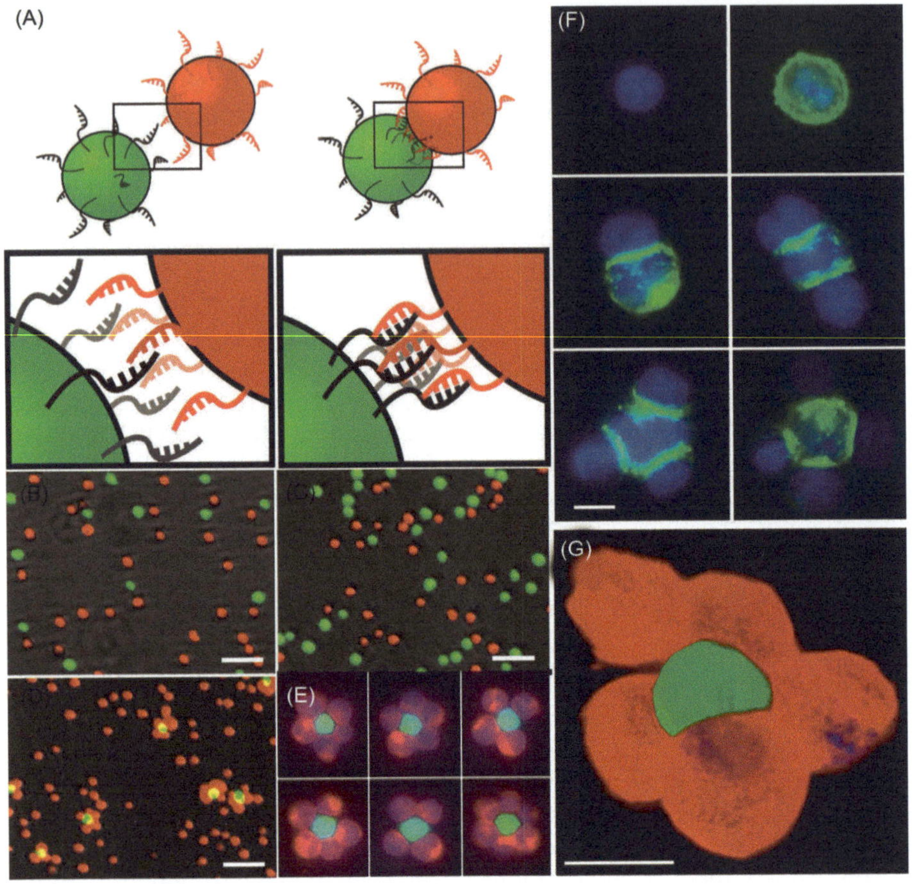

FIGURE 5.9 Oligonucleotides direct the synthesis of 3D multicellular structures with defined patterns of connectivity. (A) Cells bearing complementary cell surface oligonucleotides react to form stable cell–cell contacts. (B) Nonadherent Jurkat cells with CMFDA (fluorescein, green) or CMTPX (Texas Red, red) cytosolic stains were combined at a 1:1 ratio. (C) Green- and red-stained Jurkat cells labeled with the mismatched oligos 1-Phos and 5′-Phos were also mixed at a 1:1 ratio. (D) Green- and red-stained Jurkat cells labeled with the complementary sequences 1-Phos and 1′-Phos and combined at a 1:50 ratio. (E) Discrete structures from (D) at higher magnification. (F) Cells labeled with fluorescein-conjugated DNA (single cell, top right) were assembled with cells bearing a nonfluorescent complementary strand (single cell, top left). (G) 3D reconstruction of a single multicellular structure encapsulated in agarose by using deconvolution fluorescence microscopy. Red, Texas Red; green, fluorescein; blue, DAPI. (Scale bars: (B)–(D), 50 μm; (E)–(G), 10 μm.). *Source: Reprinted with permission from Ref. [38] © PNAS 2009.*

generate 3D spheroid assemblies and multilayered tissue-like structures (Figures 5.11 and 5.12). Tissue-like networks were also formed on the surface with geometric control. Surfaces were patterned with cell-adhesive and nonadhesive regions to generate multilayered sheets and patterned tissue structures, respectively. Ketone- and oxyamine-tailored liposomes were cultured with separate fibroblast populations, resulting in membrane fusion and subsequent presentation of chemoselective sites for oxime conjugation from the surface. Culturing these groups on a solid support ($\sim 10^5$ cells/ml) and in a layer-by-layer deposition manner gave rise

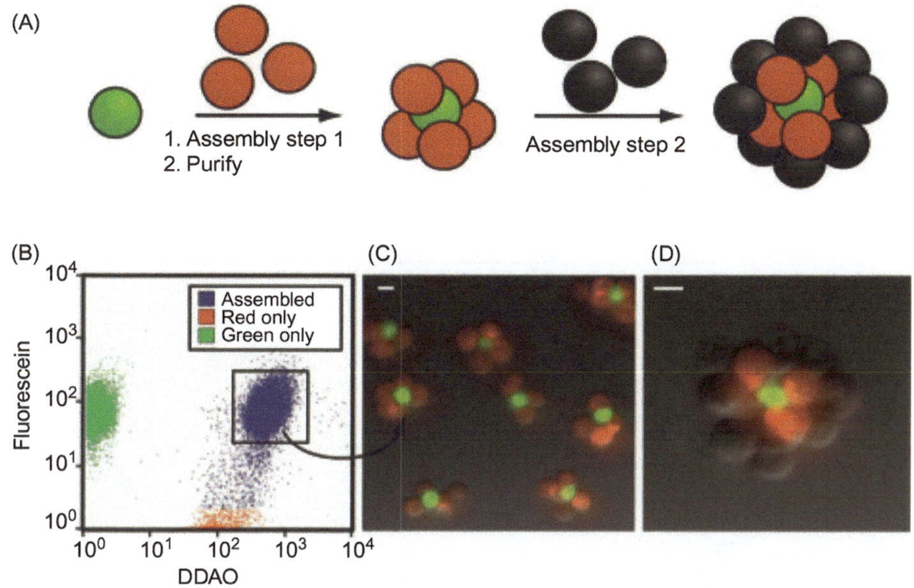

FIGURE 5.10 Fluorescence-activated cell-sorting (FACS) separates cell assemblies based on their fluorescence properties. (A) Schematic diagram of multistep assembly of cellular building blocks. (B) Jurkat cell-based structures were isolated by FACS based on red and green fluorescence proportional to the number of each cell type in the cluster. (C) Fluorescence micrograph of the isolated structures. (D) Higher order multicellular structure generated by incubation of the structures in (C) with a third population of unstained Jurkat cells bearing cell surface oligonucleotides complementary to those on the red-stained cells. Red, Texas Red; green, fluorescein. (Scale bars: 10 μm.). *Source: Reprinted with permission from Ref. [38] © PNAS 2009.*

to multilayered, tissue-like cell sheets, which were characterized by confocal microscopy. Cells cultured without the ketone and oxyamine groups did not form any multilayered structures. Spatial control on tissue formation was demonstrated using a micro-contact printing method where a variety of patterns and geometries were produced on a gold substrate. Employing SAMs and microfabrication techniques, hexadecanethiol (1 mM in ethanol) was printed on a gold surface. The surface was then backfilled with tetraethylene glycol (1 mM in ethanol, 16 h) to render the remaining regions inert to nonspecific protein absorption. Fibronectin, a cell-adhesive protein, was then added (10 mg/ml in calf bovine serum (CBS) 2 h), adhering only to the hydrophobic, patterned areas. When liposome fusion occurs to display complementary ketone and oxyamine groups from cell surfaces, multilayered 3D cell patterns were formed. The assembled tissue-like structures were generated in different shapes such as circular, bar, and squares. The strategy was also extended to form 3D cocultures of hMSCs and fibroblasts. The 3D cocultures were generated with two different cell types, in this instance, hMSCs and fibroblasts. Using this method, they retained individual cell phenotype properties that were demonstrated through differentiating hMSCs into adipocytes [49].

5.2.5 Anchoring Biomolecules on the Cell Surface Through Insertion of Hydrophobic Chain into the Cell Membrane

Although hydrophobic interactions are inherently weak in nature (in the order of

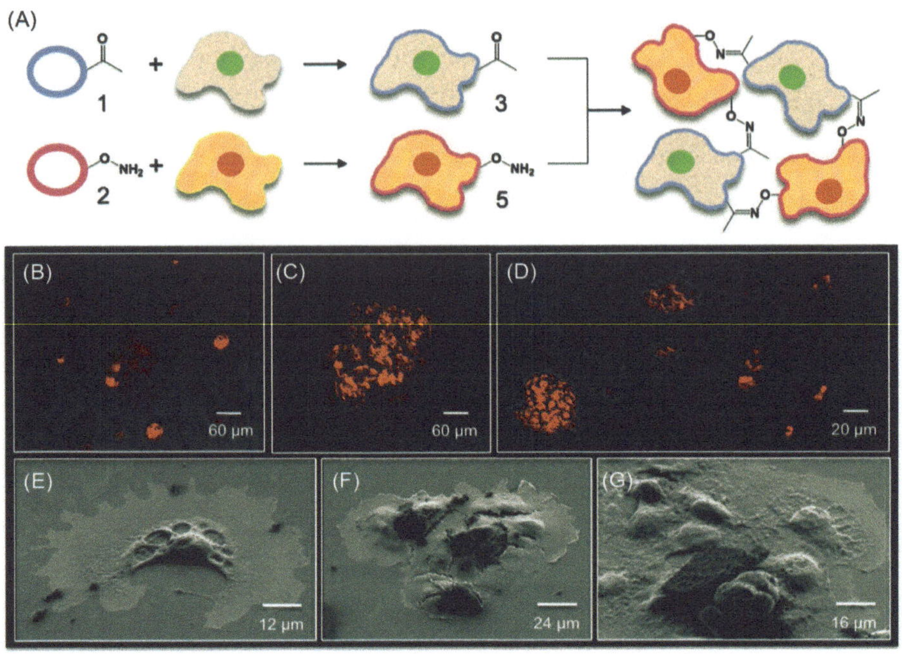

FIGURE 5.11 Fluorescent, phase contrast, and scanning electron micrographs (SEMs) describing 3D spheroid formation via liposome fusion and chemoselective cell surface tailoring. (A) Two fibroblast populations were cultured separately with ketone- or oxyamine-containing liposomes. Upon mixing these cell populations, clustering and tissue-like formation, based on chemoselective oxime conjugation, occurred. (B) Control experiments (overlay image) demonstrate no spheroid formation for cells that did not contain either ketone or oxyamine groups. (C) and (D) Two cell populations with chemoselective recognition groups were mixed to form interconnected spheroidal assemblies (overlay images). (E)–(G) Representative SEM images. *Source: Reprinted with permission from Ref. [49] © ACS 2011.*

2–5 k_BT in vacuum), they play a prominent role in many biological processes like lipid bilayer formation, protein folding, and protein–protein recognition [50]. In a seminal work by Dennis and coworkers [51] using a lipid bilayer system, MSCs were treated with palmitated protein G where the palmitate chain was incorporated into the lipid bilayer via hydrophobic interactions and protein G provides generic binding sites for antibodies (Figure 5.13). In a proof-of-concept work, intercellular cell adhesion molecule-1 (ICAM-1) was conjugated onto MSCs, which enabled the cells to bind to ICAM-1, a critical adhesion molecule expressed on activated endothelium. These chemical approaches to tailor cell surfaces with functional ligands are broad and generic and are not limited to only MSCs and the ligands mentioned herein.

Cell surface hydrophobicity and cell surface charge have been recognized as measurable physicochemical variables for evaluating cellular adhesion to surfaces. Amphiphilic polymers, such as polyethylene glycol (PEG) conjugated phospholipids (PEG-lipid) and poly(vinyl alcohol) bearing hydrophobic alkyl side chains (PVA-alkyl), have been used in cell surface modification [52–56]. In these methods, the hydrophobic alkyl chains of amphiphilic polymers are spontaneously anchored into the lipid bilayer membranes through hydrophobic interactions (Figure 5.14). These spontaneous insertions can be monitored by a surface plasmon resonance (SPR) when a

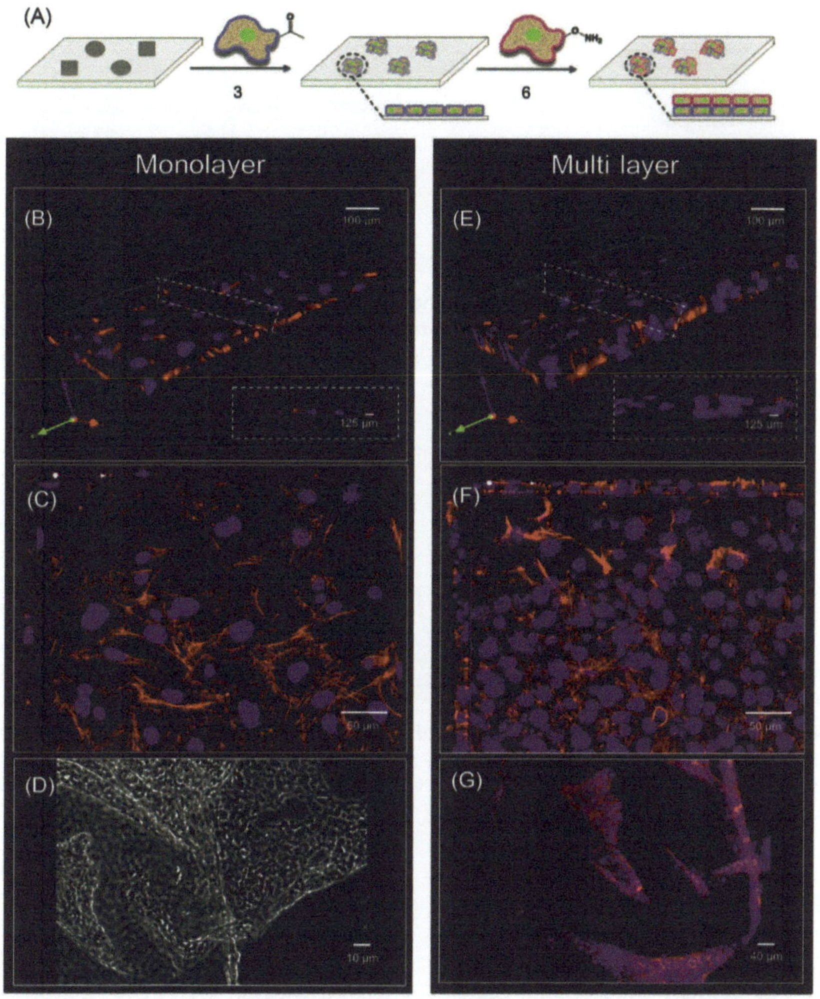

FIGURE 5.12 Schematic and images of oxime-mediated, 3D tissue-like structure formation with controlled interconnectivity. (A) Ketone- and oxyamine-containing liposomes were added to two separate fb populations, resulting in membrane fusion. Culturing these cells on substrates, alternating cell population seeding layer-by-layer, gave rise to multilayered, tissue-like cell sheets through stable oxime chemistry. (B) A 3D reconstruction and (C) confocal micrograph showing only a monolayer of cells after oxyamine-presenting cells were cultured with adhered nonfunctionalized cells. (E) A 3D reconstruction and (F) confocal micrograph of multiple cell layers. (D) and (G) Brightfield and fluorescent images of deadhered cell sheets from the surface. Cells were stained with DAPI (blue, nucleus) and phalloidin (red, actin). *Source: Reprinted with permission from Ref. [49] © ACS 2011.*

PEG-lipid solution is applied to supported lipid membranes that are formed on an SPR sensor. The spontaneous incorporation is greatly affected by the length of phospholipid alkyl chain length.

The other method is the formation of an intermediary polymer layer through hydrophobic interactions using amphiphilic polymers, such as PEG-lipid and PVA-alkyl, that bear a wide range of functional groups.

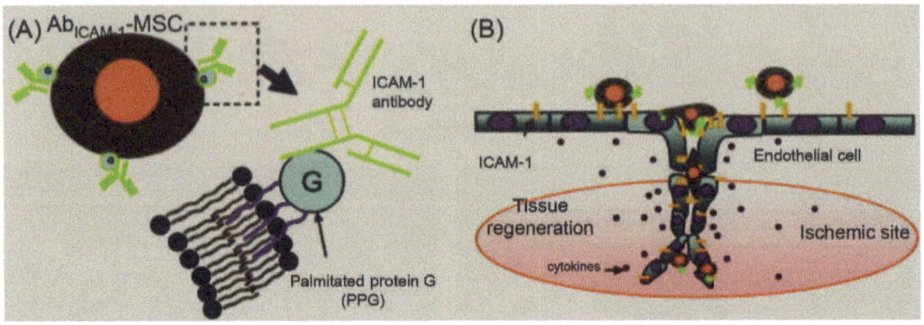

FIGURE 5.13 Diagrammatic representation of endothelial cell targeting using antibody-coated MSCs for tissue regeneration. (A) A schematic representation of the membrane structure of Ab$_{ICAM-1}$–MSC. (B) Targeting activated ECs expressing ICAM-1 abundantly in ischemic sites with Ab$_{ICAM-1}$–MSCs. *Source: Reprinted with permission from Ref. [51] © Elsevier 2010.*

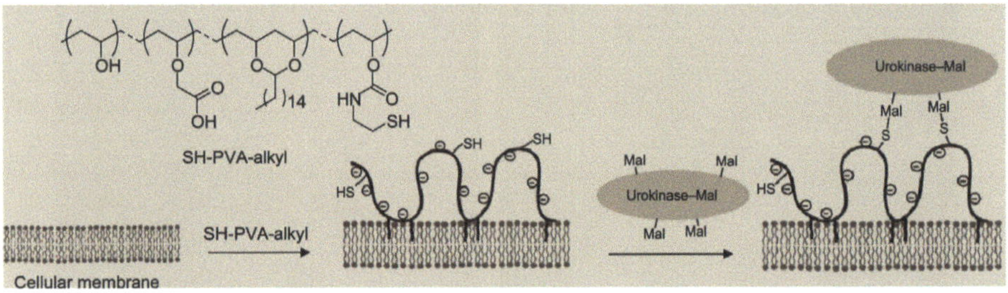

FIGURE 5.14 Schematic illustration of immobilization of urokinase on the islet surface. SH-PVA-alkyl was introduced to the cell surface by the hydrophobic interaction between alkyl side chains and the lipid bilayer of the cell membrane. Then, urokinase–Mal was immobilized on the cell surface via thiol–maleimide bonding. *Source: Reprinted with permission from Ref. [56] © Elsevier 2008.*

The functional groups that are introduced on the cell surface through amphiphilic polymers can be used for the immobilization of bioactive substances (Figure 5.14). In one example, PVA-alkyl presenting SH groups were used for the immobilization of bioactive substances displaying maleimide groups. In a similar manner, PEG-lipids presenting maleimide at the end of the PEG chain was also used for the immobilization of bioactive substances tethering SH groups without cytotoxicity.

By utilizing this approach, Totani et al. have developed a promising system that is urokinase-immobilized islets for the potential treatment of diabetes. Although transplantation of islets of Langerhans is a promising method to treat insulin-dependent type I diabetes, innate immunological response activated by blood coagulation plays a key role in the loss of islets at the beginning. Thus, to overcome this limitation and to inhibit blood coagulation on the islet surface immobilization, a plasminogen activator, urokinase, was immobilized on the islet surface via a PVA derivative with alkyl chains and thiol groups (Figure 5.14). When the PVA derivative was added to an islet suspension, the alkyl side chains spontaneously anchored into the lipid bilayer membranes of islet cells. Subsequently, urokinase modified with maleimide groups was immobilized onto the islet surface by thiol/maleimide covalent bonding with the PVA layer. Urokinase-immobilized islets exhibited fibrinolytic properties, indicating that blood coagulation can be controlled on the islet surface. Furthermore, fibrinolysis activity

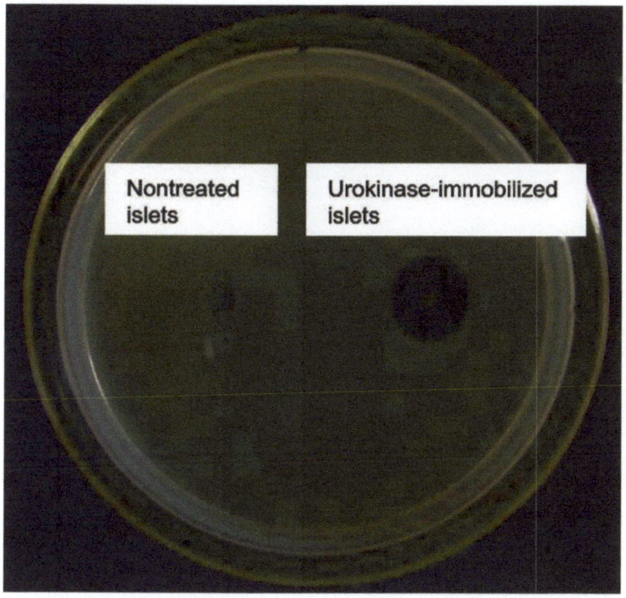

FIGURE 5.15 Fibrin plate assay of urokinase-immobilized islets. Urokinase-immobilized islets and nontreated islets (100 islets, respectively) were placed on a fibrin gel plate and incubated at 37°C for 13 h. Plasmin that is produced from plasminogen in the presence of urokinase-immobilized islets, dissolves larger area of the fibrin gel (larger black circle). *Source: Reprinted with permission from Ref. [56] © Elsevier 2008.*

of surface-immobilized urokinase was determined by the fibrin-plate-based assay (Figure 5.15). Urokinase activates plasminogen in the fibrin gel to form plasmin. Plasmin triggers a proteolytic cascade that results in fibrin gel degradation, which forms a transparent window in the gel plate. A large transparent window (2.5 cm^2) was observed around the spotted urokinase-islets, which reflected the fibrinolytic activity of urokinase immobilized on the islets. This is an elegant example of versatility and efficiency of noncovalent immobilization of biomolecules on the cell surface.

5.3 ELECTROSTATIC INTERACTIONS MEDIATED CELL SURFACE MODIFICATION

5.3.1 Engineering of Cell Surface by Harnessing Electrostatic Interactions

The cell membrane is intrinsically negatively charged due to a high composition of negatively charged lipids such as phosphatidylserine and phosphatidylinositol lipids. The possibility of constructing a thin polymer membrane on the surface of cells through electrostatic interactions between negatively charged cell surfaces and cationic polymers and then further modifying them using a layer-by-layer technique of anionic and cationic polymers [57–61] has been explored by several research groups. The ionic polymers that have been employed are poly(allylamine hydrochloride), poly(styrene sulfate), poly-L-lysine, and poly-(ethyleneimine). A layer-by-layer method is a simple and attractive technique to modify cell surfaces (Figure 5.16) [62]. The outermost layer of the polymer controls the surface properties. The thickness of the layer is also controllable by the number of polymer layers. However, most polycations, such as poly-L-lysine and polyethylene-imine, can be extremely cytotoxic and severely damage the treated cells. Although cationic polymers are efficient to interact with the cell membrane, often they rupture the cell membrane through electrostatic interactions. This leads to severe cell toxicity. To avoid cationic-polymer-associated cytotoxicity, attempts have

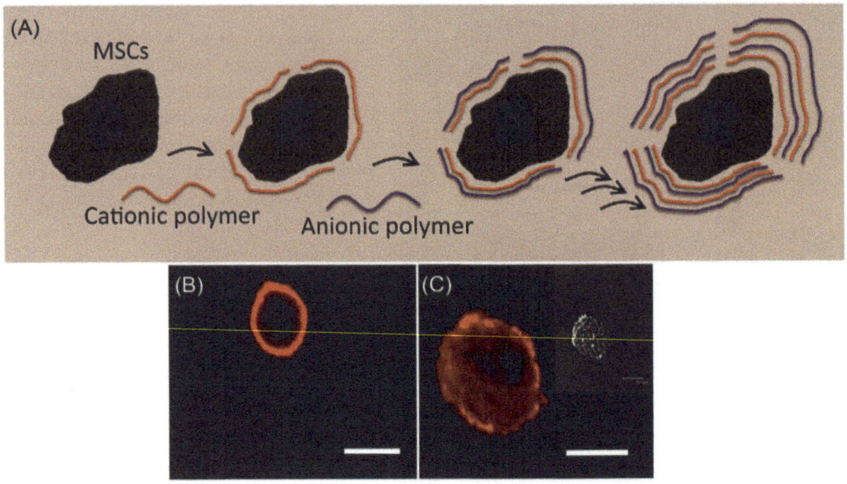

FIGURE 5.16 (A) Schematic illustrating MSCs encapsulation process using layer-by-layer assembly. Deposition of cationic PLL layer followed by deposition of anionic HA layer, and process was repeated multiple times. Confocal micrographs of MSCs encapsulated within PLL/HA at (B) 3 days postencapsulation and (C) 7 days postencapsulation. Scale bars represent 8.57 and 117 μm for panels B and C, respectively. *Source: Reprinted with permission from Ref. [62] © Wiley-VCH Verlag GMBH & Co. 2007.*

been made to minimize electrostatic interactions between cationic polymer and cell surface through modulation of charge on the polymer. However, such weak electrostatic interactions often do not yield desirable effects due to reduction of residence time of polymer on the cell surface. Thus, it is extremely important to carefully design the polymers to achieve a balance between toxicity and efficiency.

5.3.2 Noncovalent Cell Surface Modification for Bioimaging Applications

Cell surface modification can facilitate the characterization of a number of cellular processes and signal transduction pathways in terms of identifying key proteins and signaling molecules [63]. One major application of engineering cell surfaces can be in modifying the cell membrane with molecular and detection probes for subsequent tracking during postinfusion in cell therapy [64]. Chemical reporters have been incorporated to the cell membrane using the metabolic pathways [14]. Orynbayeva et al. [65] engineered chromatic polydiacetylene (PDA) polymer patches onto the cell surfaces for imaging structural perturbations of the membrane bilayer (Figure 5.17). Lipid vesicles comprising of 1,2-dimyristoyl-sn-glycerophosphoethanolamine and 1,2-dimyristoyl-sn-glycero-3-[phospho-rac-(1-glycerol)] were fused with the cell membrane, and PDA was formed *in situ* upon ultraviolet irradiation. The tethered nanopatches of PDA on the cell surface had an intense blue color. The structural perturbations of the membrane bilayer due to the conformational changes in the conjugated (ene-yne) polymer backbone led to a blue-to-red transition of the conjugated PDA. During this blue-to-red color transition, the initial nonfluorescent PDA also emitted fluorescence at 560 and 640 nm. The PDA nanopatches can therefore be used to visualize the cell bilayer perturbations induced by, for example, lidocaine, polymyxin-B, cholesterol, and oleic acid and to screen toxic pesticides and other environmentally toxic small molecules.

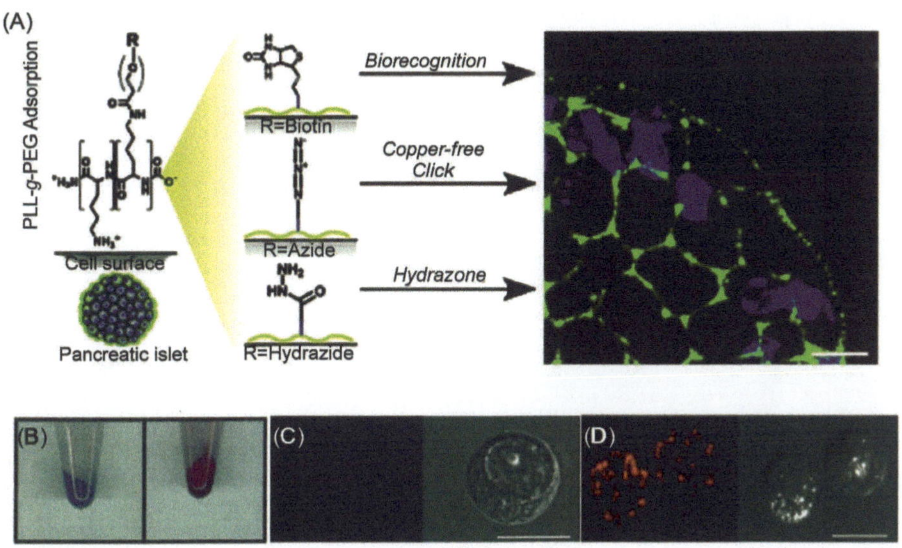

FIGURE 5.17 (A) Cell surface engineering with PLL-g-PEG copolymers. (B) Photographs of tubes containing PDA-labeled cells after sedimentation. Left tube: Blue cells before treatment; right tube: cells after incubation with oleic acid (180 μM). (C) and (D) Confocal microscopy fluorescence images (left) and transmission differential interference contrast microscopy images (right) of PDA-labeled U937 cells. (C) and (D) Cells before and after polymyxin-B (50 μM) treatment, respectively. *Source: Reprinted with permission from Refs. [65,66] © AAAS 2005 and © Wiley-VCH 2005.*

In another study, Wilson et al. [66] have used cationic graft copolymers to noncovalently engineer pancreatic islet surfaces using electrostatic interactions for subsequent labeling and visualization (Figure 5.17). Specifically, poly(L-lysine)-graft-poly(ethylene glycol) (PLL-g-PEG) copolymers where the polymer backbone was terminally functionalized with biotin, hydrazide, and azide moieties were used to reengineer the cell surfaces with streptavidin-, aldehyde-, and cyclooctyne-labeled probes.

5.3.3 Noncovalent Cell Surface Engineering to Manipulate Cell Biological Fate

Externally applied chemical moieties and materials can be engineered to cell surfaces to control cell biology. The idea is to stimulate and reversibly control different cell signaling pathways by physical tools that are normally biochemically regulated. This can be achieved by linking different molecular probes to the cell surface. Often native cell surface is not amenable to external stimuli manipulation. In such instances, transiently engineering the cell surface by immobilizing an additional agent on the cell surface would enable to manipulate the cell surface and cell fate through an external stimulus. Ingber and coworkers [67] have demonstrated the use of magnetic nanoparticles on the cell membrane to regulate cellular signaling pathways (Figure 5.18). Specifically, mast cells express membrane high-affinity IgE receptors (FcεRI) that can bind to the Fc region of IgE molecules. Multivalent antigens like dinitrophenyl (DNP) that bind to IgE can then induce FcεRI receptor

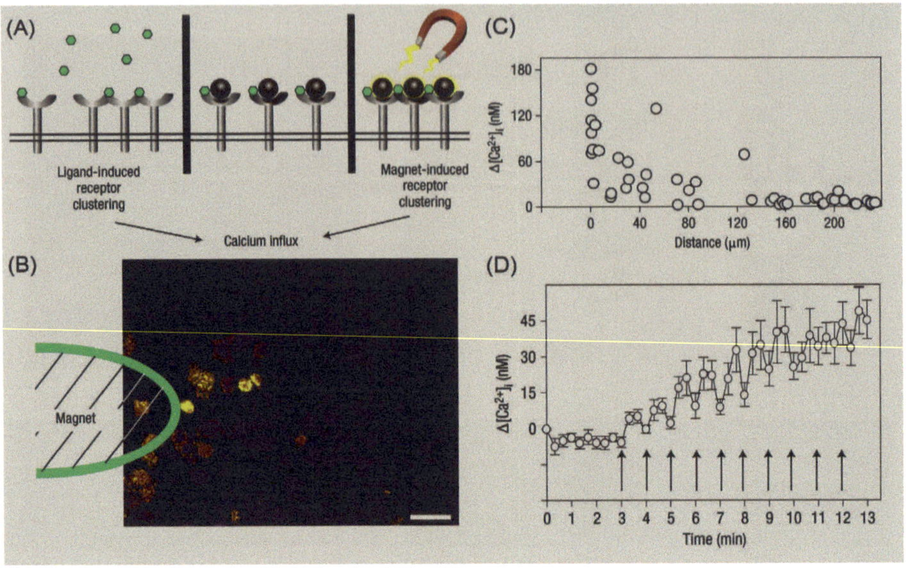

FIGURE 5.18 (A) The biochemical mechanism that stimulates downstream signaling (left) involves the binding of multivalent ligands (green hexagons) that induce oligomerization of individual IgE/FcεRI receptor complexes. In the magnetic switch, monovalent ligand-coated magnetic nanobeads (dark grey circles) bind individual IgE/FcεRI receptor complexes without inducing receptor clustering (centre). (B) The pseudocolored microfluorimetric image shows the local induction of calcium signaling (yellow) in cells near the tip of the electromagnet within 20 s of the field being applied. Scale bar represents 50 μm. (C) Quantification of peak changes in intracellular calcium relative to time 0 measured within individual cells during a 1-min pulse of applied magnetic force as a function of the distance of the tip from the cell surface. (D) Effect of a rapid cyclical magnetic stimulation regimen (40 s on, 20 s off) on intracellular calcium signaling. *Source: Reprinted with permission from Ref. [67] © NPG 2007.*

oligomerization on the cell membrane whereas monovalent antigens will fail to do so. The FcεRI receptor clustering quickly triggers an intracellular signaling response characterized by a rapid rise of cytosolic Ca^{2+}, which in turn leads to the local inflammatory response of mast cells. Instead of using an antigen ligand to induce such signaling pathway, the authors immobilized DNP-attached magnetic nanoparticles on IgE-tethered mast cells [68]. The authors ensured that only one DNP molecule was attached on each particle so that receptor clustering did not occur. When an external magnetic field is applied, magnetic nanoparticles clustered on the cell membrane, which caused the oligomerization of FcεRI receptors, a process that could be visualized by the local Ca^{2+} concentration change. Such magnetic nanoparticles mediated cellular signaling is reversible, switching on and off the magnetic field can directly correspond to a local Ca^{2+} concentration oscillation. This is an elegant example for modulation of intracellular properties through controlling an altering cell surface by an external stimulus.

5.4 ADVANTAGES AND LIMITATIONS OF THE NONCOVALENT MODIFICATION OF THE CELL SURFACE

5.4.1 Advantages of Noncovalent Modification of the Cell Surface

Cell surface engineering has primarily been a subject of molecular biology and genetic engineering. Although genetic manipulation is an efficient method to modulate the cell surface, potentially it might disrupt the inherent cell properties. Often it might lead to unforeseen and unwanted effects. Thus, selective engineering of cell surface without the need of genetic manipulation is a welcoming approach to modulate the cell surface features and properties. Typically, NHS-activated esters and cyanuric chloride derivatives are frequently used to form covalent bonds to the cell surface functional groups such as amines. In addition, maliemide derivatives can be used to react with thiol groups of membrane molecules. However, although covalent modification provides a robust functionalization approach, degree of functionalization needs to be highly optimized to avoid deactivation of membrane molecules such as proteins, lipids, and receptors. Thus, potential toxic effects may be exerted on membrane biomolecules [68]. Enzymatic treatment and metabolic introduction via sialic acid biosynthetic pathway (Bertozzi group) have been employed to add multiple functional groups such as biotin, azido, and ketone groups to living cell surfaces [14,15]. However, these technologies are limited to the introduction of specified small molecules to cells and may perturb cell physiology. Thus, noncovalent methods to transiently engineer the cell surface could overcome these limitations. Liposome fusion can serve as an important tool among many other techniques that have been previously described to modify and tailor cell surfaces with biologically relevant molecules, significantly contributing to tissue engineering fields in generating 3D tissue constructs and in the field of cell-based targeted therapy.

5.4.2 Limitations of Noncovalent Modification of the Cell Surface

Although noncovalent modification of cell surfaces does not perturb the cellular machinery of the cell, the noncovalent forces like hydrophobic and electrostatic interactions holding the molecular probes to the cell membrane are intrinsically weaker in nature. This can result in a comparatively unstable system leading to temporal sensitivity. While this approach is optimal for transient surface medication, engineering surface through noncovalent modifications might not be retained in the longer term. Thus, an extensive optimization needs to be done to increase the longevity of the engineered cell surfaces.

5.5 CONCLUSIONS AND FUTURE PERSPECTIVES

Noncovalent methods for cell surface engineering provide efficient and facile approaches toward modulating cellular functions and finding solutions to fundamental biological problems. It has potential uses in a wide range of fields of biomedicine, tissue engineering, regenerative medicine, wound healing, cell-based therapies, imaging, and diagnostics. In future, coupling multiple noncovalent forces in a single system that would enhance the longevity of the system could enable the development of more elegant and robust methods of engineering the cell surfaces with different molecular probes. Efforts should also be made in designing dynamic systems, where molecules of choice could be conjugated and released from the cell surface using external stimuli. The eventual design and manufacture of these systems would need further collaborations between biologists and engineers for such products to make it to the marketplace.

Acknowledgments

PKV thanks Department of Biotechnology, Government of India for Ramalingaswami Re-Entry Fellowship, and Institute for Stem Cell Biology and Regenerative Medicine (inStem), Bangalore for financial support.

References

[1] Gullberg D, Ekblom P. Extracellular matrix and its receptors during development. Int J Dev Biol 1995;39:845–54.

[2] Scott MD, Murad KL, Koumpouras F, Talbot M, Eaton JW. Chemical camouflage of antigenic determinants: stealth erythrocytes. Proc Natl Acad Sci USA 1997;94:7566–71.

[3] Huebsch N, Mooney DJ. Inspiration and application in the evolution of biomaterials. Nature 2009;462: 426–32.

[4] Zhao W, Karp JM. Controlling cell fate *in vivo*. Chembiochem 2009;10:2308–10.

[5] Anker JN, Hall WP, Lyandres O, Shah NC, Zhao J, Duyne PV. Biosensing with plasmonic nanosensors. Nat Mater 2008;7:442–53.

[6] Zhao W, Karp JM. Tumour targeting: nanoantennas heat up. Nat Mater 2009;8:453–4.

[7] Giepmans BNG, Adams SR, Ellisman MH, Tsien RY. The fluorescent toolbox for assessing protein location and function. Science 2006;312:217–24.

[8] Schmidt U, Günther C, Rudolph R, Böhm G. Protein and peptide delivery via engineered polyomavirus-like particles. FASEB J 2001;15:1646–8.

[9] Akin D, Ragheb K, Sturgis J, Sherman D, Burkholder K, et al. Bacteria-mediated delivery of nanoparticles and cargo into cells. Nat Nanotechnol 2007;2:441–9.

[10] Chen S, Smith LM. Photopatterned thiol surfaces for biomolecule immobilization. Langmuir 2009;25: 12275–82.

[11] Falconnet D, Csucs G, Grandin HM, Textor M. Surface engineering approaches to micropattern surfaces for cell-based assays. Biomaterials 2006;27:3044–63.

[12] Balyasnikova IV, Franco-Gou R, Mathis JM, Lesniak MS. Genetic modification of mesenchymal stem cells to express a single-chain antibody against EGFRvIII on the cell surface. J Tissue Eng Regen Med 2010;4:247–58.

[13] Saxon E. Cell surface engineering by a modified Staudinger reaction. Science 2000;287:2007–10.

[14] Prescher JA, Bertozzi CR. Chemistry in living systems. Nat Chem Biol 2005;1:13–21.

[15] Rabuka D, Forstner MB, Groves JT, Bertozzi CR. Noncovalent cell surface engineering: incorporation of bioactive synthetic glycopolymers into cellular membranes. J Am Chem Soc 2008;130:5947–53.

[16] Paulick MG, Forstner MB, Groves JT, Bertozzi CR. A chemical approach to unraveling the biological function of the glycosylphosphatidylinositol anchor. Proc Natl Acad Sci USA 2007;104:20332–7.

[17] Yun Lee D, Hee Nam J, Byun Y. Functional and histological evaluation of transplanted pancreatic islets immunoprotected by PEGylation and cyclosporine for 1 year. Biomaterials 2007;28:1957–66.

[18] Lee DY, Lee S, Nam JH, Byun Y. Minimization of immunosuppressive therapy after islet transplantation: combined action of heme oxygenase-1 and PEGylation to islet. Am J Transplant 2006;6:1820–8.

[19] Cabric S, Sanchez J, Lundgren T, Foss A, Felldin M, Kallen R, et al. Islet surface heparinization prevents the instant blood-mediated inflammatory reaction in islet transplantation. Diabetes 2007;56:2008–15.

[20] Stabler CL, Sun X-L, Cui W, Wilson JT, Haller CA, Chaikof EL, et al. Surface re-engineering of pancreatic islets with recombinant azido-thrombomodulin. Bioconjug Chem 2007;18:1713–5.

[21] Kadowaki K, Matsusaki M, Akashi M. Control of cell surface and functions by layer-by-layer nanofilms. Langmuir 2010;26:5670–8.

[22] Belardi B, O'Donoghue GP, Smith AW, Groves JT, Bertozzi CR. Investigating cell surface galectin-mediated cross-linking on glycoengineered cells. J Am Chem Soc 2012;134:9549–52.

[23] Hangauer MJ, Bertozzi CR. A FRET-based fluorogenic phosphine for live-cell imaging with the Staudinger ligation. Angew Chem Int Ed 2008;47:2394–7.

[24] Laughlin ST, Bertozzi CR. Metabolic labeling of glycans with azido sugars and subsequent glycan-profiling and visualization via Staudinger ligation. Nat Protoc 2007;2:2930–44.

[25] Karp JM, Leng Teo GS. Mesenchymal stem cell homing: the devil is in the details. Cell Stem Cell 2009;4:206–16.

[26] Rombouts WJC, Ploemacher RE. Primary murine MSC show highly efficient homing to the bone marrow but lose homing ability following culture. Leukemia 2003;17:160–70.

[27] Kumar S, Nagy TR, Ponnazhagan S. Therapeutic potential of genetically modified adult stem cells for osteopenia. Gene Ther 2010;17:105–16.

[28] Brenner S, Whiting-Theobald N, Kawai T, Linton GF, Rudikoff AG, Choi U, et al. CXCR4-transgene expression significantly improves marrow engraftment of cultured hematopoietic stem cells. Stem Cells 2004;22: 1128–33.

[29] Zhang D, Fan G-C, Zhou X, Zhao T, Pasha Z, Xu M, et al. Over-expression of CXCR4 on mesenchymal stem cells augments myoangiogenesis in the infarcted myocardium. J Mol Cell Cardiol 2008;44:281–92.

[30] Sackstein R, Merzaban JS, Cain DW, Dagia NM, Spencer JA, Lin CP, et al. *Ex vivo* glycan engineering of CD44 programs human multipotent mesenchymal stromal cell trafficking to bone. Nat Med 2008;14: 181–7.

[31] Chavakis E, Urbich C, Dimmeler S. Homing and engraftment of progenitor cells: a prerequisite for cell therapy. J Mol Cell Cardiol 2008;45:514–22.

[32] Sarkar D, Vemula PK, Teo GSL, Spelke D, Karnik R, Wee LY, et al. Chemical engineering of mesenchymal stem cells to induce a cell rolling response. Bioconjug Chem 2008;19:2105–9.

[33] Sarkar D, Vemula PK, Zhao W, Gupta A, Karnik R, Karp JM. Engineered mesenchymal stem cells with self-assembled vesicles for systemic cell targeting. Biomaterials 2010;31:5266–74.

[34] Dutta D, Pulsipher A, Luo W, Mak H, Yousaf MN. Engineering cell surfaces via liposome fusion. Bioconjug Chem 2011;22:2423–33.

[35] Nelson CM, Bissell MJ. Of extracellular matrix, scaffolds, and signaling: tissue architecture regulates development, homeostasis, and cancer. Annu Rev Cell Dev Biol 2006;22:287–309.

[36] Gillette BM, Jensen JA, Tang B, Yang GJ, Bazargan-Lari A, Zhong M, et al. *In situ* collagen assembly for integrating microfabricated three-dimensional cell-seeded matrices. Nat Mater 2008;7:636–40.

[37] Tanaka H, Murphy CL, Murphy C, Kimura M, Kawai S, Polak JM. Chondrogenic differentiation of murine embryonic stem cells: effects of culture conditions and dexamethasone. J Cell Biochem 2004;93: 454–62.

[38] Gartner ZJ, Bertozzi CR. Programmed assembly of 3-dimensional microtissues with defined cellular connectivity. Proc Natl Acad Sci USA 2009;106:4606–10.

[39] Albrecht DR, Underhill GH, Wassermann TB, Sah RL, Bhatia SN. Probing the role of multicellular organization in three-dimensional microenvironments. Nat Methods 2006;3:369–75.

[40] Gray DS, Tan JL, Voldman J, Chen CS. Dielectrophoretic registration of living cells to a microelectrode array. Biosens Bioelectron 2004;19:771–80.

[41] Odde DJ, Renn MJ. Laser-guided direct writing of living cells. Biotechnol Bioeng 2000;67:312–8.

[42] Nahmias Y, Odde DJ. Micropatterning of living cells by laser-guided direct writing: application to fabrication of hepatic-endothelial sinusoid-like structures. Nat Protoc 2006;1:2288–96.

[43] Barron JA, Krizman DB, Ringeisen BR. Laser printing of single cells: statistical analysis, cell viability, and stress. Ann Biomed Eng 2005;33:121–30.

[44] Inaba R, Khademhosseini A, Suzuki H, Fukuda J. Electrochemical desorption of self-assembled monolayers for engineering cellular tissues. Biomaterials 2009;30:3573–9.

[45] Ringeisen BR, Othon CM, Barron JA, Young D, Spargo BJ. Jet-based methods to print living cells. Biotechnol J 2006;1:930–48.

[46] Chiou PY, Ohta AT, Wu MC. Massively parallel manipulation of single cells and microparticles using optical images. Nature 2005;436:370–2.

[47] Khademhosseini A, Langer R, Borenstein J, Vacanti JP. Microscale technologies for tissue engineering and biology. Proc Natl Acad Sci USA 2006;103:2480–7.

[48] Koyfman AY, Braun GB, Reich NO, Yousaf MN. Cell-targeted self-assembled DNA nanostructures. J Am Chem Soc 2009;131:14237–9.

[49] Dutta D, Pulsipher A, Luo W, Yousaf MN. Synthetic chemoselective rewiring of cell surfaces: generation of three-dimensional tissue structures. J Am Chem Soc 2011;133:8704–13.

[50] Young L, Jernigan RL, Covell DG. A role for surface hydrophobicity in protein–protein recognition. Protein Sci 1994;3:717–29.

[51] Ko IK, Kean TJ, Dennis JE. Targeting mesenchymal stem cells to activated endothelial cells. Biomaterials 2009;30:3702–10.

[52] Teramura Y, Kaneda Y, Iwata H. Islet-encapsulation in ultra-thin layer-by-layer membranes of poly (vinyl alcohol) anchored to poly(ethylene glycol)-lipids in the cell membrane. Biomaterials 2007;28: 4818–25.

[53] Teramura Y, Kaneda Y, Totani T, Iwata H. Behavior of synthetic polymers immobilized on a cell membrane. Biomaterials 2008;29:1345–55.

[54] Teramura Y, Iwata H. Islet encapsulation with living cells for improvement of biocompatibility. Biomaterials 2009;30:2270–5.

[55] Miura S, Teramura Y, Iwata H. Encapsulation of islets with ultra-thin polyion complex membrane through poly(ethylene glycol)-phospholipids anchored to cell membrane. Biomaterials 2006;27:5828–35.

[56] Totani T, Teramura Y, Iwata H. Immobilization of urokinase on the islet surface by amphiphilic poly (vinyl alcohol) that carries alkyl side chains. Biomaterials 2008;29:2878–83.

[57] Krol S, del Guerra S, Grupillo M, Diaspro A, Gliozzi A, Marchetti P. Multilayer nanoencapsulation. New approach for immune protection of human pancreatic islets. Nano Lett 2006;6:1933–9.

[58] Elbert DL, Herbert CB, Hubbell JA. Thin polymer layers formed by polyelectrolyte multilayer techniques on biological surfaces. Langmuir 1999;15:5355–62.

[59] Chanana M, Gliozzi A, Diaspro A, Chodnevskaja I, Huewel S, Moskalenko V, et al. Interaction of polyelectrolytes and their composites with living cells. Nano Lett 2005;5:2605–12.

[60] Germain M, Balaguer P, Nicolas JC, Lopez F, Esteve JP, Sukhorukov GB, et al. Protection of mammalian cell used in biosensors by coating with a polyelectrolyte shell. Biosens Bioelectron 2006;21:1566–73.
[61] Veerabadran NG, Goli PL, Stewart-Clark SS, Lvov YM, Mills DK. Nanoencapsulation of stem cells within polyelectrolyte multilayer shells. Macromol Biosci 2007;7:877–82.
[62] Decher G. Fuzzy nanoassemblies: toward layered polymeric multicomposites. Science 1997;277:1232–7.
[63] Stevens MM, George JH. Exploring and engineering the cell surface interface. Science 2005;310:1135–8.
[64] Daley GQ, Scadden DT. Prospects for stem cell-based therapy. Cell 2008;132:544–8.
[65] Orynbayeva Z, Kolusheva S, Groysman N, Gavrielov N, Lobel L, Jelinek R. Vaccinia virus interactions with the cell membrane studied by new chromatic vesicle and cell sensor assays. J Virol 2007;81:1140–7.
[66] Wilson JT, Krishnamurthy VR, Cui W, Qu Z, Chaikof EL. Noncovalent cell surface engineering with cationic graft copolymers. J Am Chem Soc 2009;131:18228–9.
[67] Mannix RJ, Kumar S, Cassiola F, Montola-Zavaya M, Feinstein E, Prentiss M, et al. Nanomagnetic actuation of receptor-mediated signal transduction. Nat Nanotechnol 2008;3:36–40.
[68] Yung CW, Fiering J, Mueller AJ, Ingber DE. Micromagnetic-microfluidic blood cleansing device. Lab Chip 2009;9:1171–7.

CHAPTER

6

Lipid-Mediated Cell Surface Engineering

Jaina M. Patel, Vincent F. Vartabedian and Periasamy Selvaraj

Department of Pathology and Laboratory Medicine, Emory University School of Medicine, Atlanta, GA, USA

OUTLINE

J.M. Karp & W. Zhao (Eds): Micro- and Nanoengineering of the Cell Surface.
DOI: http://dx.doi.org/10.1016/B978-1-45-573146-6.00006-4

6.1 INTRODUCTION

The term "cell" was first coined in 1665, in Robert Hooke's *Micrographia* [1]. Since then, breakthroughs in the field of cellular biology have given us not only a tremendous understanding of the cell and its components but also the ability to modify it through various means. In particular, our understanding of the cell membrane structure, its role in cellular function, molecules expressed on it, and their function have enabled us to engineer the cell surface for desired purposes.

In 1972, Singer and Nicholson [2] proposed the fluid mosaic model for cell membranes. The fluid mosaic model represents the cellular membrane as a fluid phospholipid bilayer decorated with proteins and oligosaccharides. This model also places an emphasis on the vast variety of structures that the phospholipid bilayer consists of, breaking the components down into three major categories: proteins, lipids, and oligosaccharides. Additionally, the lipids of the cell membrane exist as separate domains called lipid rafts. Lipid rafts are highly enriched with sphingolipids, complex glycosphingolipids, and cholesterol [3–6], and they play a vital role in cell membrane functions.

Cell membranes have been associated with a variety of cellular functions. Not only do they provide a barrier between interior cellular components and extracellular space, but cell membranes also are important for cell-to-cell communication and effector functions of the immune cell. Adhesion receptors found on the cell surface help to mediate these cellular communications as well as allow for targeted delivery of effector molecules. Additional cell surface receptors also bind to growth factors and cytokines allowing signals to be sent from outside the cell to inside further permitting communication between different cell types. These receptors selectively mediate communication between the extracellular and intracellular environment since not all molecules can pass easily through the cell membrane. These selective signals delivered by cell surface receptors determine cell survival, function, and in general, cell fate. Other cell surface expressed proteins, such as major histocompatibility complex class I (MHC I), also display internal peptide antigens on the cell surface facilitating immune surveillance by immune cells. This immune surveillance component of the cell membrane becomes important during infections, such as viral infections, as it is involved in detection and termination of sick and infected cells. Therefore, the cell membrane plays a vital role in determining the fate of the cell during disease state.

Consequently, cell membrane surfaces can be engineered to manipulate cell functions. Cell surface modifications serve as useful tools for a variety of applications such as targeting cells for drug delivery, activation, and amplification of immune cell responses, studying receptor functions, and development of cellular vaccines for tumor immunotherapy.

Genetic engineering techniques have been used to modify cell surfaces. These genetic engineering methods involve gene transfection in which the gene encoding the protein of interest is taken up within cells so that the cells translate and express the molecule of interest on the cell surface. Although this process is successful in leading to protein surface expression, many drawbacks exist: development of a culturable cell line or the availability of a large number of cells is needed; difficulty in transfecting many cell lines; the requirement of viral vectors; the amount of protein expressed by the cells is not controllable; and it is a time-consuming process. To counteract these drawbacks, surface engineering methods to exogenously place proteins on cell membranes have been developed. This phenomenon that decorates cell surfaces is known as "protein transfer" or "protein painting" [7].

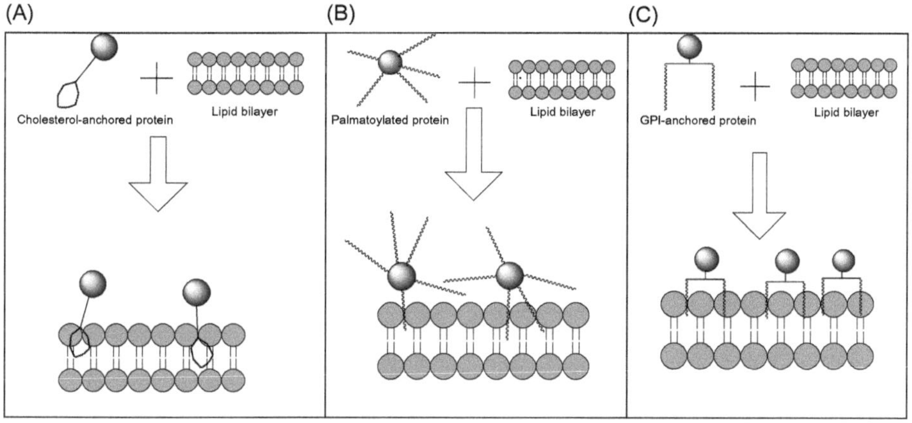

FIGURE 6.1 One-step protein transfer mediated by (A) cholesterol anchoring, (B) palmitoylation, or (C) GPI-anchoring of proteins.

One means of protein transfer, the focus of this chapter, involves attaching a lipid hydrophobic anchor to a protein and then exogenously adding the modified protein to the cells of interest. Such lipid-mediated surface engineering techniques allow for the anchored protein to be spontaneously incorporated into the cell membrane and into lipid rafts in particular [8], thus permitting the cell surface to be decorated by the protein in a controllable and time-efficient manner without any genetic modification.

Protein transfer to decorate cell surfaces can be divided into two main categories: the one-step and the two-step method. The one-step method consists of proteins that can naturally incorporate into lipid bilayers through hydrophobic chains (Figure 6.1), whereas the two-step protein transfer method requires incorporation of a scaffold onto the cell surface, which then in turn binds the protein of interest and thus allows for attachment to the cell surface (Figure 6.2). Thus, in the two-step protein transfer method, the protein of interest attaches indirectly to the cell surface. Of the one-step protein transfer methods, we will discuss cholesterol anchoring, palmitoylation modifications, glycosyl phosphatidylinositol (GPI)-anchoring, and use of hydrophobic sequences to anchor proteins on the cell membrane by protein transfer.

6.2 ONE-STEP PROTEIN TRANSFER

6.2.1 Protein Transfer Using Cholesterol-Modified Proteins

Cholesterol is an integral component of lipid rafts. With the depletion of cholesterol from cells, lipid raft domains disintegrate [6]. Cholesterol promotes lipid raft formation due to its ability to tightly pack with sphingolipids; however, other modified sterol structures that cannot tightly pack with sphingolipids disrupt lipid raft formations [9].

6.2.1.1 Exogenously Added Cholesterol Incorporates into Membranes

Cholesterol-tethering of biomolecules has been used to paint the surface of cells. Exchange between cholesterol derivatives found in lipid suspensions with those in erythrocyte ghost membranes have been detected upon incubation

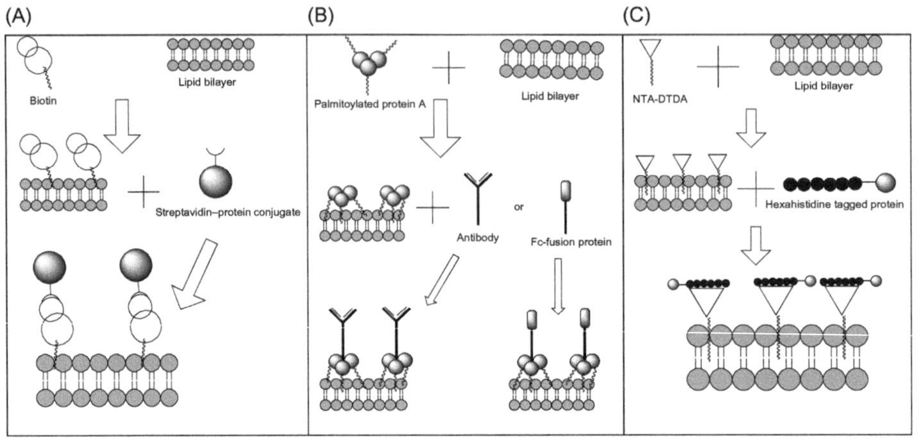

FIGURE 6.2 Two-step protein transfer mediated by (A) biotin–streptavidin, (B) pal-protein A, or (C) lipid-metal chelators.

for 24 h [10,11]. In order to further investigate the "spur" red cell phenomenon in which red blood cells from patients with spur cell anemia uptake cholesterol found in the serum, Cooper et al. [12] attempted to artificially reproduce the environment. They found that when phospholipids were sonicated with cholesterol, the cholesterol-to-phospholipid (C/P) molar ratio affected uptake of cholesterol in red cells that were incubated with the mixture. When the C/P ratio was less than 1.0, cholesterol was depleted from the red cells and when the C/P ratio was equal to 1.0, no change in cholesterol content in red cells was seen. However, when the C/P ratio was greater than 1.0, cholesterol was deposited into normal human red cell membranes and the spur cell phenomenon was detected [10–12]. As the C/P ratio was increased, increased incorporation of cholesterol in red cells was also detected in a dose-dependent manner over 24 h [12]. These results show that cholesterol can be exogenously added to incorporate into cell membranes.

6.2.1.2 Cholesterol-Tethering of Biomolecules to Membrane Surfaces

Due to the ease with which exogenously added cholesterol can incorporate into cell membranes, a method of membrane painting to modify the cell surface with biomolecules was established (Figure 6.1A). This method involves the use of cholesterol derivatives, such as *N*-(3β)-cholesterylglycine, that are tethered to biomolecules [13]. The addition of amino groups on cholesterol derivatives that decrease the cholesterol–phospholipid affinity in the donor membranes leads to enhanced incorporation of the cholesterol derivative in the accepting cell membrane vesicles [14]. Conjugation of a biomolecule, such as biotin or epitopes for anti-HA or anti-FLAG antibodies, to the cholesterol derivatives creates an amphiphilic complex. The amphiphilic nature allows for insertion of the cholesterol derivative of the biomolecule into lipid rafts, whereas the polar segment of the biomolecule remains exposed on the cell surface. Upon incubation of the cholesterol-tethered molecules for 1 h at 37°C in RPMI medium with 10% fetal bovine serum (FBS), stable dose-dependent incorporation is detected within lipid rafts of cellular membranes [13].

6.2.1.3 Application of Cholesterol-Tethering to Create Synthetic Receptors

Modification of cell surfaces by cholesterol-tethering of biomolecules allows for delivery of drugs or antitumor agents into target cells. This delivery system occurs when a cholesterol-tethered receptor bound to its ligand is internalized. Beforehand, a cholesterol derivative is tethered to the biomolecule receptor, which is then allowed to incorporate into lipid rafts of cells for 1 h. When the surrogate receptor (SR)-incorporated cells are incubated with ligand, nearly 100% of the cells bind to the ligand in a matter of 10 min. Upon receptor–ligand binding, internalization of the receptor–ligand complex was initiated, and as the concentration of SR added to the cells increased, the amount of internalization also increased upon ligand addition in a dose-dependent manner [13]. Furthermore, Hussey et al. [15] showed that internalization occurred partially within 10 min of incubation and completely within 4 h of incubation through a clathrin-mediated endocytosis pathway.

The length of the tethering linker between the cholesterol derivative and the biomolecule is imperative [16]. A longer linker allows for full insertion of the cholesterol derivative into the membrane, whereas the shorter linker prevents this. An 11-atom linker resulted in membrane insertion of a cholesterol derivative bound to biotin that was capable of internalizing its ligand, streptavidin, within 10 min [16].

Using this approach of creating synthetic cell surface receptors, Martin and Peterson [13] showed that tethering an HA epitope to a cholesterol derivative allowed for surface expression of the epitope on Jurkat cells. An anti-HA antibody complexed with a protein A-β-galactosidase conjugate bound the surface expressed HA epitope resulting in engulfment of the protein A-β-galactosidase complex. Furthermore, when the substrate to β-galactosidase was added to these Jurkat cells, fluorescence was detected intracellularly within 75% of the cells. These experiments suggest that not only can cholesterol-derived SRs be used to modify the cell surface, but also the added protein remains functionally active even when endocytosed within the cell. Similar binding and internalization studies were conducted with cholesterol-derived SRs bound to either FLAG epitopes or streptavidin ligands, such as biotin [13,16].

Cholesterol-derived SRs can be added to any cells that contain lipid rafts. Many multidrug-resistant tumors have been shown to produce increased levels of lipid raft components [16]. Therefore, the addition of cholesterol-derived SRs to these multidrug-resistant cancer cells via intratumoral injections should readily lead to the uptake and expression of the receptor on the tumor cell surfaces that contain increased lipid raft components. In this manner, many drugs or toxins can be preferentially targeted to these cancer cells as opposed to neighboring healthy cells that have fewer lipid raft components in comparison.

6.2.1.4 *Advantages/Disadvantages of Cholesterol-Tethering*

Cholesterol-tethering of biomolecules onto cell membrane surfaces provides for a simple method to express proteins on cell surfaces due to the ease with which cholesterol and cholesterol derivatives incorporate into the cell membranes. Since lipid rafts are commonly found on cells, albeit at different levels, incorporation of cholesterol-tethered biomolecules could occur on almost any cell surface. Through the production of SRs, this method can also lead to the engulfment of high molecular weight proteins, such as enzymes like A-β-galactosidase, toxins, or drugs, that remain functionally active within the cells, allowing for cell targeting via this mechanism [13,16].

However, the production of these cholesterol-tethered biomolecules is not an easy task; it involves many chemical additions and the linkage between the biomolecule and cholesterol derivative is vital for incorporation and ligand binding. Moreover, there is a possibility that lipids present in circulation may bind to these administered conjugates and block incorporation onto the defined target cells.

6.2.2 Surface Engineering with Palmitoylated Proteins

Palmitoylation is a process that involves the lipid modification of proteins in which a hydrophobic fatty acid palmitate group is added covalently to the protein of choice [17]. In the human body, palmitoyl groups are commonly found attached covalently to neurologic proteins and also play key roles in signal transduction of G protein and G protein receptors [18]. The unique reversibility of palmitoylation allows this process to play a role in not only the activation and deactivation of proteins [17,18], but also in protein localization to membranes and in interactions with other proteins [17]. In addition to its importance among biologic systems, palmitoylation is also gaining relevance as a protein-anchoring technique used for protein transfer, engineering, and vaccine creation.

In the body, palmitoylation often occurs in the cytoplasm after another fatty acid, such as a myristoyl group, has already been added [17]. However, procedures to modify proteins with palmitate *in vitro* also have been established. It has been shown that auto-palmitoylation of G proteins can be performed stoichiometrically *in vitro* via incubation with palmitoyl-CoA [19]. An additional preparation method involves using the reaction between the *N*-hydroxysuccinimide ester of palmitic acid with the protein of interest, forming a palmitate-derivatized protein [20]. Another commonly used method involves palmitoylation with palmitoyl chloride [21]. Enhancing the hydrophobicity of proteins by the addition of fatty acid palmitate groups allows for insertion of the exogenously added protein onto cell membranes (Figure 6.1B). The ability to create a more hydrophobic protein easily by palmitoylation *in vitro* makes palmitoylation an incredibly appealing technique for modifying cell surfaces by protein transfer.

Recently, human IL-2, a cytokine commonly used in anticancer therapy, was palmitoylated and protein transferred to incorporate onto cytotoxic T cells (CTLs) [22]. These protein-transferred CTLs were then injected into mice along with EG7-OVA lymphoma cells. Mice that received protein-transferred CTLs had decreased tumor growth when compared to mice that received unmodified CTLs. Furthermore, intratumoral injection of palmitoylated-IL-2 in mice with melanoma led to increased overall survival, decreased metastasis to the lungs and increased levels of splenic and tumor infiltrating IFN-γ producing T cells compared to mice that received unmodified IL-2 intratumorally [22]. Therefore, the use of palmitoylated cytokines, such as IL-2, has the clinical potential to be used as an antitumor therapy.

6.2.2.1 Palmitolyation to Create Surrogate Cell Surface Receptors

Originally, artificial receptors were incorporated onto B cells by palmitoylating antibodies. Peacock et al. [20] showed that antibodies can be palmitoylated and subsequently inserted into a cell membrane. These membrane-bound antibodies can act as receptors, and upon binding to target antigen can induce an immune response by delivering signals to the decorated cells. In the study, protein 315, a secreted antibody against dinitrophenyl (DNP) lacking the Fc region, was palmitoylated before being incubated with murine splenocytes at 37°C. The palmitoylated antibody spontaneously incorporated onto splenocyte cell membranes in cell media, specifically via the palmitate groups since nonpalmitoylated proteins did not incorporate. Incorporation of the palmitoylated antibody on cell membranes was concentration and time dependent, and the incorporated palmitoylated antibody was still able to bind the

DNP antigen and induce a polyclonal antibody response specific to DNP [20]. However, the degree of palmitoylation was also important for functional antigen binding. Increasing palmitoylation led to decreased ability of the incorporated antibody to bind antigen [20]. Furthermore, through fluorescence photobleaching recovery (FPR), palmitoylated-antibody receptors that incorporated exogenously onto B cells were found to have similar membrane properties as surface immunoglobulin (Ig) expressed naturally on B cells [23]. Therefore, exogenously decorating B cells with palmitoylated antibodies could provide insight into how surface antibodies are able to exhibit an antigen-dependent immune response.

This antibody-anchoring technology was further developed to create "SRs" on the cell surface. The establishment of SRs allows for the creation of "custom-designed killer cells" in which palmitoylated antibodies incorporated onto cell membranes provide specificity for target cells [24]. Either palmitoylated F(ab′)$_2$ against mouse Ig or intact anti-TNP antibody was incorporated onto the membrane of nylon wool nonadherent spleen cells from mice. The antimouse Ig decorated cells were then incubated with syngeneic B cells stimulated with LPS, while the spleen cells decorated with anti-TNP antibody were incubated with TNP-modified EL4 T-lymphoma cells. Both sets of decorated cells demonstrated the ability to lyse their targets, while their undecorated counterparts did not. Furthermore, neither set of decorated cells exhibited the ability to lyse cells not displaying the antigen specific for their SR, demonstrating the specificity of the SR. Of particular interest, this SR-mediated cellular cytotoxicity (SRMCC) occurred even when the SR did not have an Fc region, suggesting that SRMCC is independent of Fc receptor interactions [25]. Although SRMCC was not detected when NK cells were depleted by anti-asialo GM$_1$, SRMCC could induce lysis of NK-resistant cells. This suggests that NK cells, while not the sole effector cells, play a key role in SRMCC [24]. Thus, incorporation of palmitoylated SRs onto cells provides a means to engineer killer cell surfaces to express receptors that can specifically target cells expressing particular ligands.

6.2.3 Surface Engineering with GPI-Anchored Proteins

Proteins are expressed on the membrane of a cell through multiple mechanisms. Aside from commonly found transmembrane anchored proteins, lipid anchored proteins also exist. Lipid anchoring of proteins via glycophosphatidyl inositol (GPI) is predicted to be found in 460 proteins [26]. The GPI-anchor which anchors proteins to the outer leaflet of the cell membrane does not span the entire lipid bilayer as transmembrane proteins do [3]. This anchor consists of two or three hydrophobic fatty acid chains that insert into the outer leaflet of the lipid bilayer exposing the anchored proteins to the outside environment. The core structure of the GPI-anchor is conserved in eukaryotes and consists of a phosphoethanolamine group, followed by a glycan region, followed by a phospholipid tail. The phosphoethanolamine group is covalently attached to the C-terminal end of the GPI-anchored protein (GPI-AP) by an amide bond. The conserved GPI backbone structure is "EtN-P-Man-Man-(EtN-P)Man-GlcN-PI," where EtN is ethanolamine, P is phosphate, Man is mannose, GlcN is glucosamine, and PI is phosphatidylinositol. The glycan region can vary between species and proteins via side chain modifications; however, the core backbone is conserved [27,28]. Inositol attaches the glycan region to the phospholipid tail, which consists of two long fatty acid chains that are either saturated or unsaturated, and vary in length. This phospholipid tail is responsible for inserting into the lipid bilayer and anchoring the GPI-AP to the cell surface. In addition to serving as a membrane anchor, other functions of the GPI-anchor include signal transduction, membrane trafficking, and targeting proteins to lipid rafts [27].

The discovery and purification of the enzyme phosphatidylinositol phospholipase C (PI-PLC) from the bacteria, *Bacillus cereus*, in 1976 [29] further led to the discovery of the GPI-anchor as a means for membrane-anchoring proteins. PI-PLC treatment led to the release of many proteins from tissues, thus suggesting that these proteins are anchored via a phosphatidylinositol group. Moreover, phase separation with Triton X-114 (TX-114) can be used to characterize GPI-APs [30]. At 0°C, the nonionic detergent TX-114 forms a homogenous mixture with GPI-APs and an aqueous solution; however at 37°C, TX-114 separates away from the aqueous phase. GPI-APs are found in the TX-114 phase, whereas other transmembrane and soluble proteins are found in the aqueous phase; therefore, TX-114-mediated phase separation also verifies the presence of a GPI-anchor [30]. Upon further characterization, the structure of the GPI-anchor found in many GPI-APs became evident. However, some GPI-APs are resistant to PI-PLC and phospholipase D cleavage. These GPI-APs contain an extra fatty acid chain (palmitate) attached to the inositol ring, thus they have a total of three fatty acid chains anchoring the protein to the cell membrane [31].

6.2.3.1 *Expression of Proteins on the Cell Surface by Protein Transfer of GPI-APs*

Purified GPI-APs contain the unique property of being able to reincorporate into cell membranes and still retain their biological activity. In 1984, Medof et al. [32] showed that the purified GPI-AP, DAF, was able to reincorporate into erythrocyte membranes by simply incubating purified DAF with erythrocytes. This reconstitution of DAF prevented complement protein C4b2a activity on the erythrocytes in a DAF dose-dependent manner. Furthermore, in the acquired disorder paroxysmal nocturnal hemoglobinuria (PNH), the absence of DAF in erythrocytes from PNH patients leads the erythrocytes to be susceptible to complement-induced lysis. However, this lysis is not detected when exogenous DAF is reconstituted back into erythrocyte membranes [33]. Additionally, in PNH patients, another complement regulatory protein, homologous restriction factor (HRF), later known as CD59, is also absent from erythrocyte membranes. Incubation of purified radiolabeled HRF with erythrocytes led to insertion of HRF into erythrocyte membranes and the inserted HRF remained biologically functional in preventing complement lysis by C5b-9 *in vitro* [34–38]. Oligodendrocytes that also lacked CD59 became resistant to complement-induced lysis after exogenously incorporating purified CD59 into their membranes as well [39]. Another complement regulatory protein which behaved similarly to DAF and CD59 is C8-binding protein (C8bp). In a comparable manner, C8bp, which is also not found in erythrocytes obtained from PNH patients, can be inserted into erythrocyte membranes exogenously after purification of the protein and again inhibit complement-induced lysis of erythrocytes [40].

Additionally, adhesion molecules such as lymphocyte function-associated antigen 3 (LFA-3) are also found to be deficient on erythrocytes obtained from PNH patients [41]. LFA-3 is expressed as a GPI-anchored form in erythrocytes, whereas nucleated cells express both GPI-anchored forms and polypeptide anchored forms [42]. Consequently, partial deficiency of LFA-3 was observed in nucleated cells from PNH patients. The interaction between LFA-3 on erythrocytes and its receptor CD2 on T cells induces formation of T cell rosettes [43]; however, erythrocytes from PNH patients did not form rosettes upon incubation with CD2 expressing Jurkat cells. Interestingly, when exogenous purified GPI-anchored LFA-3 was added back to erythrocytes from PNH patients *in vitro*, rosettes were formed, suggesting that protein transfer of exogenous LFA-3 leads to stable as well as functional LFA-3 expression on the cell surface [41]. Also, exogenous incorporation of purified CD16B (Fcγ receptor III), another naturally occurring GPI-AP, onto both a human T cell leukemia and B lymphoblastoid cell lines *in vitro* demonstrated that protein transferred CD16B on nucleated cells can still retain its function and mediate endocytosis of its ligand [44]. Another GPI-AP, Thy-1, was able to exogenously incorporate onto murine Thy-1-negative cells *in vitro*. After this protein transfer, staining of the exogenously incorporated Thy-1 was similar to staining patterns of endogenously expressed Thy-1 and the lateral mobility of both endogenously and exogenously inserted Thy-1 was similar as determined by fluorescence recovery after photobleaching (FRAP) [45]. Furthermore, even nonmammalian GPI-APs, such as the membrane form of variant surface glycoprotein (mfVSG) expressed on the parasite *Trypanosoma brucei*, can also be purified and exogenously incorporated onto cell surfaces with similar lateral mobility as endogenously expressed mfVSG [46]. However, PI-PLC treatment of purified GPI-APs decreases the insertion levels of these proteins onto cell membranes suggesting that protein transfer is due to the presence of the GPI-anchor [36,41,45]. All of the above examples show that purified GPI-APs can be incorporated via the GPI-anchor into cell membranes by simply exogenously adding the purified GPI-APs and that after incorporation onto cell membranes, the biological function and lateral mobility of the protein remains similar to that of the constitutively expressed protein counterpart.

6.2.3.2 *Characteristics of GPI-AP Protein Transfer*

Protein transfer of GPI-APs onto cell membranes (Figure 6.1C) occurs in a concentration-dependent manner. As higher concentrations of GPI-APs are added to the membranes, increased incorporation occurs [32,44,47]. However, a saturation point is expected to occur in which all GPI-AP sites on the membrane will be occupied and no more GPI-APs can incorporate by protein transfer without disrupting the integrity of the membranes. GPI-anchor-mediated protein transfer also results in surface expression on 100% of the cells incubated, as shown by flow cytometry [48]. Furthermore, GPI-anchor-mediated protein transfer is time dependent. As the time for protein transfer is increased, increased expression of GPI-APs on membrane surfaces occurs. Incorporation was detected in as little as 20 min postincubation of GPI-APs with membranes and incorporation levels increased with incubation time for up to at least 4 h [32,44,47]. Medof et al. [32] noticed that incorporation of radiolabeled DAF into erythrocyte membranes occurred best at 37°C followed by 20°C with no detectable DAF incorporation occurring at 0°C. Poloso et al. [47] detected a similar phenomenon in which protein transfer occurs best on cell membrane vesicles at 37°C followed by 25°C followed by 4°C. These observations suggest the necessity of membrane mobility for the incorporation of GPI-APs. Plasma membranes are most fluid at 37°C compared to lower temperatures; therefore, the dynamics between GPI-AP micelles and the lipid bilayer occurs more readily at 37°C [49].

Optimum incorporation also occurs without the presence of certain serum proteins. Incorporation of GPI-APs is inhibited by the presence of human serum lipoproteins, such as HDL and LDL, and fatty-acid-binding proteins, such as bovine serum albumin (BSA) and orosomucoid [32,41,44]. High concentrations of ovalbumin on the other hand do not affect incorporation of GPI-APs onto cell membranes [32,44]. Once incorporated onto the cell surface by protein transfer, GPI-APs remain stably integrated into the cell membrane of erythrocytes and cannot be removed by washing even with high salt concentrations [32]. After protein transfer, GPI-APs can however be removed by solubilization with detergents, such as NP-40. However, incorporation onto live or irradiated nucleated cells leads to loss of incorporated proteins under culture conditions [48].

Protein transfer leads to the incorporation of exogenously added GPI-APs into lipid rafts of cell membranes [8,50]. After Triton X-100 lysis of cells that have been protein transferred with GPI-APs, the GPI-APs are detected in detergent insoluble microdomains as are other lipid raft specific proteins [50]. Furthermore, enhanced incorporation of green fluorescent protein (GFP)-labeled GPI-anchors into lipid rafts was also associated with the presence of lipid raft-associated molecules (cholesterol, sphingomyelin, and dipalmitoyl-phosphatidylethanolamine) further proving that protein-transferred GPI-APs associate with lipid raft structures [8]. Upon initial incorporation of GPI-APs onto cell membranes by protein transfer, the GPI-APs are widely distributed throughout the cell membrane as observed by confocal microscopy [51]. However, hours after incorporation, the GPI-APs localize into lipid raft microdomains as detected by confocal microscopy or by observing association of GPI-APs with lipid raft domains after lysis with Triton X-100 [50,51].

Similar to the erythrocyte-incorporated proteins described earlier, these protein-transferred GPI-APs expressed on nucleated cells and cell membranes retain their functional biological activity in comparison to endogenously expressed forms of the proteins [44,47,48,52]. Not only do exogenously incorporated GPI-APs retain the functional extracellular activity of the GPI-AP, but they can also retain functional intracellular activity [50,51]. Exogenously incorporated GPI-APs onto cells led to association of the GPI-AP with active intracellular kinases [50,51]. Upon antibody crosslinking of exogenously incorporated GPI-B7-1, DAF, or CD59, phosphorylation induced by active kinase activity was detected [50,51] as well as an induced transient Ca^{2+} flux signal [51].

It has also been observed that cells genetically transfected to express the GPI-anchored form of the cytokine, GM-CSF, not only expressed functional surface-bound GM-CSF that was susceptible to PI-PLC cleavage, but GM-CSF was also detected in the supernatant in a soluble form. This soluble GM-CSF was released by either shedding or proteolytic cleavage and was not released by secretion [53]. Similarly, IL-2 was detected in the supernatant of cultured cell lines transfected to express GPI-IL-2; however GPI-IL-12 was not secreted when expressed on CHO cells [54,55]. It is not clear as to how only IL-2 was released; however, proteins differ in protease cleavage sites and also may differ in shedding capabilities. Therefore, this dual ability of cytokines to not only be expressed on cell surfaces by a GPI-anchor but also to be shed into the extracellular space could lead to a cytokine gradient allowing for the infiltration and activation of immune cells.

More importantly, after protein transfer of GPI-APs onto the lipid bilayers, the newly incorporated GPI-APs remain functionally active. Although it has been stated that naturally occurring GPI-APs can be reincorporated exogenously by protein transfer and retain their biological functional activity, even recombinant GPI-APs, such as GPI-B7-1 and GPI-IL-12, after protein transfer retain their biological functional activity [47,48,52,56,57]. Protein-transferred GPI-B7-1 and GPI-IL-12 incorporated on tumor cells and tumor membrane vesicles (TMVs) retained their ability to induce T cell proliferation.

6.2.3.3 *Advantages and Limitations of GPI-AP Protein Transfer*

Modification of membranes and cell surfaces provides a wide range of application possibilities. Protein transfer allows for membrane modification in an easy, quick, and efficient manner compared to the traditional genetic engineering approaches. (1) With protein transfer, purified GPI-APs incorporate onto cell membranes in a matter of minutes with maximal incorporation occurring around 2–4 h [47], whereas genetic engineering approaches take much longer from the order of days to months to establish cultured transfected cell lines. (2) Genetic engineering makes modifying some cell lines extremely difficult especially when primary cell lines, such as bone marrow progenitors and primary cultures, are hard to transfect. Also, protein transfer can be performed on small cell numbers and does not require cell division to occur. (3) Using protein transfer, the amount of incorporation can be finely controlled by adjusting the amount of exogenously added GPI-APs. (4) Finally, multiple GPI-APs can be incorporated onto the same cell membranes by protein transfer, whereas double and triple transfections by genetic engineering are more difficult to attain [58].

However, membrane anchors are known to affect the functionality of external domains of some proteins [59]. It has been noted that PI-PLC treatment of some GPI-APs renders some antibodies nonresponsive to the protein even if the antibody binds the GPI-AP in the presence of the GPI-anchor [60–62]. Therefore, the addition of a GPI-anchor can affect the structural conformation and function of proteins. Also, if important antibody epitopes are found at the C-terminal end of a protein, the addition of the GPI-anchor at the C-terminus may alter the antigenicity of the protein. Similarly, the GPI-anchor attaches the protein close to the membrane with less than 10–14 Å space between the protein and the membrane surface as determined by fluorescence resonance energy transfer (FRET) [62,63]. This close proximity to the membrane surface may alter accessibility of important antibody-binding sites. Therefore, it is imperative that the functionality of each GPI-AP be tested before use. A bottleneck of the protein transfer method using GPI-APs is the purification of GPI-APs. Purification using immunoaffinity chromatography is the best defined manner for obtaining pure fractions of GPI-APs; however, optimization of the process for each individual GPI-AP is time-consuming. Despite these disadvantages, when optimal purification of GPI-APs that retain their biological functionality is attainable, protein-transferred GPI-APs provide an efficient method to controllably modify membrane surfaces.

6.2.3.4 Protein Transfer of Antigen-Presenting Cells with GPI-Anchored MHC Complexes

During viral infections, CTL responses are important to combat infected cells. For a CTL to become activated, the T cell has to be engaged with MHC I that expresses the viral peptide; however, many viruses downmodulate the expression of MHC molecules on the cell surface to escape from the immune system. Therefore, Huang et al. [64] constituted a method to control the amount of MHC–peptide complexes expressed on a cell by protein transfer. Their model system utilized a HLA-A2.1 MHC complexed with a hepatitis B virus (HBV) peptide that induces a dominant T cell response. They constructed a GPI-anchored form of HLA-A2.1 by attaching a GPI-anchored signal sequence to the heavy chain and contransfected Schneider S2 *Drosophila melanogaster* cells with the beta 2 microglobulin gene. Upon purification of this HLA-A2:GPI, they were able to show that the MHC complex incorporated into HLA-A-negative and HLA-B-reduced cells in a time, temperature, and concentration-dependent manner. They were able to detect incorporation by flow cytometry within 1 min of incubation with cells; however, maximal incorporation occurred after 1 h incubation at 37°C.

Additionally, incorporation was inhibited by the presence of 10% FBS, but if FBS was added after protein transfer, then no affect on cell surface expression of the MHC complex was noticed. Peptide was loaded onto the GPI-MHC molecules at 4°C overnight before protein transfer leading to stable expression of the MHC−peptide complex and MHC−peptide specific cell lysis by CTLs after protein transfer. These results showed that protein transfer with GPI-anchored MHC complexes could be used to express desired quantities of specific MHC complexes on cell surfaces. Thus, antigen-presenting cells (APCs) can also be decorated with desired MHC−peptide complexes allowing further studies of APC and T cell interactions to elucidate specific peptide-mediated T cell activation and tolerance.

6.2.3.5 Protein Transfer of GPI-APs in Tumor Immunotherapy

Another application of protein transfer using GPI-APs is in tumor immunotherapy. Cancer cells have the ability to continuously evade the immune system; therefore, finding a way to activate the immune system against these cells is key to combating tumors. One method in which tumor cells evade the immune system is through the downregulation of costimulatory molecules, such as B7-1, on tumor cell surfaces. This downregulation renders tumor-specific T cells anergic and nonresponsive to the tumor. Therefore, tumor cells have been transfected to express B7-1 in order to improve the immunogenicity of cell-based therapeutic vaccines. However, in order to transfect tumor cells with immune stimulatory molecules, the establishment of primary tumor cells lines from patients' tumor tissue is required. However, tumor cells derived from patients are extremely difficult to culture; therefore, McHugh et al. [52] proposed the use of protein transfer to express GPI-anchored B7-1 on tumor cells or on TMVs. The TMVs that are derived from the homogenization of tumor cells still contain tumor-associated antigens along with a lipid bilayer, thus they can be modified with GPI-APs by protein transfer. It was shown that TMVs derived from EG7 tumors, a murine T cell lymphoma line, could be protein transferred to express GPI-B7-1 on the surface. They found that these protein-transfer-modified TMVs could protect against challenge with EG7 tumor cells. Furthermore, B7-1-modified TMVs led to T cell proliferation and cytolytic T cell activity in vaccinated mice *ex vivo* against the parental tumor [52].

Additionally, protein transfer was also shown to be conducted on TMVs derived from surgically removed human tumor tissue. These TMVs were stably incorporated with GPI-B7-1 by protein transfer in a concentration-, time-, and temperature-dependent manner. These incorporated TMVs were functionally active in stimulating T cell proliferation [47].

Many cytokine therapies, such as the administration of the cytokine IL-2, have been used as a tumor immunotherapy in order to activate immune cells against the tumor [65]. Although cytokine administration has worked in activating the immune system, systemic toxicity also results which impedes the success of the treatment. Therefore, human TMVs modified by protein transfer to express GPI-cytokines were used as a method of delivery of not only the cytokine in a less toxic manner, but also tumor-associated antigens [47,57]. Using protein transfer, purified GPI-anchored IL-12 was incorporated onto human and murine TMVs from mammary carcinomas, renal cell carcinomas, melanoma, and B cell lymphoma cultured cell lines, and the resulting TMVs expressed functionally active IL-12 that could induce T cell proliferation and IFN-γ secretion. IL-12 also worked well in combination with CD80 or CD40 to enhance T cell proliferation and IFN-γ secretion [57].

Taken together, protein transfer of costimulatory molecules and cytokines that have been converted into GPI-anchored forms has led to a promising approach for inducing antitumor immunity. Because GPI-anchored cytokines are attached to the membrane surface, a slow-release depot of the cytokine is ensured circumventing the systemic toxicity associated with administration of soluble cytokines.

6.2.4 Protein Transfer Using Proteins with Hydrophobic Domains

Another type of one-step protein transfer involves the incorporation of hydrophobic amino acid sequences into the lipid bilayer of cell membranes. Using this protein transfer method, the coding region of the superantigen toxic shock syndrome toxin-1 (TSST1) was attached to the hydrophobic sequence of the transmembrane portion of the proto-oncogene c-erb-B-2, forming a TSST1-TM molecule [66]. This TSST1-TM also had a histidine-tag attached to the N-terminal end for purification using a nickel-agarose column. Flow cytometry analysis showed that purified TSST1-TM incorporated onto multiple tumor cell line membranes after incubation for 4 h at 37°C in cell medium that contained 2.5% fetal calf serum. Furthermore, modified tumor cells were able to induce *in vitro* human peripheral blood lymphocyte proliferation and mice vaccinated with TSST1-TM-modified mitomycin C-treated tumor cells had decreased tumor growth after challenge with parental tumor cells [66]. Therefore, TSST1-TM incorporated onto tumor cells by protein transfer remained functional and exhibited antitumor responses.

As opposed to using GPI-anchor attachments to incorporate proteins onto cell membranes, this method that attaches hydrophobic sequences to proteins allows for the expression of modified prokaryotic proteins. Attachment of GPI-anchors to proteins requires a eukaryotic processing system, whereas this method does not [67]. Therefore, prokaryotic proteins, such as TSST1, can be modified and purified using a prokaryotic system without affecting the protein conformation and processing of the original molecule. This allows for functional production and incorporation of the modified prokaryotic protein.

6.3 TWO-STEP PROTEIN TRANSFER

Two-step protein transfer can be implemented by using many different techniques, such as biotin–streptavidin, palmitoylated protein A (pal-protein A) or through the use of metal chelators (Figure 6.2).

6.3.1 Biotin–Streptavidin-Mediated Two-Step Protein Transfer

The biotin–streptavidin system has been implemented in many different ways in detection, purification, immunoassays, nucleic acid hybridization assays, DNA sequencing, etc. [68]. Biotin, or vitamin H, is found as a cofactor in many carboxylation reactions. It binds tightly to its ligand, avidin or streptavidin, with a high affinity constant [68–70]. Avidin was isolated from egg whites, whereas streptavidin was found in the bacteria *Streptomyces* [71]; however, they both form tetramers [68,69]. Since streptavidin does not bind specifically to many molecules other than biotin, and since streptavidin and biotin are highly resistant to dissociation, streptavidin is widely used to bind biotinylated proteins [68]. Using this biotin–streptavidin system, the protein transfer technology ProtEx was developed.

ProtEx is a technology designed to exogenously decorate the surface of cells through the use of the biotin–streptavidin system [70,72,73]. In this method (Figure 6.2A), cells are first biotinylated with Sulfo-NHS-LC-Biotin, which results in the attachment of biotin onto the cell membrane. Then the protein of choice, which is conjugated with core streptavidin (SA), is added to the biotinylated cells, allowing for stable interaction between streptavidin and biotin to occur and thus resulting in cell surface expression of the protein of choice [70,72,73].

An advantage of using the ProtEx system is that cell surfaces can be engineered within 2 h to stably express the protein of choice without toxicity to the cell. Due to the ability of biotin to remain stably incorporated on the cell surface for weeks [70], the ProtEx technology also allows for stable cell surface expression of the protein of choice. The extracellular domain of CD80 attached to SA (CD80-SA) was found to be stably incorporated onto biotinylated cells with a half-life of over 10–12 days *in vitro* and *in vivo* [73], whereas the kinetics of the expression of the extracellular domain of Fas ligand bound to SA (SA-FasL) was cell dependent [70]. Furthermore, the ProtEx technology allows for the protein of choice to exist as tetramers due to the nature of streptavidin which may play a beneficial role in some receptor–ligand interactions. However, for ProtEx to work, the addition of streptavidin to the protein of choice must not affect the functionality of the protein or the ability of streptavidin to bind to biotin. This restriction may limit the variety of proteins that may be used with the ProtEx technology.

Many applications of the ProtEx technology have been utilized in both autoimmune or transplantation settings, and in tumor settings. During immune homeostasis, Fas-FasL interactions play a strong role in maintaining a healthy immune environment that can fight invading organisms without being detrimental to the host. However, during many graft transplants, the immune system identifies the graft as a foreign agent and thus rejects the graft. In order to prevent graft rejection, Yolcu et al. [72] showed that by using biotin–streptavidin two-step protein transfer, FasL can be expressed on donor splenocytes and aid in preventing allogeneic pancreatic islet graft rejection. SA-FasL was added to biotinylated splenocytes and cotransplanted with islet cells into mice. Mice that received splenocytes decorated with SA-FasL had prolonged survival compared to mice that received unmodified splenocytes. The suggested mechanism behind this method was that interaction between alloreactive T cells and FasL-decorated splenocytes led to apoptosis of the Fas-expressing alloreactive T cells. Therefore, decoration of cells using SA-FasL by the biotin–streptavidin system can be used in preventing graft rejection. Bone marrow cells have also been decorated with SA-FasL in order to induce tolerance of grafts [74]. To expand the usage of SA-FasL, killer regulatory T cells were also developed using the biotin–streptavidin two-step protein transfer method. In this method, regulatory T cells decorated with SA-FasL, termed killer Tregs, were able to postpone autoimmune diseases such as diabetes [75], inflammatory colitis [76], and graft-versus-host disease [77].

On the other hand, in a tumor setting, the ProtEx technology has been used to decorate tumor cells with CD80-SA. Since the costimulation molecule, CD80 or B7-1, is often absent from many tumor cells, tumors can evade the immune system. Therefore, addition of the costimulation molecule on the tumor surface activates the immune system to recognize and attack tumor cells. Singh et al. [73] decorated human tumor cells with CD80-SA; these CD80-decorated tumor cells were then able to activate T cells to recognize and kill unmodified tumor cells in an antigen-specific manner. Furthermore, this two-step protein transfer method was used to induce protection against a B cell lymphoma line, A20, in mice. Mice were injected with A20 cells followed by irradiated A20 cells decorated with CD80-SA. These mice developed protection against tumor growth compared to control mice that received unmodified irradiated A20 cells [73]. Therefore, this ProtEx technology to decorate the cell surface through the use of the biotin–streptavidin system has proved to be an efficient method in preventing autoimmunity and graft rejection, as well as in preventing tumor growth.

6.3.2 Palmitolylated Protein A-Mediated Two-Step Protein Transfer

Although creating an SR on cells using palmitoylated antibodies as a one-step protein transfer method often is an effective approach, it is not desirable under some circumstances. Too much palmitate derivatization of antibodies, especially within the antigen-binding sites, can decrease affinity of the antibody for the specified target antigen, and the site and extent of palmitoylation can cause incorporation onto the cell membrane in a nonoptimal orientation for antigen binding. For these reasons, a need for creating another method for coating cells with specific receptors arose. In 1993, a solution was developed that involved the use of pal-protein A as an intermediate between the cell surface and the antibody [78], thus resulting in a two-step protein transfer method (Figure 6.2B). It was proposed that using unmodified antibodies bound to pal-protein A, which in turn was incorporated in the cell membrane, would yield a correctly oriented receptor. This is because protein A binds to the Fc region of the antibody [79,80] thus ensuring that the antigen-binding sites on the antibody are available for antigen binding. Furthermore, the antibody remains unmodified and thus retains the original affinity for antigen.

Using this two-step SR method, pal-protein A incorporated onto T cell hybridomas in a concentration-dependent manner, whereas non-pal-protein A did not [78]. Incorporated pal-protein A was able to bind rabbit IgG as detected by flow cytometry suggesting that palmitoylation of protein A did not inhibit incorporation or binding ability to IgG. To expand on this concept, T cell hybridoma cells decorated with pal-protein A followed by anti-mouse IgG were able to form conjugates with murine B lymphoma cells that express surface mouse IgG; these intercellular conjugates were proportional to the amount of pal-protein A/mouse IgG coating the cell; however, when nonspecific IgG was bound to pal-protein A-coated cells, decreased binding to target cells resulted [78].

Using pal-protein A for the formation of SRs is a useful method for quickly preparing cells coated with antibodies of different isotypes and specificities without having to palmitoylate each individual antibody. It also orients the antibody so that the antigen-binding sites are facing out and are able to interact with the antigen specifically. Therefore, using a two-step pal-protein A incorporation followed by antibody binding can lead to specific cell–cell interactions allowing for the creation of effector cells that can target a specific cell population.

Furthermore, this pal-protein A two-step protein transfer method can be expanded to bind Fc-fusion proteins. Using this method, the costimulation molecule, B7-1, was fused with the Fc region from the IgG1 antibody. This fusion protein can be used to "paint" the cell surfaces of cells that have been treated with pal-protein A. In this system, pal-protein A incorporated onto the cell surface in a concentration-dependent manner, whereas unmodified protein A could not [81]. Then the human B7-1 • Fcγ1 fusion protein was incubated with protein A-modified cells so that the Fc region of the fusion protein bound to protein A on the cell surface. This consequently led to surface expression of the B7-1 fusion protein in a concentration-dependent manner [81]. Furthermore, B7-1 expressed on the cell surface remained functionally active and could lead to proliferation of PHA-activated or anti-CD3-activated T cells as well secretion of IFN-γ and IL-2 upon T cell stimulation [81]. Further analysis using this two-step protein transfer showed that preassembled B7-1 • Fcγ1:pal-protein A could be incorporated onto tumor cells not only *ex vivo*, but also *in vivo* by intratumorally injecting the preassembled complex into mice that had L5178Y-R lymphomas [82]. These injected mice showed B7-1 expression on the lymphoma cells 1 h after administration as detected by flow cytometry. In another study, mice that received intratumoral adoptive transfer of *ex vivo* amplified tumor-specific T cells protein transferred with B7-1 • Fcγ1:pal-protein A had increased tumor regression compared to mice that received control-modified T cells [83]. The antitumor efficacy of this vaccine was also seen in mice that had established 4T1 (a mammary carcinoma cell line) tumors, in which prolonged survival of mice that received protein-transferred T cells was observed over mice that received control-modified T cells [83]. Moreover, three other Fcγ1-fusion-costimulator molecules, 4-1BBL, CD48, and CD40L, along with B7-1 were also produced and individually incubated with pal-protein A. These preassembled complexes were then injected intratumorally together to form a "tetra-costimulator" system of protein transfer. Mice that received this tetra-costimulator complex showed 45% tumor regression of L5178Y-R lymphomas with evidence of induction of tumor-specific CTLs [82]. These studies show that this two-step protein transfer method involving pal-protein A and a fusion between costimulator molecules and the Fc region of IgG1 can be used as an effective method to induce local expression of proteins *in vivo*.

6.3.3 Lipid-Metal Chelator-Mediated Two-Step Protein Transfer

Another approach for two-step protein transfer utilizes lipid-metal chelators (Figure 6.2C). In this system, nitrilotriacetic acid ditetradecylamine (NTA-DTDA) [84] or 3 (nitrilotriacetic acid) ditetradecylamine (NTA_3-DTDA) [85] are used. The hydrophobic chains of DTDA allow for insertion of the metal–chelator complex into cell membranes leaving the NTA head groups available for subsequent binding to hexahistidine-tagged proteins or peptides [84–87]. NTA_3-DTDA allows for an anchor that is more stable than NTA-DTDA due to the presence of three NTA head groups [85]. Using this two-step protein transfer method, two different hexahistidine-tagged molecules, B7-1 and CD40, were able to simultaneously incorporate onto tumor cells [84], and hexahistidine-tagged peptides could be used to incorporate onto lipid membranes [88,89].

The lipid-metal chelator two-step protein transfer has many applications. This method was used to study low-affinity receptor–ligand interactions [90]. Due to the fluidity of the lipid bilayer, when NTA-DTDA is incorporated onto cell surfaces and engrafted with hexahistidine-tagged CD40, B7-1, or CD4, the ligands are able to move in order to better interact with the respective receptor multimerically, thus allowing for lower-affinity receptor–ligand interactions to be further studied. Moreover, this two-step protein transfer method can be used to directly decorate tumor cell surfaces with hexahistidine-tagged costimulatory molecules, B7-1 and CD40, to induce protection from tumor challenge [84] as well as enhance the immunity of liposome complexes by engrafting hexahistidine-tagged flagellin-derived peptides onto the liposome surfaces [86]. Another approach for using this two-step protein transfer is to target both liposomes and plasmid DNA-lipoplexes to specific cell types or tumor locations *in vivo* [85,87–89]. Liposomes were targeted to dendritic cells *in vivo* by engrafting hexahistidine-tagged antibody fragments against CD11c and DEC-205 (molecules expressed on DCs) onto the liposomes to specifically target either antigens or cytokines to DCs [85]. Anticancer drugs, such as doxorubicin, can also be loaded into liposomes that are then modified with NTA_3-DTDA and a hexhistidine-tagged peptide to target the drug to NIH-3T3 cells [89]. In a similar manner, plasmid DNA-lipoplexes can also be targeted to specific cells [88] and liposomes can be targeted to the tumor site *in vivo* by using hexahistidine-tagged peptides specific for proteins expressed in the vasculature formed by tumors [87]. Therefore, the lipid-metal chelator two-step protein transfer allows for specific targeting of anticancer drugs and molecules to tumor cells.

6.4 SUMMARY

Cell surface engineering by lipid-mediated protein transfer provides an efficient and quick manner to control the expression of surface proteins on cell membranes. Both one-step and two-step lipid-mediated protein transfer methods are quick and efficient ways to engineer the cell surface. While the modified protein of interest incorporates directly in the one-step method, indirect expression of the protein of interest results in the two-step protein transfer method. Both methods allow for controllable expression of functional proteins on the cell surface. The two-step protein transfer method is useful when hydrophobic modifications of the protein of interest render the protein dysfunctional. Therefore, the two-step protein transfer can be used to maintain functionality of membrane-incorporated proteins.

Modification of cell surfaces by expressing new proteins has many advantages over genetic engineering approaches. Genetic engineering approaches have been quite successful in expressing proteins on the cell surface; however, genetic engineering requires fast dividing cell lines or a large number of cells. It is difficult to transfect primary cells especially when low numbers of cells are available. Also, many primary cells, such as immune cells and primary tumor cells, are resistant to transfection. In genetic engineering, the amount of surface expression cannot be controlled and to achieve high surface protein expression of the transfected cells, rounds of selection or sorting of high expressing cells has to be performed. Thus genetic engineering of the surface is a very time-consuming method. It also becomes difficult to express high levels of multiple proteins on the cell surface by transfection.

On the other hand, protein transfer allows for controlled expression of protein on the surface of the cell as well as on cell membrane vesicles. Protein transfer can be performed quickly within a few hours and can lead to surface expression of functional protein. Since the lipid anchor attached to the proteins can incorporate on any lipid bilayer, protein transfer results in protein expression on 100% of the cells. Also, multiple functional proteins can be incorporated simultaneously on the cell surface by protein transfer. However, each protein that undergoes protein transfer needs to be modified with lipid tails so that it can insert exogenously into the lipid bilayer of cells and membrane vesicles. This modification should not interfere with the function of the protein; therefore, each modified protein needs to be tested for functionality before use. Furthermore, division of cells that are protein transferred will lead to the diffusion of the protein among the dividing cells. Therefore, it may be beneficial to use cell membrane vesicles or irradiated and mitomycin C-treated cells with protein transfer so that cell division does not diffuse protein surface expression.

Many applications for using protein transfer exist. One of the most widely used protein transfer applications involves enhancing antitumor immunity. Many applications have been made to enhance the immunogenicity of tumor cells by protein transferring immune-stimulating molecules, such as the costimulatory molecules B7-1 and B7-2 or cytokines such as IL-12, onto tumor cells or TMVs. This modification enhances antitumor immune responses and also transforms tumor cells into APCs. Protein transfer can also be used to target cells by creating artificial receptors on membrane surfaces. Through this manner, delivery of drugs or toxins can be targeted to specified cells that express incorporated receptors on their surface. On the other hand, these artificial receptors can also be used to target the incorporated cells or lipid vesicles toward other cell types in order to stimulate either killing or activation of the target cells or induce uptake by cells, such as APCs. Furthermore, protein transfer can be used to engineer APCs to activate T cells by incorporating exogenous costimulatory or activating proteins or by incorporating more adhesion molecules on the APC cell surface. However, conversely, protein transfer can also be used to induce killing of T-cells by exogenously modifying the APC cell surface with immune inhibitory molecules.

In summary, lipid-mediated cell surface engineering can be achieved by a variety of methods. These approaches to manipulating the cell surface have the potential to be used in studying cellular receptor functions, developing delivery vehicles for drugs and designing cellular vaccines.

Acknowledgments

This work was supported by NIH grants R01 CA138993 (to P. Selvaraj) and F31 CA165632 (to J. M. Patel). The authors thank Archana V. Boopathy for critical reading of the manuscript.

References

[1] The tercentenary of Robert Hooke. Can Med Assoc J 1935;33(3):312–13.

[2] Singer SJ, Nicolson GL. The fluid mosaic model of the structure of cell membranes. Science 1972;175 (4023):720–31.

[3] Brown DA, Rose JK. Sorting of GPI-anchored proteins to glycolipid-enriched membrane subdomains during transport to the apical cell surface. Cell 1992;68 (3):533–44.

[4] Simons K, Ikonen E. Functional rafts in cell membranes. Nature 1997;387(6633):569–72.

[5] Wenz JJ, Barrantes FJ. Steroid structural requirements for stabilizing or disrupting lipid domains. Biochemistry 2003;42(48):14267–76.

[6] Ishitsuka R, Sato SB, Kobayashi T. Imaging lipid rafts. J Biochem 2005;137(3):249–54.

[7] Kooyman DL, Byrne GW, Logan JS. Glycosyl phosphatidylinositol anchor. Exp Nephrol 1998;6 (2):148–51.

[8] Legler DF, Doucey MA, Schneider P, Chapatte L, Bender FC, Bron C. Differential insertion of GPI-anchored GFPs into lipid rafts of live cells. FASEB J 2005;19(1):73–5.

[9] Xu X, London E. The effect of sterol structure on membrane lipid domains reveals how cholesterol can induce lipid domain formation. Biochemistry 2000;39 (5):843–9.

[10] Bruckdorfer KR, Edwards PA, Green C. Properties of aqueous dispersions of phospholipid and cholesterol. Eur J Biochem 1968;4(4):506–11.

[11] Bruckdorfer KR, Graham JM, Green C. The incorporation of steroid molecules into lecithin sols, beta-lipoproteins and cellular membranes. Eur J Biochem 1968;4(4):512–8.

[12] Cooper RA, Arner EC, Wiley JS, Shattil SJ. Modification of red cell membrane structure by cholesterol-rich lipid dispersions. A model for the primary spur cell defect. J Clin Invest 1975;55 (1):115–26.

[13] Martin SE, Peterson BR. Non-natural cell surface receptors: synthetic peptides capped with *N*-cholesterylglycine efficiently deliver proteins into mammalian cells. Bioconjug Chem 2003;14(1):67–74.

[14] Kan CC, Yan J, Bittman R. Rates of spontaneous exchange of synthetic radiolabeled sterols between lipid vesicles. Biochemistry 1992;31(6):1866–74.

[15] Hussey SL, He E, Peterson BR. A synthetic membrane-anchored antigen efficiently promotes uptake of antifluorescein antibodies and associated protein a by mammalian cells. J Am Chem Soc 2001;123(50):12712–3.

[16] Hussey SL, Peterson BR. Efficient delivery of streptavidin to mammalian cells: clathrin-mediated endocytosis regulated by a synthetic ligand. J Am Chem Soc 2002;124(22):6265–73.

[17] Dunphy JT, Linder ME. Signalling functions of protein palmitoylation. Biochim Biophys Acta 1998;1436 (1–2):245–61.

[18] Ross EM. Protein modification. Palmitoylation in G-protein signaling pathways. Curr Biol 1995;5 (2):107–9.

[19] Duncan JA, Gilman AG. Autoacylation of G protein alpha subunits. J Biol Chem 1996;271(38):23594–600.

[20] Peacock JS, Londo TR, Roess DA, Barisas BG. Biologic activity of antigen receptors artificially incorporated onto B lymphocytes. J Immunol 1986;137(6):1916–23.

[21] Cheung KK, Newton A. Patterns of protein synthesis during development in *Caulobacter crescentus*. Dev Biol 1977;56(2):417–25.

[22] Chou SH, Shetty AV, Geng Y, Xu L, Munirathinam G, Pipathsouk A, et al. Palmitate-derivatized human IL-2: a potential anticancer immunotherapeutic of low systemic toxicity. Cancer Immunol Immunother 2013;62(3):597–603.

[23] Londo TR, Peacock JS, Roess DA, Barisas BG. Lateral diffusion of antigen receptors artificially incorporated onto B lymphocytes. J Immunol 1986;137(6):1924–31.

[24] Colsky AS, Stein-Streilein J, Peacock JS. Surrogate receptor-mediated cellular cytotoxicity. A method for "custom designing" killer cells of desired target specificity. J Immunol 1988;140(8):2515–9.

[25] Colsky AS, Peacock JS. Palmitate-derivatized antibodies can function as surrogate receptors for mediating specific cell–cell interactions. J Immunol Methods 1989;124(2):179–87.

[26] Elortza F, Mohammed S, Bunkenborg J, Foster LJ, Nuhse TS, Brodbeck U, et al. Modification-specific proteomics of plasma membrane proteins: identification and characterization of glycosylphosphatidylinositol-anchored proteins released upon phospholipase D treatment. J Proteome Res 2006;5(4):935–43.

[27] Maeda Y, Kinoshita T. Structural remodeling, trafficking and functions of glycosylphosphatidylinositol-anchored proteins. Prog Lipid Res 2011;50(4):411–24.

[28] Loizides-Mangold U, David FP, Nesatyy VJ, Kinoshita T, Riezman H. Glycosylphosphatidylinositol anchors regulate glycosphingolipid levels. J Lipid Res 2012;53(8):1522–34.

[29] Ikezawa H, Yamanegi M, Taguchi R, Miyashita T, Ohyabu T. Studies on phosphatidylinositol phosphodiesterase (phospholipase C type) of *Bacillus cereus*. I. purification, properties and phosphatase-releasing activity. Biochim Biophys Acta 1976;450(2):154–64.

[30] Bordier C. Phase separation of integral membrane proteins in Triton X-114 solution. J Biol Chem 1981;256(4):1604–7.

[31] Roberts WL, Myher JJ, Kuksis A, Low MG, Rosenberry TL. Lipid analysis of the glycoinositol phospholipid membrane anchor of human erythrocyte acetylcholinesterase. Palmitoylation of inositol results in resistance to phosphatidylinositol-specific phospholipase C. J Biol Chem 1988;263(35):18766–75.

[32] Medof ME, Kinoshita T, Nussenzweig V. Inhibition of complement activation on the surface of cells after incorporation of decay-accelerating factor (DAF) into their membranes. J Exp Med 1984;160(5):1558–78.

[33] Medof ME, Gottlieb A, Kinoshita T, Hall S, Silber R, Nussenzweig V, et al. Relationship between decay accelerating factor deficiency, diminished acetylcholinesterase activity, and defective terminal complement pathway restriction in paroxysmal nocturnal hemoglobinuria erythrocytes. J Clin Invest 1987;80(1):165–74.

[34] Zalman LS, Wood LM, Frank MM, Muller-Eberhard HJ. Deficiency of the homologous restriction factor in paroxysmal nocturnal hemoglobinuria. J Exp Med 1987;165(2):572–7.

[35] Sugita Y, Nakano Y, Tomita M. Isolation from human erythrocytes of a new membrane protein which inhibits the formation of complement transmembrane channels. J Biochem 1988;104(4):633–7.

[36] Davies A, Simmons DL, Hale G, Harrison RA, Tighe H, Lachmann PJ, et al. CD59, an LY-6-like protein expressed in human lymphoid cells, regulates the action of the complement membrane attack complex on homologous cells. J Exp Med 1989;170(3):637–54.

[37] Holguin MH, Fredrick LR, Bernshaw NJ, Wilcox LA, Parker CJ. Isolation and characterization of a membrane protein from normal human erythrocytes that inhibits reactive lysis of the erythrocytes of paroxysmal nocturnal hemoglobinuria. J Clin Invest 1989;84(1):7–17.

[38] Seifert PS, Roth I, Schmiedt W, Oelert H, Okada N, Okada H, et al. CD59 (homologous restriction factor 20), a plasma membrane protein that protects against complement C5b-9 attack, in human atherosclerotic lesions. Atherosclerosis 1992;96(2–3):135–45.

[39] Wing MG, Zajicek J, Seilly DJ, Compston DA, Lachmann PJ. Oligodendrocytes lack glycolipid anchored proteins which protect them against complement lysis. Restoration of resistance to lysis by incorporation of CD59. Immunology 1992;76(1):140–5.

[40] Hansch GM, Schonermark S, Roelcke D. Paroxysmal nocturnal hemoglobinuria type III. Lack of an erythrocyte membrane protein restricting the lysis by C5b-9. J Clin Invest 1987;80(1):7–12.

[41] Selvaraj P, Dustin ML, Silber R, Low MG, Springer TA. Deficiency of lymphocyte function-associated antigen 3 (LFA-3) in paroxysmal nocturnal hemoglobinuria. Functional correlates and evidence for a phosphatidylinositol membrane anchor. J Exp Med 1987;166(4):1011–25.

[42] Dustin ML, Selvaraj P, Mattaliano RJ, Springer TA. Anchoring mechanisms for LFA-3 cell adhesion

glycoprotein at membrane surface. Nature 1987;329 (6142):846–8.
[43] Selvaraj P, Plunkett ML, Dustin M, Sanders ME, Shaw S, Springer TA. The T lymphocyte glycoprotein CD2 binds the cell surface ligand LFA-3. Nature 1987;326 (6111):400–3.
[44] Nagarajan S, Anderson M, Ahmed SN, Sell KW, Selvaraj P. Purification and optimization of functional reconstitution on the surface of leukemic cell lines of GPI-anchored Fc gamma receptor III. J Immunol Methods 1995;184(2):241–51.
[45] Zhang F, Schmidt WG, Hou Y, Williams AF, Jacobson K. Spontaneous incorporation of the glycosyl-phosphatidylinositol-linked protein Thy-1 into cell membranes. Proc Natl Acad Sci USA 1992;89(12):5231–5.
[46] Bulow R, Overath P, Davoust J. Rapid lateral diffusion of the variant surface glycoprotein in the coat of *Trypanosoma brucei*. Biochemistry 1988;27(7):2384–8.
[47] Poloso NJ, Nagarajan S, Bumgarner GW, Selvaraj P. Development of therapeutic vaccines by direct modification of cell membranes from surgically removed human tumor tissue with immunostimulatory molecules. Vaccine 2001;19(15–16):2029–38.
[48] McHugh RS, Ahmed SN, Wang YC, Sell KW, Selvaraj P. Construction, purification, and functional incorporation on tumor cells of glycolipid-anchored human B7-1 (CD80). Proc Natl Acad Sci USA 1995;92(17):8059–63.
[49] Los DA, Murata N. Membrane fluidity and its roles in the perception of environmental signals. Biochim Biophys Acta 2004;1666(1–2):142–57.
[50] Premkumar DR, Fukuoka Y, Sevlever D, Brunschwig E, Rosenberry TL, Tykocinski ML, et al. Properties of exogenously added GPI-anchored proteins following their incorporation into cells. J Cell Biochem 2001; 82(2):234–45.
[51] van den Berg CW, Cinek T, Hallett MB, Horejsi V, Morgan BP. Exogenous glycosyl phosphatidylinositol-anchored CD59 associates with kinases in membrane clusters on U937 cells and becomes Ca^{2+}-signaling competent. J Cell Biol 1995;131(3):669–77.
[52] McHugh RS, Nagarajan S, Wang YC, Sell KW, Selvaraj P. Protein transfer of glycosyl-phosphatidylinositol-B7-1 into tumor cell membranes: a novel approach to tumor immunotherapy. Cancer Res 1999;59(10):2433–7.
[53] Poloso NJ, Nagarajan S, Mejia-Oneta JM, Selvaraj P. GPI-anchoring of GM-CSF results in active membrane-bound and partially shed cytokine. Mol Immunol 2002;38(11):803–16.
[54] Nagarajan S, Selvaraj P. Glycolipid-anchored IL-12 expressed on tumor cell surface induces antitumor immune response. Cancer Res 2002;62(10):2869–74.
[55] Bozeman EN, Cimino-Mathews A, Machiah DK, Patel JM, Krishnamoorthy A, Tien L, et al. Expression of membrane anchored cytokines and B7-1 alters tumor microenvironment and induces protective antitumor immunity in a murine breast cancer model. Vaccine 2013;31(20):2449–56.
[56] Brunschwig EB, Levine E, Trefzer U, Tykocinski ML. Glycosylphosphatidylinositol-modified murine B7-1 and B7-2 retain costimulator function. J Immunol 1995;155(12):5498–505.
[57] Nagarajan S, Selvaraj P. Human tumor membrane vesicles modified to express glycolipid-anchored IL-12 by protein transfer induce T cell proliferation *in vitro*: a potential approach for local delivery of cytokines during vaccination. Vaccine 2006;24(13):2264–74.
[58] Brunschwig EB, Fayen JD, Medof ME, Tykocinski ML. Protein transfer of glycosyl-phosphatidylinositol (GPI)-modified murine B7-1 and B7-2 costimulators. J Immunother 1999;22(5):390–400.
[59] Chesla SE, Li P, Nagarajan S, Selvaraj P, Zhu C. The membrane anchor influences ligand binding two-dimensional kinetic rates and three-dimensional affinity of FcgammaRIII (CD16). J Biol Chem 2000;275(14):10235–46.
[60] Barboni E, Rivero BP, George AJ, Martin SR, Renoup DV, Hounsell EF, et al. The glycophosphatidylinositol anchor affects the conformation of Thy-1 protein. J Cell Sci 1995;108(Pt 2):487–97.
[61] Butikofer P, Malherbe T, Boschung M, Roditi I. GPI-anchored proteins: now you see 'em, now you don't. FASEB J 2001;15(2):545–8.
[62] Paulick MG, Bertozzi CR. The glycosylphosphatidylinositol anchor: a complex membrane-anchoring structure for proteins. Biochemistry 2008;47(27):6991–7000.
[63] Lehto MT, Sharom FJ. Proximity of the protein moiety of a GPI-anchored protein to the membrane surface: a FRET study. Biochemistry 2002;41(26):8368–76.
[64] Huang JH, Getty RR, Chisari FV, Fowler P, Greenspan NS, Tykocinski ML. Protein transfer of preformed MHC–peptide complexes sensitizes target cells to T cell cytolysis. Immunity 1994;1(7):607–13.
[65] Lotze MT, Chang AE, Seipp CA, Simpson C, Vetto JT, Rosenberg SA. High-dose recombinant interleukin 2 in the treatment of patients with disseminated cancer. Responses, treatment-related morbidity, and histologic findings. JAMA 1986;256(22):3117–24.
[66] Wahlsten JL, Mills CD, Ramakrishnan S. Antitumor response elicited by a superantigen-transmembrane sequence fusion protein anchored onto tumor cells. J Immunol 1998;161(12):6761–7.
[67] Englund PT. The structure and biosynthesis of glycosyl phosphatidylinositol protein anchors. Annu Rev Biochem 1993;62:121–38.
[68] Diamandis EP, Christopoulos TK. The biotin–(strept) avidin system: principles and applications in biotechnology. Clin Chem 1991;37(5):625–36.

[69] Lindqvist Y, Schneider G. Protein–biotin interactions. Curr Opin Struct Biol 1996;6(6):798–803.
[70] Singh NP, Yolcu ES, Askenasy N, Shirwan H. ProtEx: a novel technology to display exogenous proteins on the cell surface for immunomodulation. Ann NY Acad Sci 2005;1056:344–58.
[71] Green NM. Avidin and streptavidin. Methods Enzymol 1990;184:51–67.
[72] Yolcu ES, Askenasy N, Singh NP, Cherradi SE, Shirwan H. Cell membrane modification for rapid display of proteins as a novel means of immunomodulation: FasL-decorated cells prevent islet graft rejection. Immunity 2002;17(6):795–808.
[73] Singh NP, Yolcu ES, Taylor DD, Gercel-Taylor C, Metzinger DS, Dreisbach SK, et al. A novel approach to cancer immunotherapy: tumor cells decorated with CD80 generate effective antitumor immunity. Cancer Res 2003;63(14):4067–73.
[74] Askenasy EM, Shushlav Y, Sun Z, Shirwan H, Yolcu ES, Askenasy N. Engineering of bone marrow cells with fas-ligand protein-enhances donor-specific tolerance to solid organs. Transplant Proc 2011;43(9):3545–8.
[75] Kaminitz A, Yolcu ES, Stein J, Yaniv I, Shirwan H, Askenasy N. Killer Treg restore immune homeostasis and suppress autoimmune diabetes in prediabetic NOD mice. J Autoimmun 2011;37(1):39–47.
[76] Kaminitz A, Askenasy N, Yolcu ES. Immunomodulation with regulatory T cells and Fas-ligand ameliorate established inflammatory colitis. Gut 2013;62(8):1228–30.
[77] Yolcu ES, Kaminitz A, Mizrahi K, Ash S, Yaniv I, Stein J, et al. Immunomodulation with donor regulatory T cells armed with Fas-ligand alleviates graft-versus-host disease. Exp Hematol 2013;.
[78] Kim SA, Peacock JS. The use of palmitate-conjugated protein A for coating cells with artificial receptors which facilitate intercellular interactions. J Immunol Methods 1993;158(1):57–65.
[79] Kronvall G, Grey HM, Williams Jr. RC. Protein A reactivity with mouse immunoglobulins. Structural relationship between some mouse and human immunoglobulins. J Immunol 1970;105(5):1116–23.
[80] Kronvall G, Messner RP, Williams Jr. RC. Immunochemical studies on the interaction between staphylococcal protein A and gamma-G globulin. J Immunol 1970;105(6):1353–9.
[81] Chen A, Zheng G, Tykocinski ML. Hierarchical costimulator thresholds for distinct immune responses: application of a novel two-step Fc fusion protein transfer method. J Immunol 2000;164(2):705–11.
[82] Zheng G, Chen A, Sterner RE, Zhang PJ, Pan T, Kiyatkin N, et al. Induction of antitumor immunity via intratumoral tetra-costimulator protein transfer. Cancer Res 2001;61(22):8127–34.
[83] Zheng G, Liu S, Wang P, Xu Y, Chen A. Arming tumor-reactive T cells with costimulator B7-1 enhances therapeutic efficacy of the T cells. Cancer Res 2006;66(13):6793–9.
[84] van Broekhoven CL, Parish CR, Vassiliou G, Altin JG. Engrafting costimulator molecules onto tumor cell surfaces with chelator lipids: a potentially convenient approach in cancer vaccine development. J Immunol 2000;164(5):2433–43.
[85] van Broekhoven CL, Parish CR, Demangel C, Britton WJ, Altin JG. Targeting dendritic cells with antigen-containing liposomes: a highly effective procedure for induction of antitumor immunity and for tumor immunotherapy. Cancer Res 2004;64(12):4357–65.
[86] Faham A, Herringson T, Parish C, Suhrbier A, Khromykh AA, Altin JG. pDNA-lipoplexes engrafted with flagellin-related peptide induce potent immunity and anti-tumour effects. Vaccine 2011; 29(40):6911–9.
[87] Herringson TP, Altin JG. Effective tumor targeting and enhanced anti-tumor effect of liposomes engrafted with peptides specific for tumor lymphatics and vasculature. Int J Pharm 2011;411 (1–2):206–14.
[88] Herringson TP, Patlolla RR, Altin JG. Targeting of plasmid DNA-lipoplexes to cells with molecules anchored via a metal chelator lipid. J Gene Med 2009;11(11):1048–63.
[89] Herringson TP, Altin JG. Increasing the antitumor efficacy of doxorubicin-loaded liposomes with peptides anchored via a chelator lipid. J Drug Target 2011;19 (8):681–9.
[90] Van Broekhoven CL, Altin JG. A novel system for convenient detection of low-affinity receptor–ligand interactions: chelator-lipid liposomes engrafted with recombinant CD4 bind to cells expressing MHC class II. Immunol Cell Biol 2001;79(3):274–84.

CHAPTER

7

Engineering the Surface of Cells Using Biotin–Avidin Chemistry

Kawther K. Ahmed, Sean M. Geary and Aliasger K. Salem

College of Pharmacy, University of Iowa, Iowa City, IA, USA

OUTLINE

J.M. Karp & W. Zhao (Eds): Micro- and Nanoengineering of the Cell Surface.
DOI: http://dx.doi.org/10.1016/B978-1-45-573146-6.00007-6

7.1 INTRODUCTION: RATIONALE FOR ENGINEERING THE CELL SURFACE

The interplay between cells and their surroundings plays a major role in directing cell fate in terms of differentiation, adherence, spreading, and at times, the outcome of their interactions with cells of the immune system [1]. Thus, precise patterning of the cellular environment can be used to control and manipulate cell behavior. The ability to manipulate the cell surface to impart nonnative properties to cells has been used as a tool for enhancing cell therapy [2]. Apart from the benefit cell surface engineering can have in the field of cell therapy, cells can be modified for research purposes to expand our understanding of cellular processes in both healthy and diseased states [3,4]. Examples of applications for which cell function and behavior have been modified include: promotion of cell adhesion and migration for tissue engineering [1], modulating cell trafficking and tissue homing [5,6], providing autocrine growth factors for cells used in cell therapy [2,7], and prolonged persistence of modified cells to promote immune responses for cancer vaccine applications [8–11]. While genetic engineering of cells enables the introduction of proteins (such as surface receptors or cytokines), there is the issue of safety when implementing viral transduction, the generally most efficient form of transfection. In addition, transfection methods cannot be used to functionalize cells with nonproteins such as drugs or even particles [2]. Thus, there is a need for alternative methods to surface engineer cells with the desired functionality. One of these approaches being implemented and showing promise in a range of applications is cell surface engineering using biotin–avidin crosslinking. Advantages to the biotin–avidin system include that it is inherently adjustable, relatively inexpensive, simple, and nontoxic.

7.2 BIOTIN AND AVIDIN: AN OVERVIEW

The interaction between avidin and biotin was recognized as early as 1940 by Esmond Snell with the observation that chicks fed on a diet of raw egg white were deficient in biotin, regardless of its availability in their diet [12]. The discovered interaction exhibited unique strength and stability that made it become a widely used interaction in biomedical applications. The avidin–biotin complex is one of the strongest known noncovalent interactions in biological systems. The high association constant ($K_a \approx 10^{15}\,M^{-1}$ [13]) and the exceptionally high stability of the avidin–biotin crosslink [14] have made it an indispensable tool in biomedical and biotechnological applications. Avidin–biotin binding is rapid and very stable, being resistant to: high temperature, extreme pH, organic solvents, and proteolytic enzymes [14,15]. The list of applications is long and still growing due to the advances made in cell and tissue engineering. A highly utilized methodology involving the avidin–biotin interaction is cell surface protein enrichment prior to their identification with other detection techniques like western blotting and liquid chromatography–mass spectroscopy (LC–MS) [16]. Signal amplification in enzyme-linked immunosorbent assays (ELISA), immunocytochemistry, and bioimaging have also made great use of the strong affinity and stability of the avidin–biotin interaction [17–20]. More novel uses of avidin–biotin chemistry include cell surface engineering where the cell membrane is coated with biotin for applications such as tissue engineering [21] and cell–particle hybrid synthesis [22].

7.2.1 Avidin–Biotin Complex at the Structural Level

Avidin (Mr 67,000) is a tetrameric glycoprotein with each of the four identical 128 amino acid polypeptide subunits having a carbohydrate group [23]. Primary natural sources of avidin, which possess some antibacterial activity, are the egg whites of birds, amphibians, and reptiles [24]. Each subunit has been characterized by X-ray crystallography and was found to have a barrel-like shape and contain one biotin-binding site, thus one avidin molecule has the capacity to bind four biotin molecules [13,25–27]. Research involved in determining the three-dimensional structure of the biotin-binding site and binding kinetics has identified specific amino acids essential for biotin binding and stability of the tetrameric structure [13,23,27–30]. This knowledge has enabled manipulation of the structural features of avidin to tune its binding properties in terms of biotin-binding capacity and specificity (see Section 7.2.2).

Biotin, also known as vitamin H or vitamin B_7, is small relative to avidin with a molecular weight of 244 [31]. It contains a cyclic urea group and a valeric acid side chain and plays an essential role as a coenzyme for carboxylases involved in a range of metabolic processes including anabolism of fatty acids [31,32].

The strong avidin–biotin interaction is provided mainly by van der Waals forces driven by the exquisite orientation of the amino acid residues in the biotin-binding sites of the avidin molecule [15,29,33], the small size of the biotin molecule [29], and the structural relation of avidin to the cyclic ureido group of biotin [30].

7.2.2 Avidin and Avidin Analogs

One limitation to the use of natural avidin is its nonspecific adsorption to some biological molecules. Avidin has a basic isoelectric point (pI ~ 10) [25] rendering the protein positively charged at physiological pH [34] which can result in nonspecific binding to cells and negatively charged molecules. In addition, the carbohydrate content of avidin can contribute to nonspecific adsorption to lectins [35,36]. Although nonspecific binding may be stringently controlled in certain *in vitro* applications, it becomes more problematic for tissue engineering purposes and other biomedical applications. Therefore, alternatives to avidin have been investigated, with two of the most prevalently implemented being streptavidin and neutravidin [37] (other alternatives are genetically engineered avidin and streptavidin and they will be discussed later). Streptavidin and neutravidin are nonglycosylated analogs of avidin, each having a molecular weight of 60,000 [38,39]. Both have biotin-binding affinities equivalent to that of avidin; in a variety of applications they have greater specificities and have been successfully used as an alternative to avidin [25]. Streptavidin, extracted from *Streptomyces avidinii* [27], has an isoelectric point close to 6 and is therefore anionic at physiological pH. This, combined with the lack of glycosylation, contributes to the enhanced specificity of streptavidin compared to avidin [28,30,40].

Neutravidin has a near to neutral isoelectric point (pI 6.3) [25,39] and exhibits even lower nonspecific binding than streptavidin due to the lack of an RGD-mimicking sequence, RYD, found in streptavidin [41].

While the four to one binding capacity of biotin to avidin plays an important role in using avidin–biotin chemistry to synthesize cell aggregates [42], it becomes troublesome in some applications where the aggregation of biotinylated cells with the avidin-modified ligand is undesirable, such as in tissue engineering. The solution to this problem is provided by the use of monomeric or monovalent avidin or avidin analogs. These analogs are produced by inducing mutants in the genes encoding for the protein. A good portion of the genetic engineering efforts goes to limiting the protein-binding capacity to one biotin molecule per avidin molecule, however, studies generating avidin and streptavidin mutants with enhanced binding specificity [43] and reversible biotin-binding properties for affinity chromatography have also been reported [44]. These modified biotin-binding proteins have the potential to be used in controlling cell growth and spreading in tissue engineering allowing more precise control of cell–cell and cell–matrix interaction.

An avidin monomer was synthesized by Kulomaa and coworkers by inducing a mutation to change one of the amino acid residues at the subunits interface. However, these monomers reassociated upon the addition of biotin [45]. This problem of reassociation was solved by inducing a second mutation such that two amino acids at the interface are changed in the new monomer [46]. Another group had also reported the development of stable monomeric avidin by inducing amino acid changes, although different amino acids were changed [47]. However, these stable monomers showed less affinity to binding biotin relative to the tetrameric protein and had lower stability to proteolytic enzymes. Another study has reported the synthesis of a stable streptavidin monomer that has a biotin-binding affinity higher than the previously reported monomers [48]. The synthesized monomer produced was an analog to rhizavidin, a biotin-binding protein from the bacterium *Rhizobium etli* and exists naturally as a dimer [48].

Another approach to overcome the aggregation associated with the use of tetrameric biotin-binding proteins is the use of monovalent protein developed by Alice Ting's group [18]. In monovalent avidin and streptavidin, a tetramer of three inactive subunits and only one biotin-binding subunit is the final product [49]. The process involves forming inclusion bodies in *Escherichia coli* that express active or inactive units. These units are then purified and folded with a ratio of inactive:active units of 3:1. Tetramers having three inactive units and one biotin-binding unit are then isolated using a specific tag on the active subunit to identify its multiplicity (number of active sites folded in the product) [49]. The conservation of binding affinity in the monovalent protein compared to the monomeric product was attributed to the preservation of the subunits interface at which the biotin-binding site resides [49–51].

7.2.3 Safety of the Avidin–Biotin Complex for Clinical Use

Being xenoproteins, avidin and its analogs are, not surprisingly, immunogenic [52,53]. The level of avidin-specific antibodies that are induced increases with increasing exposure to avidin and streptavidin [54]. However, the immunogenicity of avidin and streptavidin does not necessarily constitute a health hazard since 15 years of use of avidin in clinical settings has had no major side effects [55]. In addition, melanoma patients treated with two or three doses of a vaccine formulation incorporating streptavidin–biotin complexes induced, at worst, low-grade allergic reactions [53]. Furthermore, it was shown that the presence of these antibodies did not abrogate or weaken the function of the avidin–biotin complex [55]. Nevertheless, it has been suggested that the immunogenicity of avidin can be reduced by pegylation [56].

7.3 METHODS FOR ENGINEERING CELL SURFACES WITH AVIDIN–BIOTIN COMPLEXES

The literature is rich with different approaches and variable protocols for decorating cell surfaces with biotin. However, they can be categorized generally into covalent and noncovalent binding of biotin to cell surfaces (Figure 7.1).

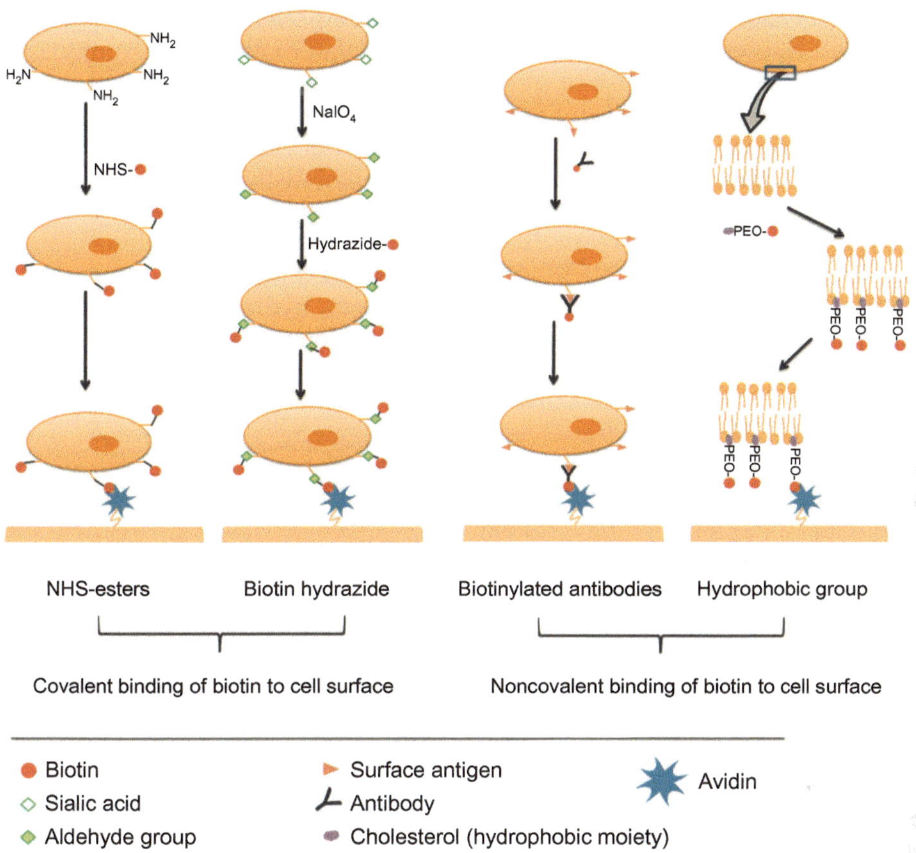

FIGURE 7.1 Methods for engineering cell surfaces with avidin–biotin complexes.

7.3.1 Covalent Binding of Biotin to Cell Surfaces

Biotin can be covalently bound to cell surfaces through its valeric acid side chain. To elaborate, the carboxylate functional group of the valeric acid side chain can be activated to bind different native or induced residues on the cell surface. Different activated groups are available for substrate biotinylation such as *N*-hydroxysuccinimide (NHS) ester, hydrazide, BMCC (1-biotinamido-4-[4′-(maleimidomethyl) cyclohexane-carboxamido]butane), and HPDP (*N*-[6-(biotinamido)hexyl]-3′-(2′-pyridyldithio) propionamide) [57]. However, NHS and hydrazide biotin conjugates are the reagents most commonly used for cell surface biotinylation and thus only these two reagents will be discussed.

7.3.1.1 NHS Esters

The valeric side chain of the biotin molecule is not required for avidin binding; therefore, the carboxylic group can be activated with carbodiimides so as to bind to amine groups present on the cell surface. Alternatively, an acylating active group like NHS can be conjugated to the valeric acid side chain to bind cell surface amines [57,58]. These biotinylating reagents (activated biotin) are commercially available as NHS–biotin, which is water insoluble, and the water-soluble form sulfo-NHS biotin [57], which is cell membrane impermeable [59]. One limitation to the use of biotin alone or NHS–biotin is the possible unavailability of the conjugated biotin for avidin binding due to steric hindrance by adjacent cell membrane proteins. In such circumstances, in order for biotin to effectively bind to avidin, a spacer is required between the NHS ester group and the valeric acid side chain [57]. NHS-LC-biotin and sulfo-LC-NHS biotin have 6-aminocaproic acid spacers that provide greater length between the modified cell surface and the biotin [57]. The method for cell surface biotinylation with NHS esters is easy and involves incubating cells (up to 2.5×10^7/ml) with a solution of the biotinylating reagent (0.5 mg/ml) for approximately 30 min at room temperature in a protein-free isotonic solution [53,60–62]. A range of cell types (adherent and nonadherent) have been biotinylated by this method and no significant effect on cell viability or cell adhesion properties (for adherent cells) was observed [53,59,61,62].

7.3.1.2 Biotin Hydrazides

Hydrazides bind spontaneously to aldehyde and ketone groups to form covalent hydrazone linkages [57]. Although aldehydes and ketones are not expressed on the surfaces of cells, they can be introduced by the oxidation of a diol group in a sugar moiety on the cell surface using sodium periodate [63,64]. The sugar moiety targeted in such cases is sialic acid. Sialic acids are a group of monosaccharides widely expressed on the cells of vertebrates [65]. This method of cell surface biotinylation (oxidizing sialic acid residues on the cells to aldehydes followed by incubating treated cells with biotin hydrazide) was employed to independently biotinylate HEK293 cells [22], live embryonic stem cells [66], and murine L6 myoblasts [67]. As with NHS–biotin, biotin-LC-hydrazide with a 6-aminocaproic acid spacer is also available [57]. The reported method for cell surface biotinylation involves initially incubating cells with a solution of sodium periodate ($NaIO_4$) at a concentration of 1–2 mM for 5–20 min at 4°C. Then treated cells are incubated in a solution of 5 mM biotin hydrazide, pH 6.5, for 90 min at room temperature [22,67,68]. Incubation with biotin hydrazide for 30 min at 37°C has also been reported [66]. This method was also cited as not resulting in major cell toxicities, however, a study comparing NHS–biotin and biotin hydrazide demonstrated that, at least for red blood cells, biotinylation with the former provided higher viable cell recovery and better cell surface biotinylation [69].

Generating reactive ketone groups as substrates for biotinylation has been suggested as a potential methodology for *in vivo* biotinylation. To explain, it was suggested that metabolic generation of ketones can provide a method for *in vivo* engineering of cell surfaces [70–72]. The method involves the use of cellular sialic acid synthesis machinery to metabolize an external precursor, *N*-azidoacetylmannosamine, to provide a substrate for cell surface ligation with biotin hydrazide [63]. *In vivo* engineering of cell surfaces was suggested to provide a tool for targeted drug delivery with ligand-labeled drug delivery systems or drugs [70,71]. It should be noted, however, that in the study that involved metabolic engineering of cell surfaces, the substrate was injected i.p. and cell surface ketones were evident in the spleen, heart, liver, and kidney of the mice [73].

7.3.2 Noncovalent Binding of Biotin to Cell Surfaces

Covalent binding ensures strong association of biotin to cell surfaces; however, noncovalent biotinylation has also been reported and shown to be effective. However, to the best of our knowledge, no direct comparison between the two approaches has been reported in terms of biotinylation efficiency and effect on cell viability.

7.3.2.1 Cell Surface Biotinylation via Biotinylated Antibodies

This methodology has been mainly used to biotinylate cells for engineering cell surfaces with particles and involves incubating cells with biotinylated antibodies at 4°C for 25–60 min in cell culture media [74]. Concentrations of antibodies used ranged between 4 and 5 μg/2–5 × 10^6 cell/ml. Cells are then washed and incubated with an avidin-modified ligand. The choice of the antibody depends on the cell type being engineered. Some examples of biotinylated antibodies used cell surface engineering via avidin–biotin linkage are anti-beta 1 (CD29) for murine melanoma (B16-F10) and prostatic cancer (RM11) cells (unpublished work from our laboratory), and anti-CD19 for B-cell lymphoma cell biotinylation [74]. The cell-bound biotin that does not complex with avidin is unlikely to have any biological consequences.

7.3.2.2 Use of Biotin-Tagged Hydrophobic Molecules to Coat Cell Surfaces

Another approach to anchor biotin to the cell surface, though not widely adopted, is through the insertion of a biotin-labeled hydrophobic moiety into the cell lipid membrane [21,75]. This approach involves the synthesis of a polyethylene oxide (PEO) polymer that is functionalized with biotin at one end and cholesterol at the other. Upon incubating this polymer with living cells, biotin is anchored via the insertion of the hydrophobic cholesterol into the lipid cell membrane. The efficiency of the cell surface biotin to subsequently bind streptavidin was found to increase with increasing polymer chain length. Similar to cholesterol, other biocompatible anchoring molecules (e.g., BAM90) have been utilized to anchor biotin on hepatic cell surfaces [21].

7.4 APPLICATIONS OF CELL SURFACE ENGINEERING USING AVIDIN–BIOTIN CHEMISTRY

As mentioned earlier, the avidin–biotin complex is widely used in different areas of biomedical and biotechnological applications. Cell surface engineering in particular is a process where the avidin–biotin complex has been exploited to tune cell behavior.

7.4.1 Spatial Control of Cell Growth on Solid Surfaces and Tissue Engineering

Spatial control of cell growth and spreading pattern is important in modulating cell properties and preserving cell phenotype [76]. In normal tissue, cell–cell and cell–ECM contact is mediated by transmembrane proteins (e.g., cadherins and integrins, respectively) [76,77]. However, these intrinsic adhesion molecules can only mediate weak cell attachment to synthetic ECM such as tissue engineering templates and solid surfaces [78]. Therefore, biomimetic surfaces coated with adhesion molecules have been developed to enhance cell adhesion and proliferation [79]. One example of applications where cell attachment needs to be augmented is the attachment of endothelial cell layers onto synthetic vascular grafts [78]. The Reichert group has utilized the avidin–biotin complex to improve the potency of synthetic vascular grafts by lining the graft lumen with a layer of endothelial cells [80–82]. Their studies have shown that the interaction between avidin and biotin is important in enhancing integrin-mediated cellular interaction with the ECM [80]. The studies involved biotinylating bovine aortic endothelial cells with sulfo-NHS-LC-biotin and using avidin to bridge biotinylated cells with a biotinylated growth surface [80,82]. The study emphasized the importance of the high affinity avidin–biotin interaction in supporting fibronectin–integrin-mediated adhesion.

The other application where spatial control of cell growth and enhancement of integrin-mediated cell adhesion becomes important is tissue engineering. In tissue engineering involving cell culturing on a polymeric templates, cell growth and spreading pattern need to be controlled and augmented to preserve cell phenotype and resemble normal tissue structures [83]. Hence a ligand–receptor patterning strategy is required to enhance cell attachment to the polymeric template. In a study carried out earlier by our group, a versatile procedure for cell growth patterning was developed [84]. The versatility stems from engineering the cell surface with biotin by oxidizing native sialic acid residues to generate reactive groups for biotin hydrazide coupling. The method was evaluated for patterning human dermal fibroblast growth on avidin-treated PLA–PEG–biotin (poly(lactic acid)–poly (ethylene glycol)–biotin) polymer surfaces. In a separate study in our group, a novel biotinylated nanotemplated degradable hydrogel was developed [85]. The fabricated hydrogel was suggested to have applications in targeted drug delivery and for cell patterning as the surface can be engineered with cell-specific receptors using the avidin–biotin interaction.

Besides cell growth on solid supports, tissue engineering involves the formation of cell aggregates between homo- and heterotypic cells as the first step toward the production of viable tissue [86]. The method of using the avidin–biotin interaction for such purposes was first described by the Kellam and Shakesheff groups where cells were aggregated into clusters for skeletal muscle engineering [42,67]. The method involved biotinylating cell surfaces with biotin hydrazide and crosslinking biotinylated cells with tetravalent avidin [68]. Cell aggregation through avidin–biotin interaction was also used to agglomerate hepatic cells and fibroblasts to form sheet-like liver tissue where the use of avidin to crosslink heterotypic biotinylated cells was shown to be effective in preserving cell viability and phenotype for nonproliferative cells such as primary cultures of hepatocytes [21].

7.4.2 Surface Conjugation of Macromolecules

The use of cells for therapeutic purposes is an expanding area of research. For example, mesenchymal stem cells have been trialed as potential therapies for immune-based diseases such as multiple sclerosis [87], whilst dendritic cell preparations have been employed as cancer vaccines such as the recently FDA-approved sipuleucel-T for prostate cancer [88]. The employment of cells for therapeutic applications requires close control of cell trafficking, proliferation, adhesion, and other microenvironmental factors [2]. Such control is often achieved by specific molecules capable of adjusting cell behavior and/or microenvironments [7]. For example, hematopoietic stem cell (HSC)-based therapy can be augmented by co-administration of a glycogen synthase kinase-3 (GSK-3) inhibitor [89] and tumor cells can be stimulated (mostly by viral transfection) to secrete certain cytokines for cancer vaccine applications [90,91]. Another approach is the stable anchoring of these moieties to cell surfaces using crosslinking strategies [92] among which is the avidin–biotin complex. For example, surface engineering tumor cells with the cytokine granulocyte macrophage colony stimulating factor (GM-CSF) has been performed for cancer vaccine purposes [53]. GM-CSF is a pleiotropic cytokine that has gained interest as a vaccine adjuvant and proved effective in cancer immunotherapy and cancer vaccines [90,93]. GM-CSF is secreted by a range of cell types, which include macrophages, T cells, endothelial cells, and fibroblasts, and is known to play a role in the recruitment and maturation of dendritic cells (the main antigen-presenting cells) [94,95]. The study involved biotinylating melanoma cells using sulfo-NHS-LC-biotin and incubating biotinylated cells with streptavidin-tagged GM-CSF [53]. This method provides a safe alternative to viral transduction of cells to promote GM-CSF expression. However, it should be noted here that the conjugated GM-CSF is not meant to be released but rather to be constantly associated with the tumor cell surface. A separate study utilized biotin–avidin crosslinking to stably bind the sialyl Lewis X (SLeX) moiety on mesenchymal stem cells [59,96]. SLeX, a cell surface carbohydrate that has a role in tumor cell metastasis and mesenchymal cell trafficking [97,98], was used as a targeting moiety to promote mesenchymal stem cell tissue homing. Cell surface modification involved cells being biotinylated using sulfo-NHS biotin followed by treating biotinylated cells with streptavidin. Biotinylated SLeX was then introduced and stable binding was provided by the biotin–streptavidin–biotin crosslinkages. Modified cells demonstrated homing responses *in vitro* [59] as well as *in vivo* [99].

7.4.3 Cell–Particle Hybrid Synthesis

Engineering cells with particulated matter using avidin–biotin chemistry is an area that has only been remotely explored. Besides the use of avidin–biotin crosslinking in studying cell–particle binding kinetics [74], it has been used to stably anchor particles onto cell surfaces in order to synthesize cell–particle hybrids. Due to the presence of drug-containing particles, these hybrids can provide a sustained drug release profile [100]. In addition, the fabricated delivery system ensures co-delivery of the particles payload and the therapeutic cells to the same target. Cell–particle hybrids that have employed the avidin–biotin crosslinking were developed for cancer vaccine purposes and targeted drug delivery [22,60]. In terms of cancer vaccine applications, the assembled hybrid delivers cellular cancer antigens (provided by the cellular component) and the loaded vaccine adjuvant (provided by the particles) such as CpG to the same dendritic cell to maximize the prophylactic/therapeutic effect of the vaccine [92,101]. As a proof of principle, a cell–particle hybrid has been synthesized in our laboratory by attaching 1.4 μm PLA–PEG–biotin microparticles to biotinylated HEK 293 tumor cells using an avidin bridge [22]. Cells were biotinylated using biotin hydrazide following their treatment with sodium periodate. More recently, another cell–particle hybrid was synthesized in our laboratory by anchoring 500 nm polylactic-*co*-glycolic acid particles to murine B16-F10 melanoma cells (unpublished data). Particles were coated with streptavidin post-fabrication using carbodiimide chemistry. Cells were biotinylated using commercially available biotinylated anti-β1 integrin antibodies and the assembled hybrids were characterized using flow cytometry, laser scanning confocal microscopy, and scanning electron microscopy. The hybrids fabricated in this study are proposed to serve as potential cancer vaccines where the irradiated allogeneic cancer cells secreting GM-CSF act as an immunogen and the particles, loaded with an immune adjuvant, can act to promote a more robust antitumor immune response. Besides cancer vaccines, cell–particle hybrids can play a role in targeted drug delivery. A study has engineered human bone marrow-derived mesenchymal stem cells with 40 nm neutravidin-coated nanoparticles [60]. The fabricated cellular particle patches are suggested to be capable of targeting chemotherapeutic agents to tumor cells via the tumoritropic properties of stem cells. In this study, cells were biotinylated with sulfo-NHS-LC-biotin and the synthesized patches were characterized with laser scanning confocal microscopy and scanning electron microscopy.

7.5 CONCLUSION

Cell surface engineering is only one of many areas of research in which avidin–biotin crosslinking plays a vital role. As cell engineering technologies grow more complex, increasing numbers of researchers are relying on the strong and stable avidin–biotin interaction as their cementing material in modifying cell surfaces and building more complex hierarchal structures. The advances in genetic engineering have enabled more control over the binding properties of avidin and its analogs allowing more precise use of the avidin–biotin complex. Such advances will pave the way for more clinically relevant approaches using biotin–avidin crosslinking such as engineering cell surfaces for vaccine applications and targeted drug delivery.

References

[1] Stevens MM, George JH. Exploring and engineering the cell surface interface. Science 2005;310 (5751):1135–8.

[2] Stephan MT, Irvine DJ. Enhancing cell therapies from the outside in: cell surface engineering using synthetic nanomaterials. Nano Today 2011;6(3):309–25.

[3] Rabuka D, Forstner MB, Groves JT, Bertozzi CR. Noncovalent cell surface engineering: incorporation of bioactive synthetic glycopolymers into cellular membranes. J Am Chem Soc 2008;130(18):5947–53.

[4] Medof ME, Nagarajan S, Tykocinski ML. Cell-surface engineering with GPI-anchored proteins. FASEB J 1996;10(5):574–86.

[5] Xia L, McDaniel JM, Yago T, Doeden A, McEver RP. Surface fucosylation of human cord blood cells augments binding to P-selectin and E-selectin and enhances engraftment in bone marrow. Blood 2004;104(10):3091–6.

[6] Sackstein R. Directing stem cell trafficking via GPS. Methods Enzymol 2010;479:93–105.

[7] Stephan MT, Moon JJ, Um SH, Bershteyn A, Irvine DJ. Therapeutic cell engineering with surface-conjugated synthetic nanoparticles. Nat Med 2010;16(9):1035–41.

[8] Buchner A, Pohla H, Willimsky G, Frankenberger B, Frank R, Baur-Melnyk A, et al. Phase 1 trial of allogeneic gene-modified tumor cell vaccine RCC-26/CD80/IL-2 in patients with metastatic renal cell carcinoma. Hum Gene Ther 2010;21(3):285–97.

[9] Grille S, Moreno M, Brugnini A, Lens D, Chabalgoity JA. A therapeutic vaccine using Salmonella-modified tumor cells combined with interleukin-2 induces enhanced antitumor immunity in B-cell lymphoma. Leuk Res 2012;37(3):341–9.

[10] Gupta R, Emens LA. GM-CSF-secreting vaccines for solid tumors: moving forward. Discov Med 2010;10 (50):52–60.

[11] Kayaga J, Souberbielle BE, Sheikh N, Morrow WJ, Scott-Taylor T, Vile R, et al. Anti-tumour activity against B16-F10 melanoma with a GM-CSF secreting allogeneic tumour cell vaccine. Gene Ther 1999;6(8):1475–81.

[12] Kresge N, Simoni RD, Hill RL. The discovery of avidin by Esmond E. Snell. J Biol Chem 2004;279(41):e5.

[13] Gitlin G, Bayer EA, Wilchek M. Studies on the biotin-binding sites of avidin and streptavidin. Tyrosine residues are involved in the binding site. Biochem J 1990;269(2):527–30.

[14] Holmberg A, Blomstergren A, Nord O, Lukacs M, Lundeberg J, Uhlen M. The biotin–streptavidin interaction can be reversibly broken using water at elevated temperatures. Electrophoresis 2005;26 (3):501–10.

[15] Sano T, Vajda S, Reznik GO, Smith CL, Cantor CR. Molecular engineering of streptavidin. Ann N Y Acad Sci 1996;799:383–90.

[16] Karhemo P-R, Ravela S, Laakso M, Ritamo I, Tatti O, Mäkinen S, et al. An optimized isolation of biotinylated cell surface proteins reveals novel players in cancer metastasis. J Proteomics 2012;77:87–100.

[17] Xiao BG. The application of reagents of a biotin–avidin system in detecting anti-acetylcholine receptor antibody. Chin J Neurol Psychiatry 1989;22 (1):5–7 61

[18] Howarth M, Ting AY. Imaging proteins in live mammalian cells with biotin ligase and monovalent streptavidin. Nat Protoc 2008;3(3):534–45.

[19] Howarth M, Takao K, Hayashi Y, Ting AY. Targeting quantum dots to surface proteins in living cells with biotin ligase. Proc Natl Acad Sci USA 2005;102 (21):7583–8.

[20] Heitzmann H, Richards FM. Use of the avidin–biotin complex for specific staining of biological membranes in electron microscopy. Proc Natl Acad Sci USA 1974;71(9):3537–41.

[21] Hamon M, Ozawa T, Montagne K, Kojima N, Ishii R, Yamaguchi S, et al. Avidin–biotin-based approach to forming heterotypic cell clusters and cell sheets on a gas-permeable membrane. Biofabrication 2011;3(3):034111.

[22] Krishnamachari Y, Pearce ME, Salem AK. Self-assembly of cell–microparticle hybrids. Adv Mater 2008;20(5):989–93.

[23] DeLange RJ, Huang TS. Egg white avidin. 3. Sequence of the 78-residue middle cyanogen bromide peptide. Complete amino acid sequence of the protein subunit. J Biol Chem 1971;246(3):698–709.

[24] Elo HA, Korpela J. The occurrence and production of avidin: a new conception of the high-affinity biotin-binding protein. Comp Biochem Physiol B: Comp Biochem 1984;78(1):15–20.

[25] Nguyen TT, Sly KL, Conboy JC. Comparison of the energetics of avidin, streptavidin, neutrAvidin, and anti-biotin antibody binding to biotinylated lipid bilayer examined by second-harmonic generation. Anal Chem 2012;84(1):201–8.

[26] Laitinen OH, Hytönen VP, Nordlund HR, Kulomaa MS. Genetically engineered avidins and streptavidins. Cell Mol Life Sci 2006;63(24):2992–3017.

[27] Livnah O, Bayer EA, Wilchek M, Sussman JL. Three-dimensional structures of avidin and the avidinbiotin complex. Proc Natl Acad Sci USA 1993;90 (11):5076–80.

[28] Laitinen OH, Hytonen VP, Ahlroth MK, Pentikainen OT, Gallagher C, Nordlund HR, et al. Chicken avidin-related proteins show altered biotin-binding and physico-chemical properties as compared with avidin. Biochem J 2002;363(Pt 3):609–17.

[29] Lazaridis T, Masunov A, Gandolfo F. Contributions to the binding free energy of ligands to avidin and streptavidin. Proteins: Struct, Funct, Bioinf 2002;47 (2):194–208.

[30] Chaiet L, Wolf FJ. The properties of streptavidin, a biotin-binding protein produced by Streptomycetes. Arch Biochem Biophys 1964;106(0):1–5.

[31] Melville DB, Moyer AW, Hofmann K, du Vigneaud V. The structure of biotin: the formation of thiophenevaleric acid from biotin. J Biol Chem 1942;146(2):487–92.

[32] McKay BE, Molineux ML, Turner RW. Biotin is endogenously expressed in select regions of the rat central nervous system. J Comp Neurol 2004;473(1):86–96.

[33] Marttila AT, Hytonen VP, Laitinen OH, Bayer EA, Wilchek M, Kulomaa MS. Mutation of the important Tyr-33 residue of chicken avidin: functional and structural consequences. Biochem J 2003;369(Pt 2):249–54.

[34] Green NM. Avidin. Adv Protein Chem 1975;29:85–133.

[35] Meyer SC, Gaj T, Ghosh I. Highly selective cyclic peptide ligands for NeutrAvidin and avidin identified by phage display. Chem Biol Drug Des 2006;68(1):3–10.

[36] Marttila AT, Laitinen OH, Airenne KJ, Kulik T, Bayer EA, Wilchek M, et al. Recombinant NeutraLite avidin: a non-glycosylated, acidic mutant of chicken avidin that exhibits high affinity for biotin and low non-specific binding properties. FEBS Lett 2000;467 (1):31–6.

[37] Sperling RA, Parak WJ. Surface modification, functionalization and bioconjugation of colloidal inorganic nanoparticles. Philos Trans A Math Phys Eng Sci 2010;368(1915):1333–83.

[38] Kobayashi H, Sakahara H, Endo K, Hosono M, Yao Z-S, Toyama S, et al. Comparison of the chase effects of avidin, streptavidin, neutravidin, and avidin–ferritin on a radiolabeled biotinylated anti-tumor monoclonal antibody. Cancer Sci 1995;86(3):310–4.

[39] Vermette P, Gengenbach T, Divisekera U, Kambouris PA, Griesser HJ, Meagher L. Immobilization and surface characterization of NeutrAvidin biotin-binding protein on different hydrogel interlayers. J Colloid Interface Sci 2003;259(1):13–26.

[40] Green NM. Avidin and streptavidin. Methods Enzymol 1990;184:51–67.

[41] Alon R, Bayer EA, Wilchek M. Streptavidin contains an RYD sequence which mimics the RGD receptor domain of fibronectin. Biochem Biophys Res Commun 1990;170(3):1236–41.

[42] Bayoussef Z, Shakesheff KM. Aggregation of cells using biomaterials and bioreactors. In: Polak JM, editor. Advances in tissue engineering. London, UK: Imperial College Press; 2008.

[43] Marttila AT, Airenne KJ, Laitinen OH, Kulik T, Bayer EA, Wilchek M, et al. Engineering of chicken avidin: a progressive series of reduced charge mutants. FEBS Lett 1998;441(2):313–7.

[44] O'Sullivan VJ, Barrette-Ng I, Hommema E, Hermanson GT, Schofield M, Wu S-C, et al. Development of a tetrameric streptavidin mutein with reversible biotin binding capability: engineering a mobile loop as an exit door for biotin. PLoS ONE 2012;7(4):e35203.

[45] Laitinen OH, Marttila AT, Airenne KJ, Kulik T, Livnah O, Bayer EA, et al. Biotin induces tetramerization of a recombinant monomeric avidin. A model for protein–protein interactions. J Biol Chem 2001;276 (11):8219–24.

[46] Laitinen OH, Nordlund HR, Hytonen VP, Uotila ST, Marttila AT, Savolainen J, et al. Rational design of an active avidin monomer. J Biol Chem 2003;278 (6):4010–4.

[47] Wu SC, Wong SL. Engineering soluble monomeric streptavidin with reversible biotin binding capability. J Biol Chem 2005;280(24):23225–31.

[48] Lim KH, Huang H, Pralle A, Park S. Stable, high-affinity streptavidin monomer for protein labeling and monovalent biotin detection. Biotechnol Bioeng 2013;110(1):57–67.

[49] Howarth M, Chinnapen DJ, Gerrow K, Dorrestein PC, Grandy MR, Kelleher NL, et al. A monovalent streptavidin with a single femtomolar biotin binding site. Nat Methods 2006;3(4):267–73.

[50] Sano T, Cantor CR. Intersubunit contacts made by tryptophan 120 with biotin are essential for both strong biotin binding and biotin-induced tighter subunit association of streptavidin. Proc Natl Acad Sci USA 1995;92(8):3180–4.

[51] Chilkoti A, Tan PH, Stayton PS. Site-directed mutagenesis studies of the high-affinity streptavidin–biotin complex: contributions of tryptophan residues 79, 108, and 120. Proc Natl Acad Sci USA 1995;92(5):1754–8.

[52] Grana CM, Chinol M, De Cicco C, Bartolomei M, Cremonesi M, Bodei L, et al. Eleven-year experience with the avidin–biotin pretargeting system in glioblastoma: toxicity, efficacy and survival. Open Nucl Med J 2012;4:6.

[53] Gao J, Huang S, Li M, Luo R, Wang X, Takashima A. GM-CSF-surface-modified B16-F10 melanoma cell vaccine. Vaccine 2006;24(25):5265–8.

[54] Paganelli G, Chinol M, Maggiolo M, Sidoli A, Corti A, Baroni S, et al. The three-step pretargeting approach reduces the human anti-mouse antibody response in patients submitted to radioimmunoscintigraphy and radioimmunotherapy. Eur J Nucl Med 1997;24(3):350–1.

[55] Petronzelli F, Pelliccia A, Anastasi AM, Lindstedt R, Manganello S, Ferrari LE, et al. Therapeutic use of avidin is not hampered by antiavidin antibodies in humans. Cancer Biother Radiopharm 2010;25(5):563–70.

[56] Chinol M, Casalini P, Maggiolo M, Canevari S, Omodeo ES, Caliceti P, et al. Biochemical modifications of avidin improve pharmacokinetics and biodistribution, and reduce immunogenicity. Br J Cancer 1998;78(2):189–97.

[57] Hermanson GT. Biotinylation reagents, in Bioconjugate Techniques. Amsterdam: Elsevier Science; 2008. p. 506–545.

[58] Drotleff S, Lungwitz U, Breunig M, Dennis A, Blunk T, Tessmar J, et al. Biomimetic polymers in pharmaceutical and biomedical sciences. Eur J Pharm Biopharm 2004;58(2):385–407.

[59] Sarkar D, Vemula PK, Teo GS, Spelke D, Karnik R, Wee le Y, et al. Chemical engineering of mesenchymal stem cells to induce a cell rolling response. Bioconjug Chem 2008;19(11):2105–9.

[60] Cheng H, Kastrup CJ, Ramanathan R, Siegwart DJ, Ma M, Bogatyrev SR, et al. Nanoparticulate cellular patches for cell-mediated tumoritropic delivery. ACS Nano 2010;4(2):625–31.

[61] Mendis D, Ginon I, Louis H, McGregor JL, Poston RN. Cell biotinylation provides a sensitive and effective detection technique for cellular adhesion assays: comparison with existing methods. J Immunol Methods 2001;253(1–2):57–68.

[62] Altin JG, Pagler EB. A one-step procedure for biotinylation and chemical cross-linking of lymphocyte surface and intracellular membrane-associated molecules. Anal Biochem 1995;224(1):382–9.

[63] Saxon E, Bertozzi CR. Cell surface engineering by a modified Staudinger reaction. Science 2000;287 (5460):2007–10.

[64] Van Lenten L, Ashwell G. Studies on the chemical and enzymatic modification of glycoproteins. A general method for the tritiation of sialic acid-containing glycoproteins. J Biol Chem 1971;246(6):1889–94.

[65] Varki A, Gagneux P. Multifarious roles of sialic acids in immunity. Ann N Y Acad Sci 2012;1253:16–36.

[66] Gothard D, Roberts SJ, Shakesheff KM, Buttery LD. Controlled embryoid body formation via surface modification and avidin–biotin cross-linking. Cytotechnology 2009;61(3):135–44.

[67] De Bank PA, Kellam B, Kendall DA, Shakesheff KM. Surface engineering of living myoblasts via selective periodate oxidation. Biotechnol Bioeng 2003;81 (7):800–8.

[68] Wood I, De Bank P, Kellam B, Shakesheff K, Kendall D. Cell surface modification of l6 myoblasts via sodium periodate treatment. Eur Cells Mater 2002;4 (Suppl. 2):2.

[69] Magnani M, Chiarantini L, Mancini U. Preparation and characterization of biotinylated red blood cells. Biotechnol Appl Biochem 1994;20(Pt 3):335–45.

[70] Mahal LK, Bertozzi CR. Engineered cell surfaces: fertile ground for molecular landscaping. Chem Biol 1997;4(6):415–22.

[71] Lemieux GA, Bertozzi CR. Chemoselective ligation reactions with proteins, oligosaccharides and cells. Trends Biotechnol 1998;16(12):506–13.

[72] Mahal LK, Yarema KJ, Bertozzi CR. Engineering chemical reactivity on cell surfaces through oligosaccharide biosynthesis. Science 1997;276(5315):1125–8.

[73] Prescher JA, Dube DH, Bertozzi CR. Chemical remodelling of cell surfaces in living animals. Nature 2004;430(7002):873–7.

[74] Sarda S, Pointu D, Pincet F, Henry N. Specific recognition of macroscopic objects by the cell surface: evidence for a receptor density threshold revealed by micrometric particle binding characteristics. Biophys J 2004;86(5):3291–303.

[75] Meier W. Reversible cell aggregation induced by specific ligand – receptor coupling. Langmuir 1999;16 (3):1457–9.

[76] West JL. Protein-patterned hydrogels: customized cell microenvironments. Nat Mater 2011;10(10):727–9.

[77] Petit V, Thiery J-P. Focal adhesions: structure and dynamics. Biol Cell 2000;92(7):477–94.

[78] Walluscheck KP, Steinhoff G, Kelm S, Haverich A. Improved endothelial cell attachment on ePTFE vascular grafts pretreated with synthetic RGD-containing peptides. Eur J Vasc Endovasc Surg 1996;12(3):321–30.

[79] Lee J, Cuddihy MJ, Kotov NA. Three-dimensional cell culture matrices: state of the art. Tissue Eng B Rev 2008;14(1):61–86.

[80] Mathur AB, Chan BP, Truskey GA, Reichert WM. High-affinity augmentation of endothelial cell attachment: long-term effects on focal contact and actin filament formation. J Biomed Mater Res A 2003;66 (4):729–37.

[81] Bhat VD, Klitzman B, Koger K, Truskey GA, Reichert WM. Improving endothelial cell adhesion to vascular graft surfaces: clinical need and strategies. J Biomater Sci Polym Ed 1998;9(11):1117–35.

[82] Bhat VD, Truskey GA, Reichert WM. Using avidin-mediated binding to enhance initial endothelial cell attachment and spreading. J Biomed Mater Res 1998;40(1):57–65.

[83] Patel N, Padera R, Sanders GH, Cannizzaro SM, Davies MC, Langer R, et al. Spatially controlled cell engineering on biodegradable polymer surfaces. FASEB J 1998;12(14):1447–54.

[84] Sinclair J, Salem AK. Rapid localized cell trapping on biodegradable polymers using cell surface derivatization and microfluidic networking. Biomaterials 2006;27(9):2090–4.

[85] Clapper JD, Pearce ME, Guymon CA, Salem AK. Biotinylated biodegradable nanotemplated hydrogel networks for cell interactive applications. Biomacromolecules 2008;9(4):1188–94.

[86] Chia S-M, Lin P-C, Yu H. TGF-β1 regulation in hepatocyte-NIH3T3 co-culture is important for the enhanced hepatocyte function in 3D microenvironment. Biotechnol Bioeng 2005;89(5):565–73.

[87] Karussis D, Karageorgiou C, Vaknin-Dembinsky A, Gowda-Kurkalli B, Gomori JM, Kassis I, et al. Safety and immunological effects of mesenchymal stem cell transplantation in patients with multiple sclerosis and amyotrophic lateral sclerosis. Arch Neurol 2010;67(10):1187–94.

[88] Geary SM, Lemke CD, Lubaroff DM, Salem AK. Proposed mechanisms of action for prostate cancer vaccines. Nat Rev Urol 2013;10(3):149–60.

[89] Trowbridge JJ, Xenocostas A, Moon RT, Bhatia M. Glycogen synthase kinase-3 is an *in vivo* regulator of hematopoietic stem cell repopulation. Nat Med 2006;12(1):89–98.

[90] Dranoff G, Jaffee E, Lazenby A, Golumbek P, Levitsky H, Brose K, et al. Vaccination with irradiated tumor cells engineered to secrete murine granulocyte-macrophage colony-stimulating factor stimulates potent, specific, and long-lasting anti-tumor immunity. Proc Natl Acad Sci USA 1993;90 (8):3539–43.

[91] Sandler AD, Chihara H, Kobayashi G, Zhu X, Miller MA, Scott DL, et al. CpG oligonucleotides enhance the tumor antigen-specific immune response of a granulocyte macrophage colony-stimulating factor-based vaccine strategy in neuroblastoma. Cancer Res 2003;63(2):394–9.

[92] Shirota H, Klinman DM. CpG-conjugated apoptotic tumor cells elicit potent tumor-specific immunity. Cancer Immunol Immunother 2011;60(5):659–69.

[93] Salgia R, Lynch T, Skarin A, Lucca J, Lynch C, Jung K, et al. Vaccination with irradiated autologous tumor cells engineered to secrete granulocyte-macrophage colony-stimulating factor augments antitumor immunity in some patients with metastatic non-small-cell lung carcinoma. J Clin Oncol 2003;21(4):624–30.

[94] Tarr P. Granulocyte-macrophage colony-stimulating factor and the immune system. Med Oncol 1996;13 (3):133–40.

[95] Murphy KP. T cell mediated immunity. Janeway's immunobiology. New York, NY: Garland Science; 2011. p. 335–386

[96] Sarkar D, Vemula PK, Spelke DP, Karp JM. Cell surface conjugation of sialyl Lewis X induces a rolling response for mesenchymal stem cells. In: Bioeng Conference, 2009 IEEE 35th Annual Northeast; 2009.

[97] Sackstein R, Merzaban JS, Cain DW, Dagia NM, Spencer JA, Lin CP, et al. *Ex vivo* glycan engineering of CD44 programs human multipotent mesenchymal stromal cell trafficking to bone. Nat Med 2008;14(2):181–7.

[98] Ohyama C, Tsuboi S, Fukuda M. Dual roles of sialyl Lewis X oligosaccharides in tumor metastasis and rejection by natural killer cells. EMBO J 1999;18 (6):1516–25.

[99] Sarkar D, Spencer JA, Phillips JA, Zhao W, Schafer S, Spelke DP, et al. Engineered cell homing. Blood 2011;118(25):e184–91.

[100] Brewer E, Coleman J, Lowman A. Emerging technologies of polymeric nanoparticles in cancer drug delivery. J Nanomaterials 2011;:**2011**.

[101] Nierkens S, den Brok MH, Sutmuller RP, Grauer OM, Bennink E, Morgan ME, et al. *In vivo* colocalization of antigen and CpG [corrected] within dendritic cells is associated with the efficacy of cancer immunotherapy. Cancer Res 2008;68(13):5390–6.

CHAPTER

8

Construction and Computation with Nucleic Acids on the Cell Surface

Alexey Y. Koyfman[1] *and Norbert O. Reich*[2]

[1]Department of Biochemistry and Molecular Biology, and Sealy Center for Structural Biology and Molecular Biophysics, University of Texas Medical Branch, Galveston, TX, USA,
[2]Biomolecular Science and Engineering Program, and Department of Chemistry and Biochemistry, University of California Santa Barbara, Santa Barbara, CA, USA

OUTLINE

Cell surfaces provide a fertile ground for the construction of architectures with diverse applications. This chapter explores the capabilities of nucleic acids to perform computations on cell surface, attach nanostructures to cells, bridge cells, activate innate cellular responses, and build transmembrane channels and pores. We outline the multiple reasons researchers are interested in identifying the cell type, modifying a surface of a cell, and attaching an

J. M. Karp & W. Zhao (Eds): Micro- and Nanoengineering of the Cell Surface.
DOI: http://dx.doi.org/10.1016/B978-1-45-573146-6.00008-8

endogenous structure to the cell of choice. We also explain the basis for the emergent interest and new methods to rationally modify cell surfaces.

DNA is the preferred biological material to store and transfer genetic information. RNA can fold into intricate three-dimensional (3D) structures utilizing a number of naturally occurring structural motifs [1] and is used for enzymatic catalysis, and gene expression and regulation. Both nucleic acids show promise in bionanotechnology due to their defined length, unique base pair recognition capabilities, and possible base and backbone modifications.

In order to create sophisticated nanomachines, a complex asymmetrical structure has to be constructed. Since each position of the DNA sequence is addressable, DNA is a material that allows creating structures with broken symmetry—placing chemical groups at specific locations. In principle, any functional group can be placed anywhere in the DNA structure with high spatial precision. Chemistry is in a powerful position to enhance the capabilities of nucleic acid nanotechnology by adding tailored functionality through the selective modification of nucleic acid's bases and backbone with chemical tags.

Small molecules such as cholesterol [2], peptides [3], enzymes [4], antibodies [5,6], quantum dots [7,8], inorganic materials [9,10], and carbon nanotubes [11] have been positioned at precise locations within DNA nanostructures.

DNA origami is a complex asymmetrical structure where every nucleotide is unique and addressable (Figure 8.1A and B). DNA origami, named after the Japanese art of paper folding, is composed of a multiple kilobase single-stranded DNA scaffold, often obtained from the M13 phage genome, which is folded into a designed shape with hundreds of short oligonucleotide staple strands (Figure 8.1A) [15]. The staples are short, single-stranded oligonucleotides that are complementary to sequences in multiple locations of the single-stranded M13 phage DNA scaffold and can therefore bridge distant sites. Therefore, it is possible to staple and fold the system into any

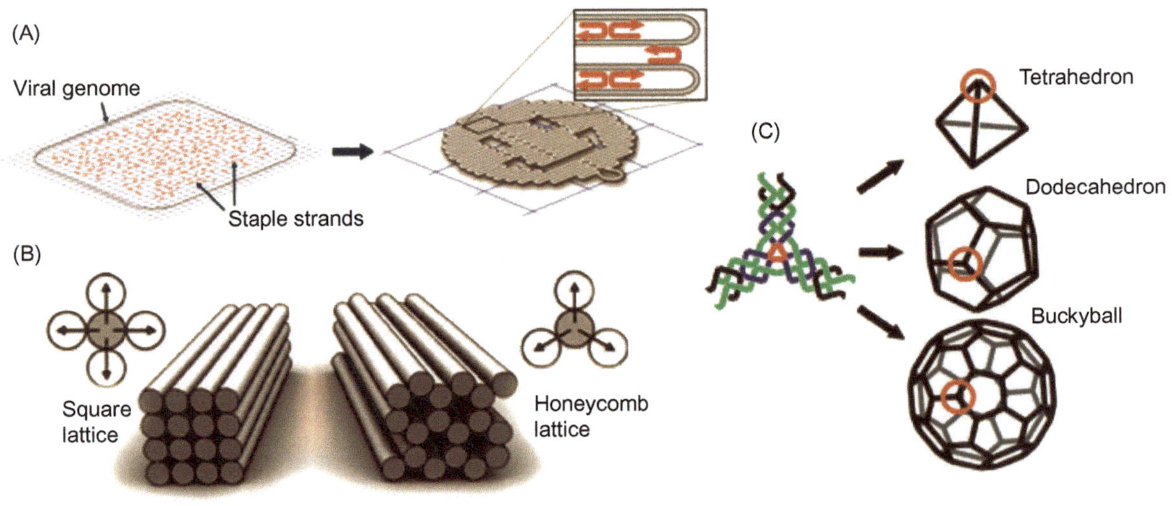

FIGURE 8.1 Construction of DNA nanostructures: (A) stapling a 2D DNA origami with hundreds of strands, (B) 3D DNA origami on square and honeycomb lattices, and (C) 3D DNA objects from three strands. Source: *Adapted from (A) Ref. [12], (B) Ref. [13], and (C) Ref. [14].*

3D shape. DNA origami structures form in a one-pot thermal annealing reaction when scaffold and staple strands are mixed. DNA origami has been assembled into a variety of 3D shapes (Figure 8.1B) [16–18] and can be used as a nanoscale template to precisely position diverse components [19–21].

An asymmetric object like DNA origami composed of hundreds of unique oligonucleotides is expensive and not always desired. To minimize the number of DNA strands and simplify the design, symmetry has to be introduced (Figure 8.1C) [14,22–24]. Icosahedral viruses have C60 symmetry and structurally cannot be as programmable as DNA origami [25]. Tetrahedral, dodecahedral, and icosohedral DNA architectures were also developed with DNA [14] (Figure 8.1C). The advantage of this approach is that the same molecule can be expressed multiple times on the surface of symmetrical objects and can be used for certain applications when you do not want to position a specific molecule in one particular location. Also, only three oligonucleotides are needed to construct an intricate structure like a DNA icosohedron (Figure 8.1C) [14].

8.1 COMPUTATION OF CELL IDENTITY

Currently it is possible to accurately identify an unknown cell type using nucleic acid computations. An easily identifiable DNA fluorescent nanorobot targets only a specific receptor or multiple receptors on the cell. Alternatively, it is possible to target a specific cell type known to be in the cellular mixture by attaching a directed nanostructure.

A cell membrane contains many cell surface proteins also known as surface antigens that are unique to each cell type. Over 200 types of cell are histologically identifiable in humans. Furthermore, various cell types express multiple cell surface markers, also known as "clusters of differentiation" (CDs) [26]. These surface antigens can be used to identify both the cell type and the functional state of the cell. Healthy and diseased cells can be also classified by inventories of their cell surface markers [27,28]. Most identified cell surface markers are not exclusively expressed on the target population of diseased cells. A marker overexpressed on the surface of cancer cells is often also expressed at a lower concentration on the surface of some normal cells. The distribution of any given protein type on a cell surface is not uniform. Cells will have very common "housekeeping" antigens and sometimes very unusual antigens [29]. Cells can be triggered to perform a certain function or behave in a certain manner upon binding by antibodies or drugs. Some drugs target disease-causing cells by binding to a specific receptor. Unfortunately, some disease-causing cells do not have unique markers and therefore drugs also bind to non-targeted healthy cells and cause "off-target" toxicities [30]. This is sometimes the case for cancer therapy utilizing antibody–drug conjugates (ADCs). For example in leukemia, both healthy and diseased subpopulations of white blood cells display surface markers that are indistinguishable by the current single-receptor antibody therapy, leading to serious complications [31] In comparison, a more practical and less risk-prone approach would simultaneously assess multiple surface receptors to pinpoint specific disease cells and enhance diagnostic accuracy [32,33].

The ability to specifically label a particular and narrow cell type subpopulation within a much larger population of related cells for the purpose of cellular elimination [34], isolation and analysis [35], and imaging [30] has many potential benefits. This challenge could be readily addressed in a direct manner if targeted subpopulations could have some unique cell surface marker against which antibodies could be raised [30].

8.1.1 Accurate Identification of the Cell Type

Accurate identification of the cell type depends on having sufficient information about what uniquely defines that cell. Although a single, cell-specific antigen is sometimes known, it is frequently necessary to evaluate multiple surface antigens to pinpoint a specific cell. Since a cell contains a number of different cell surface antigens, a molecular device that is able to bind multiple receptors will be able to pinpoint a specific cell more accurately. Four to five unique antigens would suffice to specifically identify a particular cell within a multicellular population [36].

Sergei Rudchenko, Milan Stojanovic, and coworkers at Columbia University reported in *Nature Nanotechnology* [36] that they have harnessed the power of DNA to identify a cell by performing computations directly on the cell surface. Blood cells were chosen for the experimental design since these cells are well understood [37]. Blood cells are often characterized using flow cytometry according to their different levels of expression of the CDs. Lineages and stages of differentiation in blood cells are defined by the presence or absence of multiple cell surface markers. The team of scientists designed molecular robots that can identify multiple cell surface antigens, thereby effectively labeling more specific subpopulations of cells. Each probe molecule is composed of an antibody and a synthetic DNA duplex. While the targeting antibody plays an identification role, the DNA executes the computation (Figure 8.2). Cells can be labeled in 15 min using this technique. The researchers also showed that the targeted DNA strand could be programmed to fluoresce in the cellular solution. These molecular robots can essentially label a more specific subsets of cell populations therefore allowing for more targeted therapy. Researchers could either label cells that they want to target or avoid the labeled cells.

The computation YESCD45YESCD20 would only label B cells and avoid T cells (Figure 8.2). YESCD45YESCD3 would only label T cells, and the computation YESCD3YESCD20, is a negative control, since no cells display these two markers at the same time. No labeling was observed in YESCD3YESCD20 two-step computation which indicated that the T cells did not exchange DNA oligonucleotides with B cells through diffusion or through direct physical contact between B and T cells. Rudchenko et al. present a simple method, where a number

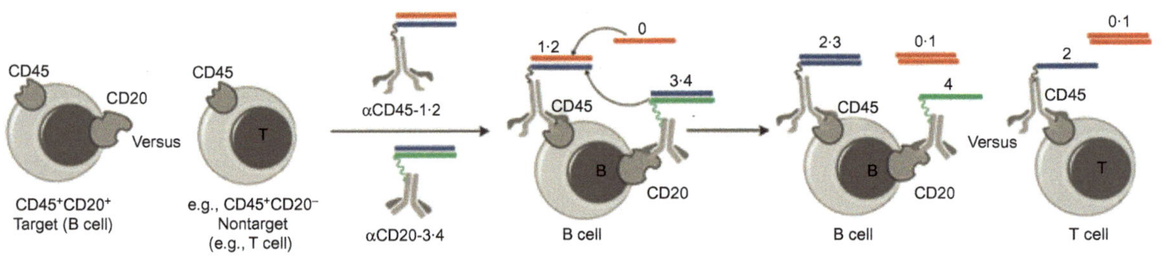

FIGURE 8.2 Computational scheme operating on a target B cell with $CD45^{+}CD20^{+}$ antigens and on nontargeted T cell with $CD45^{+}CD20^{-}$ antigens. Oligonucleotides, displayed as colored horizontal lines, are attached to antibodies, displayed as Y-shaped structures. A computation is a transfer of oligonucleotides from one antigen to another and is driven by an increased number of complementary base pairs. As the result of the computation, the unique single-stranded oligonucleotide 4 is only present on the targeted B cells. Source: *Adapted from Ref. [36].*

of straightforward components are assembled on the cell surface to compute the cell type.

8.1.2 Logic-Based Cancer Targeting and Therapy

A related DNA-based approach [38] combines the structure-switching properties of DNA aptamers with toehold-mediated strand displacement reactions. DNA Nano-Claw quantitates multiple cancer cell surface markers, produces a diagnostic signal, and autonomously induces therapeutic operations. The targeted therapeutic effect is triggered by these structure-switchable aptamer conjugates to induce photodynamic therapy (PDT) [39,40].

Both DNA-based methods are easy to reproduce and reliable because they use standard antibodies and commercially available synthetic DNA. Both systems are also modular and scalable in principle. There is a wide range of possible clinical applications if molecular robots work in mice studies and human clinical trials. For example, cancer patients could benefit from more targeted chemotherapeutics. Also, drugs for autoimmune diseases could be more specifically tailored to impact disease-causing autoimmune cells and not the immune cells needed to fight infection.

With advancement of DNA computations, molecular probes may be used to find viable bloodstream targets. However, in order to proceed to the routine *in vivo* use of molecular DNA computational robots, a few major barriers have to be addressed. DNA robots have to be distributed well within the body. Molecules with mass below 500 Da and relatively low polarity are traditionally used in medicine [41]. Unfortunately, nucleic acids do not diffuse rapidly due to a mass well above small molecules. Also DNA by itself does not cross membranes well. For other *in vivo* targets alternative delivery methods will be required.

8.2 TARGETED TRANSPORT OF A PAYLOAD TO A CELL SURFACE AND ACTIVATION OF INNATE CELLULAR RESPONSE

DNA nanostructures capable of sensing cell surface antigens were used to transport molecular payloads to cells and activate intracellular transformations.

The ability to organize materials into defined shapes is a core goal of bionanotechnology. Cell surface engineering [42] seeks to localize nanoscale materials onto cellular membranes. The defined organization of nanoscale objects on the cell surface has a number of biomedical applications. Some examples include the organization of cells into predictable architectures on surfaces [43,44] and the delivery of diverse molecules to cells [45]. Cells have been assembled into microtissues using DNA-mediated interactions [46].

8.2.1 Cellular DNA Patches Targeted to Cancer Cells

Self-assembled DNA arrays were assembled from three single-stranded DNA oligonucleotide strands into hexagonal units repeated every 30 nm (Figure 8.3A). Hexagonal DNA arrays assemble when the three strands cool down from boiling water temperature to room temperature. The size of the DNA patch is controllable with concentration of divalent magnesium ions from compact multilayers (Figure 8.3B) and into extended networks spanning several microns [23].

Self-assembled DNA arrays can be directed to the surface of cells through biotin–streptavidin (STV) interactions and specific antibody–cell surface interactions [24] (Figures 8.4 and 8.5). The versatile cargo-carrying ability of arrays is explored for directing cell–surface interactions, cell–cell bridging, and positioning multiple cells onto a DNA fabric (Figure 8.4).

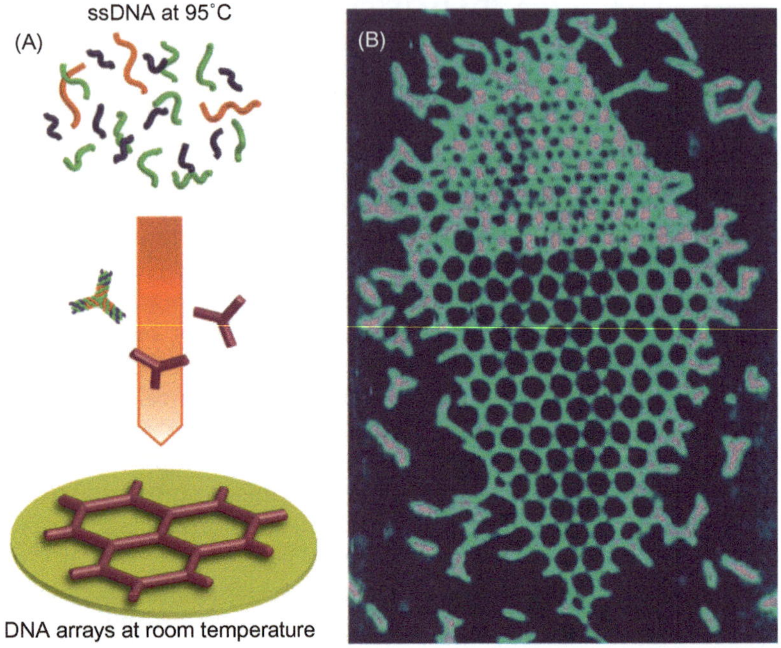

FIGURE 8.3 (A) Diagram of a DNA array assembly. As single-stranded DNA molecules (red, green, and blue) cool down, they first self-assemble into three-point star motifs and then into 2D DNA arrays (purple). (B) AFM of a DNA array.

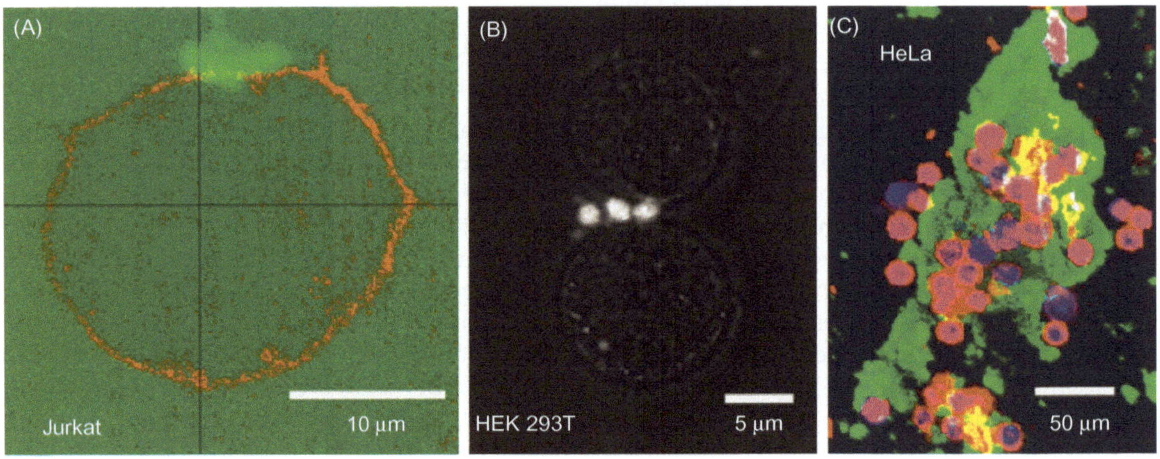

FIGURE 8.4 Self-assembled DNA arrays are attached to the surface of three different cell lines. (A) One DNA patch is attached to a Jurkat cell. (B) A DNA patch is bridging two HEK 293T cells. (C) DNA fabric enclosed multiple HeLa cells. *Source: Adapted from Ref. [24].*

We present two strategies to attach hexagonal DNA arrays to human cells. First, biotin-modified arrays were bound to biotinylated cancer cells using STV as a bridging component. Our second strategy for directing DNA arrays to cancer cells builds on the previous strategy but uses interactions between native cell-specific surface proteins and their antibodies.

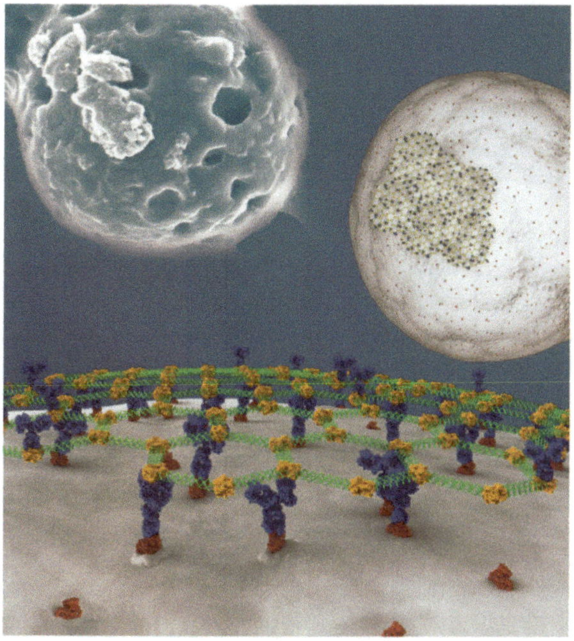

FIGURE 8.5 Self-assembled DNA arrays can be directed to the surface of cells through specific antibody–cell surface interactions. The versatile cargo-carrying ability of arrays is explored for directing cell–surface interactions, cell–cell bridging, and positioning multiple cells onto a DNA fabric. SEM of a cell containing a DNA patch is shown in the top left corner. *Source: Illustration by Peter Allen.*

This approach provides a basis for directing the assembly of distinct cell types through the use of various cell-specific surface markers. Epidermal growth factor (EGF) receptors (EGFRs) are overexpressed on the surface of many human cancer cells, at levels of up to 10^6 per cell [47,48]. We used two EGFR-expressing cell lines (HEK 293T and HeLa) and one non-expressing control cell line (Jurkat) for attachment of DNA patches through EGFR antibody. The biotin-modified arrays attached to antibodies were bound to native EGFRs using STV and antibodies as bridging components (Figure 8.5).

Micron-sized DNA–biotin–STV–Ab arrays bind to cancer cells and envelop multiple cells (Figures 8.4 and 8.5). Thus, interactions mediated by DNA assemblies as well as DNA hybridization [46] may be used to direct tissue engineering. Large numbers of antibodies would likely be required for efficient bridging of cells. Arrays optimized to uniquely bind multiple surfaces could form the framework for organizing cells on surfaces and inside 3D matrices. By incorporating gold and magnetic iron oxide nanoparticles in nucleic acid nanostructures [49], it is possible to assemble 3D tissue cultures based on magnetic levitation of cells [50]. The geometry of the cell mass can be further manipulated by spatially controlling the magnetic field. A large number of repeating units in a hexagonal DNA array provide an opportunity for loading many copies of identical or different molecules for cellular delivery or interconnection of cells. The generic biotin–STV modification allows the anchoring of a variety of biomolecules like peptides or aptamers to arrays, which might serve to identify or induce the fate of stem cells [51].

Further, since DNA networks can be systematically varied in size [15,17,23], it is possible to identify various cells by both the size and color of the fluorescent DNA patches. This could potentially be used for cancer cell screening. Having many fluorophores positioned very closely to each other on a DNA scaffold enhances sensitivity to lower intensity light used for array detection. It is possible to uniquely color fluorescent multifunctional DNA patches, identifying the location of tethered small molecules, nanoparticles, or other cargo.

DNA nanoarrays provide a number of strategies to specifically label cell surfaces with functionalized micron-sized patches, to deliver materials to cell surfaces, and to engineer cell/cell networks. The porous and periodic nature of the DNA material could be useful for tissue engineering. Functionalized nanoarrays could be used for screening cancer cells, the control of stem cell fate, and the controlled activation of immune response.

8.2.2 DNA Crosslinked Receptors Trigger Robust Signaling Responses

Antigen-mediated crosslinking of IgE bound to its receptor, FcεRI, on mast cells stimulates degranulation, phospholipid metabolism, and cytokine production leading to inflammation and allergies [52]. Spatial organization of the clustered IgE–FcεRI complexes affects the assembly of the transmembrane signaling complexes. DNA three-way junctions with tunable lengths were utilized to bring IgE–FcεRI complexes and crosslink them together [53]. Sil et al. reported that the trivalent DNA ligands triggered robust signaling responses, stimulated phosphorylation of FcεRI subunits, and lead to degranulation [53].

8.2.3 DNA Origami Barrels Activate Signaling Pathways in Target Cells

A recent paper described a method for creating tiny machines out of strands of protein and DNA [54]. Each DNA origami nanorobot measures 45 nm long × 35 nm wide, roughly the size of a virus. Both virus and DNA origami barrel can be thought of as molecular delivery trucks, virus delivering nucleic acids, and DNA barrel transporting medications to specific cells in the body. The nanoscale DNA origami barrel releases Fab antibody fragments in the presence of target cells *in vitro*.

To characterize the system, the authors delivered fluorescently labeled Fab antibody fragments to cells. DNA origami barrel showed highly specific binding to the cells that displayed the correct combination of surface antigens in mixed cell populations and whole blood leukocytes.

The constructed DNA origami barrel consists of two domains that are covalently attached in the back by single-stranded scaffold hinges (Figure 8.6). The two domains stay locked when DNA strands in the front are base paired. DNA origami barrel opens when DNA aptamer-based lock binds the antigen keys (Figure 8.6). This lock system was inspired by aptamer beacons [55] and structure-switching aptamers [56], which typically undergo target-induced switching between an aptamer–complement duplex and an aptamer–target complex (Figure 8.6). Both aptamers have to recognize their targets in order for DNA origami barrel to open and expose the inner surfaces. The DNA robot could be programmed to activate in response to a single type of key by using the same aptamer sequence in both lock sites. Alternatively, different aptamer sequences could be encoded in the locks to recognize two inputs. Both locks needed to be opened simultaneously to activate the robot. The robot remained inactive when only one of the two locks was opened. Two different aptamers specifically binding two different protein antigens were designed. This lock mechanism is thus equivalent to an AND logic gate that requires the presence of two different protein targets to be activated. When both aptamers bind their targets, a conformational change opens up the barrel and releases the cargo.

To test the generality and robustness of the aptamer-encoded logic gating, six different DNA barrels were designed using pairwise combinations of aptamer locks drawn from a set of three well-characterized aptamer sequences: 41t, against platelet-derived growth factor (PDGF) [57]; TE17 [58]; and sgc8c, against protein tyrosine kinase 7 [59].

Various leukemia cells opened DNA origami barrels only with the right aptamer locks. No DNA barrels were activated on Burkitt's lymphoma cells (Ramos). An acute myeloid leukemia cell line (Kasumi-1) activated all robots. An

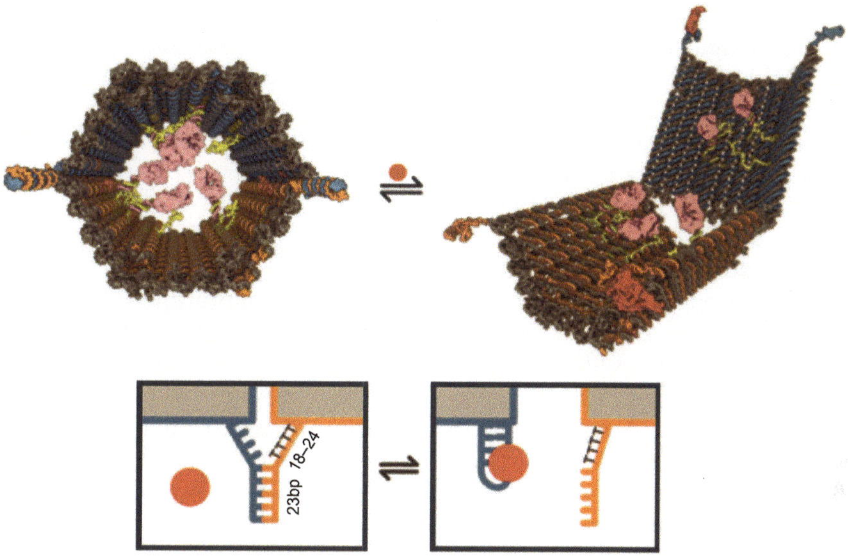

FIGURE 8.6 Aptamer-gated DNA nanorobot design. (Top left) Closed DNA nanorobot with a protein payload. Blue DNA aptamer strand and orange partially complementary strand lock the nanorobot through base pairing. Flexible hinges are located on the other side of the nanorobot. (Top right) DNA nanorobot is opened with a red antigen. (Bottom) Aptamer lock mechanism consisting of a blue DNA aptamer and a partially complementary orange strand. Red antigen key unzips DNA duplex. Source: *Adapted from Ref. [54].*

aggressive lymphocytic NK-type leukemia (NKL) cell line [60] activated DNA barrels with two 41t locks, two TE17 locks, and with one 41t and one TE17 lock. Jurkat cell line, isolated from acute T cell leukemia and acute lymphoblastic leukemia (CCRF-CEM), activated DNA barrels with two TE17 locks, two sgc8c locks, as well as one TE17 lock and one sgc8c lock. Neuroblastoma cell line (SH-SY5Y) activated DNA barrels with two sgc8c locks.

DNA origami barrels were also designed to activate signaling pathways in target cells. DNA barrels were loaded with Fab antibody fragments known to bind human CD33 and human CDw328 and induce growth arrest in leukemic cells [61]. Upon recognition of the surface antigen PDGF on cells from a patient with aggressive lymphocytic NKL, the barrel structure opened, allowing the antibody payload to bind to cell surface receptors and stopped the growth of the targeted NKL cells.

DNA origami barrels were also loaded with a combination of antibodies against a T cell surface receptor and flagellin, the protein making up bacterial flagella. Flagellin is known to promote T cell activation [62]. DNA nanorobots collected flagellin from the solution environment of the cells and induced activation of T cells.

Much work remains to take these ideas toward successful drug delivery. For example, DNA origami barrels might be unstable in the bloodstream in the presence of nucleases and other enzymes. It might be necessary to chemically modify the DNA backbone to increase resistance to degradation. Also, aptamer-encoded locks may lose specificity and efficiency in protein-rich serum. In addition, delivery into cells is required for many intracellular targets. DNA strands and large DNA nanostructures have difficult penetrating biological membranes. In order to cross the membrane and increase cellular uptake, amphiphilic molecules or peptides could be displayed on the surface of DNA origami barrels. Another concern is that DNA origami is a complex structural molecule that requires hundreds of unique oligonucleotides to assemble. The cost to synthesize sufficient amounts of customized long-scaffold DNA has to decrease. On the other hand, it may be necessary to minimize the number of DNA strands and simplify the design to increase the production of the nanostructure as a medicine [14,22–24].

Currently, we treat cancer with chemical and nuclear weapons: chemotherapy and radiation. Maybe we need to switch to smarter, more targeted therapeutics. The aim of smart drug delivery systems is to administer smaller drug doses to patients while offering improved therapeutic efficiency and fewer side effects compared with conventional drug delivery methods. Chemotherapy might shut down cancer cells but also shuts down all sorts of vital processes in the human body that we want to keep going. The toxic chemotherapy molecules lead to hair loss, nausea, and other side effects. By contrast, nanorobots could deliver medication directly to cancerous tumors. Nanorobots this small can interact directly with the surface of individual cells. The amount of medicine would be drastically decreased, since it is being delivered directly to target cells.

8.3 DNA CHANNELS AND PORES SPANNING LIPID MEMBRANE

Various kinds of DNA origami channels and pores were constructed and inserted into the lipid membrane using chemically attached hydrophobic molecules. DNA origami transmembrane channel and pores conduct an electrical current proportional to the potential that is placed across the membrane. The channel is similar to what is observed in many naturally occurring protein pores and shows stochastic gating—conformational fluctuations between open and closed states. Attaching transmembrane nanostructures to cells will provide a new type of sensing device and help manipulate or augment cell functions.

8.3.1 Transmembrane DNA Channel

A transmembrane channel made entirely from DNA and DNA-bound cholesterol moieties was constructed [2]. The inspiration for the design of the DNA transmembrane channel came from the bacterial protein pore α-hemolysin [63,64]. *Streptococcus* bacteria excretes α-hemolysin and perforates a target cell in order to leach iron through the resulting channel for the bacterium to consume [65]. Similar to α-hemolysin structure, the DNA channel structure consists of a channel that spans the lipid bilayer and a barrel-shaped cap that binds to the surface of the membrane.

The channel consists of two parts: a stem that penetrates through a lipid membrane and a barrel-shaped cap that attaches through cholesterol molecules to the membrane. The channel comprises six DNA helices that form a hollow tube. The interior of this tube acts as a transmembrane channel with a diameter of 2 nm and a length of 42 nm. The stem of the channel runs through both stem and barrel (Figure 8.7A). Adhesion to the lipid bilayer is mediated by 26 cholesterol molecules that are attached to the surface of the barrel (Figure 8.7A).

Transmission electron microscopy (TEM) images of purified DNA channels with unilamellar lipid vesicles confirmed the membrane binding (Figure 8.7B). Synthetic DNA channels bind to lipid bilayer membranes in the desired orientation in which the cholesterol-modified face of the barrel forms a tight contact with the membrane and the stem appears to protrude into the lipid bilayer (Figure 8.7A and B).

8.3.2 Charge-Neutralized Transmembrane DNA Pores

A DNA origami nanopore composed of six interconnected DNA duplexes was constructed [66]. The DNA-based nanopore structurally mimics the amphiphilic nature of protein pores in order to insert into the membrane. Membrane proteins feature an outer hydrophobic surface. DNA nanobarrels have to overcome the unfavorable energetic interaction between the hydrophobic environment of the membrane and the hydrophilic, negatively charged phosphate groups in the outer pore wall. To ensure insertion into the membrane, the pore was chemically modified with alkyl groups, which mask the negatively charged oligonucleotide backbone (Figure 8.7C). This modification allows the insertion of the DNA pore into the hydrophobic environment of the lipid membrane. The DNA barrel contains the 2.2-nm-long outer hydrophobic ring (Figure 8.7C, magenta), which matches the thickness of the lipid bilayer. A total of 72 alkyl-phosphorothioates groups (12 per each duplex) make up the hydrophobic belt. DNA origami nanopores, inserted into lipid bilayers, were tested using single channel current analysis for ionic current. After membrane insertion, DNA nanopores supported a steady transmembrane flow of ions through a single pore.

A total of 26 cholesterol molecules were placed in the strategic positions on the surface of the DNA barrel [2] (Figure 8.7A). A total of 72 alkyl-phosphorothioates groups make up the hydrophobic belt of the DNA origami pore [66] (Figure 8.7C). In order to simplify nanopore design and minimize chemical intervention, a smaller number of chemical tags with greater hydrophobicity were used for membrane anchoring [67]. Only two porphyrin-based hydrophobic tags achieve the otherwise energetically unfavorable anchoring of the negatively charged DNA nanopore into the hydrophobic core of lipid bilayers. Porphyrin has a large aromatic core and a van der Waals surface area approximately 12 times higher than ethane. The aromatic porphyrin tags can also be powerful visualization tags since they are chromophores with a fluorescence emission at 656 nm. Moreover, inserting porphyrins into lipid membranes leads to fluorescence shift and confirms membrane anchoring.

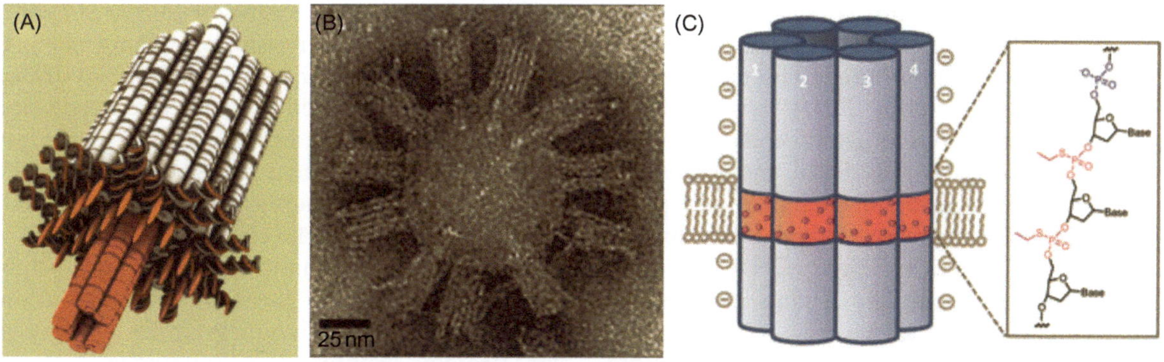

FIGURE 8.7 Synthetic DNA origami transmembrane channel and pore. (A) Schematic illustration of the DNA channel formed by 54 DNA helices. Cylinders indicate double-stranded DNA. Transmembrane domain is colored red. Orange ellipsoids indicate cholesterol-modified oligonucleotides that bind to single-stranded DNA adaptor strands. (B) TEM of a unilamellar lipid vesicle bound to 11 DNA origami channels through cholesterols. (C) DNA origami nanopore composed of six interconnected duplexes represented as cylinders. Phosphates of the DNA backbone are substituted by charge-neutral phosphorothioate ethyl groups (inset). *Source: (A) and (B) Adapted from Ref. [2] and (C) Ref. [66].*

The engineered DNA transmembrane channels and pores demonstrated electrical conductivity using an integrated chip-based setup. The synthetic DNA channel was further engineered to display gating behavior similar to natural ion channels. The mutant was designed by incorporating a single-stranded heptanucleotide protruding inside the transmembrane tube shown in red in Figure 8.7. The transmembrane current gating behavior depended on fine structural details of the synthetic DNA channel. It is challenging to modify the geometry of natural biological pores made from protein. On the other hand, the geometry of synthetic DNA origami objects and their chemical properties can be tailored for custom sensing applications of desired materials [66–69]. The possibilities with this system are endless, because one can easily vary the sequence and the corresponding structure, analogous to site-directed mutagenesis in biology. The DNA transmembrane channel was used to study unzipping of DNA hairpins and unfolding of guanine quadruplexes [70,71]. Single-stranded DNA molecules fit through the designed 2-nm central pore of the DNA channel, whereas larger DNA complexes such as hairpins or quadruplexes did not. Hairpins and quadruplexes had to unzip or unfold to pass through the DNA channel and the kinetics of the unfolding was monitored. The channel can also distinguish differing DNA lengths. DNA origami nanopores were used for the detection of λ-DNA molecules [68]. DNA molecules can be used to create designer channels and pores with adaptable diameter, shape, and surface functionality. Thus, synthetic DNA channels can be custom made as sensing devices for desired analyte molecules.

Synthetic DNA channels spanning lipid membranes are a crucial first step toward harnessing ion flux for creating sophisticated nanodevices. The inspiration for DNA channels can be found in nature by the rich functional diversity of membrane machines such as rotary motors, ion pumps, and transport proteins. The transmembrane channel and nanopore designs can be utilized as artificial pores with tunable voltage-gating and ion-selective permeation properties. Possibly, a DNA machine can be engineered to actively transport matter across a membrane by utilizing DNA-based motors [72–74]. Synthetic DNA channels could be engineered to have antimicrobial properties and used for molecular detection applications, and DNA sequencing, and research into nanofluidics and controlled transmembrane transport. DNA nanostructures can be used for sophisticated drug delivery applications in medical therapies targeting specific cell types and inserting a transmembrane pore for drug delivery much like a virus would [75,76,82].

8.4 ADVANCES IN MICROSCOPY

The biological processes occurring on the surface and inside the cell are complex and dynamic. In order to achieve a comprehensive understanding of the molecular mechanisms underlying these biological processes, both temporal and spatial resolution are required.

Cryo-electron tomography (cryoET) [77] provides the spatial resolution. Using cryoET, 3D volumes of near-native frozen hydrated structures are obtained at molecular resolution (Figure 8.8A). Recent cryoET advances vividly demonstrate inner structures of the flagella [79,80] and attachments of viruses to cells (Figure 8.8A) [75,77,81,82].

On the other hand, fluorescence light microscopy provides the temporal resolution. It provides time-resolved dynamics of fluorescent proteins of interest in living cells.

Recent advances in fluorescence light microscopy have further improved the spatial resolution by an order of magnitude [81–84]. Super-resolution fluorescence microscopy imaging, which applies switchable probes, as

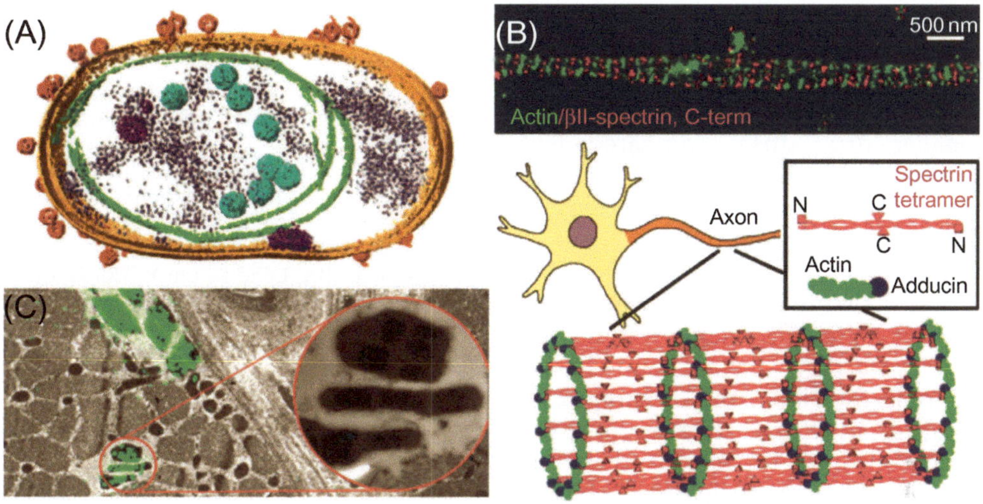

FIGURE 8.8 Advances in microscopy. (A) Multiple cyanophages (red) infect a cyanobacteria in the early stage. Infected cell is resolved by cryoET. (B) Arrangement of actin, spectrin, and adducin proteins in axons is resolved by super-resolution. (C) Both light microscopy and electron microscopy data can be captured in one instrument using correlative microscopy. *Source: Adapted from (A) Ref. [77], (B) Ref. [78], and (C) FEI brochure: Breaking frontiers in correlative light and electron microscopy.*

in PALM and STORM [85,86] has enabled a better structural understanding of the elaborate arrangement of cellular components (Figure 8.8B) [78].

Combining and integrating fluorescence light microscopy and electron microscopy (termed correlative microscopy) is an emerging powerful approach to visualize dynamic cellular processes, since it unites the strengths of the two complimentary imaging modalities (Figure 8.8C) [87,88]. On one hand, correlative microscopy provides temporal resolution and dynamic information available from optical microscopy. On the other hand, spatial resolution and molecular-level details are provided by cryoET (Figure 8.8C).

Advances in emerging techniques like cryoET, super-resolution, and correlative microscopy will help us pinpoint interactions of nucleic acid nanostructures with cell membrane and elucidate the intricate arrangement of structural components on the cell surface.

8.5 CONCLUSION

In this chapter, we summarize the most recent advances in the engineering and computing on the cell surface. In particular, we focus on translational research with the most promising potential applications and emerging imaging techniques.

In order to create more targeted therapeutics, it is necessary to clearly identify the target. Diseased cells often have a number of overexpressed cell markers. Unfortunately these markers are sometimes expressed in healthy cells. In order to correctly identify a diseased cell, more than one marker has to be located on the cell. Recently reported DNA-based computational methods for cellular identification are modular, scalable, and use standard antibodies and synthetic DNA. However, in order to proceed as a therapeutic drug, cell targeted DNA robots have to be able to find their target first. We need to switch to smarter, more targeted therapeutics in order to treat diseases with fewer side effects. The nanoscale machines can interact directly with the surfaces of individual cells of interest and deliver smaller drug doses compared with conventional medicine. In order to increase the production and decrease the cost, it may be beneficial to simplify the design of a nanostructure.

There are still a number of challenges before successful drug delivery is viable. We have to address major barriers like large molecular mass, stability in the blood, and crossing cellular membranes. Chemical stability of DNA nanostructures has to be improved in the presence of bloodstream nucleases and other digestive enzymes. It might be necessary to chemically modify the DNA backbone to increase resistance to degradation. In order to cross the membrane and increase cellular uptake, amphiphilic molecules or peptides could be displayed on the surface of DNA origami objects. For some targets fully or partially eliminating the negative charge of the DNA backbone [66,67] is an option. For other targets, alternative delivery and modification methods will be required.

A treasure trove of the natural membrane machines like rotary and ion pumps, and transport proteins provides the inspiration for DNA channels and pores. Synthetic DNA origami channels spanning lipid membranes can be engineered to harness ion flux with tunable voltage-gating and ion-selective properties. Artificial DNA pores can be designed to actively transport payload across the lipid bilayer. DNA nanostructures are about the size of a virus and can be used for sophisticated drug delivery applications. Currently it is possible to recognize a specific cell type, insert a transmembrane pore, and deliver the cargo. Synthetic DNA channels can also be custom made with adaptable diameter, shape, and surface functionality for detecting desired analyte molecules.

References

[1] Chworos A, Severcan I, Koyfman AY, Weinkam P, Oroudjev E, Hansma HG, et al. Building programmable jigsaw puzzles with RNA. Science 2004;306:2068–72.

[2] Langecker M, Arnaut V, Martin TG, List J, Renner S, Mayer M, et al. Synthetic lipid membrane channels formed by designed DNA nanostructures. Science 2012;338:932–6.

[3] Williams BA, Lund K, Liu Y, Yan H, Chaput JC. Self-assembled peptide nanoarrays: an approach to studying protein-protein interactions. Angew Chem Int Ed 2007;46:3051–4.

[4] Wilner OI, Weizmann Y, Gill R, Lioubashevski O, Freeman R, Willner I. Enzyme cascades activated on topologically programmed DNA scaffolds. Nat Nanotechnol 2009;4:249–54.

[5] He Y, Tian Y, Ribbe AE, Mao C. Antibody nanoarrays with a pitch of approximately 20 nanometers. J Am Chem Soc 2006;128:12664–5.

[6] Li H, LaBean TH, Kenan DJ. Single-chain antibodies against DNA aptamers for use as adapter molecules on DNA tile arrays in nanoscale materials organization. Org Biomol Chem 2006;4:3420–6.

[7] Höfner G, Wanner KT. Competitive binding assays made easy with a native marker and mass spectrometric quantification. Angew Chem 2003;120:5235–7. (2003 not 2008)

[8] Sharma J, Ke Y, Lin C, Chhabra R, Wang Q, Nangreave J, et al. DNA-tile-directed self-assembly of quantum dots into two-dimensional nanopatterns. Angew Chem Int Ed 2008;47:5157–9.

[9] Koyfman AY, Braun G, Magonov S, Chworos A, Reich NO, Jaeger L. Controlled spacing of cationic gold nanoparticles by nanocrown RNA. J Am Chem Soc 2005;127:11886–7.

[10] Le JD, Pinto Y, Seeman NC, Musier-Forsyth K, Taton TA, Kiehl RA. DNA-templated self-assembly of metallic nanocomponent arrays on a surface. Nano Lett 2004;4:2343–7.

[11] Maune HT, Han SP, Barish RD, Bockrath M, Iii WA, Rothemund PW, et al. Self-assembly of carbon nanotubes into two-dimensional geometries using DNA origami templates. Nat Nanotechnol 2010;5:61–6.

[12] Sanderson K. What to make with DNA origami. Nature 2010;464:158–9.

[13] Castro CE, Kilchherr F, Kim DN, Shiao EL, Wauer T, Wortmann P, et al. A primer to scaffolded DNA origami. Nat Methods 2011;8(3):221–9.

[14] He Y, Ye T, Su M, Zhang C, Ribbe AE, Jiang W, et al. Hierarchical self-assembly of DNA into symmetric supramolecular polyhedra. Nature 2008;452:198–201.

[15] Rothemund PWK. Folding DNA to create nanoscale shapes and patterns. Nature 2006;440:297–302.
[16] Ke Y, Sharma J, Liu M, Jahn K, Liu Y, Yan H. Scaffolded DNA origami of a DNA tetrahedron molecular container. Nano Lett 2009;9:2445–7.
[17] Douglas SM, Dietz H, Liedl T, Högberg B, Graf F, Shih WM. Self-assembly of DNA into nanoscale three-dimensional shapes. Nature 2009;459:414–8.
[18] Dietz H, Douglas SM, Shih WM. Folding DNA into twisted and curved nanoscale shapes. Science 2009;325:725–30.
[19] Kuzuya A, Kimura M, Numajiri K, Koshi N, Ohnishi T, Okada F, et al. Precisely programmed and robust 2D streptavidin nanoarrays by using periodical nanometer-scale wells embedded in DNA origami assembly. Chem Bio Chem 2009;10:1811–5.
[20] Voigt NV, Tørring T, Rotaru A, Jacobsen MF, Ravnsbaek JB, Subramani R, et al. Single-molecule chemical reactions on DNA origami. Nat Nanotechnol 2010;5:200–3.
[21] Stephanopoulos N, Liu M, Tong GJ, Li Z, Liu Y, Yan H, et al. Immobilization and one-dimensional arrangement of virus capsids with nanoscale precision using DNA origami. Nano Lett 2010;10:2714–20.
[22] He Y, Chen Y, Liu H, Ribbe AE, Mao C. Self-assembly of hexagonal DNA two-dimensional (2D) arrays. J Am Chem Soc 2005;127(35):12202–3.
[23] Koyfman AY, Magonov SN, Reich NO. Self-assembly of DNA arrays into multilayer stacks. Langmuir 2009;25:1091–6.
[24] Koyfman AY, Braun GB, Reich NO. Cell-targeted self-assembled DNA nanostructures. J Am Chem Soc 2009;131:14237–9.
[25] Blum AS, Soto CM, Wilson CD, Cole JD, Kim M, Gnade B, et al. Cowpea mosaic virus as a scaffold for 3-D patterning of gold nanoparticles. Nano Lett 2004;4(5):867–70.
[26] De Silva AP, Uchiayam S. Molecular logic and computing. Nat Nanotechnol 2007;2:399–410.
[27] Vickaryous MK, Hall BK. Human cell type diversity, evolution, development, and classification with special reference to cells derived from the neural crest. Biol Rev 2006;81:425–55.
[28] Rosen JM, Jordan CT. The increasing complexity of the cancer stem cell paradigm. Science 2009;324:1670–3.
[29] da Cunha JP, Galante PA, de Souza JE, de Souza RF, Carvalho PM, Ohara DT, et al. Bioinformatics construction of the human cell surfaceome. Proc Natl Acad Sci USA 2009;106:16752–7.
[30] Ichise M, Harris PE. Imaging of b-cell mass and function. J Nucl Med 2010;51:1001–4.
[31] Ikeda A, Shankar DB, Watanabe M, Tamanoi F, Moore TB, Sakamoto KM. Molecular targets and the treatment of myeloid leukemia. Mol Genet Metab 2006;88(3):216–24. Epub. 2006 May 5.
[32] Zhou H, Jiao P, Yang L, Li X, Yan B. Enhancing cell recognition by scrutinizing cell surfaces with a nanoparticle array. J Am Chem Soc 2011;133:680–2.
[33] Kluza E, van der Schaft DW, Hautvast PA, Mulder WJ, Mayo KH, Griffioen AW, et al. Synergistic targeting of alphavbeta3 integrin and galectin-1 with heteromultivalent paramagnetic liposomes for combined MR imaging and treatment of angiogenesis. Nano Lett 2010;10:52–8.
[34] Weiner LM, Surana R, Wang S. Monoclonal antibodies: versatile platforms for cancer immunotherapy. Nat Rev Immunol 2010;10:317–27.
[35] Grützkau A, Radbruch A. Small but mighty: how the MACS-technology based on nanosized superparamagnetic particles has helped to analyze the immune system within the last 20 years. Cytometry A 2010;77(7):643–7. Available from: http://dx.doi.org/doi:10.1002/cyto.a.20918.
[36] Rudchenko M, Taylor S, Pallavi P, Dechkovskaia A, Khan S, Butler Jr VP, et al. Autonomous molecular cascades for evaluation of cell surfaces. Nat Nanotech 2013;8:580–6.
[37] Bendall SC, Simonds EF, Qiu P, Amir el-AD, Krutzik PO, Finck R, et al. Single-cell mass cytometry of differential immune and drug responses across a human hematopoietic continuum. Science 2011;332:687–96.
[38] You M, Peng L, Shao N, Zhang L, Qiu L, Cui C, et al. DNA "Nano-Claw": logic-based autonomous cancer targeting and therapy. JACS 2014;136(4):1256–9.
[39] Castano AP, Mroz P, Hamblin MR. Photodynamic therapy and anti-tumour immunity. Nat Rev Cancer 2006;6(7):535–45.
[40] Han D, Zhu G, Wu C, Zhu Z, Chen T, Zhang X, et al. Engineering a cell-surface aptamer circuit for targeted and amplified photodynamic cancer therapy. ACS Nano 2013;7:2312–9.
[41] Leeson PD, Springthorpe B. The influence of drug-like concepts on decision-making in medicinal chemistry. Nat Rev Drug Discov 2007;6:881–90.
[42] Rabuka D, Forstner MB, Groves JT, Bertozzi CR. Noncovalent cell surface engineering: incorporation of bioactive synthetic glycopolymers into cellular membranes. J Am Chem Soc 2008;130:5947–53.
[43] Stevens MM, George JH. Exploring and engineering the cell surface interface. Science 2005;310:1135–8.
[44] Hsiao SC, Crow AK, Lam WA, Bertozzi CR, Fletcher DA, Francis MB. DNA-coated AFM cantilevers for the investigation of cell adhesion and the patterning of live cells. Angew Chem Int Ed 2008;47:8473–7.
[45] Luo D, Saltzman WM. Synthetic DNA delivery systems. Nat Biotechnol 2000;18:33–7.

[46] Gartner ZJ, Bertozzi CR. Programmed assembly of 3-dimensional microtissues with defined cellular connectivity. Proc Natl Acad Sci USA 2009;106:4606–10.

[47] Carpenter G, Cohen S. Epidermal growth factor. Annu Rev Biochem. 1979;48:193–216.

[48] Bremer E, Samplonius DF, van Genne L, Dijkstra MH, Kroesen BJ, de Leij LF, et al. Simultaneous inhibition of epidermal growth factor receptor (EGFR) signaling and enhanced activation of tumor necrosis factor-related apoptosis-inducing ligand (TRAIL) receptor-mediated apoptosis induction by an scFv:sTRAIL fusion protein with specificity for human EGFR. J Biol Chem 2005;280(11):10025–33. Epub. 2005 Jan 11.

[49] Koyfman AY, Braun G, Magonov S, Chworos A, Reich NO, Jaeger L. Controlled spacing of cationic gold nanoparticles by nanocrown RNA. J Am Chem Soc 2005;127:11886–7.

[50] Souza GR, Molina JR, Raphael RM, Ozawa MG, Stark DJ, Levin CS, et al. Three-dimensional tissue culture based on magnetic cell levitation. Nat Nanotechnol 2010;5(4):291–6. Available from: http://dx.doi.org/doi:10.1038/nnano.2010.23. Epub. 2010 Mar 14.

[51] Ding S, Wu TY, Brinker A, Peters EC, Hur W, Gray NS, et al. Synthetic small molecules that control stem cell fate. Proc Natl Acad Sci USA 2003;100:7632–7.

[52] Rivera J, Gilfillan AM. Molecular regulation of mast cell activation. J Allergy Clin Immunol 2006;117:1214–25.

[53] Sil D, Lee JB, Luo D, Holowka D, Baird B. Trivalent ligands with rigid DNA spacers reveal structural requirements for IgE receptor signaling in RBL mast cells. ACS Chem Biol 2007;2(10):674–84.

[54] Douglas SM, Bachelet I, Church GM. A logic-gated nanorobot for targeted transport of molecular payloads. Science 2012;335:831–4.

[55] Hamaguchi N, Ellington A, Stanton M. Aptamer beacons for the direct detection of proteins. Anal Biochem 2001;294:126–31.

[56] Nutiu R, Li Y. Structure-switching signaling aptamers. J Am Chem Soc 2003;125:4771–8.

[57] Green LS, Jellinek D, Jenison R, Ostman A, Heldin CH, Janjic N. Inhibitory DNA ligands to platelet-derived growth factor B-chain. Biochemistry 1996;35:14413–24.

[58] Tang Z, Shangguan D, Wang K, Shi H, Sefah K, Mallikratchy P, et al. Selection of aptamers for molecular recognition and characterization of cancer cells. Anal Chem 2007;79:4900–7.

[59] Huang YF, Shangguan D, Liu H, Phillips JA, Zhang X, Chen Y, et al. Molecular assembly of an aptamer-drug conjugate for targeted drug delivery to tumor cells. Chem Bio Chem 2009;10:862–8.

[60] Robertson MJ, Cochran KJ, Cameron C, Le JM, Tantravahi R, Ritz J. Characterization of a cell line, NKL, derived from an aggressive human natural killer cell leukemia. Exp Hematol 1996;24:406–15.

[61] Vitale C, Romagnani C, Puccetti A, Olive D, Costello R, Chiossone L, et al. Surface expression and function of p75/AIRM-1 or CD33 in acute myeloid leukemias: engagement of CD33 induces apoptosis of leukemic cells. Proc Natl Acad Sci USA 2001;98: 5764–9.

[62] Ye Z, Lee CM, Sun GW, Gan YH. Burkholderia pseudomallei infection of T cells leads to T-cell costimulation partially provided by flagellin. Infect Immun 2008;76:2541–50.

[63] Song L, Hobaugh MR, Shustak C, Cheley S, Bayley H, Gouaux JE. Structure of staphylococcal alpha-hemolysin, a heptameric transmembrane pore. Science 1996;274:1859–66.

[64] Hall AR, Scott A, Rotem D, Mehta KK, Bayley H, Dekker C. Hybrid pore formation by directed insertion of α-haemolysin into solid-state nanopores. Nat Nanotechnol 2010;5(12):874–7.

[65] Kowalczyk SW, Blosser TR, Dekker C. Biomimetic nanopores: learning from and about nature. Trends Biotechnol 2011;29:607–14.

[66] Burns JR, Stulz E, Howorka S. Self-assembled DNA nanopores that span lipid bilayers. Nano Lett 2013;13:2351–6.

[67] Burns JR, Göpfrich K, Wood JW, Thacker VV, Stulz E, Keyser UF, et al. Lipid-bilayer-spanning DNA nanopores with a Bifunctional porphyrin. AnchorAngew Chem Int Ed 2013;52:12069–72.

[68] Bell NA, Engst CR, Ablay M, Divitini G, Ducati C, Liedl T, et al. DNA origami nanopores. Nano Lett 2012;12:512–7.

[69] Wei R, Martin TG, Rant U, Dietz H. DNA origami gatekeepers for solid-state nanopores. Angew Chem Int Ed 2012;51:4864–7.

[70] Vercoutere W, Winters-Hilt S, Olsen H, Deamer D, Haussler D, Akeson M. Rapid discrimination among individual DNA hairpin molecules at single-nucleotide resolution using an ion channel. Nat Biotechnol 2001;19:248–52.

[71] Sauer-Budge AF, Nyamwanda JA, Lubensky DK, Branton D. Unzipping kinetics of double-stranded DNA in a nanopore. Phys Rev Lett 2003;90:238101.

[72] Wickham SF, Bath J, Katsuda Y, Endo M, Hidaka K, Sugiyama H, et al. A DNA-based molecular motor that can navigate a network of tracks. Nat Nanotechnol 2012;7(3):169–73.

[73] Omabegho T, Sha R, Seeman NC. A bipedal DNA Brownian motor with coordinated legs. Science 2009;324(5923):67–71.

[74] Venkataraman S, Dirks RM, Rothemund PW, Winfree E, Pierce NA. An autonomous polymerization motor

powered by DNA hybridization. Nat Nanotechnol 2007;2(8):490–4.
[75] Liu X, Zhang Q, Murata K, Baker ML, Sullivan MB, Fu C, et al. Structural changes in a marine podovirus associated with release of its genome into Prochlorococcus. Nat Struct Mol Biol 2010;17:830–6.
[76] Dai W, Fu C, Raytcheva D, Flanagan J, Khant HA, Liu X. Visualizing virus assembly intermediates inside marine cyanobacteria. Nature 2013;502:707–10.
[77] Leis A, Rockel B, Andrees L, Baumeister W. Visualizing cells at the nanoscale. Trends Biochem Sci 2009;34:60–70.
[78] Xu K, Zhong G, Zhuang X. Actin, spectrin, and associated proteins form a periodic cytoskeletal structure in axons. Science 2013;339:452–6.
[79] Koyfman AY, Schmid MF, Gheiratmand L, Fu CJ, Khant HA, Huang D, et al. Structure of Trypanosoma brucei flagellum accounts for its bihelical motion. PNAS 2011;108(27):11105–8.
[80] Sun E, He J, Zhuang X. Live cell imaging of viral entry. Curr Opin Virol 2013;3(1):34–43.
[81] Jones SA, Shim SH, He J, Zhuang X. Fast, three-dimensional super-resolution imaging of live cells. Nat Methods 2011;8(6):499–508.
[82] Fernandez-Suarez M, Ting AY. Fluorescent probes for super-resolution imaging in living cells. Nat Rev Mol Cell Biol 2008;9:929–43.
[83] Huang B, Bates M, Zhuang X. Super-resolution fluorescence microscopy. Annu Rev Biochem 2009;78:993–1016.
[84] Schermelleh L, Heintzmann R, Leonhardt H. A guide to super-resolution fluorescence microscopy. J Cell Biol 2010;190:165–75.
[85] Betzig E, Patterson GH, Sougrat R, Lindwasser OW, Olenych S, Bonifacino JS, et al. Imaging intracellular fluorescent proteins at nanometer resolution. Science 2006;313:1642–5.
[86] Rust MJ, Bates M, Zhuang X. Sub-diffraction-limit imaging by stochastic optical reconstruction microscopy (STORM). Nat Methods 2006;3:793–5.
[87] Zhang P. Correlative cryo-electron tomography and optical microscopy of cells. Curr Opin Struct Biol 2013;23:763–70.
[88] Fridman K, Mader A, Zwerger M, Elia N, Medalia O. Advances in tomography: probing the molecular architecture of cells. Nat Rev Mol Cell Biol 2012;13:736–42.

CHAPTER

9

Cell Surface Enzymatic Engineering-Based Approaches to Improve Cellular Therapies

Ayman F. Abuelela, Kosuke Sakashita and Jasmeen S. Merzaban

Biological and Environmental Sciences and Engineering, King Abdullah University of Science and Technology, Thuwal, Kingdom of Saudi Arabia

OUTLINE

J. M. Karp & W. Zhao (Eds): Micro- and Nanoengineering of the Cell Surface.
DOI: http://dx.doi.org/10.1016/B978-1-45-573146-6.00009-X

9.1 INTRODUCTION

Cell migration is central to many physiological and pathological functions *in vivo* from immunity to metastasis of cancer cells, to the homing of therapeutic stem cells to a target tissue, and tissue repair. In some cases, it would be desirable to promote migration of particular cells whereas in others inhibition of migration would be preferable. An understanding of the molecules involved in this migration paradigm will permit better development of technologies for therapies.

This chapter will start by introducing the principle of migration of cells to tissues/organs and how this process is controlled by a specific sugar decoration, mainly sialyl Lewis x/a ($sLe^{x/a}$), on cell surface proteins/lipids. This key epitope is created through the activity of specific glycosyltransferases (GTs). When these enzymes are stimulated to become expressed and/or active in a particular cell or are introduced *ex vivo*, this will result in migration of that cell to specific locations within the body where counter receptors to the epitope, the selectins, are expressed.

The chapter will focus on how controlling the expression and/or activity of these GTs makes it possible to (i) improve the migration of therapeutic cells (i.e., stem cells) to desired locations within the body in order to optimize therapeutic outcomes of disease, and also (ii) inhibit the migration of harmful cells (i.e., circulating tumor cells, CTCs) in order to control metastasis or inflammation. In this chapter, the role of other enzymes in the modification of cell surface architecture related to refining cell function(s) to benefit therapy will also be considered.

9.1.1 Multistep Paradigm of Cellular Migration

Nature has developed an extremely efficient mechanism for the delivery of circulating cells to specific sites in the body in immune processes such as inflammation and injury. An interesting analogy exemplifying cell movement within the blood stream is offered by Ehrhardt et al. [1], which states "Comparable to a human being plunged into a roaring river, a leukocyte (cell in flow) is exposed to high shears within the mainstream of the blood." In whichever scenario, arrest and exit from the flow represent a particularly complex task for human or cell. The emigration of cells out of flow (such as in the blood circulation) is a sophisticated process controlled by various

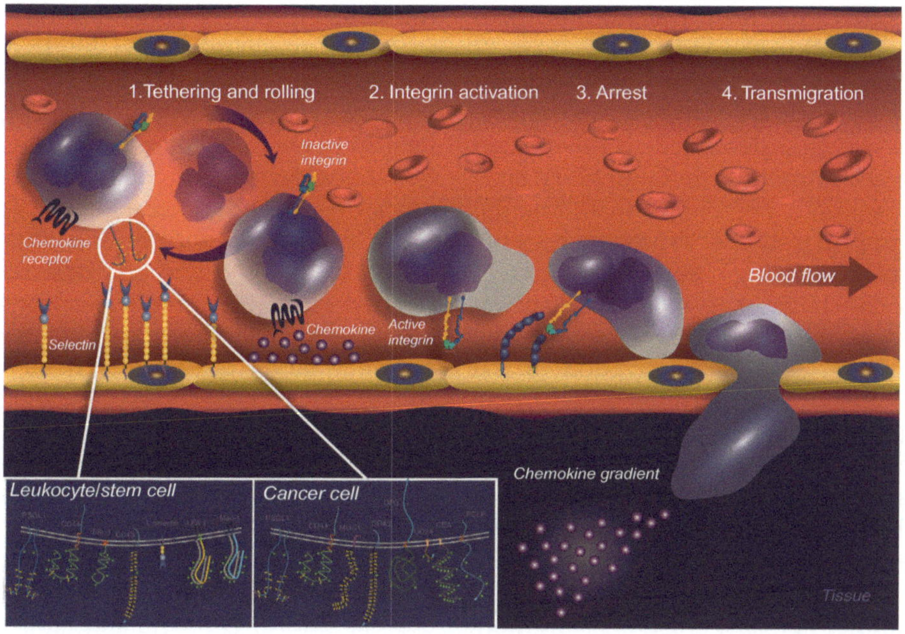

FIGURE 9.1 The multistep paradigm of cell migration highlighting the main interactions that take place between the cell (leukocyte, stem cell, CTC) in flow and the endothelial cells lining the blood vessels of an organ or tissue (such as the bone marrow). Step 1, "tethering and rolling" is mediated by the selectins and their ligands and acts to slow down the cell. Step 2, "integrin activation" is controlled by the expression of chemokines on the surface of the endothelial cells finding their appropriate chemokine receptors on the surface of the cell in flow triggering G-protein-coupled "inside-out" activation of integrins. This activation results in conformational changes of the integrin(s) expressed on the cell in flow and the subsequent high affinity binding of the integrin(s) to their CAM (cellular adhesion molecules) ligand(s) on the endothelium leading to Step 3 "firm adhesion." The interaction of all of these molecules with their appropriate receptors culminates at Step 4, the transmigration of the cell in flow into the intended tissue or organ. Each step in this process is a prerequisite for the next; however, the initial step, mediated by the selectins and their ligands, is crucial to the overall migration paradigm. Common selectin glycoprotein ligands identified on leukocytes, stem cells, and cancer cells are depicted in the inset; O-glycans are shown as yellow squiggly lines (~) and N-glycans are shown as green hexagons (⬢).

adhesion molecules. These comprise, in particular, selectins (E, P, and L), chemokines, and integrins, which function in a coordinated multistep cascade (Figure 9.1) [2–4]. The first essential event in cellular recruitment involves shear-resistant adhesion of flowing cells onto the endothelial surface, this being most efficiently mediated by the selectins and their respective ligands [5,6]. These interactions initially tether the cell in flow to the vessel wall and, in the context of vascular shear flow, cause the tethered cell to roll along the endothelium lining the inside of the vessel. Selectin-dependent tethering and rolling (Step 1) thus decreases cellular velocity below that of the prevailing hemodynamic stream and brings the cell into close physical proximity with the vessel wall. This process facilitates engagement of specific cell-borne chemokine receptors to pertinent chemokines present in the perivascular areas [7–9], thereby triggering activation through inside-out signal transduction events [10–13] leading to increased adhesiveness of integrin family members (Step 2). Adhesive engagements between the activated cell integrins and their cognate endothelial cell counter receptors then lead to arrest and firm adhesion of the cell to the vessel wall (Step 3) and, ultimately, the process of transmigration (Step 4) [14] to reach the extravascular space. This process was first described in the context of leukocyte recruitment to inflammatory sites and has subsequently been applied to describe other biological migration activities such as the movement of circulating hematopoietic stem progenitor cells (HSPCs) to the bone marrow during bone marrow transplants, and in the metastasis of the CTC. The migration of the cell out of circulation and into the target tissue or organ depends on each of these distinct classes of molecules acting sequentially, each step being a prerequisite for the next. In fact, it has been shown in numerous biological scenarios that the selectins, as the most effective mediators [15] of Step 1 interactions, are essential for recruitment of cells altogether and may be the gatekeepers of the multistep paradigm [16]. In fact, it is suggested that the selectins may even be able to substitute for the absence of chemokine receptor signaling events, in the case that chemokines are not present [17,18].

9.1.1.1 *The Selectins*

The selectins are a family of three C-type (Ca^{2+}-dependent) lectins expressed exclusively by bone marrow-derived cells and endothelial cells: L-selectin (expressed on *L*eukocytes and HSPCs), E-selectin (expressed on *E*ndothelial cells), and P-selectin (expressed on *P*latelets and endothelial cells). Due to their expression on the endothelium, E-selectin and P-selectin are often referred to as the "vascular selectins" whereas L-selectin is termed the "leukocyte selectin." These molecules are type I transmembrane glycoproteins sharing a similar topology which comprises an amino-terminal lectin-like domain, an epidermal growth factor (EGF)-like domain, a variable number of consensus repeats (CRs; also known as complement regulatory proteins or "sushi" domains), a single membrane spanning domain, and a carboxy-terminal cytoplasmic domain (Figure 9.2A) [4,6]. The lectin domain primarily mediates ligand binding with minimal contribution provided by the EGF-like domain [19–22]. The number (and hence length) of sushi domains facilitates the recognition between selectins and their ligands as it allows a selectin to extend beyond the negatively charged cloud of the glycocalyx. The selectins bind to specialized carbohydrate determinants (Figure 9.2B), comprising sialofucosylations containing an α(2,3)-linked sialic acid substitution on galactose, and an α(1,3)-linked fucose modification on *N*-acetylglucosamine, prototypically displayed as the terminal tetrasaccharide sialyl Lewis x (sLe^x; or also to its isomer sLe^a) [16,23]. These determinants, also known as "CD15s," may be displayed on either a protein scaffold (i.e., a glycoprotein) or a lipid scaffold (i.e., a glycolipid), and are recognized by mAbs such as CSLEX-1, KM93, and HECA-452 [24,25]. Although additional structural modifications, principally involving sulfation, increase the binding affinity of P-selectin and L-selectin to sLe^x, no known modifications are needed for optimal binding of E-selectin [26,27]. These characteristics of engaging rapidly and with high tensile strength to their ligands help to make the selectins the most important initiators of adhesion under flow. A brief introduction to each of the selectins is provided below.

L-selectin (CD62L) is expressed constitutively by all myeloid cells, virtually all B cells and naïve T cells, a subpopulation of memory T cells, NK cells, and both early and mature hematopoietic cells in the bone marrow. Unlike the other selectins, however, L-selectin is not expressed on endothelial cells [28]. Functionally, L-selectin is critical for the homing of naïve lymphocytes to high endothelial venules (HEVs) of secondary lymphoid organs [3,27]. L-selectin-dependent lymphocyte rolling on HEV requires localization of L-selectin to the tips of the microvilli characteristic of the surfaces of lymphocytes and other leukocytes, allowing for optimal leukocyte–endothelial

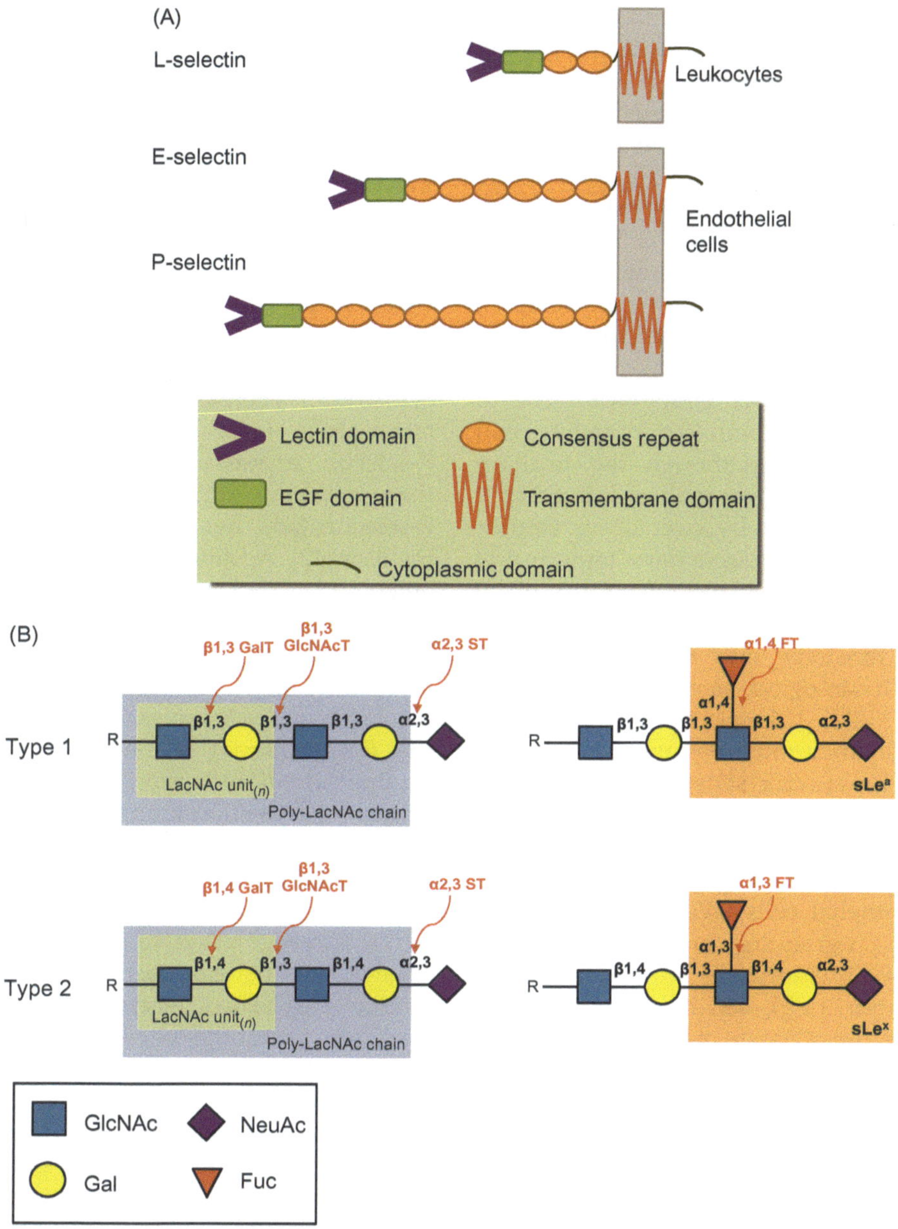

FIGURE 9.2 Selectins and their carbohydrate ligands. (A) Selectins are type 1 transmembrane cell surface proteins comprised of five different domains: lectin-binding domain, EGF-like domain, varying numbers of CRs (sushi domains), transmembrane domain, and a short cytoplasmic tail. The lectin-binding domain binds the carbohydrate $sLe^{a/x}$ containing ligands while the varying sushi domains help to extend the selectin beyond the glycocalyx of the cell surface. (B) GTs involved in the formation of sLe^{a} and sLe^{x} structures on selectin ligands. Although type 1 and type 2 lactosamine [Galβ1,4GlcNAc or Galβ1,3GlcNAc] structures tend to be common to many glycans, the expression of the GTs responsible for capping the galactose (Gal,) of the lactosamine with sialic acid (NeuAc,)—*ST*—and the terminal *N*-acetylglucosamine (GlcNAc,) of the lactosamine with fucose (Fuc,)—*FT*—are often correlated with cells that migrate or home to tissues where selectins are expressed.

interactions. Mapping of L-selectin domains by mAbs has revealed that the NH_2 terminal nine amino acids are critical for ligand binding. The EGF-like and the two short sushi domains maintain the spatial conformation and are also important for ligand binding [27]. HEV-borne L-selectin ligands are collectively referred to as peripheral node addressins (PNAds) and include the glycoproteins GlyCAM-1 (glycosylation-dependent cell adhesion molecule-1), CD34, MAdCAM-1 (mucosal vascular addressin cell adhesion molecule-1), podocalyxin, Sgp200, endomucin, and endoglycan. All of these PNAds have mucin-like domains that carry O-glycans [27]. Although the primary role of L-selectin is in promoting lymphocyte homing, its expression by circulating neutrophils may facilitate the secondary tethering of these cells to an already rolling neutrophil at an inflammatory site (secondary tethering) [29,30]; this, however, may be a rare event of which the physiological significance has been questioned [31,32]. If L-selectin is to recognize them, these ligands must be modified by various GTs to give specific sialylated, fucosylated, and sulfated oligosaccharides.

P-selectin (CD62P) is a glycoprotein that is constitutively expressed within secretory granules of endothelial cells (Weibel–Palade bodies) and in α-granules of platelets. Upon activation by inflammatory mediators (histamine, thrombin, LPS), secretory vesicles fuse rapidly with the plasma membrane allowing expression of P-selectin on the surface of the platelet or endothelial cells within minutes [6]. Thus, P-selectin is often involved in early leukocyte recruitment during an inflammatory response. P-selectin is also constitutively expressed at low levels on the endothelium of thymus [33], lung and choroid plexus microvessels [34], bone marrow microvasculature [35,36], in postcapillary venules in skin [37], and on peritoneal macrophages [38]. The transcription of P-selectin can be induced by cytokines such as interleukin (IL)-4 and IL-13 [39,40]. Once expressed on the surface of endothelial cells, P-selectin is rapidly internalized by endocytosis [41]. In addition to mediating the homing of immune cells to inflammatory sites, P-selectin-mediated interactions are also involved in controlling T-cell progenitor (circulating thymic progenitor, CTP) migration from the bone marrow to the site of T-cell development, the thymus [33].

It is important to note that differences appear to exist between the responses of human compared with mouse endothelium to inflammatory cytokines. Whereas IL-1 and tumor necrosis factor-alpha (TNF-α) are potent inducers of P-selectin expression in mouse endothelium, these cytokines only induce the expression of E-selectin (and not P-selectin) in human endothelium [42]. Additionally, experiments in transgenic mice comparing human and mouse P-selectin genes have demonstrated that following injection of TNF-α, the expression of human P-selectin mRNA decreases whereas mouse P-selectin mRNA increases [43]. This suggests that E-selectin may be the more dominant selectin in human inflammation and disease and brings into question the extrapolation to human disease of the emphasis on P-selectin ligand interactions related to mouse models.

E-selectin (CD62E) is a glycosylated glycoprotein that recognizes diverse groups of glycoconjugates on hematopoietic and cancer cells. Unlike P-selectin, E-selectin is not presynthesized in endothelial cells but, rather, is transcriptionally regulated by mediators including TNF-α, IL-1β, IFN-γ, and LPS. However, exceptions to this rule exist including the observations that vascular beds within the skin [37,44] and bone marrow [36,45] constitutively express E-selectin. Expression can occur as early as 2 h after stimulation and decline within 24 h [46]. Despite this delayed expression (as compared to P-selectin), E-selectin overlaps with P-selectin temporarily enhancing leukocyte recruitment during an inflammatory response and functionally leading to a dramatic decrease in the rolling velocity (slow rolling), and enhancing the probability of adhesion [47] and the subsequent steps of the cascade (including integrin activation) [48]. Even following the downregulation of P-selectin, E-selectin continues to enhance leukocyte–endothelial interactions leading to slower rolling velocities and improved adhesion [49]. When the stimulus wanes, E-selectin becomes internalized and degraded in lysosomes [50]. Physiologically, E-selectin-mediated interactions are involved in controlling leukocyte recruitment to injury and inflammatory sites [51], steady-state, tissue-specific homing in cutaneous tropism of skin-homing T cells [52,53], HSPC entry into bone marrow [45,52,53], as well as metastasis of CTCs.

9.1.2 E-Selectin Ligands in Therapy and Disease

Emphasis on E-selectin-mediated interactions and the regulation of migration of *human* cells in flow is crucial to understanding many physiological phenomena in human disease and therapy since: (i) E-selectin responds differently from P-selectin to stimuli such as inflammatory cytokines, (ii) E-selectin contributes dramatically to decreasing the velocity of the cells in flow for an extended time, and (iii) E-selectin is constitutively expressed in some vascular beds such as the bone marrow, this being important both for an understanding of how to enhance migration of blood stem cells to bone marrow and also for how to inhibit the metastasis of CTCs to this site.

9.1.2.1 E-Selectin Binds Specific Selectin Ligands on HSPCs to Guide them to the Bone Marrow

The vascular endothelial cells within the bone marrow constitutively express E-selectin and vascular cell adhesion molecule-1 (VCAM-1) [36]. Numerous studies in mice, including intravital microscopy studies, have demonstrated the importance of E-selectin, VCAM-1, and stromal cell-derived factor 1 (SDF-1) in mediating the entry of cells in flow (such as HSPCs and immune cells found in the blood) to the bone marrow [54–56]. These findings have established that homing to bone marrow involves a multistep cascade initiated by E-selectin binding to its ligands on the HSPCs to mediate tethering and rolling, followed by SDF-1 binding to CXCR4 (on HSPC) leading to activation of VLA-4 (on HSPC), resulting in firm adhesion mediated by VLA-4 binding to VCAM-1 and subsequent transmigration. The Step 1 effectors of this cascade, E-selectin and its ligands, play a fundamental role in the recruitment of circulating cells to the bone marrow as demonstrated by many studies and reviewed by Sackstein [16].

Studies in both humans and mice indicate that E-selectin is constitutively expressed on marrow endothelial cells [36,54,57], and intravital studies have revealed that HSPC migration to marrow occurs at specialized microvascular beds expressing E-selectin [54]. These and several other independent lines of evidence [45,58–62] have highlighted a central role for E-selectin-dependent interactions in HSPC recruitment to marrow. E-selectin-dependent binding of HSPC to marrow microvessels, either concurrent with transmigration or subsequently within "vascular niches," could affect hematopoietic activity and with differential effects depending on the target glycoprotein ligand(s). Importantly, it is known that ligation of P-selectin glycoprotein ligand 1 (PSGL-1) can suppress hematopoiesis in mouse [63] and human [64] HSPCs. This fact, together with evidence that non-PSGL-1 and E-selectin ligands can also trigger apoptosis and growth inhibition of HSPCs from human and mouse [64], highlights a broader role for E-selectin receptor/ligand interactions in hematopoiesis. Indeed, recent work by Winkler et al. demonstrated that E-selectin is a crucial component of the vascular HSC niche that actively induces hematopoietic stem cell (HSC) proliferation and if E-selectin is removed from the equation, HSC quiescence is maintained [65]. This suggests that in the absence of a specific E-selectin-mediated proliferative signal, greater numbers of HSCs remain dormant.

Control of this process may be through the expression upon "metabolic activation" [66] of a specific E-selectin ligand with an ability to bind E-selectin and induce the observed proliferation and commitment.

9.1.2.2 E-Selectin/E-Selectin Ligand Interactions Involved in CTC Metastasis, Regulation, and Maintenance

Metastasis is the culmination of an elaborate cascade of events in which a cancer cell breaks away from the primary tumor, enters blood vessels, and travels through the body until it finally extravasates through the vessels of a distant organ to establish a secondary colony. In order to successfully establish a metastatic clone at a distant site, any CTCs must survive the targeted abuse present in the blood stream that may lead to induction of apoptosis or necrosis as well as avoid elimination by immune cells. CTCs that possess these enhanced survival capabilities (and have overcome this "microevolutionary" process) will generate metastatic colonies in distant organs and even "self-seed" [67] the primary site with more hostile tumor cells. Research focused on elucidating the molecular mediators used by CTCs to adhere to the endothelial cells lining vessels at a target site, i.e., bone marrow, may lead to therapies that can be used to dissuade metastasis.

Endothelial cells expressing E-selectin, such as the bone marrow or activated inflamed vessels, are able to capture CTCs displaying E-selectin ligands on their surface, and initiate the adhesion/metastasis cascade that leads to invasion of the tumor cell (Figure 9.1). These ligands have also been implicated in cancer stem cell regulation and maintenance relating to epithelial-to-mesenchymal transition (EMT) and other events necessary for metastatic growth. E-selectin-dependent interactions have been established in promoting metastasis of many cancers including breast, pancreatic, colon, and prostate [68–72]. $sLe^{x/a}$ epitopes are significantly upregulated on cancer cells compared with normal cells and this may play a role in promoting CTC adhesion during metastasis [68,72,73]. This increased expression of selectin ligands is due, in part, to altered regulation of the GTs responsible for creating the $sLe^{x/a}$ epitope [74,75]. Thus, it is necessary to identify not only the glycoproteins and glycolipids presenting sialo-fucosylated glycans but also the GTs responsible in the upregulated expression of selectin ligands in order to target their downregulation. The role of E-selectin ligands and relevant GTs will be discussed in the following section.

9.1.3 Role of GTs in Controlling Cell Migration

As illustrated in Figure 9.1, the migration of cells to a specified site is dependent on the interaction of a number of ligands and receptor pairs expressed both on the surface of the migrating cell and on the surface of the endothelium, these working together in a coordinated multistep fashion to orchestrate the delivery of cells to a specific site. Although many molecules are involved in organizing the steps involved in this cascade of events, the selectin binding to its ligand(s) literally "gets the ball (cell) rolling" and on its way to enter a tissue/organ. This initial interaction is the prerequisite for the remaining steps in the cascade and is dependent on the careful glycosylation of selectin ligands. The glycan structure most common for the recognition of ligands by selectins is the sLe^x (or its stereoisomer sLe^a) addition to the N-/O-glycan on the protein (Figure 9.2). GTs expressed in the ER and Golgi act in an assembly line fashion by adding one glycan at a time to a growing carbohydrate chain whereby the product of one

GT reaction becomes the substrate for the following GT. All GTs are type II transmembrane-Golgi/ER localized enzymes with a short NH_2 terminal tail in the cytosol followed by a transmembrane domain, which is linked to the catalytic domain in the Golgi/ER lumen by a short stem region. Generally, each GT uses only one of several sugar nucleotide substrates (including UDP-Galactose (Gal), UDP-Glucose (Glc), UDP-*N*-acetylgalactosamine (GalNAc), UDP-*N*-acetylglucosamine (GlcNAc), GDP-fucose (Fuc), GDP-mannose (Man), UDP-xylose (Xyl), or CMP-sialic acid (SA)). Each GT also uses one and sometimes more glycan acceptor substrates and, as they are stereospecific, these enzymes are only capable of forming one specific glycosidic bond (i.e., either an α or β anomer) [76]. Glycan synthesis is the consequence of a series of ordered, GT-dependent events that are characterized by glycan polymer elongation or branching. In general, it is the suite of GTs expressed by a cell that will dictate the glycan structures made. However, the expression of specific GTs does not permit prediction of the glycan structures that will be synthesized. Many factors can affect the activity and ability of the GTs to create a given structure such as the availability of sugar nucleotides, the relative location of the GT in the Golgi stacks in relation to the substrates that it can recognize, and even the possibility of modifications of the glycan product by naturally expressed glycosidases [5,77]. In the context of the glycan structures that confer selectin counter receptor activity on a cell, the GT suite will therefore participate in the control of selectin-dependent leukocyte recruitment. Although all three selectins recognize the sLe^x epitope, each selectin requires particular fine structural details of this epitope or of the peptide backbone to confer optimal binding affinities.

In order for selectin ligands to be formed, GTs responsible for creating the $sLe^{a/x}$ structures must be endogenously expressed and active. The native display of these glycan structures on selectin ligands requires the expression and activity of various GTs and sulfotransferases, including the action of α1,3- or α1,4-fucosyltransferases, α2,3-sialyltransferases, β1,4-galactosyltransferases, and β1,6-*N*-acetylglucosaminyltransferases on either a protein scaffold (i.e., N/O-glycans on glycoproteins) or a lipid scaffold (i.e., glycolipids). Specifically, this specialized carbohydrate motif comprises sialofucosylations containing an α(2,3)-linked sialic acid substitution on galactose, and an α(1,4)-linked (for sLe^a) or α(1,3)-linked (for sLe^x) fucose modification on *N*-acetylglucosamine, of either a type 1 (for sLe^a) or type 2 (for sLe^x) lactosamine unit, creating a terminal tetrasaccharide sLe^a or sLe^x structure, respectively (Figure 9.2). Although E-selectin binds both sLe^a and sLe^x, it binds sLe^x with a higher avidity [78] and, with the exception of human metastatic cancer cells, most human hematopoietic cells and "normal" stem cells prototypically express only type 2 lactosamine units [79,80]. Understanding the expression and regulation of GTs in the creation of E-selectin ligands will guide efforts to manipulate the migration of human cells in circulation. Indeed, knowledge of which GTs are required to create E-selectin ligands has allowed researchers to manipulate cells in order to create ligands on cells *ex vivo* with the intention of helping direct the migration of therapeutic cells (such as HSPCs) to specific tissues (such as the bone marrow) that express the corresponding selectins [59,79]. Such work will be discussed in the following section.

9.1.3.1 GTs Involved in Creation of E-Selectin Ligands

For the most part, the enzymes responsible for forming E-selectin ligands are similar to those responsible for both L-selectin and P-selectin ligand formation in that they require sLe^x structures. However, unlike L- and P-selectin ligands, the sLe^x cap can occur on a number of different glycans from N-glycans to glycolipids. In addition, analysis of FX^{null} mice (mice with a null mutation of the FX locus that encodes an enzyme within the *de novo* pathway for GDP-fucose synthesis) by Lowe's group suggests that E-selectin ligands are much more sensitive than P-selectin ligands to loss of fucose, this further highlighting the importance of fucosylation to the formation of functional E-selectin ligands [81].

9.1.3.2 FTs and E-Selectin Ligand Formation

α1,3-fucosylation is imperative for E-selectin ligand activity and a number of human fucosyltransferases (FTs) have been identified that are able to modify terminal lactosamines. These include FT-III, FT-IV, FT-V, FT-VI, FT-VII, and FT-IX [82,83]. FT-III to FT-VI are all able to modify either sialylated or unsialylated type 2 lactosamines resulting in sLe^x or Le^x (Figure 9.2) structures, respectively, while FT-III (and FT-V) is also able to modify type 1 lactosamines resulting in sLe^a or Le^a.

Analyses of mice with targeted mutations in FT genes reveal that only two of these enzymes are required to form functional selectin counter receptor activity in leukocytes. These studies all support the observation that FT-VII and FT-IV cooperate in the synthesis of fucosylated glycans required for all leukocyte selectin ligand activity [84–86]. Specifically $FT\text{-}VII^{null}$ mice have leukocytosis and their blood leukocytes have reduced E- and P-selectin ligand activity that effectively compromises the recruitment of leukocytes to inflammatory sites in *in vivo* mouse models of inflammation, such as the thioglycollate-induced peritonitis model [85] and the contact hypersensitivity model (CHS) [84,86]. However, residual recruitment is still evident (particularly with myeloid cells) and is due to FT-IV activity. CHS response studies performed in mice deficient in FT-IV, FT-VII, or both enzymes revealed that, although lymphocyte (Th1 and Tc1) recruitment was absent in $FT\text{-}VII^{null}$ animals, both enzymes needed to be knocked out to inhibit the recruitment of myeloid cells. A subtle decrease in the myeloid cell CHS response was seen in $FT\text{-}IV^{null}$ mice whereas lymphocyte recruitment was unaffected. Overall, these observations support the hypothesis that FT-VII dependent fucosylation of selectin ligands accounts for most of the selectin-dependent adhesion in leukocytes especially within the lymphocyte fraction. The dependence of selectin ligand formation on FT-VII is also paralleled in studies with human T lymphocytes [87–89]. Indeed, studies suggest that FT-IV acts preferentially on glycolipids whereas FT-VII acts on glycoproteins [90] and, interestingly, gangliosides are able to mediate tethering and rolling [91–93] on endothelium. However, glycoproteins are better at facilitating firm adhesion under physiological flow conditions [93] which could also be related to signaling through the glycoprotein selectin ligands to activate integrins directly [17]. Recent studies focused on assessing knockdowns of FT-IV and -VII in human cells suggest that an additional α1,3-fucosyltransferase, FT-IX, is important in decorating E-selectin ligands on human leukocytes but not mouse [94,95]. This is significant since the selectin ligands, particularly those that bind E-selectin, vary between different leukocyte cell populations and among species [96].

9.1.3.3 ST3GalTs and E-Selectin Ligands

α2,3-sialic acid linkages to terminal galactose residues are crucial in the formation of sLe^x epitopes. The contribution of sialylation to the formation of selectin ligands is well documented as sialidase treatment of leukocytes under static conditions *in vitro* completely abolishes binding of P-selectin and E-selectin to selectin ligands [97]. This linkage is catalyzed by the activity of the sialyltransferases (STs) ST3GalT-I, -II, -III, -IV, -V, and -VI. Only ST3GalT-III, -IV, and -VI have been shown to sialylate type 2 lactosamines and ST3GalT-IV and ST3GalT-VI are implicated in E-selectin (and P-selectin) ligand formation [97–100]. *In vivo* inflammatory studies of ST3GalT-IV^{null} mice reveal a mild but significant defect in selectin ligand function, including a mild reduction in E-selectin-dependent rolling, an increase in E-selectin-dependent rolling velocity, a decrease in L-selectin-dependent rolling during inflammation (but no effect on L-selectin-dependent rolling on HEVs [101]), and no contribution to P-selectin ligands [97,102]. Since sialylation is so crucial to selectin binding but yet the phenotype of the ST3GalT-IV^{null} is so mild, it is suggested that other STs, particularly ST3GalT-VI, contribute to selectin ligand formation. With the generation of the ST3GalT-VI^{null} and ST3GalT-IV/ST3GalT-VI double knockout mouse, the role of these STs appears to be critical for the generation of functional selectin ligands, especially in the case of the formation of P-selectin ligands (ST3GalT-VI is involved in E-selectin-mediated capturing but not in E-selectin-dependent rolling) [100]. Implications for the involvement of additional STs involved in selectin ligand synthesis and function are still warranted, especially for E-selectin and L-selectin.

9.1.3.4 C2GnTs and Functional E-Selectin Ligands

C2GnT (core 2 β(1,6)-*N*-acetylglucosaminyltransferase) isoenzymes (C2GnT-I, -II, -III) create the core 2 O-glycan branch [103] by adding GlcNAc to the Galβ1,3GalNAc core 1 structure expressed on serine or threonine residues. Capping of the core 2 branch with a sLe^x moiety creates the P-selectin counter receptor. Analyses of C2GnT-I^{null} mice, however, show that it is the isoenzyme essential for PSGL-1 modifications on leukocytes [104–106]. Despite the essential role of C2GnT-I in P-selectin ligand formation, C2GnT-I^{null} mice possess a relatively mild phenotype, showing only a partial reduction in selectin ligands and essentially no effect on lymphocyte homing [104], reflecting the possible redundancy of the C2GnT isoenzymes. It has been suggested that C2GnT-II and/or C2GnT-III contribute to selectin ligand formation and cell trafficking [107,108]. C2GnT-I and C2GnT-III are expressed primarily by lymphocytes while C2GnT-II is associated with goblet cell mucin production in the intestinal epithelium. A single knockout of any of these enzymes and double knockout of C2GnT-I and C2GnT-III had no effect on the trafficking of lymphocytes in an inflammatory mouse model of the large intestine. However, mice deficient in C2GnT-I and C2GnT-II or all 3 C2GnT enzymes were not able to overcome a helminth infection, suggesting that C2GnT-dependent modifications on lymphocytes may play a role in migration of lymphocytes (to the large intestine) [108].

In the context of E-selectin ligands, a number of studies indicate that cells deficient in C2GnT-I are able to bind to E-selectin, although at a slightly reduced level compared with wild-type cells [104,107,109]. These studies reveal that 75–80% of E-selectin binding is maintained in the absence of C2GnT-I, suggesting that either the isoenzymes are compensating for the loss of C2GnT-I or that other E-selectin ligands (besides PSGL-1) are dominant mediators of this interaction.

9.2 USE OF ENZYMES TO MODIFY CELL SURFACE CARBOHYDRATES ON PROTEINS AND LIPIDS TO ENHANCE MIGRATION TO TISSUES

Direct injection of therapeutic cells into target tissue sites is only practical for a limited number of applications including pancreatic beta islet transplants through the portal vein into the liver or intracoronary infusions of myoblasts into damaged heart muscles. For the vast majority of therapeutic cellular-based therapies that are targeting systemic multifocal sites of injured or inflamed tissues or organs, intravenous transfer is the routine method of administrating cells. Therefore, it is crucial that the efficient homing of donor cells to the site of intended action be ensured. Since only a subset of intravenously injected cells may engraft in the intended location due to the absence of the key homing molecules on infused cells, the development of methods for improving therapeutic cell trafficking is a high priority. One relatively simple cell engineering method uses the GT enzymatic machinery responsible for creating elaborate glycan structures on cell surface proteins to program the *in vivo* trafficking of systemically delivered cells. Through the introduction of GTs important for creating the key carbohydrate structure, sialyl Lewis x/a (sLe^x and sLe^a), on the cell surface, infused cells can then be attracted by the cognate selectin counter receptor expressed on the endothelium and direct its subsequent migration. These GTs can be introduced genetically using expression vectors containing the GT gene of interest, induced through cytokine signaling mechanisms, or using recombinant active GTs to add on glycans *ex vivo*.

9.2.1 Genetic Approaches for Modification of Cell Surface Glycoconjugates

It is possible to introduce into cells genes that are required to achieve the formation of the glycoconjugates on the surface of cells that render selectin ligand binding.

Among the 13 family members recognized in the human genome, eight FTs (FT-III, FT-IV, FT-V, FT-VI, FT-VII, FT-IX, FT-X, and FT-XI) are reported to possess the alpha-1,3-fucosyltransferase activity which potentially could confer formation of the sLe^x structure with distinctive substrate specificities [110,111]. In particular, it has been shown that FT-III, FT-IV, FT-V, FT-VI, and FT-VII are involved in the formation of the selectin ligand and by inhibiting the expression of these FTs in various cell types including leukocytes [95], prostate cancer cells [71], and gastric cancer cells [112], selectin binding is inhibited. Consequently, FTs can be considered as good candidate genes to introduce to therapeutic cells and enhance sLe^x formation. The mode of introduction of FTs into cells could be either transient or stable. Since the formation of the sLe^x moiety is a prerequisite only in the initial step of the transition of cells to a tissue, transient expression of FT could adequately satisfy this requirement. The benefit of the transient expression strategy compared with stable expression through viral vector-mediated gene integration is that the former approach avoids genetic disturbance of the genome and so minimizes the likelihood of spontaneous oncogenesis.

Successful physical approaches for the delivery of plasmid vectors into cells for transient expression have been reported using lipofection [113,114] or electroporation [115,116], in both cases with good transfection efficiency, high viability and recovery of transfected cells, and without any significant effect on capability for differentiation or proliferation.

One potentially intriguing approach to overcome the relatively lower efficiency of transient transfection is the use of a nonintegrating version of foamy virus-based vector [117], which has been shown to transduce stem cells rather effectively [118]. Although this vector system is still under development, it could permit a high efficiency by virtue of being a virus-based transduction while at the same time keeping the genome of the target cells intact.

Stable expression of exogenous genes can be achieved using genomic integration mediated by virus-based vectors. Various viral vector systems have been used for transduction of cells including lentivirus-, retrovirus-, adenovirus-, adeno-associated virus-, or baculovirus-based vectors to achieve long-term and stable expression of exogenous genes [119,120].

Although *ex vivo* enzymatic treatments are generally viewed as being more desirable in terms of simplicity and efficiency, it would seem reasonable to employ a viral vector approach when the introduction of transgenes or correction of a genomic mutation is required.

9.2.2 Cytokine and Growth Factor Induced Expression of GTs Within Cells

Several cytokines are implicated in the regulation of the key GTs involved in selectin ligand synthesis of leukocytes and other circulating cells, such as hematopoietic stem/progenitor cells, as they undergo differentiation and specialization.

In *T cells*, expression and regulation of GTs important in forming P- and E-selectin ligands require antigen-stimulated activation [121–124]. Following this finding, a number of studies focused on defining how key cytokines involved in the activation of T cells affect the expression of GTs that create sLe^x epitopes on E- and P-selectin ligands. *In vitro* activation studies using cytokines that induce the polarization of CD4$^+$ and CD8$^+$ T cells toward activated T_H1 (T_C1) or T_H2 (T_C2) subsets demonstrated that the GTs C2GnT-I, FT-VII and ST3GalT-IV and -VI are differentially regulated depending on cytokines used. T_H1 cytokines, including IL-12, -2, -15, and TGF-β tend to enhance the expression of C2GnTs [107,125,126], FT-VII [127–131], and ST3GalTs [127,132] and induce binding to E-selectin and P-selectin, while T_H2 cytokines, such as IL-4 and -5, repress their expression. Based on their potent impact *in vitro*, these cytokines were expected to modulate homing receptor formation under *in vivo* conditions. Although these cytokines appeared to regulate C2GnT-I *in vitro*, there was no effect on the selectin ligand formation and proliferative responses in the absence of these cytokines *in vivo* [133].

Other signals have been implicated in the regulation of selectin ligand formation, such as the vitamin A and D derivatives. A recent study illustrated that exposure of human T cells to physiological levels of retinoic acid (RA) and 1,25D(3) completely eliminated E-selectin ligand activity and subsequent skin-homing abilities, whereas the inactive/precursor forms of vitamins A and D had little inhibitory action [134]. Analysis of GTs involved in forming sLe^x structures revealed that FT-VII and ST3GalT-III were significantly reduced upon exposure to these vitamin derivatives. This finding is intriguing as RA is currently being used successfully as a differentiation therapy for the treatment of acute myeloid leukemia (AML) subtype 3. In addition to inducing the differentiation of the diseased leukemic cells, it may also influence the ability of the leukemic blasts/stem cells to migrate to the bone marrow where E-selectin is expressed by inhibiting the GTs involved in creating E-selectin ligands.

In *hematopoietic stem/progenitor cells and myeloid cells*, G-CSF (granulocyte-colony stimulating factor) significantly enhanced the expression of sLe^x on HSPCs and their subsequent ability to bind E-selectin and home *in vivo*. This was accompanied by an associated increase in the expression of the GTs, ST3GalT-IV, FT-IV, and FT-VII [135]. Studies on human CD34$^+$ cord blood stem cells exposed to dimethyl-prostaglandin E2 (dm-PGE2) have demonstrated that this treatment increases HSPC numbers and enhances their homing [136–138]. However, to date, this enhanced homing has not been correlated with measurements of GT or sLex expression or selectin binding.

Immune blood cells, such as lymphocytes and macrophages, are known to secrete inflammatory factors that could potentially influence *cancer cell* metastatic progression. In particular, the proinflammatory cytokines, TNF-α, and IL-6 have been reported to be elevated in the blood serum of patients diagnosed with advanced stage breast cancer and this has been correlated with an increase in the number and size of metastatic sites [139]. These cytokines have also been shown to promote the growth and invasiveness of breast, colon, and prostate cancer cells [140–142]. In addition, cancer cell chemokine receptors have also been implicated in playing a role in identifying the destination of metastases [143]. It can be inferred that these cytokines lead to enhanced expression of GTs involved in the creation of sLe^x epitopes on ligands of these cancer cells, which results in enhanced adhesion to selectins [140]. This process will be discussed later.

9.2.3 Use of Recombinant GTs *ex vivo*

GTs expressed in the ER and Golgi act in an assembly line fashion by adding one glycan at a time to a growing carbohydrate chain decorating a particular glycoprotein. Depending on which GTs are expressed in a particular cell, a diverse set of structures can be made. In order for selectin ligands to be formed, GTs responsible for creating the sLe^x structure need to be endogenously expressed and active, as outlined in Figure 9.2. The knowledge of which GTs are required to create particular structures allows researchers to manipulate cells in order to create ligands on cells' surface *ex vivo* with the intention of helping to direct the migration of therapeutic cells to specific tissues that express the corresponding selectins [59,79,144].

The microvasculature of human and murine bone marrow constitutively expresses E-selectin and, as described earlier, extensive research has determined that HSPC migration to the bone marrow depends on the expression of ligands that bind E-selectin. Both mouse and human HSPCs express a number of E-selectin ligands. Studies to identify the glycoprotein ligands revealed that on human $CD34^+$ HSPCs, at least three different ligands contributed to E-selectin binding. These include the hematopoietic cell E-/L-selectin ligand (HCELL; sLe^x-decorated glycoform of CD44 that binds E-selectin and L-selectin), cutaneous lymphocyte antigen (CLA; sLe^x-decorated glycoform of PSGL-1 that binds E-selectin), and a sLe^x-decorated glycoform of CD43 [96]. In contrast, on mouse HSPCs only CLA and CD43 were identified as ligands suggesting that the modest E-selectin binding detected in mouse HSPCs compared with human HSPCs is due to the absence of HCELL. Indeed, inducing the expression of HCELL on mouse HSPCs leads to significant enhancement of E-selectin binding in mouse HSPCs [96]. HCELL expression was achieved using an innovative tool to custom engineer desirable cell surface glycoforms (i.e., sLe^x) on therapeutically beneficial cells (i.e., stem cells) using recombinant GTs to transform the cell surface glycans temporarily in order to enhance their migration to specific sites (i.e., E-selectin on bone marrow microvasculature).

Selectin ligands must be α1,3-fucosylated to form the terminal glycan sLe^x determinant. Often stem cells are inadequately α1,3-fucosylated, this being the case with MSCs [59], HSPCs isolated from cord blood [61,144], as well as with neural stem cells (unpublished data), and so exhibit homing defects. Indeed *ex vivo* treatment of stem cells (principally human MSCs and HSPCs derived from human cord blood) with FTs, particularly FT-VI and FT-VII, increases cell surface sLe^x determinants, boosts binding to E-selectin and P-selectin, and enhances homing and engraftment into bone marrow of immunocompromised mice [59,144–146]. By generating a comprehensive set of tools (i.e., GTs) for *ex vivo* design of sLe^x structures on the surface of a therapeutic cell, strategies to enhance the migration of these cells for the treatment of diseases can be further anticipated.

For many years, scientists have been interested in understanding and characterizing the complex nature of carbohydrates displayed on cell membranes and the GTs responsible for creating these molecules. Early studies focused on isolating small amounts of GTs from natural sources and using these enzymes in reactions that included radioactive donor sugar nucleotides to rebuild cell surface structures following their removal using glycosidases; the proteins that resulted were then identified by SDS-PAGE [147,148]. These studies also helped to characterize carbohydrate-dependent binding interactions such as those between pathogens and host cells. The use of GTs to manipulate glycans became much more practical with advancements in recombinant DNA technologies that allowed for the expression of proteins and enzymes on a large scale. The potential for creating libraries of recombinant GTs for fine tuning glycans on cells or proteins brings extensive possibilities, such as enhancing the stability of proteins, creating enhanced bioactive/humanized mAbs [149–151], and directing the movement of cells within the body.

The use of recombinant GTs to modify cells *ex vivo* has been elegantly formulated and discussed by Sackstein [59,79,152–154]. He outlines the guiding principles (summarized in Table 9.1) and methodological details (Figure 9.3) required for the custom engineering of cell surface glycans of a cell population of interest, particularly stem cells, using his "GPS" (glycosyltransferase-programmed stereosubstitution) technology [152,153]. Any GT enzyme can be used to create the glycan of interest using this technique but special attention must be given to ensure that the enzyme assay conditions (storage buffer, reaction buffer) are optimized to maintain cell viability and avoid phenotypic effects (such as the differentiation) on cell(s). One major concern is that many GTs require divalent cations for their catalytic activity [156]. Accordingly, the recombinant human FT-VI enzyme and reaction conditions specifically developed by Sackstein effectively catalyzed α(1,3)-fucosylation at physiologic pH in the absence of divalent cations [59].

TABLE 9.1 Guiding Principles Developed by Sackstein [152,153] to Apply GPS Technology to Cells

Strategy for Custom Engineering Cell Surface Glycans
I. Identify target glycoconjugate "acceptor"
II. Use appropriate enzyme(s) that result in stereospecific carbohydrate epitopes
III. Confirm the creation of the target modification biochemically and functionally *in vitro* and *in vivo*

Many mammalian proteins have been expressed and prepared as active recombinant proteins [157–159]. A number of different expression systems could be used (bacterial cells, mammalian cell lines, baculovirus-infected insect cells, yeast cells, and even silk worm) to express these GTs but, since several reports have shown that N-glycosylation is necessary, higher order systems are preferred for optimal enzymatic activity; in fact, bacteria-expressed FTs exhibit no detectable activity [160,161]. As bacteria-based systems would not work and production of large amounts of the required enzymes is necessary, Sackstein et al. chose to use the methanotrophic yeast *Pichia pastoris* expression system. This system offers the advantages of generating a high yield of recombinant proteins and an ability for the proteins to be secreted into the medium which facilitates downstream preparation of the enzyme [162]. Recent efforts have been made to investigate the expression and isolation of GTs from silkworm. This is a clear option for the preparation of recombinant enzymes, which cannot be expressed functionally in bacterial expression systems. The silkworm approach uses pupal or larval individuals of a lepidopteran insect, *Bombix mori*, and a baculovirus-based vector for exogenous gene expression

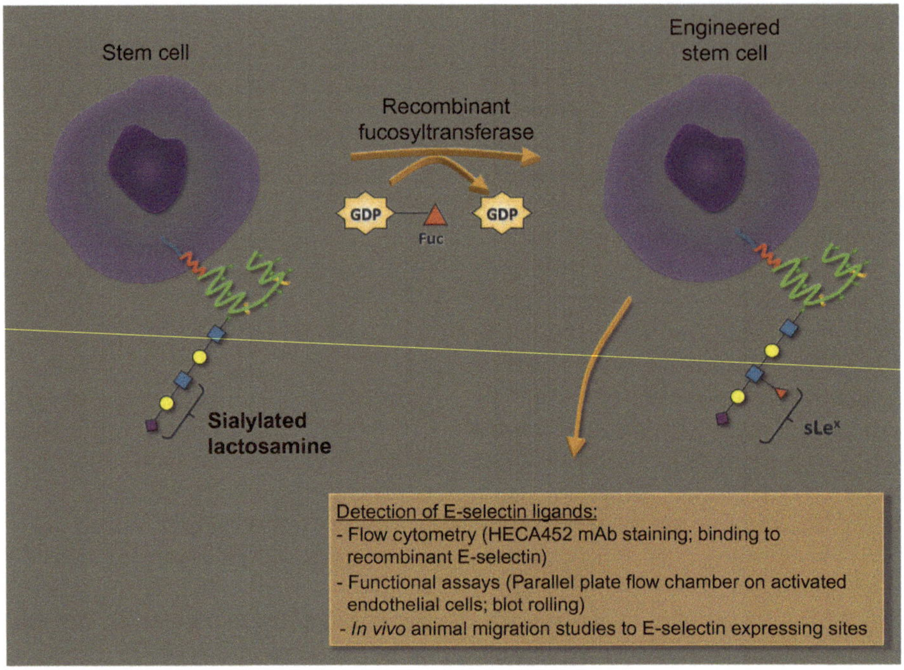

FIGURE 9.3 *Ex vivo* GT treatment of cells. Preparations of recombinant GTs, such as FT-VI [59,144] and -VII [146], have been used to treat stem cells. In this example, the FT-VI enzyme is illustrated, along with the reaction conditions needed to catalyze α(1,3)-exofucosylation efficiently (near-physiologic pH; absence of divalent cations, i.e. Mn^{2+}), by adding fucose from a nucleotide sugar donor, GDP-Fuc (guanosine 5′-diphospho-β-L-fucose), to a sialylated lactosamine structure on glycoproteins or glycolipids in order to create the sLe^x structure critical to guide the migration of cells to E-selectin-expressing endothelial beds such as the bone marrow [59,152]. Once the structure is formed, a number of functional assays are performed to determine the efficacy of the enzymatic treatment including parallel plate flow chamber studies and blot rolling assays [152,155].

[163]. This system brings several benefits including optimal folding and posttranslational modifications resulting in proper functionality of the many recombinant proteins, an extremely high yield, great amenability to scaling-up, and a short time period for production of the protein. Although few laboratories currently have the necessary technology, several companies offer a commercial protein expression service, which uses this system.

9.2.3.1 Glycans and Their Importance in Mediating Engraftment

Once HSPCs have used their newly acquired homing molecules to enter the bone marrow, successful transplantation then requires engraftment of the stem cells within their particular niche. Implications for the activity of GTs depend not only on improving the homing of therapeutic cells to their niche but also on the successful engraftment of these cells. Normally HSCs are located within specific regions in the bone marrow that dictate the behavior of these cells. The complex interplay of cells and molecules helps to regulate HSPC quiescence, division, and differentiation. Various cell types have been identified that contribute to the HSPC niche, including various mesenchymal stem cells ($CD146^+$ perivascular MSCs [164], $nestin^+$ MSCs [165], and leptin $receptor^+$ perivascular MSCs [166]), CXCL12 abundant reticular (CAR) cells, and MSC-derived osteoblast lineage cells [167–169]. In addition, endothelial cells form an overlapping HSC niche [166,170]. These niches also appear to be further regulated by local sympathetic nerves [171]. A model has been proposed whereby the "osteoblastic" niche (endosteal surface of bone) harbors and maintains HSCs in a quiescent state, whereas the "vascular" niche (marrow vessels) promotes HSC proliferation and differentiation [65,172,173]. However, many recent studies have challenged this model [174] stating that the osteoblastic niche is actually extensively vascularized [55,175], and that since the most potent and quiescent HSCs prefer to reside in hypoxic [176,177], poorly vascularized regions, these may not reside in the "osteoblastic" niche of the bone marrow. A particularly interesting study has demonstrated that E-selectin, expressed on around 20% of bone marrow endothelial cells (BMECs), directly promotes HSC proliferation and, in the absence of this selectin, HSC quiescence and self-renewal potential increases [65]. These findings suggest that, in addition to directing the migration of HSPCs to the bone marrow, E-selectin—through a currently undefined ligand—also plays a key role in regulating HSC engraftment and fate. Currently, quite which E-selectin ligand(s) on the HSC mediate this proliferation remains unknown as studies focused on knocking down the key E-selectin ligands identified to date do not present the same HSC phenotype [65,96]. Further studies are warranted to identify these ligands.

9.3 MODIFICATION OF ENZYME ACTIVITY LEADS TO CHANGES ON CELL SURFACE STRUCTURES THAT DETER MIGRATION AND METASTASIS

Metastatic colonization, outgrowth at a distant site, starts with the interaction between CTCs and endothelial cells lining lymph and postcapillary vessels. This interaction involves many adhesion molecules and molecular pathways, such as selectins, integrins, intercellular adhesion molecules (ICAMs), and chemokines with their receptors. These molecules work collectively in a process that obeys the multistep paradigm of HSPC recruitment to bone or leukocytes to sites of inflammation (Figure 9.1). Although still debated, "hematopoietic mimicry" or the "homing concept of metastasis" hypothesis states that at Step 1 the circulating cell, moving under high shear force of circulation, uses selectin-mediated low affinity reversible rolling interactions with endothelial cells expressing E-selectin and P-selectin [68,178,179]. This interaction facilitates extravasation of CTCs and determines organ-specific metastasis depending on the tumor type and metastatic environment. Selectins are the receptors for many glycoconjugate selectin ligands.

9.3.1 Selectin–Selectin Ligand Axis in Metastasis

The role of the selectin–selectin ligand axis in promoting cancer metastasis is supported by a number of findings described below and is summarized in Figure 9.4.

Many types of metastatic epithelial tumors, including breast [74,180,181], prostate [182,183], colon [184–187], lung [188,189], and pancreatic carcinomas [190], show a substantial ability to roll on endothelial cell layers natively expressing E-selectin. Although this interaction alone is not sufficient to explain the organ specificity of metastasis, it is the initial interaction and the primary adhesion that directs organ-specific metastasis. Lewis lung cancer cells tend to metastasize preferentially to liver where endothelium shows upregulation of E-selectin expression [191,192]. Additionally, prostate and breast carcinomas tend to metastasize to bone, where E-selectin is constitutively expressed and, interestingly, highly metastatic cells show a stronger interaction with bone marrow endothelium compared with weakly metastatic carcinomas [181–183,191,192].

Experiments in mouse models deficient in or overexpressing one of the selectins indicate a direct correlation with selectin expression and metastatic potential. For example, Biancone et al. [193] demonstrated a reversion of metastasis from lung to liver when E-selectin was upregulated *in vivo*. Another study showed that metastasis is inhibited upon immunoglobulin E-selectin blockade [194]. Moreover, studies in P-selectin and L-selectin deficient mice displayed impaired metastatic potential [195,196]. A growing body of evidence supports the proposal that inflammation potentiates cancer dissemination and progression. Immune cells lying in a tumor microenvironment both at the primary site and at premetastatic sites release proinflammatory cytokines that not only induce the upregulation of selectin expression on endothelial monolayers but also upregulate the enzymes responsible for carbohydrate ligands (GTs) on

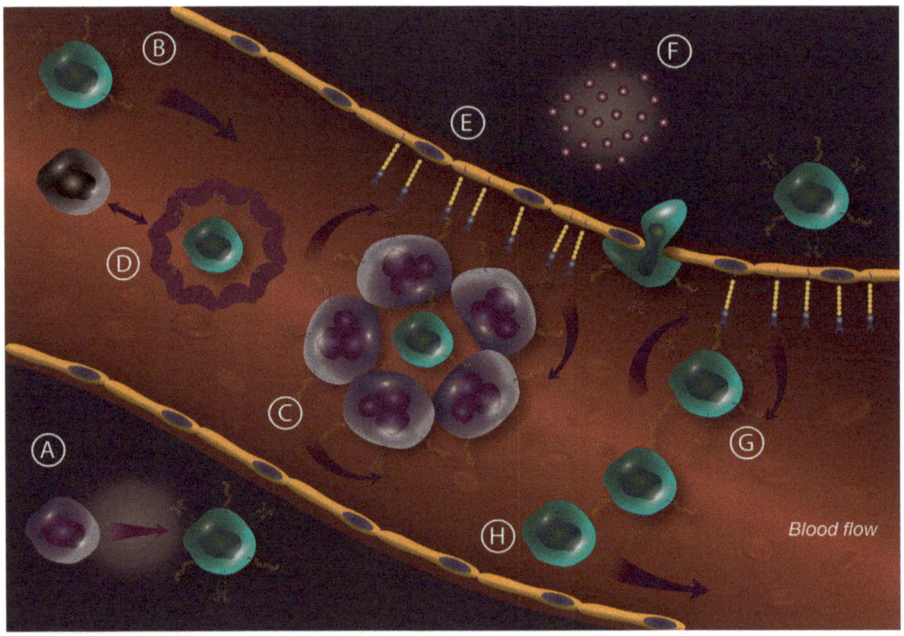

FIGURE 9.4 Selectins mediate tumor progression and metastasis. At the primary site, a prometastatic transition program confers higher migratory abilities of the tumor cell including increased expression of GTs involved in the creation of sLe$^{x/a}$ (A). Migratory tumor cells invade the blood circulation searching for a proper metastatic niche at a distant site to form the secondary tumor mass; in selectin-dependent metastasis, tissues that express selectin(s) capture selectin ligand-expressing CTCs allowing tissue tropism of metastasis (B). In the circulation, selectin-expressing leukocytes and platelets (D) coat the tumor cells to form metastatic emboli (C) that help the tumor cell evade the recognition and attack by immune cells. Moreover, metastatic emboli bridge an interaction between tumor cells and selectin-expressing endothelial cells through selectin ligand-expressing leukocytes (E). Premetastatic niche induces selectin expression on the surrounding endothelial linings by releasing inflammatory cytokines (F). Metastatic colonization commences when the CTCs encounter selectin-mediated capture and rolling allowing downstream adhesion and extravasation events (G). Cells devoid of selectin ligands lack this ability to traffic, which decreases their ability to survive in the circulation leading to a lower metastatic potential (H).

cancer cells. For instance, E-selectin expression in colorectal tissues is directly proportional to inflammatory reactions at the lesion–environment interface [191,192,197], and this may explain the sensitivity of colon cancer to anti-inflammatory therapy [198].

Selectins support the formation of metastatic emboli. L-selectin and P-selectin expressed on leukocytes and platelets, respectively, bind selectin ligand glycoconjugates and soluble mucins on the surface of CTCs forming metastatic emboli [199]. Metastatic emboli support metastasis in two ways: first, they allow better interaction with E-selectin-expressing endothelium through ligand-expressing leukocytes, and secondly, the coat of platelets is thought to offer a shield against the immune system [195,200].

Carcinogenesis is associated with abnormal GT expression [73] and, hence, typically a shift toward increased $sLe^{x/a}$ expression [201–203] (as outlined below). In fact, levels of sLe^x within tumors correlate with malignancy and patient survival [75,204–208].

Selectin ligand binding, perturbation in GT expression, and subsequent $sLe^{x/a}$ expression are correlated with prometastatic modifications such as EMT [209]. EMT is a rapid transformation process involved in embryogenesis, wound healing, and invasiveness of cancer cells, within which a polarized epithelial cell undergoes cellular and biochemical changes to transform into a mesenchymal phenotype. Mesenchymal phenotype is characterized by loss of polarization, increased invasiveness, high motility, elevated resistance to apoptosis, and expression of extracellular matrix components [210–214]. Cancer cells undergoing this transformation, often at the tumor host interface, tend to invade and metastasize to surrounding and distant tissues leading to life-threatening manifestations of cancer progression [215–223]. A study by Sakuma et al. [209] showed that EMT induction in colon cancer cell lines using EGF and bFGF caused elevated expression of sLe^x and sLe^a as well as E-selectin binding. These changes were linked to upregulation of ST and FT expression. Pinho et al. [224] reported that there is a negative correlation between sLe^x and E-cadherin expression in two mammary tumor models. The loss of the latter is a well-known marker for EMT.

9.3.1.1 Selectin Ligands and Metastasis

To date, many selectin ligands are known to be expressed on carcinoma cells, including PSGL-1 [182,225], CD44v [181,185,226–228], podocalyxin-like protein (PCLP) [229], death receptor-3 (DR-3) [186], gangliosides [93], mucin 1 (MUC1) [180,230,231], CD43 [232,233], and Mac-2BP [234] (Figure 9.1). PSGL-1 is the most characterized selectin ligand at the cellular, molecular, and functional levels in addition to being the only ligand currently known to be able to bind all three selectins. PSGL-1 is a type 1 transmembrane $sLe^{x/a}$-bearing mucin-like glycoprotein of 402 or 412 amino acids [235,236]. Moore et al. [237] first identified PSGL-1 as a P-selectin ligand on myeloid cells including neutrophils and HL-60 cells. PSGL-1 is a disulfide homodimer with apparent molecular weights of 240 kD (dimer) and 120 kD (monomer) on SDS-PAGE. The extracellular domain is rich in serine, threonine, and proline, and consists of 15 or 16 decameric repeats extended linearly due to the presence of multiple serine/threonine O-glycosylated sites that straighten the backbone of mucin proteins [238]. Three N-terminal tyrosine amino acids at positions 5, 7, and 10 are arranged in a consensus sequence that favors tyrosine sulfation. The cytoplasmic domain of PSGL-1 has 69 amino acids with signaling functionality but with no consensus sequence for phosphorylation [235]. PSGL-1 is expressed on hematopoietic cells and cancer cells. $CD34^+$ HSPCs and myeloid cells at different states of maturation (except erythrocytes and megakaryocytes) and all peripheral blood leukocytes express PSGL-1 [239,240]. Expression of PSGL-1 on platelets has been reported by Frenette et al., although remains debated [241]. PSGL-1 is also expressed by leukemic cells [242] and carcinoma cells such as colon [225] and prostate cancer cells [182]. PSGL-1 binds to E-selectin in a different way than to L-selectin and P-selectin. Both sialylation and α1,3-fucosylation, that appear on core 2 O-glycan chains, are required for E-selectin binding but sulfation (of either sLe^x moiety or tyrosine) is not required [243]. The binding to E-selectin of PSGL-1 expressed by CTCs implies a functional role in bone tropism of metastatic cancer cells. Dimitroff et al. demonstrated that PSGL-1, expressed on the surface of the bone-metastatic cell line MDA-PCa-2b, supports rolling on BMECs. This rolling was abolished by using neutralizing antibodies against E-selectin, suggesting that E-selectin/E-selectin ligand interactions mediate this adhesion. Neuraminidase treatment of the MDA-PCa-2b cells also impaired rolling indicating the importance of sialic acid in mediating this interaction and subsequent rolling. Performing a Western blot analysis on the membrane fraction of MDA-PCa-2b for the expression of sLex (detected by HECA-452 mAb) revealed reactivity with CD44 in addition to PSGL-1 [182,183].

Many reports have characterized the binding of CD44 to E-selectin on HSCs, leukocytes, and CTCs. CD44 is known to direct the recruitment of lymphocytes toward inflammatory sites through its interaction with matrix HA (hyaluronin) and endothelial cell expressed E-selectin, and a similar interaction is thought to direct the metastasis of many cancer types [80]. The E-selectin binding form of CD44 is known as HCELL. HCELL is expressed on the surface of HSPCs, human leukemia cells and has recently been located on colon cancer and breast cancer cells. Unlike mucin-like glycoproteins, CD44 bears sLe$^{x/a}$ structures on N-linked glycan branches [181,185,226–228,244].

MUC1 is a large heavily glycosylated transmembrane protein with an extracellular region comprising a variable number of tandem repeats that include 20 conserved amino acids (HGVTSAPDTRPAPGSTAPPA) each with five potential sites of O-glycosylation (as underlined). In normal cells, MUC1 is expressed on apical surfaces and glycosylated with core 2 branched O-glycans and lactosamine extensions, whereas in cancer cells MUC1 is expressed on the whole surface and decorated with aberrantly truncated O-glycosylated carbohydrate antigens. MUC1 expressed on breast cancer cells tends to present sialylated and unsialylated core 1 structures instead of core 2 structures. It has long been known that ST expression and activity is higher in breast cancer cells than in normal breast tissue. Thus, the exposure of truncated glycan structures may be due to sialylation of core 1 GalNAc or lactosamine structures leading to a blockade of further extension and expression of T and Tn antigens [245]. The MUC1 interaction with E-selectin was first described in colon cancer cells by Zhang et al. [231]; MUC1 from a colon cancer cell line or colon cancer patient serum was shown to be able to inhibit the binding of leukemic cells to E-selectin-expressing cells. MUC1, expressed by colon cancer cells, was reported to bind E-selectin better than CD43 from the same cells under flow conditions [230]. A recent study by Geng et al. [180] used E-selectin and ICAM-coated microtubes as a model for microvascular endothelium to show that the MUC1-expressing metastatic breast cancer cell line, ZR-75-1, was able to roll and adhere to the walls of the microtube under flow conditions, so implicating MUC1 as a ligand for E-selectin on breast cancer metastatic cells. CD24, another mucin-like glycoprotein, was also identified as an E-selectin and P-selectin ligand in breast cancer cell lines (KS and MCF-7) [246,247].

More recently, a number of novel ligands have been identified as E-selectin ligands on cancer cells including PCLP on colon carcinoma cells [229], Mac-2BP on breast cancer cells [234], and DR-3 on colon cancer [186]. In addition to these glycoproteins, gangliosides have also been shown to play a role as E-selectin ligands in breast cancer cell lines [93]. Mac-2BP is a highly glycosylated protein whose expression is correlated with cancer metastasis. Shirure et al. were able to immunoprecipitate E-selectin ligands from metastatic breast cancer (ZR-75-1) lysate using recombinant E-selectin and identified Mac-2BP as a novel ligand through MS analysis. Gene silencing of Mac-2BP significantly reduced the binding of ZR-75-1 cells to E-selectin-expressing human umbilical vein endothelial cell (HUVEC) cells in flow assays [234]. All these studies highlighted the role of selectin/selectin ligand interaction in supporting metastasis. Since it is the glycan decorations, primarily the expression of $sLe^{x/a}$ on these proteins (and lipids), that mediate this interaction of the CTC with the endothelial E-selectin, the contribution of GTs to migration and metastasis-supporting programs such as EMT needs to be better understood.

9.3.1.2 Perturbation in the Expression of GTs in Metastasis

It is widely accepted that abnormalities in glycosylation are closely associated with malignant transformation and progression [201–203]. Accordingly, highly metastatic carcinomas (e.g., colon, breast, and prostate) exhibit high levels of $sLe^{x/a}$ expression [74,180–183,185,186,225]. $sLe^{x/a}$ biosynthesis involves the concerted action of four GTs [68] and, hence, any dysregulation in GT expression and function has a direct impact on selectin-mediated metastasis. For instance, C2GnT is overexpressed in colorectal adenocarcinoma [248,249]. Lung carcinoma cells were found to acquire a colonizing phenotype upon FT-VII overexpression [188,189,250] and, in colon carcinoma, FucT-III gene silencing caused downregulation of sLe^x expression and subsequent inhibition of tumor proliferation and metastasis to the liver [251,252]. Matsuura et al. [74] investigated the expression of FTs and STs in breast cancer samples and cell lines and found that FT-VI and -III are significantly elevated, and that by expressing these FTs in breast cancer cells normally negative for sLe^x and sLe^a expression (MCF-7), rolling events are amplified on IL-1β activated HUVECs. One decade later, Barthel et al. reported similar results, i.e., elevated expression of FucT-III and -VI in the metastatic prostate cancer cell line, MDA-PCa-2b, compared with nonmetastatic cell lines. When PC3 cells, a prostate cancer cell line that is normally negative for E-selectin binding, were transfected with FucT-III, -VI or -VII, the cells gained sLe^x expression and the ability to roll on E-selectin-expressing endothelial cells and to metastasize to bone [71,201].

Moreover, it was demonstrated that the most highly expressed ST in patient samples of breast carcinoma was ST3Gal-III. This ST is responsible for sLe^a biosynthesis and its overexpression was associated with shorter overall survival [253,254]. Correspondingly, ST expression in metastasizing mammary tumors is higher compared with levels in nonmetastasizing tumors [255]. A recent report has indicated that induction of EMT, a prometastatic transformation program, was accompanied by elevation in levels of ST3Gal-I/-III/-IV and FT-III which are responsible for terminal capping of lactosamine units with sialic acid and fucose residues to form $sLe^{x/a}$ moieties on colon cancer cells [209]. In combination, these studies indicate the fundamental role of GTs as modifiers of surface glycome features in supporting cancer metastasis and also reveal the potential significance of GT inhibitors as basic research and therapeutic tools.

9.3.2 Metabolic Inhibition of GTs

In common with all glycans, $sLe^{x/a}$ is not directly encoded by the genome and its upregulation is achieved through overexpression of GTs responsible for its biosynthesis.

The tetrasaccharide sLe$^{x/a}$ requires four GT enzymes to catalyze its biosynthesis: *N*-acetylglucosaminyltransferase, β-galactosyltransferase, α-fucosyltransferase, and α-sialyltransferase. A detailed biosynthesis pathway of sLe$^{x/a}$ and an explanation of the differences in glycan structure recognized by the three different selectins are outlined in Figure 9.2. Glycoside bond synthesis by the Leloir pathway involves the transfer of a sugar from an activated sugar nucleotide donor, nucleoside diphospho sugar (NDP sugar), or a nucleoside monophospho sugar (NMP sugar), to the hydroxyl group of an aglycone acceptor, catalyzed by the action of a GT enzyme. GT action may either have an overall retaining or inverting effect on the configuration at the anomeric center of the donor sugar [156]. sLe$^{x/a}$ acts as a recognition marker for selectin and is involved in diverse biological processes including cell–cell adhesion, cell growth, cell differentiation, morphogenesis, leukocyte trafficking, and tumor dissemination [46,256]. In addition, aberration of glycan structure to sLe$^{x/a}$ and expression occurs in various diseases including inflammatory disorders and tumor metastasis [68,204,257]. Consequently, GT inhibitors as therapeutic tools are promising in targeting diseases associated with surface glycoproteins such as inflammatory diseases and cancer metastasis. Alternatively, inhibitors of GTs provide valuable information about the role of glycans and glycoconjugates in biological processes through functional experiments in which the glycosylation of surface markers is perturbed and the cell surface glycoproteome is altered. Inhibitors are also used to study the active sites of GT enzymes and the molecular mechanisms of their action. Due to their biological and therapeutic significance, numerous inhibitors have been developed to target GTs.

9.3.2.1 GT Inhibitors and Primers

Natural GT inhibitors include papulacandin B [258], polyoxin D [259], tunicamycin [260], castanospermine [261], swainsonine [262], deoxymannojirimycin [263], and *N*-butyldeoxynojirimycin [264]. Natural inhibitors may cause extensive modifications in glycoconjugate structures such that it is difficult to study or control a specific interaction or enzyme; this introduces the necessity for the design and synthesis of selective agents. Synthetic GT inhibitors can be divided into two classes (refer to Figure 9.5). The first comprises substrate analogs which cause dead-end inhibition whereby a modified substrate is used to mimic one or more of the enzyme substrate(s) in order to block the enzyme or compete with natural substrates and terminate or delay glycan chain initiation or elongation. This class is further separated into three classes: donor substrate analogs (mimetics) of the donor sugar, nucleoside diphosphate, nucleoside monophosphate, or a diphosphate moiety. Because the mammalian system has only nine sugar nucleotide donors, the selectivity of GTs is thought to be associated with the identification of the acceptor rather than with the donor substrate. Within the group of donor analogs, diphosphate mimetics are thought to confer even less selectivity than sugar-modified donor analogs. In order to fulfill the selectivity requirement of GT inhibitors, acceptor mimetics have been developed. Acceptor substrate analogs compete with natural acceptors or block the active site of the enzyme to prevent the glycosyl transfer to the natural acceptor. Acceptor analogs usually have modifications at the hydroxyl group. The group may be deoxygenated, derivatized, or replaced with a halogen or other functionalities. Bisubstrate analogs have two covalently linked motifs mimicking both the acceptor and donor substrates. These analogs are also referred to as transition state analogs because they mimic the transition state of the reaction, and they are thought to be the most active and selective inhibitors of GTs. The second class includes GT primers and metabolic decoys. In contrast to the dead-end GT inhibitors, GT

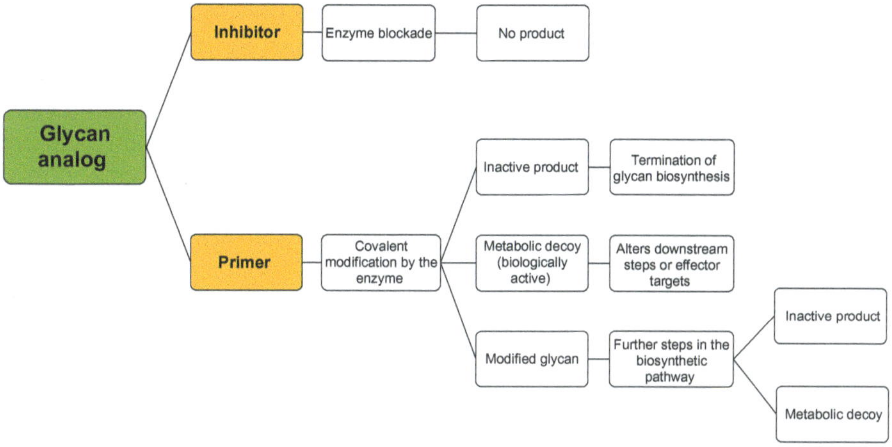

FIGURE 9.5 Mode of action of GT modulators. Glycan analogs modulate the action of GTs through either inhibitory or priming activity. GT inhibitors are enzyme substrate analogs that mimic one or more of the substrate analogs to block the enzyme or compete with natural substrate leading to termination of natural glycan biosynthesis pathway. GT primers prime the enzyme activity using irrelevant substrate leading to consumption of the enzyme activity in the production of irrelevant products and diversion from the natural glycan biosynthesis pathway.

primers prime the enzyme activity using irrelevant substrates leading to the consumption of free active enzyme and the production of irrelevant products. Products of the priming reaction may be either inactive or metabolic decoys, or further modified with downstream enzymes to produce metabolic decoys capable of inhibiting a target pathway or ligand. GT synthetic inhibitors will be discussed in more detail in the following section. For simplicity, the first class will be designated as "inhibitors" and the second class as "primers."

9.3.2.2 GT Substrate-Analogous Inhibitors

As this chapter is focused on selectin ligand prototype sLe$^{x/a}$, this section in turn will discuss inhibitors against the key GTs involved in its biosynthesis (Figure 9.2) and with which sLe$^{x/a}$ expression and/or selectin ligand binding were reported to be affected: C2GnT, FT, and ST.

C2GnT inhibitors: C2GnT (β(1,3)-galactosyl-*O*-glycosyl-glycoprotein β(1,6)-*N*-acetylglucosaminyltransferase; EC 2.4.1.102) is an inverting GT that catalyzes the transfer of a GlcNAc residue from UDP-αGlcNAc to β(1,6) linkage on core 1 structure (Galβ1,3GalNAcα-O-pp) to form core 2 branching, one of the major mucin-type O-glycan core structures [265]. An inhibitor to C2GnT GT was developed based on a deoxygenated analog of the enzyme acceptor (Galβ1,3,6-deoxy-GalNAcα-O$(CH_2)_8COOCH_3$) which produced moderate inhibition compared with the corresponding acceptor (Galβ1,3GalNAcα-O$(CH_2)_8COOCH_3$) *in vitro* [266]. A recent study showed that 5-thio-GlcNAc could act as a donor substrate analog against the C2GnT enzyme. This inhibitor is active *in vitro* and the first to be reported as active *in vivo*. Although it caused a decrease in levels of cellular *O*-GlcNAc, the surface glycan profile did not change. *In vivo* activity is thought to be due to peracetylation of 5T-GlcNAc, which overcomes the problem of poor diffusion of charged analogs. The peracetylated prodrug is converted to the 5T-GlcNAc by the intracellular estrases and, in turn, is converted to UDP-5T-GlcNAc through the salvage pathway [267].

FUT inhibitors: α1,3-fucosyltransferase (FucT-IV, -V, -VI, -VII; EC 2.4.1.152) is an inverting enzyme that catalyzes the transfer of the fucosyl moiety from GDP-fucose to the 3-OH of group of *N*-acetyllactosamine to give a Lewis X trisaccharide. α1,3/4-fucosyltransferase (EC 2.4.1.65) transfers a fucose sugar from GDP-fucose to the 4-OH of Galβ1,3GlcNAc moiety to give a Lewis A trisaccharide. FTs can act on sialylated LacNAc to give sLe$^{x/a}$ tetrasaccharides [268].

As for 5T-GlcNAc, the same research group developed 5-thio-fucose and evaluated its activity against sLe^{x} expression and selectin binding activity. Peracetylated 5T-Fuc was taken up by the relevant cells and converted through a salvage pathway into a sugar nucleotide donor analog, GDP-5T-Fuc that, in turn, inhibited the addition of fucose by the FT enzyme. 5T-Fuc was able to inhibit sLe^{x} expression in HepG2 cells and impaired adhesion of the cells to selectin-coated surfaces, and activated endothelial monolayer cells [269]. A cell-permeable acetylated analog of a fucose bearing a fluorine atom near the endocyclic oxygen, 2-fluoro-fucose (2F-Fuc), was reported to exhibit global inhibition of FTs. After being deacetylated intracellularly, this fluorinated analog is converted to the corresponding nucleotide sugar substrate GDP-2-fluoro-Fuc by the salvage pathway and is able to inhibit sLe^{x} formation in a human myeloid cell line and, subsequently, to hinder E-selectin- and P-selectin-mediated rolling on coated surfaces [270]. A novel study investigated the oral bioavailability of 2F-Fuc in a murine model. 2F-Fuc inhibited protein fucosylation, neutrophil sLe^{x} expression, and selectin-mediated adhesion during oral administration with no apparent toxicity in mice over the course of 3 weeks. Inhibitory activities were reversed upon discontinuing the drug administration. 2F-Fuc was also found to inhibit tumor outgrowth in colon and renal cancer metastasis mouse models after oral administration. This study supports the development and potential use of clinically effective GT inhibitors [179,180] in the treatment of cancer and prevention of metastasis.

ST inhibitors: The α2,3-sialyltransferase enzyme (ST3Gal) is a retaining enzyme that catalyzes the transfer of sialic acid from cytidine-5′-monophospho-*N*-acetylneuraminic acid (CMP-Neu5Ac) to the terminal Galβ1,4GlcNAc (EC 2.4.99.6) or Galβ1,3GlcNAc. Donor-analogous substrates were developed to mimic the common donor substrate, CMP-Neu5Ac, by replacing the anomeric carbon-linked oxygen atom with an ethyl bridge. This analog caused significant inhibition as it probably prevented the nucleophilic attack by the enzyme basic group [271]. Recently, using a cell-permeable acetylated analog of a sialic acid bearing a fluorine atom near the endocyclic oxygen, a global inhibition of ST activity was reported. After being deacetylated intracellularly, this fluorinated analog is converted to the corresponding nucleotide sugar substrate, CMP-2-fluoro-NeuAc, by the salvage pathway. This analog is then able to inhibit sLe^{x} formation in a human myeloid cell line as well as E-selectin- and P-selectin-mediated rolling on coated surfaces. It is thought that the observed inhibition is due to both enzyme and feedback inhibition of *de novo* synthesis of CMP-Neu5Ac by the salvage pathway and, to date, this is the only α2,3-sialyltransferase analog that has been found to produce inhibitory effects *in vivo* [270,272]. Acceptor-analogous inhibitors were reported to inhibit rat liver α2,3-sialyltransferase activity *in vitro* using a trisaccharide analog with a modified 3′-OH. The modifications include deoxygenation, fluorination, amination, or methoxy derivatization [273]. The 3′-deoxy lactosamine derivative was also reported, but with no inhibition activity *in vitro* [266]. Bisubstrate analogs containing the donor CMP-Neu5Ac mimetic and galactose, lactose or lactosamine acceptor have been developed and found to produce weak inhibitory effects against α2,3-sialyltransferase in the case of galactose- and lactose-containing acceptors, whereas they produced potent effects in the case of lactosamine-containing analogs [271,274].

9.3.2.3 GT Primers and Metabolic Decoys

Traditional GT inhibitors are analogs with alterations in specific functionalities that affect the thermodynamics and kinetics of the reaction. These analogs can still bind to the enzyme, but cannot undergo covalent modifications to give reaction products (dead-end inhibition). Conversely, priming activity is achieved when the modified analog binds the active enzyme and undergoes covalent modification to yield modified soluble glycan. Modified glycans can be inactive and thus inhibitory by only the consumption of the active enzyme in an extraneous pathway diverting glycan synthesis from endogenous glycoconjugates (see Figure 9.5). Otherwise, the modified glycan can be biologically active so that it acts as a metabolic decoy, or further modified by the biosynthesis pathway enzymes to a metabolic decoy that can inhibit the glycoconjugate formation or block an effector target. Many of the well-known inhibitors are inactive *in vivo* probably due to poor permeation into cells and subcellular compartments such as the Golgi apparatus where most glycan formation takes place. Primers can circumvent this problem by conjugating the substrate to a hydrophobic aglycone that enables passive diffusion through the lipophilic cell membrane without affecting the binding kinetics to the target enzyme. Conjugating specific aglycones may also facilitate the isolation of reaction products for further study or create probes for the study of active sites. Also, heterogeneity between cell types in glycosylation patterns can be studied through analysis of the primed products. The monosaccharide glycoside paranitrophenol α-GalNAc (pNp α-GalNAc) was used to inhibit mucin glycosylation through priming modified lactosamine formation and extension in colon cancer and lymphoid tumor cells [275,276]. PNp α-GalNAc is an acceptor analog for β1,3-galactosyltransferase that transfers a Gal residue from UDP-Gal to GalNAc to form core 1 structure. The core 1 structure (Gal β1,3GalNAc) acts as a molecular scaffold for mucin decorations with glycan side chains (i.e., core 2 branching, lactosamine extension, and $sLe^{x/a}$ formation). PNp α-GalNAc could enter viable cells, probably due to the hydrophobic aryl group, and consume glycosylation machinery from mucin decoration toward the priming reaction. The aryl group was also used to isolate and secrete the priming reaction products Galβ1,3GalNAcα-pNp and Galβ1,3(Galβ1,4GlcNAcβ-6)GalNAcα-pNp, and to study the glycosylation patterns in these cells. Two independent groups who used GalNAcα-*O*-benzyl as a primer for such pathways [277–279] also reported similar results. Likewise, several hydrophobic glycosides of GlcNAc, GlcNAcα-*O*-benzyl, GlcNAcβ-*O*-benzyl, GlcNAcβ-*O*-phenyl, and GlcNAcβ-*O*-pnp could act as primers for polylactosamine chain synthesis [280]. More recently, fluorinated peracetylated GalNAc was reported to be incorporated in selectin ligands expressed by human promyelocytic leukemia cells (HL-60) suggesting a metabolic priming activity and inhibition of glycan chain elongation through the blockade of carbon 4 with a fluorine atom, so preventing the addition of either fucose or galactose sugars. Addition of 4F-GalNAc to the culture medium caused a significant decrease in sLe^x expression and leukocyte adhesion *in vivo* [281]. Another study confirmed the activity of 4F-GlcNAc in inhibiting sLe^x expression and selectin binding. However, this research showed that there was no incorporation of 4F-GlcNAc into the growing glycan chain, suggesting that the main mechanism of 4F-GlcNAc activity lies in interference with UDP-GlcNAc biosynthesis and not in the priming and incorporation of 4F-GlcNAc [282].

Disaccharide glycosides can also be used as primers for GTs. GlcNAcβ1,3Galβ-*O*-naphthalenemethanol (GlcNAcβ1,3Galβ-O-NM) and

GlcNAcβ1,4Galβ-O-NM inhibited sLe^x expression in leukemia, colon cancer, and lung cancer models, and subsequently reduced E-selectin-dependent cell adhesion to activated endothelial cells [283–286]. Furthermore, it was shown that the treatment of colon and lung adenocarcinoma with the peracetylated disaccharide primer $(Ac)_6$GlcNAcβ1,3Galβ-O-NM reduced the potential for lung colonization in immunodeficient mice. Acetylated derivatives are thought to be more readily bioavailable to cells and, hence, more active *in vivo* as the polar hydroxyl groups are masked. Acetyl groups are then removed by the action of intracellular and membrane bound esterases [287]. 4-deoxy derivative of $(Ac)_6$GlcNAcβ1,3Galβ-O-NM acted as an acceptor-analogous substrate against β1,4-galactosyltransferase and inhibited experimental tumor metastasis by Lewis lung carcinoma *in vivo* [288].

The development of a diverse group of GT inhibitors as modulators of surface selectin glycoconjugates holds great promise for the treatment of metastatic tumors and inflammatory disorders. Although this field is several decades old, most of the inhibitors known lack activity *in vivo* due to their polar nature and underprivileged cell permeability. Recently, several reports have indicated that this problem can be avoided by using the prodrug technology whereby peracetylated derivatives show enhanced uptake. Peracetylated prodrugs are easily deacetylated intracellularly to give the active form of the drug. Another approach is to use bulky hydrophobic aglycones to confer hydrophobic properties on the inhibitor candidate. Nevertheless the development of GT inhibitors with acceptable bioavailability remains a goal in the development of clinically effective antimetastasis therapeutic agents. On the other hand, since both physiological and pathological cell players share the sLe^x decoration, increased efforts should be directed toward to the development of selective agents that target-specific GT isoforms that are differentially expressed in CTCs but not in normal immune cells.

9.4 USE OF ENZYMES TO IMPROVE THE THERAPEUTIC FUNCTION OF CELLS THROUGH MODIFICATION OF THE CELL SURFACE

Other types of enzymes have been used or are potentially available for *ex vivo* enzymatic treatment of the cell surface with the aim of therapeutic applications. One such example is the transient proteolytic treatment of MSCs to improve the efficiency of delivery to the target sites [289,290]. When a systemic infusion is conducted for the delivery of MSCs, one major obstacle is the natural tendency of the injected cells to accumulate initially in the lung, and this may lower the efficiency of delivery of the MSCs to the target sites. Use of pronase rather than the standard use of trypsin for the preparation of the cell suspension results in increased lung clearance of MSCs and a better delivery to the site of inflammation, without any apparent functional defect, presumably due to the enhanced cleavage of cell surface proteins involved in adhesion and migration.

In another example, O-sialoglycoprotein endopeptidase (OSGE) treatment has been shown to enhance the functionality of aged CD4 T cells [291]. In humans and mice, the function of the antibody-mediated immune system becomes limited with age, there being lowered clonal diversity of antibodies and inadequate formation of long-term memory [292], due in part to age-related defects in CD4 T cells, especially a malfunction in the interaction between T cells and antigen presenting cells (APCs) [293]. The treatment of aged CD4 cells with OSGE *ex vivo* restored the age-dependent loss of function, possibly through a mechanism that removes glycoproteins resulting in increased access to the CD28 molecules for interaction with APCs [291].

9.5 CONCLUSION

This chapter has discussed in detail that the adhesive recruitment of both therapeutic stem cells as well as detrimental metastatic cancer cells are initiated by glycoprotein–selectin interactions facilitating cell rolling on endothelium. The formation of the key glycan responsible for mediating these interactions, sLe^x, is controlled by the precise activity of a family of ER/Golgi localized enzymes, the GTs. Through the identification and manipulation of these enzymes, migration can be effectively controlled either to enhance or inhibit it. In applications dealing with enabling a remedial cell to migrate to a predetermined site *in vivo*, the sLe^x structure needs to be created by the sequential activity of GTs under conditions that do not alter the viability, metabolism, or physiology of that cell, such as through the application of "GPS" technology. Multiple therapeutic avenues can then be sought including the guiding of stem cells or cell-based drug delivery systems to selectin expressing endothelial cells intrinsic to autoimmune or inflammatory diseases. In contrast, for applications where the focus is to inhibit migration/metastasis of detrimental cancer cells or immune cells in inflammatory disease, this interaction between selectins and glycans can be perturbed by inhibiting the activity of these enzymes. Since every family of GT has numerous members, identifying unique enzymes that are responsible for contributing to sLe^x formation in these harmful cells, but not in "normal" cells, and engineering selective inhibitors to these unique GTs, the migration of the harmful cells could potentially be targeted. Most of the inhibitors known to date lack this specificity and in addition lack activity *in vivo* due to their polar nature and underprivileged cell permeability. Although recent approaches have helped to overcome membrane permeabilization issues, the development of inhibitors with acceptable bioavailability and specificity remains a goal in the development of clinically effective antimetastasis therapeutic agents.

References

[1] Ehrhardt C, Kneuer C, Bakowsky U. Selectins—an emerging target for drug delivery. Adv Drug Deliv Rev 2004;56:527–49.

[2] Butcher EC. Leukocyte-endothelial cell recognition: three (or more) steps to specificity and diversity. Cell 1991;67:1033–6.

[3] Butcher EC, Picker LJ. Lymphocyte homing and homeostasis. Science 1996;272:60–6.

[4] Springer TA. Traffic signals for lymphocyte recirculation and leukocyte emigration: the multistep paradigm. Cell 1994;76:301–14.

[5] Lowe JB. Glycosylation in the control of selectin counter-receptor structure and function. Immunol Rev 2002;186:19–36.

[6] Vestweber D, Blanks JE. Mechanisms that regulate the function of the selectins and their ligands. Physiol Rev 1999;79:181–213.

[7] Kunkel EJ, Butcher EC. Chemokines and the tissue-specific migration of lymphocytes. Immunity 2002;16:1–4.

[8] Rossi D, Zlotnik A. The biology of chemokines and their receptors. Annu Rev Immunol 2000;18:217–42.

[9] Sallusto F, Mackay CR, Lanzavecchia A. The role of chemokine receptors in primary, effector, and memory immune responses. Annu Rev Immunol 2000;18:593–620.

[10] Cinamon G, Grabovsky V, Winter E, Franitza S, Feigelson S, Shamri R, et al. Novel chemokine functions in lymphocyte migration through vascular endothelium under shear flow. J Leukoc Biol 2001;69:860–6.

[11] Harris ES, McIntyre TM, Prescott SM, Zimmerman GA. The leukocyte integrins. J Biol Chem 2000;275:23409–12.

[12] van Kooyk Y, Figdor CG. Avidity regulation of integrins: the driving force in leukocyte adhesion. Curr Opin Cell Biol 2000;12:542–7.

[13] Woodside DG, Liu S, Ginsberg MH. Integrin activation. Thromb Haemost 2001;86:316–23.

[14] Weber C. Novel mechanistic concepts for the control of leukocyte transmigration: specialization of integrins, chemokines, and junctional molecules. J Mol Med 2003;81:4–19.

[15] Lawrence MB, Springer TA. Leukocytes roll on a selectin at physiologic flow rates: distinction from and prerequisite for adhesion through integrins. Cell 1991;65:859–73.

[16] Sackstein R. The lymphocyte homing receptors: gatekeepers of the multistep paradigm. Curr Opin Hematol 2005;12:444–50.

[17] Thankamony SP, Sackstein R. Enforced hematopoietic cell E- and L-selectin ligand (HCELL) expression primes transendothelial migration of human mesenchymal stem cells. Proc Natl Acad Sci USA 2011;108:2258–63.

[18] Yago T, Shao B, Miner JJ, Yao L, Klopocki AG, Maeda K, et al. E-selectin engages PSGL-1 and CD44 through a common signaling pathway to induce integrin alphaLbeta2-mediated slow leukocyte rolling. Blood 2010;116:485–94.

[19] Gibson RM, Kansas GS, Tedder TF, Furie B, Furie BC. Lectin and epidermal growth factor domains of P-selectin at physiologic density are the recognition unit for leukocyte binding. Blood 1995;85:151–8.

[20] Kansas GS, Saunders KB, Ley K, Zakrzewicz A, Gibson RM, Furie BC, et al. A role for the epidermal growth factor-like domain of P-selectin in ligand recognition and cell adhesion. J Cell Biol 1994;124(4):609–18.

[21] Wenzel K, Felix S, Kleber FX, Brachold R, Menke T, Schattke S, et al. E-selectin polymorphism and atherosclerosis: an association study. Hum Mol Genet 1994;3:1935–7.

[22] Wenzel K, Hanke R, Speer A. Polymorphism in the human E-selectin gene detected by PCR-SSCP. Hum Genet 1994;94:452–3.

[23] Polley MJ, Phillips ML, Wayner E, Nudelman E, Singhal AK, Hakomori S, et al. CD62 and endothelial cell-leukocyte adhesion molecule 1 (ELAM-1) recognize the same carbohydrate ligand, sialyl-Lewis x. Proc Natl Acad Sci USA 1991;88:6224–8.

[24] Alon R, Feizi T, Yuen CT, Fuhlbrigge RC, Springer TA. Glycolipid ligands for selectins support leukocyte tethering and rolling under physiologic flow conditions. J Immunol 1995;154:5356–66.

[25] Fuhlbrigge RC, Kieffer JD, Armerding D, Kupper TS. Cutaneous lymphocyte antigen is a specialized form of PSGL-1 expressed on skin-homing T cells. Nature 1997;389:978–81.

[26] Leppanen A, White SP, Helin J, McEver RP, Cummings RD. Binding of glycosulfopeptides to P-selectin requires stereospecific contributions of individual tyrosine sulfate and sugar residues. J Biol Chem 2000;275:39569–78.

[27] Rosen SD. Ligands for L-selectin: homing, inflammation, and beyond. Annu Rev Immunol 2004;22:129–56.

[28] Tedder TF, Steeber DA, Chen A, Engel P. The selectins: vascular adhesion molecules. FASEB J 1995;9:866–73.

[29] Bargatze RF, Kurk S, Butcher EC, Jutila MA. Neutrophils roll on adherent neutrophils bound to cytokine-induced endothelial cells via L-selectin on the rolling cells. J Exp Med 1994;180:1785–92.

[30] Walcheck B, Moore KL, McEver RP, Kishimoto TK. Neutrophil–neutrophil interactions under hydrodynamic shear stress involve L-selectin and PSGL-1. A mechanism that amplifies initial leukocyte accumulation of P-selectin *in vitro*. J Clin Invest 1996;98:1081–7.

[31] Kunkel EJ, Chomas JE, Ley K. Role of primary and secondary capture for leukocyte accumulation *in vivo*. Circ Res 1998;82:30–8.

[32] Mitchell DJ, Li P, Reinhardt PH, Kubes P. Importance of L-selectin-dependent leukocyte–leukocyte interactions in human whole blood. Blood 2000;95:2954–9.

[33] Rossi FMV, Corbel SY, Merzaban JS, Carlow DA, Gossens K, Duenas J, et al. Recruitment of adult thymic progenitors is regulated by P-selectin and its ligand PSGL-1. Nat Immunol 2005;6:262–70.

[34] Kivisakk P, Mahad DJ, Callahan MK, Trebst C, Tucky B, Wei T, et al. Human cerebrospinal fluid central memory $CD4^+$ T cells: evidence for trafficking through choroid plexus and meninges via P-selectin. Proc Natl Acad Sci USA 2003;100:8389–94.

[35] Jacobsen K, Kravitz J, Kincade PW, Osmond DG. Adhesion receptors on bone marrow stromal cells: *in vivo* expression of vascular cell adhesion molecule-1 by reticular cells and sinusoidal endothelium in normal and gamma-irradiated mice. Blood 1996;87:73–82.

[36] Schweitzer KM, Drager AM, van der Valk P, Thijsen SF, Zevenbergen A, Theijsmeijer AP, et al. Constitutive expression of E-selectin and vascular cell adhesion molecule-1 on endothelial cells of hematopoietic tissues. Am J Pathol 1996;148:165–75.

[37] Weninger W, Ulfman LH, Cheng G, Souchkova N, Quackenbush EJ, Lowe JB, et al. Specialized contributions by alpha(1,3)-fucosyltransferase-IV and FucT-VII during leukocyte rolling in dermal microvessels. Immunity 2000;12:665–76.

[38] Tchernychev B, Furie B, Furie BC. Peritoneal macrophages express both P-selectin and PSGL-1. J Cell Biol 2003;163:1145–55.

[39] Yao L, Pan J, Setiadi H, Patel KD, McEver RP. Interleukin 4 or oncostatin M induces a prolonged increase in P-selectin mRNA and protein in human endothelial cells. J Exp Med 1996;184:81–92.

[40] Woltmann G, McNulty CA, Dewson G, Symon FA, Wardlaw AJ. Interleukin-13 induces PSGL-1/P-selectin-dependent adhesion of eosinophils, but not neutrophils, to human umbilical vein endothelial cells under flow. Blood 2000;95:3146–52.

[41] Hattori R, Hamilton KK, Fugate RD, McEver RP, Sims PJ. Stimulated secretion of endothelial von Willebrand factor is accompanied by rapid redistribution to the cell surface of the intracellular granule membrane protein GMP-140. J Biol Chem 1989;264:7768–71.

[42] Pan J, Xia L, McEver RP. Comparison of promoters for the murine and human P-selectin genes suggests species-specific and conserved mechanisms for transcriptional regulation in endothelial cells. J Biol Chem 1998;273:10058–67.
[43] Liu Z, Miner JJ, Yago T, Yao L, Lupu F, Xia L, et al. Differential regulation of human and murine P-selectin expression and function *in vivo*. J Exp Med 2010;207:2975–87.
[44] Keelan ET, Licence ST, Peters AM, Binns RM, Haskard DO. Characterization of E-selectin expression *in vivo* with use of a radiolabeled monoclonal antibody. Am J Physiol 1994;266:H278–90.
[45] Frenette PS, Subbarao S, Mazo IB, von Andrian UH, Wagner DD. Endothelial selectins and vascular cell adhesion molecule-1 promote hematopoietic progenitor homing to bone marrow. Proc Natl Acad Sci USA 1998;95:14423–8.
[46] Kansas GS. Selectins and their ligands: current concepts and controversies. Blood 1996;88:3259–87.
[47] Kunkel EJ, Ley K. Distinct phenotype of E-selectin-deficient mice. E-selectin is required for slow leukocyte rolling *in vivo*. Circ Res 1996;79:1196–204.
[48] Kunkel EJ, Dunne JL, Ley K. Leukocyte arrest during cytokine-dependent inflammation *in vivo*. J Immunol 2000;164:3301–8.
[49] Hwang JM, Yamanouchi J, Santamaria P, Kubes P. A critical temporal window for selectin-dependent $CD4^+$ lymphocyte homing and initiation of late-phase inflammation in contact sensitivity. J Exp Med 2004;199:1223–34.
[50] Subramaniam M, Koedam JA, Wagner DD. Divergent fates of P- and E-selectins after their expression on the plasma membrane. Mol Biol Cell 1993;4:791–801.
[51] Bevilacqua MP, Stengelin S, Gimbrone Jr. MA, Seed B. Endothelial leukocyte adhesion molecule 1: an inducible receptor for neutrophils related to complement regulatory proteins and lectins. Science 1989;243: 1160–5.
[52] Picker LJ, Michie SA, Rott LS, Butcher EC. A unique phenotype of skin-associated lymphocytes in humans. Preferential expression of the HECA-452 epitope by benign and malignant T cells at cutaneous sites. Am J Pathol 1990;136:1053–68.
[53] Berg EL, Yoshino T, Rott LS, Robinson MK, Warnock RA, Kishimoto TK, et al. The cutaneous lymphocyte antigen is a skin lymphocyte homing receptor for the vascular lectin endothelial cell-leukocyte adhesion molecule 1. J Exp Med 1991;174:1461–6.
[54] Sipkins DA, Wei X, Wu JW, Runnels JM, Cote D, Means TK, et al. *In vivo* imaging of specialized bone marrow endothelial microdomains for tumour engraftment. Nature 2005;435:969–73.
[55] Lo Celso C, Fleming HE, Wu JW, Zhao CX, Miake-Lye S, Fujisaki J, et al. Live-animal tracking of individual haematopoietic stem/progenitor cells in their niche. Nature 2009;457:92–6.
[56] Sackstein R. The bone marrow is akin to skin: HCELL and the biology of hematopoietic stem cell homing. J Invest Dermatol 2004;122:1061–9.
[57] Frenette PS, Mayadas TN, Rayburn H, Hynes RO, Wagner DD. Susceptibility to infection and altered hematopoiesis in mice deficient in both P- and E-selectins. Cell 1996;84:563–74.
[58] Mazo IB, Gutierrez-Ramos JC, Frenette PS, Hynes RO, Wagner DD, von Andrian UH. Hematopoietic progenitor cell rolling in bone marrow microvessels: parallel contributions by endothelial selectins and vascular cell adhesion molecule 1. J Exp Med 1998;188:465–74.
[59] Sackstein R, Merzaban JS, Cain DW, Dagia NM, Spencer JA, Lin CP, et al. *Ex vivo* glycan engineering of CD44 programs human multipotent mesenchymal stromal cell trafficking to bone. Nat Med 2008;14:181–7.
[60] Katayama Y, Hidalgo A, Furie BC, Vestweber D, Furie B, Frenette PS. PSGL-1 participates in E-selectin-mediated progenitor homing to bone marrow: evidence for cooperation between E-selectin ligands and alpha4 integrin. Blood 2003;102:2060–7.
[61] Hidalgo A, Weiss LA, Frenette PS. Functional selectin ligands mediating human $CD34^+$ cell interactions with bone marrow endothelium are enhanced postnatally. J Clin Invest 2002;110:559–69.
[62] Dimitroff CJ, Lee JY, Rafii S, Fuhlbrigge RC, Sackstein R. CD44 is a major E-selectin ligand on human hematopoietic progenitor cells. J Cell Biol 2001;153:1277–86.
[63] Levesque JP, Zannettino AC, Pudney M, Niutta S, Haylock DN, Snapp KR, et al. PSGL-1-mediated adhesion of human hematopoietic progenitors to P-selectin results in suppression of hematopoiesis. Immunity 1999;11:369–78.
[64] Winkler IG, Snapp KR, Simmons PJ, Levesque JP. Adhesion to E-selectin promotes growth inhibition and apoptosis of human and murine hematopoietic progenitor cells independent of PSGL-1. Blood 2004;103:1685–92.
[65] Winkler IG, Barbier V, Nowlan B, Jacobsen RN, Forristal CE, Patton JT, et al. Vascular niche E-selectin regulates hematopoietic stem cell dormancy, self renewal and chemoresistance. Nat Med 2012;18:1651–7.
[66] Manfredini R, Zini R, Salati S, Siena M, Tenedini E, Tagliafico E, et al. The kinetic status of hematopoietic stem cell subpopulations underlies a differential expression of genes involved in self-renewal, commitment, and engraftment. Stem Cells 2005;23:496–506.

[67] Kim MY, Oskarsson T, Acharyya S, Nguyen DX, Zhang XH, Norton L, et al. Tumor self-seeding by circulating cancer cells. Cell 2009;139:1315–26.
[68] Barthel SR, Gavino JD, Descheny L, Dimitroff CJ. Targeting selectins and selectin ligands in inflammation and cancer. Expert Opin Ther Targets 2007;11:1473–91.
[69] Geng Y, Marshall JR, King MR. Glycomechanics of the metastatic cascade: tumor cell–endothelial cell interactions in the circulation. Ann Biomed Eng 2012;40: 790–805.
[70] Khatib AM, Fallavollita L, Wancewicz EV, Monia BP, Brodt P. Inhibition of hepatic endothelial E-selectin expression by C-raf antisense oligonucleotides blocks colorectal carcinoma liver metastasis. Cancer Res 2002;62:5393–8.
[71] Barthel SR, Wiese GK, Cho J, Opperman MJ, Hays DL, Siddiqui J, et al. Alpha 1,3 fucosyltransferases are master regulators of prostate cancer cell trafficking. Proc Natl Acad Sci USA 2009;106:19491–6.
[72] Burdick MM, Henson KA, Delgadillo LF, Choi YE, Goetz DJ, Tees DF, et al. Expression of E-selectin ligands on circulating tumor cells: cross-regulation with cancer stem cell regulatory pathways? Front Oncol 2012;2:103.
[73] Kannagi R, Izawa M, Koike T, Miyazaki K, Kimura N. Carbohydrate-mediated cell adhesion in cancer metastasis and angiogenesis. Cancer Sci 2004;95:377–84.
[74] Matsuura N, Narita T, Hiraiwa N, Hiraiwa M, Murai H, Iwase T, et al. Gene expression of fucosyl- and sialyl-transferases which synthesize sialyl Lewis x, the carbohydrate ligands for E-selectin, in human breast cancer. Int J Oncol 1998;12:1157–64.
[75] Renkonen J, Paavonen T, Renkonen R. Endothelial and epithelial expression of sialyl Lewis(x) and sialyl Lewis(a) in lesions of breast carcinoma. J Int Cancer 1997;74:296–300.
[76] Czlapinski JL, Bertozzi CR. Synthetic glycobiology: exploits in the Golgi compartment. Curr Opin Chem Biol 2006;10:645–51.
[77] Gadhoum SZ, Sackstein R. CD15 expression in human myeloid cell differentiation is regulated by sialidase activity. Nat Chem Biol 2008;4:751–7.
[78] Tyrrell D, James P, Rao N, Foxall C, Abbas S, Dasgupta F, et al. Structural requirements for the carbohydrate ligand of E-selectin. Proc Natl Acad Sci USA 1991;88:10372–6.
[79] Sackstein R. Glycosyltransferase-programmed stereosubstitution (GPS) to create HCELL: engineering a roadmap for cell migration. Immunol Rev 2009;230: 51–74.
[80] Jacobs PP, Sackstein R. CD44 and HCELL: preventing hematogenous metastasis at step 1. FEBS Lett 2011; 585:3148–58.
[81] Smith PL, Myers JT, Rogers CE, Zhou L, Petryniak B, Becker DJ, et al. Conditional control of selectin ligand expression and global fucosylation events in mice with a targeted mutation at the FX locus. J Cell Biol 2002;158:801–15.
[82] Becker DJ, Lowe JB. Fucose: biosynthesis and biological function in mammals. Glycobiology 2003;13: 41R–53R.
[83] Ma B, Simala-Grant JL, Taylor DE. Fucosylation in prokaryotes and eukaryotes. Glycobiology 2006;16: 158R–84R.
[84] Smithson G, Rogers CE, Smith PL, Scheidegger EP, Petryniak B, Myers JT, et al. Fuc-TVII is required for T helper 1 and T cytotoxic 1 lymphocyte selectin ligand expression and recruitment in inflammation, and together with Fuc-TIV regulates naive T cell trafficking to lymph nodes. J Exp Med 2001;194:601–14.
[85] Maly P, Thall A, Petryniak B, Rogers CE, Smith PL, Marks RM, et al. The alpha(1,3)fucosyltransferase Fuc-TVII controls leukocyte trafficking through an essential role in L-, E-, and P-selectin ligand biosynthesis. Cell 1996;86:643–53.
[86] Homeister JW, Thall AD, Petryniak B, Maly P, Rogers CE, Smith PL, et al. The alpha(1,3)fucosyltransferases FucT-IV and FucT-VII exert collaborative control over selectin-dependent leukocyte recruitment and lymphocyte homing. Immunity 2001;15:115–26.
[87] Knibbs RN, Craig RA, Natsuka S, Chang A, Cameron M, Lowe JB, et al. The fucosyltransferase FucT-VII regulates E-selectin ligand synthesis in human T cells. J Cell Biol 1996;133:911–20.
[88] Knibbs RN, Craig RA, Maly P, Smith PL, Wolber FM, Faulkner NE, et al. Alpha(1,3)-fucosyltransferase VII-dependent synthesis of P- and E-selectin ligands on cultured T lymphoblasts. J Immunol 1998;161:6305–15.
[89] Bengtson P, Lundblad A, Larson G, Pahlsson P. Polymorphonuclear leukocytes from individuals carrying the G329A mutation in the alpha 1,3-fucosyltransferase VII gene (FUT7) roll on E- and P-selectins. J Immunol 2002;169:3940–6.
[90] Huang MC, Laskowska A, Vestweber D, Wild MK. The alpha (1,3)-fucosyltransferase Fuc-TIV, but not Fuc-TVII, generates sialyl Lewis X-like epitopes preferentially on glycolipids. J Biol Chem 2002;277: 47786–95.
[91] Burdick MM, McCaffery JM, Kim YS, Bochner BS, Konstantopoulos K. Colon carcinoma cell glycolipids, integrins, and other glycoproteins mediate adhesion to HUVECs under flow. Am J Physiol Cell Physiol 2003;284:C977–87.
[92] Nimrichter L, Burdick MM, Aoki K, Laroy W, Fierro MA, Hudson SA, et al. E-selectin receptors on human leukocytes. Blood 2008;112:3744–52.

[93] Shirure VS, Henson KA, Schnaar RL, Nimrichter L, Burdick MM. Gangliosides expressed on breast cancer cells are E-selectin ligands. Biochem Biophys Res Commun 2011;406:423–9.

[94] Mondal N, Buffone Jr. A, Neelamegham S. Distinct glycosyltransferases synthesize E-selectin ligands in human vs. mouse leukocytes. Cell Adh Migr 2013;7: 288–92.

[95] Buffone Jr. A, Mondal N, Gupta R, McHugh KP, Lau JT, Neelamegham S. Silencing alpha1,3-fucosyltransferases in human leukocytes reveals a role for FUT9 enzyme during E-selectin-mediated cell adhesion. J Biol Chem 2013;288:1620–33.

[96] Merzaban JS, Burdick MM, Gadhoum SZ, Dagia NM, Chu JT, Fuhlbrigge RC, et al. Analysis of glycoprotein E-selectin ligands on human and mouse marrow cells enriched for hematopoietic stem/progenitor cells. Blood 2011;118:1774–83.

[97] Ellies LG, Sperandio M, Underhill GH, Yousif J, Smith M, Priatel JJ, et al. Sialyltransferase specificity in selectin ligand formation. Blood 2002;100:3618–25.

[98] Okajima T, Fukumoto S, Miyazaki H, Ishida H, Kiso M, Furukawa K, et al. Molecular cloning of a novel alpha2,3-sialyltransferase (ST3Gal VI) that sialylates type II lactosamine structures on glycoproteins and glycolipids. J Biol Chem 1999;274:11479–86.

[99] Sasaki K, Watanabe E, Kawashima K, Sekine S, Dohi T, Oshima M, et al. Expression cloning of a novel Gal beta (1–3/1–4) GlcNAc alpha 2,3-sialyltransferase using lectin resistance selection. J Biol Chem 1993;268:22782–7.

[100] Yang WH, Nussbaum C, Grewal PK, Marth JD, Sperandio M. Coordinated roles of ST3Gal-VI and ST3Gal-IV sialyltransferases in the synthesis of selectin ligands. Blood 2012;120:1015–26.

[101] Sperandio M, Frommhold D, Babushkina I, Ellies LG, Olson TS, Smith ML, et al. Alpha 2,3-sialyltransferase-IV is essential for L-selectin ligand function in inflammation. Eur J Immunol 2006;36:3207–15.

[102] Kono M, Ohyama Y, Lee YC, Hamamoto T, Kojima N, Tsuji S. Mouse beta-galactoside alpha 2,3-sialyltransferases: comparison of *in vitro* substrate specificities and tissue specific expression. Glycobiology 1997;7:469–79.

[103] Schachter H, Brockhausen I. The biosynthesis of branched O-glycans. Symp Soc Exp Biol 1989;43:1–26.

[104] Ellies LG, Tsuboi S, Petryniak B, Lowe JB, Fukuda M, Marth JD. Core 2 oligosaccharide biosynthesis distinguishes between selectin ligands essential for leukocyte homing and inflammation. Immunity 1998;9:881–90.

[105] Snapp KR, Heitzig CE, Ellies LG, Marth JD, Kansas GS. Differential requirements for the O-linked branching enzyme core 2 beta1-6-N-glucosaminyltransferase in biosynthesis of ligands for E-selectin and P-selectin. Blood 2001;97:3806–11.

[106] Sperandio M, Thatte A, Foy D, Ellies LG, Marth JD, Ley K. Severe impairment of leukocyte rolling in venules of core 2 glucosaminyltransferase-deficient mice. Blood 2001;97:3812–9.

[107] Merzaban JS, Zuccolo J, Corbel SY, Williams MJ, Ziltener HJ. An alternate core 2 {beta}1,6-*N*-acetylglucosaminyltransferase selectively contributes to P-selectin ligand formation in activated CD8 T cells. J Immunol 2005;174:4051–9.

[108] Mullaly SC, Oudhoff MJ, Min PH, Burrows K, Antignano F, Rattray DG, et al. Requirement for core 2 O-glycans for optimal resistance to helminth infection. PLoS ONE 2013;8:e60124.

[109] Huang MC, Zollner O, Moll T, Maly P, Thall AD, Lowe JB, et al. P-selectin glycoprotein ligand-1 and E-selectin ligand-1 are differentially modified by fucosyltransferases Fuc-TIV and Fuc-TVII in mouse neutrophils. J Biol Chem 2000;275:31353–60.

[110] de Vries T, Knegtel RM, Holmes EH, Macher BA. Fucosyltransferases: structure/function studies. Glycobiology 2001;11:119R–28R.

[111] Mollicone R, Moore SE, Bovin N, Garcia-Rosasco M, Candelier JJ, Martinez-Duncker I, et al. Activity, splice variants, conserved peptide motifs, and phylogeny of two new alpha1,3-fucosyltransferase families (FUT10 and FUT11). J Biol Chem 2009;284:4723–38.

[112] Padro M, Cobler L, Garrido M, de Bolos C. Down-regulation of FUT3 and FUT5 by shRNA alters Lewis antigens expression and reduces the adhesion capacities of gastric cancer cells. Biochim Biophys Acta 2011;1810:1141–9.

[113] Boura JS, Santos FD, Gimble JM, Cardoso CM, Madeira C, Cabral JM, et al. Direct head-to-head comparison of cationic liposome-mediated gene delivery to mesenchymal stem/stromal cells of different human sources: a comprehensive study. Hum Gene Ther Methods 2013;24:38–48.

[114] Madeira C, Mendes RD, Ribeiro SC, Boura JS, Aires-Barros MR, da Silva CL, et al. Nonviral gene delivery to mesenchymal stem cells using cationic liposomes for gene and cell therapy. J Biomed Biotechnol 2010;2010:735349.

[115] Lim JY, Park SH, Jeong CH, Oh JH, Kim SM, Ryu CH, et al. Microporation is a valuable transfection method for efficient gene delivery into human umbilical cord blood-derived mesenchymal stem cells. BMC Biotechnol 2010;10:38.

[116] Madeira C, Ribeiro SC, Pinheiro IS, Martins SA, Andrade PZ, da Silva CL, et al. Gene delivery to

human bone marrow mesenchymal stem cells by microporation. J Biotechnol 2011;151:130–6.

[117] Deyle DR, Li Y, Olson EM, Russell DW. Nonintegrating foamy virus vectors. J Virol 2010;84: 9341–9.

[118] Wiktorowicz T, Peters K, Armbruster N, Steinert AF, Rethwilm A. Generation of an improved foamy virus vector by dissection of cis-acting sequences. J Gen Virol 2009;90:481–7.

[119] Dwyer RM, Khan S, Barry FP, O'Brien T, Kerin MJ. Advances in mesenchymal stem cell-mediated gene therapy for cancer. Stem Cell Res Ther 2010;1:25.

[120] Shakhbazov AV, Kosmacheva SM, Kartel NA, Potapnev MP. Gene therapy based on human mesenchymal stem cells: strategies and methods. Tsitol Genet 2010;44:76–82.

[121] Kaufmann M, Blaser C, Takashima S, Schwartz-Albiez R, Tsuji S, Pircher H. Identification of an alpha2,6-sialyltransferase induced early after lymphocyte activation. Int Immunol 1999;11:731–8.

[122] Damle NK, Klussman K, Dietsch MT, Mohagheghpour N, Aruffo A. GMP-140 (P-selectin/CD62) binds to chronically stimulated but not resting $CD4^+$ T lymphocytes and regulates their production of proinflammatory cytokines. Eur J Immunol 1992;22:1789–93.

[123] Piller F, Piller V, Fox RI, Fukuda M. Human T-lymphocyte activation is associated with changes in O-glycan biosynthesis. J Biol Chem 1988;263: 15146–50.

[124] Picker LJ, Treer JR, Ferguson-Darnell B, Collins PA, Bergstresser PR, Terstappen LW. Control of lymphocyte recirculation in man. II: Differential regulation of the cutaneous lymphocyte-associated antigen, a tissue-selective homing receptor for skin-homing T cells. J Immunol 1993;150:1122–36.

[125] Carlow DA, Corbel SY, Williams MJ, Ziltener HJ. IL-2, -4, and -15 differentially regulate O-glycan branching and P-selectin ligand formation in activated CD8 T cells. J Immunol 2001;167:6841–8.

[126] Lim YC, Xie H, Come CE, Alexander SI, Grusby MJ, Lichtman AH, et al. IL-12, STAT4-dependent up-regulation of $CD4^+$ T cell core 2 beta-1,6-*N*-acetylglucosaminyltransferase, an enzyme essential for biosynthesis of P-selectin ligands. J Immunol 2001;167:4476–84.

[127] Blander JM, Visintin I, Janeway Jr. CA, Medzhitov R. Alpha(1,3)-fucosyltransferase VII and alpha(2,3)-sialyltransferase IV are up-regulated in activated CD4 T cells and maintained after their differentiation into Th1 and migration into inflammatory sites. J Immunol 1999;163:3746–52.

[128] Nakayama F, Teraki Y, Kudo T, Togayachi A, Iwasaki H, Tamatani T, et al. Expression of cutaneous lymphocyte-associated antigen regulated by a set of glycosyltransferases in human T cells: involvement of alpha1, 3-fucosyltransferase VII and beta1,4-galactosyltransferase I. J Invest Dermatol 2000;115:299–306.

[129] Wagers AJ, Waters CM, Stoolman LM, Kansas GS. Interleukin 12 and interleukin 4 control T cell adhesion to endothelial selectins through opposite effects on alpha1, 3-fucosyltransferase VII gene expression. J Exp Med 1998;188:2225–31.

[130] Lim YC, Henault L, Wagers AJ, Kansas GS, Luscinskas FW, Lichtman AH. Expression of functional selectin ligands on Th cells is differentially regulated by IL-12 and IL-4. J Immunol 1999;162:3193–201.

[131] van Wely CA, Blanchard AD, Britten CJ. Differential expression of alpha3 fucosyltransferases in Th1 and Th2 cells correlates with their ability to bind P-selectin. Biochem Biophys Res Commun 1998;247: 307–11.

[132] Underhill GH, Zisoulis DG, Kolli KP, Ellies LG, Marth JD, Kansas GS. A crucial role for T-bet in selectin ligand expression in T helper 1 (Th1) cells. Blood 2005;106:3867–73.

[133] Carlow DA, Williams MJ, Ziltener HJ. P-selectin ligand in activated CD8 T cells proceeds in absence of IL2 and IL12Rp40 *in vivo*. J Immunol 2005;174.

[134] Yamanaka KI, Dimitroff CJ, Fuhlbrigge RC, Kakeda M, Kurokawa I, Mizutani H, et al. Vitamins A and D are potent inhibitors of cutaneous lymphocyte-associated antigen expression. J Allergy Clin Immun 2008;121: 148–57.

[135] Dagia NM, Gadhoum SZ, Knoblauch CA, Spencer JA, Zamiri P, Lin CP, et al. G-CSF induces E-selectin ligand expression on human myeloid cells. Nat Med 2006;12:1185–90.

[136] Hoggatt J, Singh P, Sampath J, Pelus LM. Prostaglandin E-2 enhances hematopoietic stem cell homing, survival, and proliferation. Blood 2009; 113:5444–55.

[137] North TE, Goessling W, Walkley CR, Lengerke C, Kopani KR, Lord AM, et al. Prostaglandin E2 regulates vertebrate haematopoietic stem cell homeostasis. Nature 2007;447:1007–11.

[138] Goessling W, Allen RS, Guan X, Jin P, Uchida N, Dovey M, et al. Prostaglandin E2 enhances human cord blood stem cell xenotransplants and shows long-term safety in preclinical nonhuman primate transplant models. Cell Stem Cell 2011;8:445–58.

[139] Berberoglu U, Yildirim E, Celen O. Serum levels of tumor necrosis factor alpha correlate with response to neoadjuvant chemotherapy in locally advanced breast cancer. Int J Biol Markers 2004;19:130–4.

[140] Geng Y, Chandrasekaran S, Hsu JW, Gidwani M, Hughes AD, King MR. Phenotypic switch in blood: effects of pro-inflammatory cytokines on breast cancer cell aggregation and adhesion. PLoS ONE 2013;8: e54959.
[141] Bates RC, Mercurio AM. Tumor necrosis factor-alpha stimulates the epithelial-to-mesenchymal transition of human colonic organoids. Mol Biol Cell 2003; 14:1790–800.
[142] Lee SO, Lou W, Hou M, de Miguel F, Gerber L, Gao AC. Interleukin-6 promotes androgen-independent growth in LNCaP human prostate cancer cells. Clin Cancer Res 2003;9:370–6.
[143] Muller A, Homey B, Soto H, Ge N, Catron D, Buchanan ME, et al. Involvement of chemokine receptors in breast cancer metastasis. Nature 2001;410:50–6.
[144] Xia L, McDaniel JM, Yago T, Doeden A, McEver RP. Surface fucosylation of human cord blood cells augments binding to P-selectin and E-selectin and enhances engraftment in bone marrow. Blood 2004;104:3091–6.
[145] Robinson SN, Simmons PJ, Thomas MW, Brouard N, Javni JA, Trilok S, et al. *Ex vivo* fucosylation improves human cord blood engraftment in NOD-SCID IL-2Rgamma(null) mice. Exp Hematol 2012;40:445–56.
[146] Wan X, Sato H, Miyaji H, McDaniel JM, Wang Y, Kaneko E, et al. Fucosyltransferase VII improves the function of selectin ligands on cord blood hematopoietic stem cells. Glycobiology 2013;23:1184–91.
[147] Whiteheart SW, Passaniti A, Reichner JS, Holt GD, Haltiwanger RS, Hart GW. Glycosyltransferase probes. Methods Enzymol 1989;179:82–95.
[148] Palcic MM. Glycosyltransferases in glycobiology. Methods Enzymol 1994;230:300–16.
[149] Warnock D, Bai X, Autote K, Gonzales J, Kinealy K, Yan B, et al. *In vitro* galactosylation of human IgG at 1 kg scale using recombinant galactosyltransferase. Biotechnol Bioeng 2005;92:831–42.
[150] Zheng K, Bantog C, Bayer R. The impact of glycosylation on monoclonal antibody conformation and stability. MAbs 2011;3:568–76.
[151] Anthony RM, Nimmerjahn F, Ashline DJ, Reinhold VN, Paulson JC, Ravetch JV. Recapitulation of IVIG anti-inflammatory activity with a recombinant IgG Fc. Science 2008;320:373–6.
[152] Sackstein R. Directing stem cell trafficking via Gps. Method Enzymol 2010;479:93–105.
[153] Sackstein R. Engineering cellular trafficking via glycosyltransferase-programmed stereosubstitution. Ann N Y Acad Sci 2012;1253:193–200.
[154] Sackstein R. Glycoengineering of HCELL, the human bone marrow homing receptor: sweetly programming cell migration. Ann Biomed Eng 2012;40:766–76.
[155] Sackstein R, Fuhlbrigge R. The blot rolling assay: a method for identifying adhesion molecules mediating binding under shear conditions. Methods Mol Biol (Clifton, NJ) 2006;341:217–26.
[156] Lairson LL, Henrissat B, Davies GJ, Withers SG. Glycosyltransferases: structures, functions, and mechanisms. Annu Rev Biochem 2008;77:521–55.
[157] Gallet PF, Vaujour H, Petit JM, Maftah A, Oulmouden A, Oriol R, et al. Heterologous expression of an engineered truncated form of human Lewis fucosyltransferase (Fuc-TIII) by the methylotrophic yeast *Pichia pastoris*. Glycobiology 1998;8:919–25.
[158] Stacke C, Ziegelmuller P, Hahn U. Comparison of expression systems for human fucosyltransferase IX. Eur J Cell Biol 2010;89:35–8.
[159] Auge C, Malleron A, Tahrat H, Marc A, Goergen JL, Cerutti M, et al. Outstanding stability of immobilized recombinant alpha(1 -> 3/4)-fucosyltransferases exploited in the synthesis of Lewis a and Lewis x trisaccharides. Chem Commun 2000;:2017–8.
[160] Baboval T, Koul O, Smith FI. N-glycosylation site occupancy of rat alpha-1,3-fucosyltransferase IV and the effect of glycosylation on enzymatic activity. Biochim Biophys Acta 2000;1475:383–9.
[161] Morais VA, Costa MT, Costa J. N-glycosylation of recombinant human fucosyltransferase III is required for its *in vivo* folding in mammalian and insect cells. Biochim Biophys Acta 1619;133–138:2003.
[162] Li P, Anumanthan A, Gao XG, Ilangovan K, Suzara VV, Duzgunes N, et al. Expression of recombinant proteins in *Pichia pastoris*. Appl Biochem Biotechnol 2007;142:105–24.
[163] Kato T, Kajikawa M, Maenaka K, Park EY. Silkworm expression system as a platform technology in life science. Appl Microbiol Biotechnol 2010;85:459–70.
[164] Sacchetti B, Funari A, Michienzi S, Di Cesare S, Piersanti S, Saggio I, et al. Self-renewing osteoprogenitors in bone marrow sinusoids can organize a hematopoietic microenvironment. Cell 2007;131:324–36.
[165] Mendez-Ferrer S, Michurina TV, Ferraro F, Mazloom AR, Macarthur BD, Lira SA, et al. Mesenchymal and haematopoietic stem cells form a unique bone marrow niche. Nature 2010;466:829–34.
[166] Ding L, Saunders TL, Enikolopov G, Morrison SJ. Endothelial and perivascular cells maintain haematopoietic stem cells. Nature 2012;481:457–62.
[167] Calvi LM, Adams GB, Weibrecht KW, Weber JM, Olson DP, Knight MC, et al. Osteoblastic cells regulate the haematopoietic stem cell niche. Nature 2003;425:841–6.

[168] Raaijmakers MH, Mukherjee S, Guo S, Zhang S, Kobayashi T, Schoonmaker JA, et al. Bone progenitor dysfunction induces myelodysplasia and secondary leukaemia. Nature 2010;464:852–7.
[169] Visnjic D, Kalajzic Z, Rowe DW, Katavic V, Lorenzo J, Aguila HL. Hematopoiesis is severely altered in mice with an induced osteoblast deficiency. Blood 2004;103:3258–64.
[170] Butler JM, Nolan DJ, Vertes EL, Varnum-Finney B, Kobayashi H, Hooper AT, et al. Endothelial cells are essential for the self-renewal and repopulation of Notch-dependent hematopoietic stem cells. Cell Stem Cell 2010;6:251–64.
[171] Yamazaki S, Ema H, Karlsson G, Yamaguchi T, Miyoshi H, Shioda S, et al. Nonmyelinating Schwann cells maintain hematopoietic stem cell hibernation in the bone marrow niche. Cell 2011;147:1146–58.
[172] Trumpp A, Essers M, Wilson A. Awakening dormant haematopoietic stem cells. Nat Rev Immunol 2010;10:201–9.
[173] Wilson A, Laurenti E, Oser G, van der Wath RC, Blanco-Bose W, Jaworski M, et al. Hematopoietic stem cells reversibly switch from dormancy to self-renewal during homeostasis and repair. Cell 2008;135:1118–29.
[174] Kiel MJ, Morrison SJ. Uncertainty in the niches that maintain haematopoietic stem cells. Nat Rev Immunol 2008;8:290–301.
[175] Xie Y, Yin T, Wiegraebe W, He XC, Miller D, Stark D, et al. Detection of functional haematopoietic stem cell niche using real-time imaging. Nature 2009;457:97–101.
[176] Levesque JP, Winkler IG, Hendy J, Williams B, Helwani F, Barbier V, et al. Hematopoietic progenitor cell mobilization results in hypoxia with increased hypoxia-inducible transcription factor-1 alpha and vascular endothelial growth factor A in bone marrow. Stem Cells 2007;25:1954–65.
[177] Parmar K, Mauch P, Vergilio JA, Sackstein R, Down JD. Distribution of hematopoietic stem cells in the bone marrow according to regional hypoxia. Proc Natl Acad Sci USA 2007;104:5431–6.
[178] McIntyre TM, Prescott SM, Weyrich AS, Zimmerman GA. Cell–cell interactions: leukocyte-endothelial interactions. Curr Opin Hematol 2003;10:150–8.
[179] Okeley NM, Alley SC, Anderson ME, Boursalian TE, Burke PJ, Emmerton KM, et al. Development of orally active inhibitors of protein and cellular fucosylation. Proc Natl Acad Sci USA 2013;110:5404–9.
[180] Geng Y, Yeh K, Takatani T, King MR. Three to tango: MUC1 as a ligand for both E-selectin and ICAM-1 in the breast cancer metastatic cascade. Front Oncol 2012;2:76.
[181] Zen K, Liu DQ, Guo YL, Wang C, Shan J, Fang M, et al. CD44v4 is a major E-selectin ligand that mediates breast cancer cell transendothelial migration. PLoS ONE 2008;3:e1826.
[182] Dimitroff CJ, Descheny L, Trujillo N, Kim R, Nguyen V, Huang W, et al. Identification of leukocyte E-selectin ligands, P-selectin glycoprotein ligand-1 and E-selectin ligand-1, on human metastatic prostate tumor cells. Cancer Res 2005;65:5750–60.
[183] Dimitroff CJ, Lechpammer M, Long-Woodward D, Kutok JL. Rolling of human bone-metastatic prostate tumor cells on human bone marrow endothelium under shear flow is mediated by E-selectin. Cancer Res 2004;64:5261–9.
[184] Burdick MM, Bochner BS, Collins BE, Schnaar RL, Konstantopoulos K. Glycolipids support E-selectin-specific strong cell tethering under flow. Biochem Biophys Res Commun 2001;284:42–9.
[185] Burdick MM, Chu JT, Godar S, Sackstein R. HCELL is the major E- and L-selectin ligand expressed on LS174T colon carcinoma cells. J Biol Chem 2006; 281:13899–905.
[186] Gout S, Morin C, Houle F, Huot J. Death receptor-3, a new E-Selectin counter-receptor that confers migration and survival advantages to colon carcinoma cells by triggering p38 and ERK MAPK activation. Cancer Res 2006;66:9117–24.
[187] Tremblay PL, Auger FA, Huot J. Regulation of transendothelial migration of colon cancer cells by E-selectin-mediated activation of p38 and ERK MAP kinases. Oncogene 2006;25:6563–73.
[188] Martin-Satue M, de Castellarnau C, Blanco J. Overexpression of alpha(1,3)-fucosyltransferase VII is sufficient for the acquisition of lung colonization phenotype in human lung adenocarcinoma HAL-24Luc cells. Br J Cancer 1999;80:1169–74.
[189] Martin-Satue M, Marrugat R, Cancelas JA, Blanco J. Enhanced expression of alpha(1,3)-fucosyltransferase genes correlates with E-selectin-mediated adhesion and metastatic potential of human lung adenocarcinoma cells. Cancer Res 1998;58:1544–50.
[190] Iwai K, Ishikura H, Kaji M, Sugiura H, Ishizu A, Takahashi C, et al. Importance of E-selectin (ELAM-1) and sialyl Lewis(a) in the adhesion of pancreatic carcinoma cells to activated endothelium. Int J Cancer 1993;54:972–7.
[191] Brodt P, Fallavollita L, Bresalier RS, Meterissian S, Norton CR, Wolitzky BA. Liver endothelial E-selectin mediates carcinoma cell adhesion and promotes liver metastasis. Int J Cancer 1997;71: 612–9.
[192] Khatib AM, Kontogiannea M, Fallavollita L, Jamison B, Meterissian S, Brodt P. Rapid induction of cytokine

and E-selectin expression in the liver in response to metastatic tumor cells. Cancer Res 1999;59:1356–61.

[193] Biancone L, Araki M, Araki K, Vassalli P, Stamenkovic I. Redirection of tumor metastasis by expression of E-selectin *in vivo*. J Exp Med 1996; 183:581–7.

[194] Mannori G, Santoro D, Carter L, Corless C, Nelson RM, Bevilacqua MP. Inhibition of colon carcinoma cell lung colony formation by a soluble form of E-selectin. Am J Pathol 1997;151:233–43.

[195] Borsig L, Wong R, Feramisco J, Nadeau DR, Varki NM, Varki A. Heparin and cancer revisited: mechanistic connections involving platelets, P-selectin, carcinoma mucins, and tumor metastasis. Proc Natl Acad Sci USA 2001;98:3352–7.

[196] Borsig L, Wong R, Hynes RO, Varki NM, Varki A. Synergistic effects of L- and P-selectin in facilitating tumor metastasis can involve non-mucin ligands and implicate leukocytes as enhancers of metastasis. Proc Natl Acad Sci USA 2002;99:2193–8.

[197] Nakamura S, Ohtani H, Watanabe Y, Fukushima K, Matsumoto T, Kitano A, et al. *In situ* expression of the cell adhesion molecules in inflammatory bowel disease. Evidence of immunologic activation of vascular endothelial cells. Lab Invest 1993;69:77–85.

[198] Kakiuchi Y, Tsuji S, Tsujii M, Murata H, Kawai N, Yasumaru M, et al. Cyclooxygenase-2 activity altered the cell-surface carbohydrate antigens on colon cancer cells and enhanced liver metastasis. Cancer Res 2002;62:1567–72.

[199] Kim YJ, Borsig L, Han HL, Varki NM, Varki A. Distinct selectin ligands on colon carcinoma mucins can mediate pathological interactions among platelets, leukocytes, and endothelium. Am J Pathol 1999;155:461–72.

[200] Nieswandt B, Hafner M, Echtenacher B, Mannel DN. Lysis of tumor cells by natural killer cells in mice is impeded by platelets. Cancer Res 1999;59:1295–300.

[201] Barthel SR, Gavino JD, Wiese GK, Jaynes JM, Siddiqui J, Dimitroff CJ. Analysis of glycosyltransferase expression in metastatic prostate cancer cells capable of rolling activity on microvascular endothelial (E)-selectin. Glycobiology 2008;18:806–17.

[202] Fukuda M, Hiraoka N, Yeh JC. C-type lectins and sialyl Lewis X oligosaccharides. Versatile roles in cell–cell interaction. J Cell Biol 1999;147:467–70.

[203] Zipin A, Israeli-Amit M, Meshel T, Sagi-Assif O, Yron I, Lifshitz V, et al. Tumor–microenvironment interactions: the fucose-generating FX enzyme controls adhesive properties of colorectal cancer cells. Cancer Res 2004;64:6571–8.

[204] Dube DH, Bertozzi CR. Glycans in cancer and inflammation—potential for therapeutics and diagnostics. Nat Rev Drug Discov 2005;4:477–88.

[205] Hoff SD, Matsushita Y, Ota DM, Cleary KR, Yamori T, Hakomori S, et al. Increased expression of sialyl-dimeric LeX antigen in liver metastases of human colorectal carcinoma. Cancer Res 1989;49:6883–8.

[206] Nakamori S, Furukawa H, Hiratsuka M, Iwanaga T, Imaoka S, Ishikawa O, et al. Expression of carbohydrate antigen sialyl Le(a): a new functional prognostic factor in gastric cancer. J Clin Oncol 1997;15:816–25.

[207] Nakamori S, Kameyama M, Imaoka S, Furukawa H, Ishikawa O, Sasaki Y, et al. Involvement of carbohydrate antigen sialyl Lewis(x) in colorectal cancer metastasis. Dis Colon Rectum 1997;40:420–31.

[208] Renkonen R, Mattila P, Majuri ML, Rabina J, Toppila S, Renkonen J, et al. *In vitro* experimental studies of sialyl Lewis x and sialyl Lewis a on endothelial and carcinoma cells: crucial glycans on selectin ligands. Glycoconj J 1997;14:593–600.

[209] Sakuma K, Aoki M, Kannagi R. Transcription factors c-Myc and CDX2 mediate E-selectin ligand expression in colon cancer cells undergoing EGF/bFGF-induced epithelial–mesenchymal transition. Proc Natl Acad Sci USA 2012;109:7776–81.

[210] Greenburg G, Hay ED. Epithelia suspended in collagen gels can lose polarity and express characteristics of migrating mesenchymal cells. J Cell Biol 1982; 95:333–9.

[211] Greenburg G, Hay ED. Cytodifferentiation and tissue phenotype change during transformation of embryonic lens epithelium to mesenchyme-like cells *in vitro*. Dev Biol 1986;115:363–79.

[212] Greenburg G, Hay ED. Cytoskeleton and thyroglobulin expression change during transformation of thyroid epithelium to mesenchyme-like cells. Development 1988;102:605–22.

[213] Kalluri R. EMT: when epithelial cells decide to become mesenchymal-like cells. J Clin Invest 2009; 119:1417–9.

[214] Kalluri R, Weinberg RA. The basics of epithelial–mesenchymal transition. J Clin Invest 2009; 119:1420–8.

[215] Aktas B, Tewes M, Fehm T, Hauch S, Kimmig R, Kasimir-Bauer S. Stem cell and epithelial–mesenchymal transition markers are frequently overexpressed in circulating tumor cells of metastatic breast cancer patients. Breast Cancer Res 2009;11: R46.

[216] Blick T, Hugo H, Widodo E, Waltham M, Pinto C, Mani SA, et al. Epithelial mesenchymal transition traits in human breast cancer cell lines parallel the CD44(hi/)CD24 (lo/ −) stem cell phenotype in human breast cancer. J Mammary Gland Biol Neoplasia 2010;15:235–52.

[217] Brabletz T, Jung A, Spaderna S, Hlubek F, Kirchner T. Opinion: migrating cancer stem cells—an integrated concept of malignant tumour progression. Nat Rev Cancer 2005;5:744–9.
[218] Kong D, Banerjee S, Ahmad A, Li Y, Wang Z, Sethi S, et al. Epithelial to mesenchymal transition is mechanistically linked with stem cell signatures in prostate cancer cells. PLoS ONE 2010;5:e12445.
[219] Kong D, Wang Z, Sarkar SH, Li Y, Banerjee S, Saliganan A, et al. Platelet-derived growth factor-D overexpression contributes to epithelial–mesenchymal transition of PC3 prostate cancer cells. Stem Cells 2008;26:1425–35.
[220] Morel AP, Lievre M, Thomas C, Hinkal G, Ansieau S, Puisieux A. Generation of breast cancer stem cells through epithelial–mesenchymal transition. PLoS ONE 2008;3:e2888.
[221] Santisteban M, Reiman JM, Asiedu MK, Behrens MD, Nassar A, Kalli KR, et al. Immune-induced epithelial to mesenchymal transition *in vivo* generates breast cancer stem cells. Cancer Res 2009;69:2887–95.
[222] Thomson S, Buck E, Petti F, Griffin G, Brown E, Ramnarine N, et al. Epithelial to mesenchymal transition is a determinant of sensitivity of non-small-cell lung carcinoma cell lines and xenografts to epidermal growth factor receptor inhibition. Cancer Res 2005;65:9455–62.
[223] Zhau HE, Odero-Marah V, Lue HW, Nomura T, Wang R, Chu G, et al. Epithelial to mesenchymal transition (EMT) in human prostate cancer: lessons learned from ARCaP model. Clin Exp Metastasis 2008;25:601–10.
[224] Pinho SS, Reis CA, Gartner F, Alpaugh ML. Molecular plasticity of E-cadherin and sialyl Lewis X expression, in two comparative models of mammary tumorigenesis. PLoS ONE 2009;4.
[225] Goetz DJ, Ding H, Atkinson WJ, Vachino G, Camphausen RT, Cumming DA, et al. A human colon carcinoma cell line exhibits adhesive interactions with P-selectin under fluid flow via a PSGL-1-independent mechanism. Am J Pathol 1996;149:1661–73.
[226] Hanley WD, Burdick MM, Konstantopoulos K, Sackstein R. CD44 on LS174T colon carcinoma cells possesses E-selectin ligand activity. Cancer Res 2005;65:5812–7.
[227] Hanley WD, Napier SL, Burdick MM, Schnaar RL, Sackstein R, Konstantopoulos K. Variant isoforms of CD44 are P- and L-selectin ligands on colon carcinoma cells. FASEB J 2006;20:337–9.
[228] Napier SL, Healy ZR, Schnaar RL, Konstantopoulos K. Selectin ligand expression regulates the initial vascular interactions of colon carcinoma cells: the roles of CD44v and alternative sialofucosylated selectin ligands. J Biol Chem 2007;282:3433–41.
[229] Thomas SN, Schnaar RL, Konstantopoulos K. Podocalyxin-like protein is an E-/L-selectin ligand on colon carcinoma cells: comparative biochemical properties of selectin ligands in host and tumor cells. Am J Physiol Cell Physiol 2009;296:C505–13.
[230] Fernandez-Rodriguez J, Dwir O, Alon R, Hansson GC. Tumor cell MUC1 and CD43 are glycosylated differently with sialyl-Lewis a and x epitopes and show variable interactions with E-selectin under physiological flow conditions. Glycoconj J 2001;18:925–30.
[231] Zhang K, Baeckstrom D, Brevinge H, Hansson GC. Secreted MUC1 mucins lacking their cytoplasmic part and carrying sialyl-Lewis a and x epitopes from a tumor cell line and sera of colon carcinoma patients can inhibit HL-60 leukocyte adhesion to E-selectin-expressing endothelial cells. J Cell Biochem 1996;60:538–49.
[232] Alcaide P, King SL, Dimitroff CJ, Lim YC, Fuhlbrigge RC, Luscinskas FW. The 130-kDa glycoform of CD43 functions as an E-selectin ligand for activated Th1 cells *in vitro* and in delayed-type hypersensitivity reactions *in vivo*. J Invest Dermatol 2007;127:1964–72.
[233] Matsumoto M, Atarashi K, Umemoto E, Furukawa Y, Shigeta A, Miyasaka M, et al. CD43 functions as a ligand for E-Selectin on activated T cells. J Immunol 2005;175:8042–50.
[234] Shirure VS, Reynolds NM, Burdick MM. Mac-2 binding protein is a novel e-selectin ligand expressed by breast cancer cells. PLoS ONE 2012;7:e44529.
[235] Sako D, Chang XJ, Barone KM, Vachino G, White HM, Shaw G, et al. Expression cloning of a functional glycoprotein ligand for P-selectin. Cell 1993;75:1179–86.
[236] Veldman GM, Bean KM, Cumming DA, Eddy RL, Sait SN, Shows TB. Genomic organization and chromosomal localization of the gene encoding human P-selectin glycoprotein ligand. J Biol Chem 1995;270:16470–5.
[237] Moore KL, Stults NL, Diaz S, Smith DF, Cummings RD, Varki A, et al. Identification of a specific glycoprotein ligand for P-selectin (CD62) on myeloid cells. J Cell Biol 1992;118:445–56.
[238] Li FG, Erickson HP, James JA, Moore KL, Cummings RD, McEver RP. Visualization of P-selectin glycoprotein ligand-1 as a highly extended molecule and mapping of protein epitopes for monoclonal antibodies. J Biol Chem 1996;271:6342–8.
[239] Laszik Z, Jansen PJ, Cummings RD, Tedder TF, McEver RP, Moore KL. P-selectin glycoprotein ligand-1 is broadly expressed in cells of myeloid,

lymphoid, and dendritic lineage and in some nonhematopoietic cells. Blood 1996;88:3010–21.

[240] Moore TM, Khimenko P, Adkins WK, Miyasaka M, Taylor AE. Adhesion molecules contribute to ischemia and reperfusion-induced injury in the isolated rat lung. J Appl Physiol 1995;78:2245–52.

[241] Frenette PS, Denis CV, Weiss L, Jurk K, Subbarao S, Kehrel B, et al. P-Selectin glycoprotein ligand 1 (PSGL-1) is expressed on platelets and can mediate platelet–endothelial interactions *in vivo*. J Exp Med 2000;191:1413–22.

[242] Kappelmayer J, Kiss A, Karaszi E, Veszpremi A, Jako J, Kiss C. Identification of P-selectin glycoprotein ligand-1 as a useful marker in acute myeloid leukaemias. Br J Haematol 2001;115:903–9.

[243] Li F, Wilkins PP, Crawley S, Weinstein J, Cummings RD, McEver RP. Post-translational modifications of recombinant P-selectin glycoprotein ligand-1 required for binding to P- and E-selectin. J Biol Chem 1996;271:3255–64.

[244] Dimitroff CJ, Lee JY, Schor KS, Sandmaier BM, Sackstein R. Differential L-selectin binding activities of human hematopoietic cell L-selectin ligands, HCELL and PSGL-1. J Biol Chem 2001;276:47623–31.

[245] Brockhausen I. Mucin-type O-glycans in human colon and breast cancer: glycodynamics and functions. EMBO Rep 2006;7:599–604.

[246] Aigner S, Ramos CL, Hafezi-Moghadam A, Lawrence MB, Friederichs J, Altevogt P, et al. CD24 mediates rolling of breast carcinoma cells on P-selectin. FASEB J 1998;12:1241–51.

[247] Myung JH, Gajjar KA, Pearson RM, Launiere CA, Eddington DT, Hong S. Direct measurements on CD24-mediated rolling of human breast cancer MCF-7 cells on E-selectin. Anal Chem 2011;83:1078–83.

[248] Shimodaira K, Nakayama J, Nakamura N, Hasebe O, Katsuyama T, Fukuda M. Carcinoma-associated expression of core 2 beta-1,6-*N*-acetylglucosaminyltransferase gene in human colorectal cancer: role of O-glycans in tumor progression. Cancer Res 1997;57:5201–6.

[249] St Hill CA, Farooqui M, Mitcheltree G, Gulbahce HE, Jessurun J, Cao Q, et al. The high affinity selectin glycan ligand C2-O-sLex and mRNA transcripts of the core 2 beta-1,6-*N*-acetylglucosaminyltransferase (C2GnT1) gene are highly expressed in human colorectal adenocarcinomas. BMC Cancer 2009;9:79.

[250] Ogawa J, Inoue H, Koide S. Expression of alpha-1,3-fucosyltransferase type IV and VII genes is related to poor prognosis in lung cancer. Cancer Res 1996;56: 325–9.

[251] Hiller KM, Mayben JP, Bendt KM, Manousos GA, Senger K, Cameron HS, et al. Transfection of alpha (1,3)fucosyltransferase antisense sequences impairs the proliferative and tumorigenic ability of human colon carcinoma cells. Mol Carcinog 2000;27:280–8.

[252] Weston BW, Hiller KM, Mayben JP, Manousos GA, Bendt KM, Liu R, et al. Expression of human alpha (1,3)fucosyltransferase antisense sequences inhibits selectin-mediated adhesion and liver metastasis of colon carcinoma cells. Cancer Res 1999;59:2127–35.

[253] Recchi MA, Harduin-Lepers A, Boilly-Marer Y, Verbert A, Delannoy P. Multiplex RT-PCR method for the analysis of the expression of human sialyltransferases: application to breast cancer cells. Glycoconj J 1998;15:19–27.

[254] Recchi MA, Hebbar M, Hornez L, Harduin-Lepers A, Peyrat JP, Delannoy P. Multiplex reverse transcription polymerase chain reaction assessment of sialyltransferase expression in human breast cancer. Cancer Res 1998;58:4066–70.

[255] Bernacki RJ, Kim U. Concomitant elevations in serum sialytransferase activity and sialic acid content in rats with metastasizing mammary tumors. Science 1977;195:577–80.

[256] Varki A. Selectin ligands: will the real ones please stand up? J Clin Invest 1997;100:S31–5.

[257] Munro JM. Endothelial–leukocyte adhesive interactions in inflammatory diseases. Eur Heart J 1993;14 (**Suppl. K**):72–7.

[258] Perez P, Garcia-Acha I, Duran A. Effect of papulacandin B on the cell wall and growth of *Geotrichum lactis*. J Gen Microbiol 1983;129:245–50.

[259] Cabib E, Kang MS, Au-Young J. Chitin synthase from *Saccharomyces cerevisiae*. Methods Enzymol 1987;138:643–9.

[260] Lehrman MA. Biosynthesis of *N*-acetylglucosamine-P-P-dolichol, the committed step of asparagine-linked oligosaccharide assembly. Glycobiology 1991;1:553–62.

[261] Saul R, Chambers JP, Molyneux RJ, Elbein AD. Castanospermine, a tetrahydroxylated alkaloid that inhibits beta-glucosidase and beta-glucocerebrosidase. Arch Biochem Biophys 1983;221:593–7.

[262] Goss PE, Baptiste J, Fernandes B, Baker M, Dennis JW. A phase I study of swainsonine in patients with advanced malignancies. Cancer Res 1994;54:1450–7.

[263] Bischoff J, Kornfeld R. The effect of 1-deoxymannojirimycin on rat liver alpha-mannosidases. Biochem Biophys Res Commun 1984;125:324–31.

[264] Platt FM, Neises GR, Dwek RA, Butters TD. *N*-butyldeoxynojirimycin is a novel inhibitor of glycolipid biosynthesis. J Biol Chem 1994;269:8362–5.

[265] Schwientek T, Yeh JC, Levery SB, Keck B, Merkx G, van Kessel AG, et al. Control of O-glycan branch formation. Molecular cloning and characterization of a novel thymus-associated core 2 beta1, 6-*N*-acetylglucosaminyltransferase. J Biol Chem 2000;275:11106–13.

[266] Hindsgaul O, Kaur KJ, Srivastava G, Blaszczyk-Thurin M, Crawley SC, Heerze LD, et al. Evaluation of deoxygenated oligosaccharide acceptor analogs as specific inhibitors of glycosyltransferases. J Biol Chem 1991;266:17858–62.

[267] Gloster TM, Zandberg WF, Heinonen JE, Shen DL, Deng L, Vocadlo DJ. Hijacking a biosynthetic pathway yields a glycosyltransferase inhibitor within cells. Nat Chem Biol 2011;7:174–81.

[268] Grabenhorst E, Nimtz M, Costa J, Conradt HS. *In vivo* specificity of human alpha1,3/4-fucosyltransferases III–VII in the biosynthesis of LewisX and Sialyl LewisX motifs on complex-type N-glycans. Coexpression studies from bhk-21 cells together with human beta-trace protein. J Biol Chem 1998;273:30985–94.

[269] Zandberg WF, Kumarasamy J, Pinto BM, Vocadlo DJ. Metabolic inhibition of sialyl-Lewis X biosynthesis by 5-thiofucose remodels the cell surface and impairs selectin-mediated cell adhesion. J Biol Chem 2012;287:40021–30.

[270] Rillahan CD, Antonopoulos A, Lefort CT, Sonon R, Azadi P, Ley K, et al. Global metabolic inhibitors of sialyl- and fucosyltransferases remodel the glycome. Nat Chem Biol 2012;8:661–8.

[271] Izumi M, Wada K, Yuasa H, Hashimoto H. Synthesis of bisubstrate and donor analogues of sialyltransferase and their inhibitory activities. J Org Chem 2005;70:8817–24.

[272] Burkart MD, Vincent SP, Duffels A, Murray BW, Ley SV, Wong CH. Chemo-enzymatic synthesis of fluorinated sugar nucleotide: useful mechanistic probes for glycosyltransferases. Bioorg Med Chem 2000;8:1937–46.

[273] Van Dorst JA, Tikkanen JM, Krezdorn CH, Streiff MB, Berger EG, Van Kuik JA, et al. Exploring the substrate specificities of alpha-2,6- and alpha-2,3-sialyltransferases using synthetic acceptor analogues. Eur J Biochem 1996;242:674–81.

[274] Hinou H, Sun XL, Ito Y. Systematic syntheses and inhibitory activities of bisubstrate-type inhibitors of sialyltransferases. J Org Chem 2003;68:5602–13.

[275] Kuan SF, Byrd JC, Basbaum C, Kim YS. Inhibition of mucin glycosylation by aryl-N-acetyl-alpha-galactosaminides in human colon cancer cells. J Biol Chem 1989;264:19271–7.

[276] Zhuang D, Grey A, Harris-Brandts M, Higgins E, Kashem MA, Dennis JW. Characterization of O-linked oligosaccharide biosynthesis in cultured cells using paranitrophenyl alpha-D-GalNAc as an acceptor. Glycobiology 1991;1:425–33.

[277] Gouyer V, Leteurtre E, Delmotte P, Steelant WF, Krzewinski-Recchi MA, Zanetta JP, et al. Differential effect of GalNAcalpha-O-bn on intracellular trafficking in enterocytic HT-29 and Caco-2 cells: correlation with the glycosyltransferase expression pattern. J Cell Sci 2001;114:1455–71.

[278] Huet G, Hennebicq-Reig S, de Bolos C, Ulloa F, Lesuffleur T, Barbat A, et al. GalNAc-alpha-O-benzyl inhibits NeuAcalpha2-3 glycosylation and blocks the intracellular transport of apical glycoproteins and mucus in differentiated HT-29 cells. J Cell Biol 1998;141:1311–22.

[279] Huet G, Kim I, de Bolos C, Lo-Guidice JM, Moreau O, Hemon B, et al. Characterization of mucins and proteoglycans synthesized by a mucin-secreting HT-29 cell subpopulation. J Cell Sci 1995;108(**Pt 3**):1275–85.

[280] Neville DC, Field RA, Ferguson MA. Hydrophobic glycosides of *N*-acetylglucosamine can act as primers for polylactosamine synthesis and can affect glycolipid synthesis *in vivo*. Biochem J 1995;307(**Pt 3**):791–7.

[281] Marathe DD, Buffone Jr. A, Chandrasekaran EV, Xue J, Locke RD, Nasirikenari M, et al. Fluorinated per-acetylated GalNAc metabolically alters glycan structures on leukocyte PSGL-1 and reduces cell binding to selectins. Blood 2010;115:1303–12.

[282] Barthel SR, Antonopoulos A, Cedeno-Laurent F, Schaffer L, Hernandez G, Patil SA, et al. Peracetylated 4-fluoro-glucosamine reduces the content and repertoire of N- and O-glycans without direct incorporation. J Biol Chem 2011;286:21717–31.

[283] Brown JR, Fuster MM, Whisenant T, Esko JD. Expression patterns of alpha 2,3-sialyltransferases and alpha 1,3-fucosyltransferases determine the mode of sialyl Lewis X inhibition by disaccharide decoys. J Biol Chem 2003;278:23352–9.

[284] Sarkar AK, Brown JR, Esko JD. Synthesis and glycan priming activity of acetylated disaccharides. Carbohydr Res 2000;329:287–300.

[285] Sarkar AK, Fritz TA, Taylor WH, Esko JD. Disaccharide uptake and priming in animal-cells—inhibition of sialyl-Lewis-X by acetylated Gal-beta-1-] 4glcnac-beta-O-naphthalenemethanol. Proc Natl Acad Sci USA 1995;92:3323–7.

[286] Sarkar AK, Rostand KS, Jain RK, Matta KL, Esko JD. Fucosylation of disaccharide precursors of sialyl LewisX inhibit selectin-mediated cell adhesion. J Biol Chem 1997;272:25608–16.

[287] Fuster MM, Brown JR, Wang L, Esko JD. A disaccharide precursor of sialyl Lewis X inhibits metastatic potential of tumor cells. Cancer Res 2003;63: 2775–81.

[288] Brown JR, Yang F, Sinha A, Ramakrishnan B, Tor Y, Qasba PK, et al. Deoxygenated disaccharide analogs as specific inhibitors of beta 1-4-galactosyltransferase 1 and selectin-mediated tumor metastasis. J Biol Chem 2009;284:4952–9.

[289] Kerkela E, Hakkarainen T, Makela T, Raki M, Kambur O, Kilpinen L, et al. Transient proteolytic modification of mesenchymal stromal cells increases lung clearance rate and targeting to injured tissue. Stem Cells Transl Med 2013;2:510–20.

[290] Nystedt J, Anderson H, Tikkanen J, Pietila M, Hirvonen T, Takalo R, et al. Cell surface structures influence lung clearance rate of systemically infused mesenchymal stromal cells. Stem Cells 2013;31:317–26.

[291] Perkey E, Miller RA, Garcia GG. *Ex vivo* enzymatic treatment of aged CD4 T cells restores cognate T cell helper function and enhances antibody production in mice. J Immunol 2012;189:5582–9.

[292] Kumar R, Burns EA. Age-related decline in immunity: implications for vaccine responsiveness. Expert Rev Vaccines 2008;7:467–79.

[293] Haynes L, Eaton SM. The effect of age on the cognate function of $CD4^+$ T cells. Immunol Rev 2005;205:220–8.

CHAPTER

10

Cell Microencapsulation for Tissue Engineering and Regenerative Medicine

Jian Du and Kevin J. Yarema

Department of Biomedical Engineering and the Translational Tissue Engineering Center, The Johns Hopkins University, Baltimore, MD, USA

OUTLINE

J.M. Karp & W. Zhao (Eds): Micro- and Nanoengineering of the Cell Surface.
DOI: http://dx.doi.org/10.1016/B978-1-45-573146-6.00010-6

10.1 INTRODUCTION

Cell-based therapies have been attempted sporadically for well over a hundred years and became firmly established in the medical mainstream in 1968 when the first successful bone marrow transplantations were performed in humans [1]. During the same era, early efforts to treat diabetes with transplanted cells also flourished while in recent years, advances in stem cell biology have led to a rapidly increasing profusion of cellular therapies [2,3]. Throughout the past five decades, a primary obstacle to the clinical implementation of these otherwise exciting therapies has been avoidance of immune rejection of the implanted cells while secondary challenges, such as retention of the implanted cells at the desired location in the body, also persist. As will be outlined in this chapter, cell encapsulation technology offers a solution to these problems, as well as providing additional advantages such as ease of handling, a highly hydrated tissue-like environment for cell and tissue growth, and the ability to form complex structures *in vivo*. Emerging opportunities to include additional materials (growth factors, cells, drugs, even adsorbents, and magnetic materials) or other design features (e.g., topographical cues [4]) that control cell fate [5,6] are now complementing cell-based therapies and helping to propel them into the forefront of emerging therapeutic advances.

Cell encapsulation, a strategy that aims to physically isolate an implanted cell mass from the outside environment within the confines of a semipermeable membrane barrier [7–9], was originally inspired by the need to overcome a major obstacle associated with the transplantation of nonautologous cells or tissue, which is graft rejection [10,11]. To avoid graft rejection, transplant recipients currently are obliged to take immunosuppressive drugs for extended periods of time, which can result in severe side effects that include increased susceptibility to infections [12], negative effects on bone and cardiovascular function [13,14], and even cancer due to compromised immunosurveillance [14]. The immunosuppressive drugs also interact with other medicines and affect their metabolism and action causing wide-ranging medical difficulties [15]. The increasing number of options for achieving cell immunoisolation in microcapsules (as shown in Figure 10.1) negates the requirement for long-term administration of immunosuppressive agents and is opening the door for a growing number of cell-based therapies.

Before describing state of the art cell encapsulation applications, a brief history of this field begins with an early description of basic requirements for the technology by Chang in the early 1950s [16]. A primary requirement of microcapsule-surrounding membranes was to ensure that they were amenable to the diffusion of nutrients (e.g., glucose) and other molecules such as oxygen and growth factors essential for cell survival [17]. Twenty years later, Lim and Sun [18] presented the first implantable alginate–poly(L-lysine) microcapsules harboring rat islet cells, that naturally secreted insulin, for the treatment of diabetes. In the following three decades, the permselective capsule environment has been optimized for supporting cellular metabolism, proliferation, differentiation, and cellular morphogenesis.

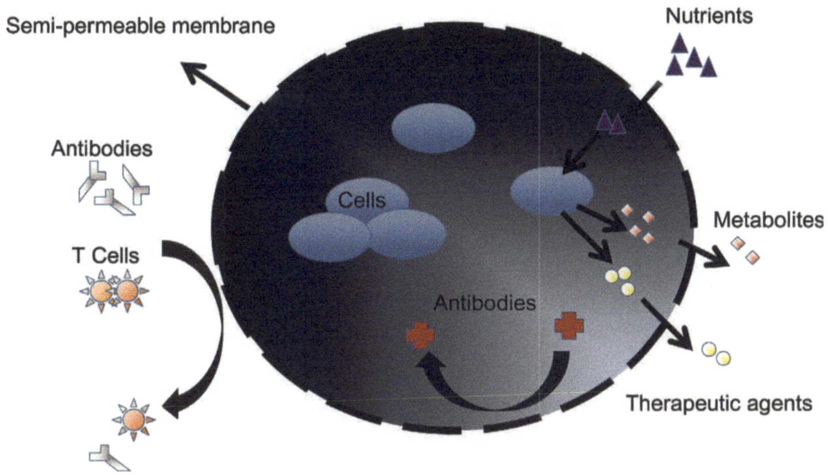

FIGURE 10.1 Schematic illustration of cell microencapsulation.

Today's ability to combine cells and polymer scaffolds by using microencapsulation methods creates "living cell medicines" that provide long-term drug delivery has opened new doors in the use of allografts. For example, encapsulated cells can deliver therapeutic proteins continuously for prolonged time periods of days, weeks, or in some cases even months. These features made this technology particularly attractive for islet transplantation for treatment of diabetes [19–22]. Moreover, cell encapsulation allows the use of cells from a variety of sources such as primary or stem cells, or genetically engineered cells which can be modified to express any desired protein *in vivo* without the modification of the host's genome [9].

Recent advances in biotechnology, molecular biology, nanotechnology, and polymer chemistry are now opening up further exciting possibilities in this field, making increasingly sophisticated applications of cell encapsulation technologies feasible such as forming functional new tissues. For example, the Elisseeff group has developed cell compatible, chemically functionalized hydrogels that can function as "glues" to encapsulate cells for cartilage reconstruction [23–25]. The robustness of this technique is evidenced by its ability to paste living cells onto a beating heart, as has been done in nascent efforts to achieve cell-based cardiovascular regeneration [26]. These encapsulation strategies not only protect the implanted cells from the host immune system, but also ensure cell localization and retention at a defined site in the body to an extent not yet possible even by using emerging glycoengineering techniques that guide cell migration [27,28].

This chapter describes the historical development and principles behind polymeric artificial cells, and discusses the potential of cell capsulation systems for efficiently utilizing the functions of transplanted cells or tissues. In our view, an understanding of the current status and limitations of cell implantation technology will enable the development of new strategies and increasingly successful cell-based transplantation therapies. Accordingly, this chapter provides an overview of encapsulation strategies and followed by a description of various biocompatible polymers used in this field. Finally, a survey of long-standing disease targets—primarily diabetes—is given followed by a synopsis of emerging efforts to use encapsulated cells to treat an expanded repertoire of conditions that include cancer, CNS disorders, and liver disease.

10.2 ENCAPSULATION REQUIREMENTS AND STRATEGIES

10.2.1 Purpose and Requirements of Cell Encapsulation

To achieve the promise inherent in cell-based therapies, a capsule must satisfy a stringent set of dichotomous requirements. The capsule membrane, to serve as an immunoisolation device, must be able to keep various components of the immune system away from the living cells while at the same time allowing nutrients such as glucose and oxygen, as well as proteins such as growth factors, to pass through without much impedance. The membrane must also be biocompatible to the host and to the cells it encloses and when capsulation occurs, it must be cell friendly. At the same time, however, the membrane must be sufficiently robust to survive handling, transplantation, and the hostile environment inside the human body. Furthermore, in many cases, a biodegradable membrane is useful, but care must be taken to ensure that its degradation products do not adversely affect the encapsulated cells [29]. In addition, the diameter and surface charge of the capsule also needs to be taken into consideration because they also influence immune response and nutrient diffusion [30].

10.2.2 Categories of Cell Capsules

Numerous strategies have been pursued over the years to physically isolate cells from the immune system to avoid immunorejection. Immunoisolational capsules can be categorized very broadly as macrocapsules, which are usually millimeter-sized flat-sheet and hollow-core fibers or spheres or as microcapsules comprised of small spherical vehicles and conformally coated tissues. Regarding microcapsules [31], their smaller size (ranging from 100 to 500 μm) is particularly advantageous from a mass transport perspective, offering optimal surface-to-volume ratio for macromolecule (e.g., protein) and nutrient (e.g., glucose or oxygen) diffusion thereby enhancing cell viability. The small size of the capsules also allows their implantation in close contact to the bloodstream where oxygen transfer into their interiors is optimized further, which is critical for applications that require the long-term functionality of the enclosed cells [32]. Moreover, microcapsules are typically more durable than macrocapsules and are difficult to disrupt mechanically [17].

10.2.3 Classification of Cell Microencapsulation

10.2.3.1 Matrix-Core/Shell Microcapsules

Microcapsules can be classified into three categories (Figure 10.2): matrix-core/shell microcapsules, liquid-core/shell microcapsules, and cells-core/shell microcapsules (or conformal coating) [29]. Matrix-core/shell microcapsules are produced by placing polymer/cell droplets in a gelling solution. Basically, there are two methods to prepare the matrix-core/shell microcapsules. One method produces covalently crosslinked capsules, which attain an equilibrium swelling state that depends on the polymer–water interaction parameter and the crosslink density [33]. This type of capsule has significant advantages in terms of stability; however, the toxicity of the reactants and the time required to complete chemical crosslinking can be incompatible with cell survival. The other method involves the production of "reversible" or "physical" gels where the constituent polymer networks are held together by various degrees of molecular entanglements and secondary forces including ionic charges, hydrogen bonding, or hydrophobic interactions. All of these interactions are reversible and can be disrupted by

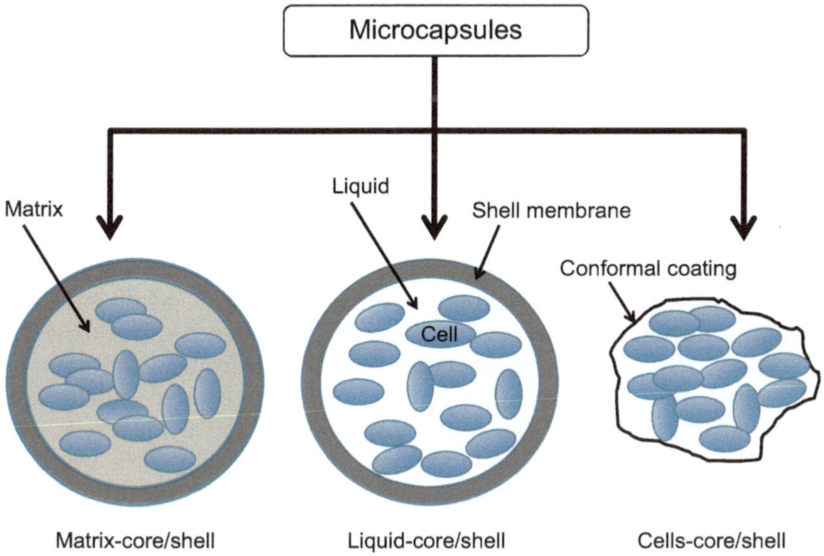

FIGURE 10.2 The three main categories of cell microcapsules.

changes in physical conditions or application of stress [33]. This class of microcapsules has been well studied due to its several advantages that include the maintenance of the cells in an aqueous environment where they are in contact with a soft and biocompatible material and for the long-term release of therapeutic molecules. Despite ability of the hydrogel matrix to support facile cell growth, the *in vivo* stability of electrostatic and hydrogen bonds maintaining the structure of this type of microcapsule is limited and the mechanical resistance of capsules also remains a concern [34].

10.2.3.2 Liquid-Core/Shell Microcapsules

Many technical improvements in microcapsule design and production have been introduced over the years. For example, the hydrogel network found within a capsule can be tuned to promote cell growth and protein production by including a liquid or semiliquid environment that supports higher diffusion of gases and nutrients than possible in a gel [35,36]. As a result, cells show increased survival and protein production. Liquid-core/shell microcapsules were first manufactured by dropping a cell/gelling solution into a polymer bath and have been pursued with different types of hollow polymeric microcapsules such as polymersomes, multilayered capsules, and hollow microspheres. Methods to fabricate this type of microcapsule have grown increasingly sophisticated beyond the self-assembly of microgels in droplets [37] to now include layer-by-layer polyelectrolyte deposition [38], interfacial polymerization [39], precipitation by phase separation [40], surface polymerization [41], copolymer vesicles [42], multiphase microfluidics [43], and a combination of molecular self-assembly and precipitation [44].

10.2.3.3 Cells-Core/Shell Microcapsules

Both of the techniques mentioned above (matrix-core and liquid-core shells) typically start with a scheme to generate a controlled-size droplet, followed by an interfacial process to stabilize the droplet and to obtain a solid microcapsule membrane around the droplet. Aside from these long-established approaches, new techniques are emerging that overcome shortcomings of the existing methods. For example, core/shell microcapsules that can be generated by conformal coating of cells been proposed [45,46]. Conformal coating constitutes a special case of microencapsulation where the term describes a method of forming a barrier directly around a small cell mass or a small piece of tissue to eliminate unutilized space in a microcapsule core. By closely surrounding the cell mass with the encapsulation membrane, this approach provides improved mass transport between the capsule exterior and the cell mass and also minimizes implant size by increasing the effectiveness of cell packing. The thickness of the poly(ethylene glycol) (PEG) layer formed on the cell surface in this strategy can be as thin as several nanometers thereby drastically reducing total graft volume compared with conventional microcapsules.

10.3 MATERIALS USED FOR CELL ENCAPSULATION

10.3.1 Hydrogels

One of the key issues in the design of a microencapsulation strategy is the choice of a suitable material for delivering cells to the host or immobilizing them for long-term delivery of molecules that they produce to the surrounding tissue. In the last few decades, the field of cell microencapsulation has greatly benefited from advances and optimization of biomaterials used to elaborate the capsule [47]. Microcapsules are almost exclusively produced from hydrogel-forming polymers because these materials hold a number of appealing features. For example, hydrogels are comprised of 3D networks of hydrophilic polymers, which provide a highly hydrated microenvironment for embedded cells that can also be designed to present biochemical, cellular, and physical stimuli that guide cellular processes including cell attachment, growth, and differentiation [48]. The unique properties of hydrogels that make them a material of choice for biomedical applications include minimal interfacial tension with surrounding biological fluids, permeation, and diffusion of low molecular weight compounds. Hydrogels also match the flexibility of human tissue, due to their water content, which reduces mechanical and frictional irritation to tissues and cells [49].

An important advantage of hydrogel-based microcapsules is a variety of administration options; in particular, they can be administered as implantable forms (preformed scaffolds) or injectable forms (*in situ* formed scaffolds). Unlike tissue engineering approaches that utilize prefabricated scaffolds, injectable systems have garnered great interest as a unique therapeutic method for administration to difficult-to-reach areas of the body using minimally invasive procedures. Another advantage is the ability of implanted microcapsules to conform to any shape irrespective of the defect geometry [50]. Overall, incorporating microencapsulated cells within hydrogel-based scaffolds improves administration options and local retention at the desired site within the body while reducing posttransplantation inflammation [51].

10.3.2 Natural Polymers

A variety of natural and synthetic polymers listed in Table 10.1 have been used for cell microencapsulation. Natural polymers that form semipermeable membranes suitable for cell encapsulation include polysaccharides (e.g., alginate, chitosan, hyaluronic acid (HA), and agarose) and extracellular matrix (ECM) proteins (e.g., collagen, gelatin, and fibrin); a sampling of these materials will be discussed in more detail in the following sections.

10.3.2.1 Alginate

Alginates are a frequently employed biomaterial for cell immobilization due to their

TABLE 10.1 Biomaterials for Cell Encapsulation

Material	Encapsulated Cells	Application	References
NATURE POLYMERS			
Alginate	Mesenchymal stem cells	Bone tissue engineering	[52]
	Pancreatic islet	Diabetes	[18,53]
	Fibroblast	Cardiac disease	[54]
	Embryonic stem cells	Diabetes	[55]
	Neonatal porcine choroid plexus cells	Neurodegenerative diseases	[56]
Alginate–agarose	BHK fibroblast and C2C12 myoblast		[57]
	Feline kidney cells	Subsieve-size capsules	[58]
Alginate–chitosan	Mesenchymal stem cells	Cartilage regeneration	[59]
	Chondrocytes	Bone tissue engineering	[60]
Alginate–collagen	Mesenchymal stem cells	Bone tissue engineering	[61]
Alginate–gelatin	Hepatocytes	Liver tissue engineering	[62]
Alginate–poly(L-lysine)	HepG2/C3A cells	Liver tissue engineering	[63]
	Chromaffin cells	Neuropathic pain	[64]
	CHO cells	Cardiovascular disease	[65]
	Porcine aortic endothelial cells	Cancer	[66]
	Genetically modified hamster kidney cells	Cancer	[67]
	Mouse genetically engineered myoblasts	Cancer	[68]
Alginate–poly(L-ornithine)	Pancreatic islet	Diabetes	[69]
	Choroid plexus	Huntington's disease	[70]
Chitosan	PC12 cells and R208F cells		[71]
	Chondrocytes	Orthopedics	[72]
	C2C12 myoblasts	Myocardial infarction	[73]
	Mesenchymal stem cells	Cartilage regeneration	[74]
Collagen	Mesenchymal stem cells	Cartilage regeneration	[75]
	Mesenchymal stem cells	Bone tissue engineering	[76]
Hyaluronic acid (HA)	3T3 fibroblasts		[77]
	Chondrocytes	Cartilage regeneration	[78]
	Adipose-derived stem cells	Adipogenic differentiation	[79]
	Chick dorsal root ganglia	Neuronal injury	[80]
	Embryonic stem cells or fibroblasts		[81]

(Continued)

TABLE 10.1 (Continued)

Material	Encapsulated Cells	Application	References
SYNTHETIC POLYMERS			
Polyethersulfone	C2C12 myoblasts	Focal epilepsies	[82]
Polyethylene glycol (PEG)	Chondrocytes	Cartilage regeneration	[83,84]
	Osteoblasts	Bone tissue engineering	[85]
	Mesenchymal stem cells	Bone tissue engineering	[86]
	Aortic valvular interstitial cells	Cartilage regeneration	[87]
	Neural cells	Neuronal regeneration	[88]
Poly(2-hydroxyethyl methacrylate)	3T3 fibroblasts CHO-K1	CNS diseases	[89]
Poly(lactic-co-glycolic acid)	P19 embryonic carcinoma cells	Nerve tissue engineering	[90]
Poly(vinyl alcohol)	Islets	Diabetes	[91].

abundance, easy gelling properties, and apparent biocompatibility; the significant breadth of literature available (reviewed in detail elsewhere [17,92]) drives home how important this compound has now become for cell and tissue encapsulation. Despite continuing relevance to the present day, alginate was first discovered in the late nineteenth century by Edward Stanford [93] and commercial production began in the 1920s through the harvest and processing of cell wall components of various species of marine brown algae [94]. This material was used in the pioneering study by Chang in 1964 [16] that laid a foundation for encapsulation as a strategy for immunoprotection of transplanted cells and the subsequent groundbreaking work of Lim and Sun in 1980 [18]. In the three decades since then, alginate has become a ubiquitous matrix material for cell, tissue, and macromolecule encapsulation with applications in many aspects of tissue engineering and regenerative medicine.

Alginates are linear block polymers consisting of α-L-guluronic acid (G block) and β-D-mannuronic acid (M block). The gelation of alginate from a solution containing divalent cations results in the formation of the calcium junctions of GG-GG, MG-GG, and MG-MG between alginate molecules. Calcium and barium are the most commonly used as the complexation ions to produce the gelled structure because of their selectivity and cooperative binding to the G blocks. The G and M content varies with the alginate source and can affect gel properties such as mechanical strength, biocompatibility, and permeability [95]. The main limitation of using alginate as an encapsulation matrix is poor mechanical stability in the presence of monovalent and certain divalent cations commonly found in cell culture media or a physiological milieu, which can have an adverse effect on hydrogel structure by destabilizing the polyelectrolyte interactions involved in gel integrity [36]. To increase the stability and to reduce the wall permeability (for immunoisolation) of the microcapsules, alginate microcapsules are often coated first with a polycation and then with another layer of alginate. Various polycations have been used for this purpose including poly-L-lysine (PLL), poly-L-ornithine (PLO), chitosan, lactose modified chitosan, and photopolymerized materials [96].

Cell adhesion cues have been incorporated into the alginate network either through the addition of Arg–Gly–Asp (RGD), a cell adhesion oligopeptide [97,98], or by copolymerizing alginate with gelatin as culture time increased and complete degradation of the gel occurred in 5 weeks [62]. In recent work, Liu et al. [99] used crosslinked gelatin and alginate derivatives to prepare cell sheets and spherical tissues wrapped in living cell sheaths. The microcapsules not only supported growth of the enclosed cells into spherical tissues but also provided a cell adhesive outer surface. Light-activated gelation of alginate was recently employed to encapsulate rat bone marrow cells through the release of calcium chloride from light-sensitive liposomes suspended in the alginate–cell solution [100].

10.3.2.2 Chitosan

Chitosan, a derivative of the extremely abundant chitin polysaccharide, has structural characteristics similar to mammalian glycosaminoglycans (GAGs) and has attracted much attention as an alternative polycation in preparing microcapsules. Chitosan is a biocompatible, biodegradable, and nontoxic cationic linear polysaccharide composed of β-(1-4)-linked D-glucosamine and *N*-acetyl-D-glucosamine units and is obtained by the extensive deacetylation of chitin by hydrolyzing the amino-acetyl groups using NaOH [101]. Chitosan can form hydrogels by ionic interactions [102] or through chemical crosslinking [103] with the bifunctional agent glutaraldehyde and it degrades via enzymatic hydrolysis [104].

Due to its weak mechanical properties and lack of bioactivity, chitosan is often combined with other materials, such as poly(lactic-co-glycolic acid) (PLGA) [105] to achieve more desirable mechanical properties for encapsulation. Chitosan has also been used to coat capsules to enhance stability and modify permeability as an alternative to PLL because it is regarded as being more biocompatible [106]. Under dilute acid conditions ($pH < 6$), chitosan is positively charged and water soluble, while at physiological pH, chitosan is neutral and hydrophobic, leading to the formation of a solidified, physically crosslinked hydrogel. The addition of polyol salts enables encapsulation of cells at neutral pH, where gelation becomes temperature dependent [102]. The addition of the adhesion-promoting RGD tripeptide has been necessary to promote cell attachment to chitosan, allowing myoblast and cardiomyocytes to be successfully encapsulated in crosslinked gel made from this polysaccharide [73]. In a separate study, polylysine was grafted to chitosan to enhance the microenvironment for neural cell growth. When fetal mouse cortical cells were encapsulated in chitosan-*g*-polylysine gels, neurite extensions were observed in the 3D gel cultures [107]. Examples that illustrate the versatility of chitosan include the use of this material for microencapsulation of human red blood cells [108], HepG2 cells [109], BHK-21 cells [110], chondrocytes [111], bone marrow stem cells [112], and PC12 and R208F cells [71].

10.3.2.3 Hyaluronic Acid

HA, or hyaluronan, is a natural GAG that is nonimmunogenic in its high molecular weight form and also nonadhesive [113]. HA is a linear polysaccharide composed of repeating disaccharide units of *N*-acetyl-D-glucosamine and D-glucuronic acid and is found in many tissues, such as skin, cartilage, and the vitreous humor. HA is a component of the ECM in mammalian connective tissues and participates in many important biological processes during wound repair [81]. Colloids of HA can be gelled by prior chemical modification of hyaluronan with thiols [114], methacrylates [115], or tyramines [116], or can be crosslinked *in situ* using formaldehyde or divinyl sulphone. One of the key advantages of using HA gels for tissue engineering is that their degradation can be mediated by hyaluronidase, an enzyme secreted by a multitude of mammalian cell types [117]. These properties make HA an interesting biomaterial for cell-based scaffolds and several cell types (chondrocytes, fibroblasts, and murine embryonic stem cells (ESCs)) encapsulated into either HA/alginate beads [118] or HA-microfabricated devices have been observed to proliferate and behave correctly [81].

10.3.2.4 Collagen

Collagen gels, as well as other mammalian-derived protein-based polymers covered in the next section, are effective matrices for cellular growth because these materials contain many cell-signaling domains present in the *in vivo* ECM [119]. Collagen type I is one of the most studied materials for tissue engineering in part because of its low immunogenicity [120]. An advantage of collagen gels is that they can be created through natural means without chemical modifications; the resulting collagen matrices, however, do have drawbacks including mechanical weakness and low resistance to the enzymatic degradation. Various strategies, therefore, have been pursued to improve the stability and mechanical properties of collagen, including chemical (glutaraldehyde and water-soluble carbodiimide) [121], physical (UV irradiation and dehydrothermal) [122], and enzymatic [123] crosslinking approaches. As an example of the promise held by collagen microcapsule technology, self-assembled collagen microspheres provide superior hyaline cartilage reproduction for the microencapsulation of human mesenchymal stem cells (hMSCs) based on the stability of these devices, their ability to be injected *in vivo*, and their provision of a protective, growth- and migration-supporting environment for hMSC both *in vitro* and *in vivo* [75].

10.3.2.5 Protein-based gels

Proteins are polymers of amino acids that form structures attractive for cell encapsulation including β-sheet-rich fibrils that can be nanometers in width and micrometers in length that also further self-organize and entangle to form 3D hydrogels under appropriate conditions [124,125]. As an alternative to natural proteins, synthetic peptides hold great promise in designing hydrogel carriers with a wealth of bioactive signals programmed directly into the hydrogel matrix. For example, one could envision incorporation of cell adhesion moieties, biochemical cues to promote tissue deposition, and specific enzyme-sensitive sequences for cell-mediated degradation [126]. Several groups have explored synthetic peptides that are relatively short ~15- to 20-mer amino acid sequences designed to self-assemble into a hydrogel for cell encapsulations. One example was an amphiphilic peptide containing positively charged lysine residues that was demonstrated to successfully encapsulate mesenchymal stem cells (MSCs) [127]. Another example is provided by elastin-like peptides (ELPs) used to encapsulate human adipose-derived adult stem cells [128] and as a preosteoblastic cell carrier for bone regeneration [129].

10.3.3 Synthetic Polymers

10.3.3.1 General Considerations

In general, natural polymers, including those discussed above, have attractive features that provide them with therapeutic relevance. For example, natural polymers, as well as their degradation products, are usually not toxic or carcinogenic, nor do they elicit adverse reproductive and developmental effects and in most cases entrapment of living cells in matrices made from natural polymers can be accomplished under physiological conditions. In addition to natural biomaterials, however, many synthetic polymers as well as inorganic compounds have also been explored for cell encapsulation [130]. The mechanical and chemical properties of synthetic polymers can be tailored for different applications without having the immunogenicity-related concerns of some naturally occurring polymers (e.g., low molecular weight HA), which constitutes an important advantage of these materials over their natural counterparts [130].

Although synthetic materials provide researchers with large flexibility in material design, one drawback is that they do not have an intrinsic mechanism for interacting with cells and cell attachment typically relies on nonspecific adhesion, which tends to be weak [131]. This limitation, which restricts the use of synthetic materials in applications that require defined control over cell–matrix interactions, can be overcome in some cases by functionalizing these materials with bioactive molecules such as the RGD tripeptide adhesion motif. A number of synthetic hydrogels, e.g., poly (ethylene glycol), poly(acrylic acid), poly (vinyl alcohol), poly(*N*-isopropyl acrylamide), and their derivatives have been developed and successfully employed for use in tissue engineering and encapsulation of cells [49,132,133]. PEG will be discussed in more detail below as a well-studied representative of synthetic polymers used for cell encapsulation.

10.3.3.2 Poly(Ethylene Glycol)

PEG has been the most widely used synthetic material to create macromeric polymers for cell encapsulation. PEG hydrogels can be formed via the photoinitiated chain polymerization of functionalized PEG macromers, specifically dimethacrylated PEG. Photoencapsulation within these gels has been shown to be cytocompatible with multiple cell types, including fibroblasts, chondrocytes [83], osteoblasts [85], mesenchymal stem cells [86], aortic valvular interstitial cells [87], and neural cells [88]. Even though PEG gels are inherently cell repellent [117], certain cell types such as chondrocytes and osteoblasts encapsulated in PEG hydrogels survive without the addition of any adhesion or biological-signaling molecules [134,135]. However, many cell types are attachment dependent and require cell adhesion sites to survive necessitating chemical modification strategies to incorporate various peptides or other signaling molecules into the gels that reduce their cell-repulsion behavior. For example, hMSCs encapsulated in PEG hydrogels require the addition of a cell adhesion component, such as the oligopeptide sequence RGD, for cell survival [136]. Alternatively, strategies have developed natural hydrogel environments by mimicking the native ECM, which not only promote cell–matrix interactions with the hydrogel but also sequester important growth factors that can augment tissue growth [137].

PEG hydrogels are very stable *in vitro* and are refractory to degradation in biological systems. This drawback—when biodegradation of an implanted material is advantageous—has been addressed by exploiting the versatile nature of PEG polymer chemistry. For example, the incorporation of copolymers enables the network structure and stability of PEG-based materials to be modified conveniently [138]. In particular, polylactic acid (PLA) and poly(caprolactone) (PCL) have been widely used in poly(α-hydroxy esters) for creating biodegradable PEG macromers for cell encapsulation because degradation profiles can be controlled by tuning chemistry and size of the degradable block [139]; in one example, Martens et al. [140] demonstrated that biodegradation time was enhanced when PEG hydrogels contained labile PLA linkages. In another strategy, the degradation of PCL-*b*–PEG-*b*–PCL hydrogels was temporally controlled by exogenously delivering lipase to match tissue growth for cartilage tissue engineering [141].

10.4 THERAPEUTIC APPLICATIONS OF ENCAPSULATED CELLS

Although challenges remain, considerable progress has been made during the past decade toward understanding the biological and technological requirements for successful transplantation of encapsulated cells in experimental animal models. These studies have provided a foundation for clinical trials with many of the long-standing translational efforts directed toward treating diabetes. Additional promising therapeutic options are currently on the horizon for cancer, neurodegenerative diseases, and liver failure. Examples of therapy directed to each of these endpoints are provided in the following sections and are listed in Table 10.2.

10.4.1 Diabetes

10.4.1.1 Cell Encapsulation Technologies were Pioneered by Efforts to Treat Diabetes

It has long been recognized that diabetic patients, who suffer from a plethora of secondary complications such as blindness, neuropathy, and end-stage renal failure, benefit from insulin treatment. Relevant to this chapter, difficulties in obtaining and administering insulin provided impetus to many of the historical developments in the field of cell encapsulation dating back to 1964 when Chang [16] proposed the idea of using ultrathin polymer membrane microcapsules for the immunoprotection of transplanted cells and introduced the term "artificial cells" to define the concept of bioencapsulation. Based on efforts started in the early 1970s that employed alginates for the immobilization of insulin-secreting pancreatic islet cells (PICs), which remained morphologically and functionally intact for as long as 4 months *in vitro* [18], this approach subsequently was implemented *in vivo* when it was used to xenograft islet cells in a rat model of diabetes [149]. Importantly, the microencapsulated PICs had a superior ability to maintain normoglycemia in animal models compared to nonencapsulated cells; for example, in a study conducted in streptozotocin-induced diabetic Wistar Lewis rats, encapsulated PICs maintained normoglycemia for 3 weeks compared to only 6–8 days with nonencapsulated cells [149]. More recently, Luca et al. [150] immobilized Sertoli cells in alginate-based microcapsules and transplanted the newly formed functional islets β-cells into nonobese diabetic mice thereby reversing the disease.

10.4.1.2 Expanding the Types of Cells Available for Treating Diabetes

Intensified insulin treatment, in particular, the delivery of insulin-secreting PICs as discussed above has proven promising for the treatment of type 1 diabetes [151]. Unfortunately, transplantation of whole human pancreases, isolated islets, or PICs into patients has been severely hampered by the scarcity of cadaveric human donor organs, which has mandated a search for insulin producing cells/tissue source alternatives. Taking one direction toward overcoming this hurdle, Abalovich et al. [152] performed a preclinical study investigating the potential of encapsulated pig islet cells for xenotransplantation. Type 1 diabetic dogs transplanted with encapsulated PICs experienced a significant reduction (20–80%) in insulin necessity after transplantation. Elliott et al. [144] showed that porcine islet cells survived and insulin production was maintained in a human patient 10 years after transplant of pig islet cells. In a second direction, recent progress in stem cell biology has provided promising leads for the conversion of progenitor cells into functionally competent, insulin-secreting cells. Currently, several studies have reported that MSCs [153], ESCs [55], and recently induced

TABLE 10.2 Clinical Studies of Encapsulated Cells

Biomaterials	Therapeutic Agents	Application	Results/Status	References
Alginate	Human islets	Diabetes	Insulin independence with tight glycemic control was demonstrated after 9 months	[142]
	Parathyroid tissue	Hypocalcaemia	2 patient, parathyroid allotransplantation without immunosuppression	[143]
	Diabecell®; neonatal porcine islets; insulin	Diabetes	1 patient; 9.5 years functionality tested A new opened investigation of the safety and effectiveness of Diabecell® started from 2009 and will complete on 2014	[144] Clinical trial number NCT00940173 www. clinicaltrials.gov
Cellulose	Pancreatic cancer cells, ifosfamide	Cancer	2 independent clinical trials, entering phase III trials	http://www. stockmarketmediagroup.com/ nuvilex-inc-will-enter-phase-iii-trials-with-advantages-using-cell-encapsulation/
Gelatin	Human retinal pigment epithelium, spheramine	Parkinson's disease	The 1-year, open-label, single center study evaluated the safety and efficacy of gelatin microcarrier	[65,145,146]
Polyethersulfone	Engineered kidney cells	Huntington's disease	6 patients, phase I study evaluated the safety and tolerability of the procedure	[147]
	Engineered baby hamster kidney cells	Amyotrophic lateral sclerosis	6 patients, neurotrophic factors can be continuously delivered within the CSF of humans by an *ex vivo* gene therapy approach	[148]
Polyethylene glycol (PEG)	Novocell®; allogeneic human islets; insulin	Diabetes	A single center phase I/II study, encapsulated islet allografts can be implanted safely in the subcutaneous tissues without the use of long-term immunosuppression	Clinical trial number NCT00260234 www.clinicaltrials.gov

pluripotent stem cells (iPSCs) [154] can differentiate into insulin producing cells potentially useful for the treatment of diabetes. Tuch et al. [155] have proposed replacing pancreatic β-cells of diabetic recipients with pancreatic progenitors derived from human ESCs (hESCs).

10.4.1.3 Diabetes Leads the Way in Clinical Translation of Cell Encapsulation Therapies

Translation of microencapsulated cell systems, as evidenced by preclinical trials, now is under way (listed in full in Table 10.2) and has often focused on diabetes. Encapsulated islet transplantation was first performed in 1999 in a 38-year-old male patient who had severe diabetes [20]. In 2000, Elliott et al. [19] transplanted encapsulated porcine islets into diabetic patients and confirmed no PERV infection in patients. Calafiore et al. [21] subsequently performed human islets microencapsulation using alginate and transplanted them into 10 T1DM patients in Italy in an approach approved by the Italian Ministry of Health [21] and an Australia group transplanted barium-alginate encapsulated human islets into four T1DM patients without immunosuppression and followed them for over 2 years in a phase I study. Similarly, Tuch et al. [22] investigated the transplantation of barium-alginate microcapsules containing human islet cells in four type 1 diabetic patients and demonstrated the safety of this method, with little C-peptide—which reflects the level of blood sugars—detected, normal renal function, negligible cytokine release, and no major infection detected during the trial. Although insulin independence was not achieved, glycemic control was improved, with a reduction in insulin daily requirement. To quickly summarize clinical testing, a high level of safety has been demonstrated with an absence of adverse events in four phase I trials showing that intraperitoneally infused microencapsulated human islets can be considered safe for up to 3 years [22,156]; conversely, efficacy remains marginal.

10.4.1.4 Status of Clinic Translation

Important lessons have been learned and precedents set from efforts to clinically translate cell encapsulation therapies to treat diabetes; in particular, this approach provides protection from immunorejection and facilitates the long-term health and function of the transplanted cells. Although the (yet to be achieved) Holy Grail of specific induction of transplant tolerance would be ideal, *in vivo* testing has demonstrated cell encapsulation is a promising technology for achieving sufficient immunoprotection to allow β-cell replacement [157,158]. The early studies to date have confirmed that an encapsulation approach limits undesirable immunological responses to the nonhost islet cells and increases the access of cell replacement therapy to a broader patient base by minimizing the need for chronic administration of immunosuppressive drugs. Furthermore, the use of a macroencapsulation device will enable retrievability of the entire graft, which may be a desirable feature for products using hESC-derived cells as a potential safety benefit, allowing removal of the cells if complications do arise [159]. Overall, as the examples of clinical testing of cell encapsulation therapies outlined in the previous paragraph illustrates, hurdles remain to make this therapeutic approach reproducible and effective. On the other hand, a high degree of safety has been achieved, making this approach attractive to treat an expanded set of diseases; examples include cancer, CNS disorders, and liver disease as discussed next.

10.4.2 Cancer

Cell encapsulation technologies originally developed for the treatment of acquired and genetic diseases such as diabetes now are also proving to be attractive for the treatment of a variety of solid tumors. An advantage for using encapsulated cells for the treatment of cancers (and also other diseases) is that therapeutic molecules can be delivered in a sustained manner from implanted cells as the cells continuously produce the desired therapeutic products (e.g., endostatin as discussed below), which would otherwise have short residence times in the body.

Several microcapsule-based strategies are emerging to introduce and maintain immunoprotected cells in the body that secrete a variety of therapeutic products for the treatment of cancer. One approach is based on the idea that without angiogenesis, most solid tumors cannot grow past a critical size because of inadequate tissue oxygenation and nutrient supply; accordingly, strategies are being pursued based on encapsulation of cells that secrete anti-angiogenic proteins such as endostatin [67,160] and angiostatin [161]. In one study, genetically engineered HEK293 cells were encapsulated in alginate particles for the local production of endostatin for treatment of malignant brain tumors in a rat model [160]. In a follow-up study, tumor growth was reduced by 35% in treated animals that had received C6 glioma spheroids implanted either ectopically or orthotopically and, interestingly, tumor cell invasion into the surrounding tissue was also inhibited [162]. In a second set of studies, baby hamster kidney (BHK) cells were stably transfected with a human endostatin expression vector and were encapsulated in alginate–PLL microcapsules for long-term delivery of endostatin. A single local injection of encapsulated endostatin-secreting cells in nude mice subcutaneously xenografted with the human glioma cell line (U-87MG) resulted in a 72% reduction in tumor weight 21 days posttreatment [67]. Cells genetically modified to secrete angiostatin have also been encapsulated and implanted in tumor bearing mice resulting in a reduction of tumor volume by 70–80% in one study [161].

In a different approach, cell-based therapies have been designed to stimulate an immune response against tumors by encapsulating cells genetically modified to produce tumor necrosis factor α (TNF-α) [163] or interleukin-6 (IL-6) [164]. The encapsulated cells that were capable of releasing therapeutically significant amounts of functional TNF-α were intratumorally implanted into athymic nude mice tumors. The treated tumors showed extensive apoptosis and necrosis in response to TNF-α production resulting in significant tumor regression [163]. In another study, Chinese hamster ovary (CHO) cells producing IL-6 were encapsulated in alginate. This study demonstrated the viability of using cell encapsulation technology to generate short-term high levels of active circulating and intra-hepatic cytokines and also raised the possibility of modifying specific signal transduction cascades that contribute to tumor progression [164].

In a third approach, cells have been employed that over-express enzymes such as the cytochrome P450 enzyme to activate chemotherapeutic agents or prodrugs such as ifosfamide; such gene-directed enzyme prodrug therapy is one of the most encouraging approaches for cancer therapy [165]. The first demonstration of this method used genetically modified epithelial cells, which was later reproduced using HEK293 cells [166]. In addition, not only ifosfamide but also cyclophosphamide and related agents can be activated by the cytochrome P450 enzyme. Therefore, this strategy provides several options to treat tumors. Artificial cell encapsulated genetically modified cells expressing prodrug-activating enzymes, such as cytochrome P450, have been studied as a potential treatment for inoperable pancreatic carcinoma in patients [166,167]. These encapsulated cells are given intra-arterially so that they are located in the pancreatic carcinoma where they activate the drug ifosfamide. Therefore, the local concentration of activated ifosfamide is much higher than in the rest of the body, thus minimizing systemic toxicity and increasing local efficacy. Another approach involved polymeric artificial cells containing genetically modified myoblast cells that secrete IL-2 linked to the Fv region of a humanized antibody with affinity to tumors over-expressing the oncogene ERBB2; animal testing with these encapsulated cells delayed tumor progression and prolonged survival [168]. Finally, encapsulation of stem cells holds promise for cancer therapy. For example, Goren et al. [169] showed that encapsulated MSCs that were genetically engineered to secrete hemopexin-like protein (PEX), in alginate–PLL microcapsules. These microcapsules were implanted in a model of human glioblastoma reducing tumor growth, blood vessel formation, and tumor cell proliferation while increasing tumor cell apoptosis.

10.4.3 Central Nervous System

Targeted delivery of active biomolecules to the central nervous system (CNS) has therapeutic potential for a variety of disorders. Systemic drug delivery is inefficient because of poor passage of conventional therapeutic molecules through the blood–brain barrier while emerging cell and gene-based modalities also require alternative methods of administration. Cell microencapsulation technology provides an innovative strategy to deliver endogenous or engineered peptides/proteins directly to the diseased brain. Implantation of encapsulated producer cells ensures local *de novo* drug delivery, continuous and long-term release, and the possibility to combine several therapeutic agents [170].

10.4.3.1 Parkinson's Disease

Initial proof of concept data for treating Parkinson's disease (PD) with encapsulated cells came from studies in rodents and primates that demonstrated that encapsulated PC12 cells could be implanted into the striatum and deliver L-3,4-dihydroxyphenylalanine (L-DOPA) and dopamine in sufficient quantities to induce behavioral improvements [171,172]. Encapsulated PC12 cells survived *in vivo* for up to 6 months, while maintaining a typical morphology and production of catecholamines [172]. Other cell types, including encapsulated bovine adrenal chromaffin cells, survived for 1.5 years with continued production of catecholamines and met-enkephalin [173]. The most widely employed cell-based approach for the treatment of PD has entailed the administration of encapsulated choroid plexus (CP) cells in conjunction with glial-cell-derived neurotrophic factor (GDNF) or vascular endothelial growth factor (VEGF) secreting cells [174–176]. In one study using this strategy, encapsulated GDNF-secreting fibroblasts were implanted into the striatum of 6-hydroxydopamine (6-OHDA) injured rats. Long-term expression of GDNF, up to 6 months, was detected along with behavioral improvement and tissue biocompatibility of the capsules [175]. In another study, porcine CP cells were encapsulated in alginate–PLO–alginate capsules and implanted into the striatum of 6-OHDA lesioned rats. Results showed that the presumed neurorestorative proteins secreted by the encapsulated cells improved motor behavior and nigrostriatal dopaminergic activity [177].

10.4.3.2 Alzheimer's Disease

Alzheimer's disease (AD) is a progressive disorder with a successive decline in cognitive functions. One early symptom is episodic memory loss. The progressive memory loss in AD patients is associated with neurodegeneration and dysfunction of the central cholinergic [34]. Positive effects of nerve growth factor (NGF) on cell survival and neurological functionality have been demonstrated in several animal studies. Initial studies demonstrated that implants of encapsulated NGF-secreting BHK cells prevented cholinergic neuron loss following aspiration of the fimbria/fornix. Control-implanted animals had an extensive loss (88%) of ChAT-positive cholinergic neurons ipsilateral to the lesion that was prevented by NGF cell implants (14% loss) [178]. Nagahara's [179] study indicated that sustained local delivery of NGF to the cholinergic basal forebrain can arrest and even reverse the degeneration of the basal forebrain cholinergic neurons involved in the cognitive decline in AD. The immobilization of genetically engineered cells to produce VEGF, ciliary neurotrophic factor (CNTF) or glucagon-like peptide-1 (GLP-1) have also been employed for the treatment of AD. Spuch et al. [170] implanted microencapsulated fibroblast cells releasing VEGF into a transgenic (APP/PS1) mouse model of AD with degenerative alterations in the microvasculature. The study reported increased angiogenesis, decreased presence of Aβ and tau protein, less apoptosis, and protection of the cognitive behavior in the APP/PS1 mice after the implantation of the microcapsules.

When encapsulated cells secreting CNTF were implanted intra-cerebroventricularly into mice expressing the mutant amyloid precursor protein, a significant improvement in cognitive function occurred [180]. In other studies, encapsulated MSCs expressing GLP-1 were evaluated in an AD double transgenic murine model for their ability to GLP-1 decrease amyloid deposition and A-β-induced toxicity [181]. Results suggested a reduction in A-β-induced toxicity *in vitro* along with anti-inflammatory and neuroprotective properties. Most recently, the authors describe the first clinical trial with encapsulated cell biodelivery implants that deliver NGF to the cholinergic basal forebrain with the intention of halting the degeneration of cholinergic neurons and the associated cognitive decline in patients with AD [182]. Follow-up MRI at 3 and 12 months postimplantation showed no evidence of inflammation or device displacement. At 12 months, implants were successfully retrieved and low but persistent NGF secretion was detected in half of the patients.

10.4.4 Liver Disease

Hepatic diseases, including acute liver failure, chronic liver disease, and congenital metabolic liver disease, require the restoration of liver function. In recent years, bone marrow stem cells have been used as feeder cells to enhance hepatocyte viability. Extending this concept from cell culture experiments, rat and immortalized human hepatocytes (IHHs) now can be encapsulated in 400 μm alginate–PLL–alginate membranes, stored using cryopreservation, and then later be transplanted into xenogeneic recipients with liver failure and sustain liver metabolic functions [183]. Unfractionated rat bone marrow stem cells coencapsulated with hepatocytes significantly enhances the viability of hepatocytes compared to encapsulation alone for time periods from 7 weeks to up to 4 months [184,185]. In a test of efficacy, rat bone marrow stem cells were encapsulated and then transplanted intraperitoneally into a 90% hepatectomy-induced acute liver failure rat model. After 2 weeks, survival of animals that received implants containing bone marrow stem cells was significantly prolonged comparable to the cohort that was transplanted with hepatocytes only [185]. Mechanistically, when coencapsulated with bone marrow stem cells, the hepatocytes attach to one another whereas most remained unattached when encapsulated in the absence of the bone marrow cells. Based on the need for cell–cell interactions for hepatocytes to function optimally, the bone marrow stem cells contribute to positive *in vivo* outcomes at least in part by promoting hepatocyte function.

10.5 CHALLENGES AND FUTURE PERSPECTIVES

In the past 50 years since the idea of cell-based therapies was given impetus by early attempts to encapsulate cells before implantation into a patient to thwart immunorejection, considerable progress has been made as has been outlined in this chapter. However, there are remaining hurdles to surmount before this technology achieves its full potential that include overcoming technological and biological limitations, as well as conforming to ethical, political, and regulatory obstacles. A major focus of this chapter was on the "materials science" aspects of cell encapsulation-based therapies where considerable progress has been made over the past decades. However, there are still limitations in the production of microcapsules of defined size, certain size ranges, shape, and properties at a scale suitable for commercial application and under controlled environmental conditions. In addition, polymers used to encapsulate cells still hinder the development and application of cell-based therapies because these materials are often of natural origin and are highly variable in chemical composition and physical properties. It is therefore difficult to prepare reproducible encapsulation devices because of batch-to-batch variations in factors such as molecular weight distributions, a wide range of viscosities, purities, and the degree of chemical substitution. In many cases, synthetic polymers suffer from the same problems.

Another important consideration for clinical success of cell encapsulation therapies is an appropriate and reproducible source of functional cells. In this regard, the potential use of allogeneic versus xenogeneic cells has provoked important social and ethical debates [7]. Ultimately for a cell encapsulation strategy to be successful, a number of practical factors must be considered in addition to the fundamental design criteria described above. The process must be easily scaled up for manufacturing; marketable; accepted by surgeons, patients, and health-care providers; and receive FDA approval [92]. Nevertheless, the accelerating pace of technical and biological advances, together with the increasing experience in the field, offers hope that that full potential of bioactive cell encapsulation will reach fruition in the coming years.

Acknowledgments

Funding for this contribution was provided by the National Institutes of Health (R01CA112314 and R01AR054005).

References

[1] Starzl TE. History of clinical transplantation. World J Surg 2000;24:759–82.
[2] Tabar V, Studer L. Pluripotent stem cells in regenerative medicine: challenges and recent progress. Nat Rev Genet 2014;15:82–92.
[3] Emmert MY, Hitchcock RW, Hoerstrup SP. Cell therapy, 3D culture systems and tissue engineering for cardiac regeneration. Adv Drug Deliv Rev 2013; pii: S0169-409X(13)00290-1. [Epub ahead of print].
[4] Han LH, Tong X, Yang F. Photo-crosslinkable PEG-based microribbons for forming 3D macroporous scaffolds with decoupled niche properties. Adv Mater 2014;26(11):1757–62.
[5] Coburn J, Gibson M, Bandalini PA, Laird C, Mao HQ, Moroni L, et al. Biomimetics of the extracellular matrix: an integrated three-dimensional fiber–hydrogel composite for cartilage tissue engineering. Smart Struct Syst 2011;7:213–22.
[6] Lim SH, Liu XY, Song H, Yarema KJ, Mao HQ. The effect of nanofiber-guided cell alignment on the preferential differentiation of neural stem cells. Biomaterials 2010;31:9031–9.
[7] Orive G, Hernandez RM, Gascon AR, Calafiore R, Chang TM, De Vos P, et al. Cell encapsulation: promise and progress. Nat Med 2003;9:104–7.
[8] Chang TMS. Therapeutic applications of polymeric artificial cells. Nat Rev Drug Disc 2005;4:221–35.
[9] Hernandez RM, Orive G, Murua A, Pedraz JL. Microcapsules and microcarriers for *in situ* cell delivery. Adv Drug Deliv Rev 2010;62:711–30.
[10] Foster JL, Williams G, Williams LJ, Tuch BE. Differentiation of transplanted microencapsulated fetal pancreatic cells. Transplantation 2007;83:1440–8.
[11] LaRosa DF, Rahman AH, Turka LA. The innate immune system in allograft rejection and tolerance. J Immunol 2007;178:7503–9.
[12] Mueller NJ. New immunosuppressive strategies and the risk of infection. Transpl Infect Dis 2008;10:379–84.
[13] Salvadori M, Bertoni E. Side effects and toxicity of immunosuppressive agents. G Ital Nefrol 2003;20:490–502.
[14] Boots JM, Christiaans MH, van Hooff JP. Effect of immunosuppressive agents on long-term survival of renal transplant recipients: focus on the cardiovascular risk. Drugs 2004;64:2047–73.
[15] Lake KD, Canafax DM. Important interactions of drugs with immunosuppressive agents used in transplant recipients. J Antimicrob Chemother 1995;36:11–22.
[16] Chang TMS. Semipermeable microcapsules. Science 1964;146:524–5.

[17] Uludag H, De Vos P, Tresco PA. Technology of mammalian cell encapsulation. Adv Drug Deliv Rev 2000;42:29–64.

[18] Lim F, Sun AM. Microencapsulated islets as bioartificial endocrine pancreas. Science 1980;210:908–10.

[19] Stamler J, Brown IJ, Daviglus ML, Chan QEN, Miura K, Okuda N, et al. Dietary glycine and blood pressure: the international study on macro/micronutrients and blood pressure. Am J Clin Nutr 2013;98:136–45.

[20] Soon-Shiong P. Treatment of type I diabetes using encapsulated islets. Adv Drug Deliv Rev 1999;35:259–70.

[21] Calafiore R, Basta G, Luca G, Lemmi A, Racanicchi L, Mancuso F, et al. Standard technical procedures for microencapsulation of human islets for graft into nonimmunosuppressed patients with type 1 diabetes mellitus. Transplant Proc 2006;38:1156–7.

[22] Tuch BE, Keogh GW, Williams LJ, Wu W, Foster JL, Vaithilingam V, et al. Safety and viability of microencapsulated human islets transplanted into diabetic humans. Diabetes Care 2009;32:1887–9.

[23] Yang F, Williams CG, Wang DA, Lee H, Manson PN, Elisseeff J. The effect of incorporating RGD adhesive peptide in polyethylene glycol diacrylate hydrogel on osteogenesis of bone marrow stromal cells. Biomaterials 2005;26:5991–8.

[24] Strehin I, Nahas Z, Arora K, Nguyen T, Elisseeff J. A versatile pH sensitive chondroitin sulfate–PEG tissue adhesive and hydrogel. Biomaterials 2010;31:2788–97.

[25] Simson JA, Strehin IA, Allen BW, Elisseeff JH. Bonding and fusion of meniscus fibrocartilage using a novel chondroitin sulfate bone marrow tissue adhesive. Tissue Eng Part A 2013;19:1843–51.

[26] Chang CY, Chan AT, Armstrong PA, Luo HC, Higuchi T, Strehin IA, et al. Hyaluronic acid–human blood hydrogels for stem cell transplantation. Biomaterials 2012;33:8026–33.

[27] Thankamony SP, Sackstein R. Enforced hematopoietic cell E- and L-selectin ligand (HCELL) expression primes transendothelial migration of human mesenchymal stem cells. Proc Natl Acad Sci USA 2011;108:2258–63.

[28] Sackstein R. Glycoengineering of HCELL, the human bone marrow homing receptor: sweetly programming cell migration. Ann Biomed Eng 2012;40:766–76.

[29] Rabanel JM, Banquy X, Zouaoui H, Mokhtar M, Hildgen P. Progress technology in microencapsulation methods for cell therapy. Biotechnol Prog 2009;25:946–63.

[30] de Vos P, de Haan BJ, Kamps JA, Faas MM, Kitano T. Zeta-potentials of alginate–PLL capsules: a predictive measure for biocompatibility? J Biomed Mat Res A 2007;80:813–9.

[31] Sakai S, Kawakami K. Development of subsieve-size capsules and application to cell therapy. Adv Exp Med Biol 2010;670:22–30.

[32] de Vos P, Spasojevic M, Faas MM. Treatment of diabetes with encapsulated islets. Ad Exp Med Biol 2010;670:38–53.

[33] Hennink WE, van Nostrum CF. Novel crosslinking methods to design hydrogels. Adv Drug Deliv Rev 2012;64:223–36.

[34] Eriksdotter-Jonhagen M, Linderoth B, Lind G, Aladellie L, Almkvist O, Andreasen N, et al. Encapsulated cell biodelivery of nerve growth factor to the basal forebrain in patients with Alzheimer's disease. Dement Geriatr Cogn Dis 2012;33:18–28.

[35] Helmlinger G, Netti PA, Lichtenbeld HC, Melder RJ, Jain RK. Solid stress inhibits the growth of multicellular tumor spheroids. Nat Biotechnol 1997;15:778–83.

[36] Breguet V, Gugerli R, von Stockar U, Marison IW. CHO immobilization in alginate/poly-L-lysine microcapsules: an understanding of potential and limitations. Cytotechnology 2007;53:81–93.

[37] Dinsmore AD, Hsu MF, Nikolaides MG, Marquez M, Bausch AR, Weitz DA. Colloidosomes: selectively permeable capsules composed of colloidal particles. Science 2002;298:1006–9.

[38] Grigoriev DO, Bukreeva T, Mohwald H, Shchukin DG. New method for fabrication of loaded micro- and nanocontainers: emulsion encapsulation by polyelectrolyte layer-by-layer deposition on the liquid core. Langmuir 2008;24:999–1004.

[39] Kobaslija M, McQuade DT. Polyurea microcapsules from oil-in-oil emulsions via interfacial polymerization. Macromolecules 2006;39:6371–5.

[40] Shulkin A, Stover HDH. Photostimulated phase separation encapsulation. Macromolecules 2003;36:9836–9.

[41] Schmidt AM. The synthesis of magnetic core-shell nanoparticles by surface-initiatied ring-opening polymerization of epsilon-caprolactone. Macromol Rapid Commun 2005;26:93–7.

[42] Korobko AV, Jesse W, van der Maarel JRC. Encapsulation of DNA by cationic diblock copolymer vesicles. Langmuir 2005;21:34–42.

[43] Shah RK, Kim JW, Agresti JJ, Weitz DA, Chu LY. Fabrication of monodisperse thermosensitive microgels and gel capsules in microfluidic devices. Soft Matter 2008;4:2303–9.

[44] Zhao H, Chen JF, Zhao Y, Jiang L, Sun JW, Yun J. Hierarchical assembly of multilayered hollow microspheres from an amphiphilic pharmaceutical molecule of azithromycin. AdvMater 2008;20:3682–6.

[45] Khademhosseini A, May MH, Sefton MV. Conformal coating of mammalian cells immobilized onto magnetically driven beads. Tissue Eng 2005;11:1797–806.

[46] Wilson JT, Cui WX, Chaikof EL. Layer-by-layer assembly of a conformal nanothin PEG coating for intraportal islet transplantation. Nano Lett 2008;8:1940–8.
[47] Orive G, Tam SK, Pedraz JL, Halle JP. Biocompatibility of alginate–poly-L-lysine microcapsules for cell therapy. Biomaterials 2006;27:3691–700.
[48] Vermonden T, Fedorovich NE, van Geemen D, Alblas J, van Nostrum CF, Dhert WJA, et al. Photopolymerized thermosensitive hydrogels: Synthesis, degradation, and cytocompatibility. Biomacromolecules 2008;9:919–26.
[49] Trongsatitkul T, Budhlall BM. Multicore-shell PNIPAm-co-PEGMa microcapsules for cell encapsulation. Langmuir 2011;27:13468–80.
[50] Tran RT, Gyawali D, Nair P, Yang J. Biodegradable injectable systems for bone tissue engineering. In: Sharma SK, Mudhoo A, eds. A Handbook of Applied Biopolymer Technology: Synthesis, degradation and applications. Royal Society of Chemistry (London, UK); 2011.
[51] Acarregui A, Pedraz JL, Blanco FJ, Hernandez RM, Orive G. Hydrogel-based scaffolds for enclosing encapsulated therapeutic cells. Biomacromolecules 2013;14:322–30.
[52] Moshaverinia A, Ansari S, Chen C, Xu X, Akiyama K, Snead ML, et al. Co-encapsulation of anti-BMP2 monoclonal antibody and mesenchymal stem cells in alginate microspheres for bone tissue engineering. Biomaterials 2013;34:6572–9.
[53] Weir GC. Islet encapsulation: advances and obstacles. Diabetologia 2013;56:1458–61.
[54] Hunt NC, Shelton RM, Henderson DJ, Grover LM. Calcium–alginate hydrogel-encapsulated fibroblasts provide sustained release of vascular endothelial growth factor. Tissue Eng Part A 2013;19:905–14.
[55] Wang N, Adams G, Buttery L, Falcone FH, Stolnik S. Alginate encapsulation technology supports embryonic stem cells differentiation into insulin-producing cells. J Biotechnol 2009;144:304–12.
[56] Borlongan CV, Skinner SJ, Geaney M, Vasconcellos AV, Elliott RB, Emerich DF. Intracerebral transplantation of porcine choroid plexus provides structural and functional neuroprotection in a rodent model of stroke. Stroke 2004;35:2206–10.
[57] Orive G, Hernandez RM, Gascon AR, Igartua M, Pedraz JL. Survival of different cell lines in alginate–agarose microcapsules. Eur J Pharm Sci 2003;18:23–30.
[58] Sakai S, Hashimoto I, Kawakami K. Development of alginate–agarose subsieve-size capsules for subsequent modification with a polyelectrolyte complex membrane. Biochem Eng J 2006;30:76–81.
[59] Wang L, Rao RR, Stegemann JP. Delivery of mesenchymal stem cells in chitosan/collagen microbeads for orthopedic tissue repair. Cell Tiss Organs 2013;197:333–43.
[60] Wasiak I, Ciach T. Encapsulation of chondrocytes in hydrogel systems effect of chitosan viscosity and microcapsule shape. Chem Proc Eng-Inzynieria Chemiczna I Procesowa 2012;33:529–38.
[61] Perez RA, Kim M, Kim TH, Kim JH, Lee JH, Park JH, et al. Utilizing core-shell fibrous collagen–alginate hydrogel cell delivery system for bone tissue engineering. Tissue Eng Part A 2014;20:103–14.
[62] Balakrishnan B, Jayakrishnan A. Self-cross-linking biopolymers as injectable *in situ* forming biodegradable scaffolds. Biomaterials 2005;26:3941–51.
[63] Capone SH, Dufresne M, Rechel M, Fleury MJ, Salsac AV, Paullier P, et al. Impact of alginate composition: from bead mechanical properties to encapsulated HepG2/C3A cell activities for *in vivo* implantation. PLoS One 2013;8:e62032.
[64] Kim YM, Jeon YH, Jin GC, Lim JO, Baek WY. Immunoisolated chromaffin cells implanted into the subarachnoid space of rats reduce cold allodynia in a model of neuropathic pain: a novel application of microencapsulation technology. Artif Organs 2004;28:1059–66.
[65] Zhang H, Zhu SJ, Wang W, Wei YJ, Hu SS. Transplantation of microencapsulated genetically modified xenogeneic cells augments angiogenesis and improves heart function. Gene Ther 2008;15:40–8.
[66] Schuch G, Oliveira-Ferrer L, Loges S, Laack E, Bokemeyer C, Hossfeld DK, et al. Antiangiogenic treatment with endostatin inhibits progression of AML *in vivo*. Leukemia 2005;19:1312–7.
[67] Joki T, Machluf M, Atala A, Zhu J, Seyfried NT, Dunn IF, et al. Continuous release of endostatin from microencapsulated engineered cells for tumor therapy. Nat Biotechnol 2001;19:35–9.
[68] Li AA, Shen F, Zhang T, Cirone P, Potter M, Chang PL. Enhancement of myoblast microencapsulation for gene therapy. J Biomed Mater Res Part B—Appl Biomater 2006;77B:296–306.
[69] Opara EC, Mirmalek-Sani SH, Khanna O, Moya ML, Brey EM. Design of a bioartificial pancreas(+). J Investig Med 2010;58:831–7.
[70] Emerich DF, Thanos CG, Goddard M, Skinner SJ, Geany MS, Bell WJ, et al. Extensive neuroprotection by choroid plexus transplants in excitotoxin lesioned monkeys. Neurobiol Dis 2006;23:471–80.
[71] Zielinski BA, Aebischer P. Chitosan as a matrix for mammalian-cell encapsulation. Biomaterials 1994;15:1049–56.

[72] Hong Y, Song HQ, Gong YH, Mao ZW, Gao CY, Shen JC. Covalently crosslinked chitosan hydrogel: properties of *in vitro* degradation and chondrocyte encapsulation. Acta Biomater 2007;3:23–31.

[73] Yeo Y, Geng WL, Ito T, Kohane DS, Burdick JA, Radisic M. Photocrosslinkable hydrogel for myocyte cell culture and injection. J Biomed Mater Res Part B—Appl Biomater 2007;81B:312–22.

[74] Cho JH, Kim SH, Park KD, Jung MC, Yang WI, Han SW, et al. Chondrogenic differentiation of human mesenchymal stem cells using a thermosensitive poly (*N*-isopropylacrylamide) and water-soluble chitosan copolymer. Biomaterials 2004;25:5743–51.

[75] Hui TY, Cheung KMC, Cheung WL, Chan D, Chan BP. *In vitro* chondrogenic differentiation of human mesenchymal stem cells in collagen microspheres: influence of cell seeding density and collagen concentration. Biomaterials 2008;29:3201–12.

[76] Chan BP, Hui TY, Wong MY, Yip KH, Chan GC. Mesenchymal stem cell-encapsulated collagen microspheres for bone tissue engineering. Tissue Eng Part C Methods 2010;16:225–35.

[77] Burdick JA, Chung C, Jia X, Randolph MA, Langer R. Controlled degradation and mechanical behavior of photopolymerized hyaluronic acid networks. Biomacromolecules 2005;6:386–91.

[78] Chung C, Mesa J, Randolph MA, Yaremchuk M, Burdick JA. Influence of gel properties on neocartilage formation by auricular chondrocytes photoencapsulated in hyaluronic acid networks. J Biomed Mater Res A 2006;77:518–25.

[79] Flynn L, Prestwich GD, Semple JL, Woodhouse KA. Adipose tissue engineering with naturally derived scaffolds and adipose-derived stem cells. Biomaterials 2007;28:3834–42.

[80] Horn EM, Beaumont M, Shu XZ, Harvey A, Prestwich GD, Horn KM, et al. Influence of cross-linked hyaluronic acid hydrogels on neurite outgrowth and recovery from spinal cord injury. J Neurosurg Spine 2007;6:133–40.

[81] Khademhosseini A, Eng G, Yeh J, Fukuda J, Blumling 3rd J, Langer R, et al. Micromolding of photocrosslinkable hyaluronic acid for cell encapsulation and entrapment. J Biomed Mater Res Part A 2006;79:522–32.

[82] Guttinger M, Padrun V, Pralong WF, Boison D. Seizure suppression and lack of adenosine A1 receptor desensitization after focal long-term delivery of adenosine by encapsulated myoblasts. Exp Neurol 2005;193:53–64.

[83] Bryant SJ, Anseth KS. Hydrogel properties influence ECM production by chondrocytes photoencapsulated in poly(ethylene glycol) hydrogels. J Biomed Mater Res 2002;59:63–72.

[84] Rice MA, Anseth KS. Controlling cartilaginous matrix evolution in hydrogels with degradation triggered by exogenous addition of an enzyme. Tissue Eng 2007;13:683–91.

[85] Burdick JA, Anseth KS. Photoencapsulation of osteoblasts in injectable RGD-modified PEG hydrogels for bone tissue engineering. Biomaterials 2002;23: 4315–23.

[86] Nuttelman CR, Tripodi MC, Anseth KS. *In vitro* osteogenic differentiation of human mesenchymal stem cells photoencapsulated in PEG hydrogels. J Biomed Mater Res Part A 2004;68A:773–82.

[87] Masters KS, Shah DN, Leinwand LA, Anseth KS. Crosslinked hyaluronan scaffolds as a biologically active carrier for valvular interstitial cells. Biomaterials 2005;26:2517–25.

[88] Mahoney MJ, Anseth KS. Three-dimensional growth and function of neural tissue in degradable polyethylene glycol hydrogels. Biomaterials 2006;27:2265–74.

[89] Mokry J, Karbanova J, Lukas J, Paleckova V, Dvorankova B. Biocompatibility of HEMA copolymers designed for treatment of CNS diseases with polymer-encapsulated cells. Biotechnol Prog 2000;16:897–904.

[90] Nojehdehian H, Moztarzadeh F, Baharvand H, Mehrjerdi NZ, Nazarian H, Tahriri M. Effect of poly-L-lysine coating on retinoic acid-loaded PLGA microspheres in the differentiation of carcinoma stem cells into neural cells. Int J Artif Organs 2010;33:721–30.

[91] Qi M, Gu Y, Sakata N, Kim D, Shirouzu Y, Yamamoto C, et al. PVA hydrogel sheet macroencapsulation for the bioartificial pancreas. Biomaterials 2004;25:5885–92.

[92] Nicodemus GD, Bryant SJ. Cell encapsulation in biodegradable hydrogels for tissue engineering applications. Tissue Eng Part B-Rev 2008;14:149–65.

[93] Remminghorst U, Rehm BHA. Bacterial alginates: from biosynthesis to applications. Biotechnol Lett 2006;28:1701–12.

[94] Rehm BHA, Valla S. Bacterial alginates: biosynthesis and applications. Appl Microbiol Biotechnol 1997;48:281–8.

[95] Zhang WJ, He XM. Microencapsulating and banking living cells for cell-based nedicine. J Healthcare Eng 2011;2:427–46.

[96] Santos E, Zarate J, Orive G, Hernandez RM, Pedraz JL. Biomaterials in cell microencapsulation. Adv Exp Med Biol 2010;670:5–21.

[97] Rowley JA, Madlambayan G, Mooney DJ. Alginate hydrogels as synthetic extracellular matrix materials. Biomaterials 1999;20:45–53.

[98] Schmidt JJ, Jeong J, Kong H. The interplay between cell adhesion cues and curvature of cell adherent alginate microgels in multipotent stem cell culture. Tissue Eng Part A 2011;17:2687–94.

[99] Liu Y, Sakai S, Taya M. Impact of the composition of alginate and gelatin derivatives in bioconjugated hydrogels on the fabrication of cell sheets and spherical tissues with living cell sheaths. Acta Biomater 2013;9:6616–23.

[100] Smith AM, Harris JJ, Shelton RM, Perrie Y. 3D culture of bone-derived cells immobilised in alginate following light-triggered gelation. J Control Rel 2007;119:94–101.

[101] Du J, Hsieh YL. PEGylation of chitosan for improved solubility and fiber formation via electrospinning. Cellulose 2007;14:543–52.

[102] Chenite A, Chaput C, Wang D, Combes C, Buschmann MD, Hoemann CD, et al. Novel injectable neutral solutions of chitosan form biodegradable gels *in situ*. Biomaterials 2000;21:2155–61.

[103] Monteiro OAC, Airoldi C. Some studies of crosslinking chitosan–glutaraldehyde interaction in a homogeneous system. Int J Biol Macromolecules 1999;26:119–28.

[104] Varum KM, Myhr MM, Hjerde RJN, Smidsrod O. *In vitro* degradation rates of partially *N*-acetylated chitosans in human serum. Carbohydr Res 1997;299: 99–101.

[105] Li AD, Sun ZZ, Zhou M, Xu XX, Ma JY, Zheng W, et al. Electrospun chitosan-graft-PLGA nanofibres with significantly enhanced hydrophilicity and improved mechanical property. Colloids Surf B Biointerfaces 2013;102:674–81.

[106] Gaserod O, Sannes A, Skjak-Braek G. Microcapsules of alginate–chitosan. II. A study of capsule stability and permeability. Biomaterials 1999;20:773–83.

[107] Crompton KE, Goud JD, Bellamkonda RV, Gengenbach TR, Finkelstein DI, Horne MK, et al. Polylysine-functionalised thermoresponsive chitosan hydrogel for neural tissue engineering. Biomaterials 2007;28:441–9.

[108] Chandy T, Mooradian DL, Rao GH. Evaluation of modified alginate–chitosan–polyethylene glycol microcapsules for cell encapsulation. Artif Organs 1999;23:894–903.

[109] Haque T, Chen H, Ouyang W, Martoni C, Lawuyi B, Urbanska A, et al. Investigation of a new microcapsule membrane combining alginate, chitosan, polyethylene glycol and poly-L-lysine for cell transplantation applications. Int J Artif Organs 2005;28:631–7.

[110] Baruch L, Machluf M. Alginate–chitosan complex coacervation for cell encapsulation: effect on mechanical properties and on long-term viability. Biopolymers 2006;82:570–9.

[111] Marsich E, Borgogna M, Donati I, Mozetic P, Strand BL, Salvador SG, et al. Alginate/lactose-modified chitosan hydrogels: a bioactive biomaterial for chondrocyte encapsulation. J Biomed Mater Res Part A 2008;84:364–76.

[112] Paul A, Shum-Tim D, Prakash S. Investigation on PEG integrated alginate–chitosan microcapsules for myocardial therapy using marrow stem cells genetically modified by recombinant baculovirus. Cardiovasc Engin Technol 2010;1:154–64.

[113] Morra M. Engineering of biomaterials surfaces by hyaluronan. Biomacromolecules 2005;6:1205–23.

[114] McCall JD, Anseth KS. Thiol-ene photopolymerizations provide a facile method to encapsulate proteins and maintain their bioactivity. Biomacromolecules 2012;13:2410–7.

[115] Seidlits SK, Khaing ZZ, Petersen RR, Nickels JD, Vanscoy JE, Shear JB, et al. The effects of hyaluronic acid hydrogels with tunable mechanical properties on neural progenitor cell differentiation. Biomaterials 2010;31:3930–40.

[116] Desai N, Abdelhafez F, Calabro A, Falcone T. Three dimensional culture of fresh and vitrified mouse preantral follicles in a hyaluronan-based hydrogel: a preliminary investigation of a novel biomaterial for *in vitro* follicle maturation. Reprod Biol Endocrinol 2012;10:29–40.

[117] Peppas NA, Hilt JZ, Khademhosseini A, Langer R. Hydrogels in biology and medicine: from molecular principles to bionanotechnology. Adv Mater 2006;18:1345–60.

[118] Bae KH, Yoon JJ, Park TG. Fabrication of hyaluronic acid hydrogel beads for cell encapsulation. Biotechnol Prog 2006;22:297–302.

[119] Smethurst PA, Onley DJ, Jarvis GE, O'Connor MN, Knight CG, Herr AB, et al. Structural basis for the platelet–collagen interaction: the smallest motif within collagen that recognizes and activates platelet glycoprotein VI contains two glycine–proline–hydroxyproline triplets. J Biol Chem 2007;282:1296–304.

[120] Parenteau-Bareil R, Gauvin R, Berthod F. Collagen-based biomaterials for tissue engineering applications. Materials 2010;3:1863–87.

[121] Wu X, Black L, Santacana-Laffitte G, Patrick Jr. CW. Preparation and assessment of glutaraldehyde-crosslinked collagen–chitosan hydrogels for adipose tissue engineering. J Biomed Mater Res Part A 2007;81:59–65.

[122] Weadock KS, Miller EJ, Bellincampi LD, Zawadsky JP, Dunn MG. Physical crosslinking of collagen fibers: comparison of ultraviolet irradiation and dehydrothermal treatment. J Biomed Mater Res 1995;29:1373–9.

[123] Garcia Y, Hemantkumar N, Collighan R, Griffin M, Rodriguez-Cabello JC, Pandit A. *In vitro*

characterization of a collagen scaffold enzymatically cross-linked with a tailored elastin-like polymer. Tissue Eng Part A 2009;15:887–99.

[124] Gosal WS, Clark AH, Pudney PDA, Ross-Murphy SB. Novel amyloid fibrillar networks derived from a globular protein: beta-lactoglobulin. Langmuir 2002;18:7174–81.

[125] Gosal WS, Clark AH, Ross-Murphy SB. Fibrillar β-lactoglobulin gels: Part 1. Fibril formation and structure. Biomacromolecules 2004;5:2408–19.

[126] Kisiday J, Jin M, Kurz B, Hung H, Semino C, Zhang S, et al. Self-assembling peptide hydrogel fosters chondrocyte extracellular matrix production and cell division: implications for cartilage tissue repair. Proc Natl Acad Sci USA 2002;99:9996–10001.

[127] Haines-Butterick L, Rajagopal K, Branco M, Salick D, Rughani R, Pilarz M, et al. Controlling hydrogelation kinetics by peptide design for three-dimensional encapsulation and injectable delivery of cells. Proc Natl Acad Sci USA 2007;104:7791–6.

[128] Betre H, Ong SR, Guilak F, Chilkoti A, Fermor B, Setton LA. Chondrocytic differentiation of human adipose-derived adult stem cells in elastin-like polypeptide. Biomaterials 2006;27:91–9.

[129] Sargeant TD, Guler MO, Oppenheimer SM, Mata A, Satcher RL, Dunand DC, et al. Hybrid bone implants: self-assembly of peptide amphiphile nanofibers within porous titanium. Biomaterials 2008;29:161–71.

[130] Wang C, Varshney RR, Wang DA. Therapeutic cell delivery and fate control in hydrogels and hydrogel hybrids. Adv Drug Del Rev 2010;62:699–710.

[131] Chan G, Mooney DJ. New materials for tissue engineering: towards greater control over the biological response. Trend Biotechnol 2008;26:382–92.

[132] Lee KY, Mooney DJ. Hydrogels for tissue engineering. Chem Rev 2001;101:1869–79.

[133] Nuttelman CR, Rice MA, Rydholm AE, Salinas CN, Shah DN, Anseth KS. Macromolecular monomers for the synthesis of hydrogel niches and their application in cell encapsulation and tissue engineering. Prog Polymer Sci 2008;33:167–79.

[134] Bryant SJ, Durand KL, Anseth KS. Manipulations in hydrogel chemistry control photoencapsulated chondrocyte behavior and their extracellular matrix production. J Biomed Mater Res Part A 2003;67A:1430–6.

[135] Benoit DSW, Durney AR, Anseth KS. Manipulations in hydrogel degradation behavior enhance osteoblast function and mineralized tissue formation. Tiss Eng 2006;12:1663–73.

[136] Ruoslahti E. RGD and other recognition sequences for integrins. Annu Rev Cell Dev Biol 1996;12: 697–715.

[137] Cai SS, Liu YC, Shu XZ, Prestwich GD. Injectable glycosaminoglycan hydrogels for controlled release of human basic fibroblast growth factor. Biomaterials 2005;26:6054–67.

[138] Yu L, Ding JD. Injectable hydrogels as unique biomedical materials. Chem Soc Rev 2008;37:1473–81.

[139] Metters AT, Bowman CN, Anseth KS. A statistical kinetic model for the bulk degradation of PLA-b–PEG-b–PLA hydrogel networks. J Phys Chem B 2000;104:7043–9.

[140] Martens PJ, Bryant SJ, Anseth KS. Tailoring the degradation of hydrogels formed from multivinyl poly (ethylene glycol) and poly(vinyl alcohol) macromers for cartilage tissue engineering. Biomacromolecules 2003;4:283–92.

[141] Rice MA, Sanchez-Adams J, Anseth KS. Exogenously triggered, enzymatic degradation of photopolymerized hydrogels with polycaprolactone subunits: experimental observation and modeling of mass loss behavior. Biomacromolecules 2006;7:1968–75.

[142] Soon-Shiong P, Heintz RE, Merideth N, Yao QX, Yao Z, Zheng T, et al. Insulin independence in a type 1 diabetic patient after encapsulated islet transplantation. Lancet 1994;343:950–1.

[143] Hasse C, Klock G, Schlosser A, Zimmermann U, Rothmund M. Parathyroid allotransplantation without immunosuppression. Lancet 1997;350:1296–7.

[144] Elliott RB, Escobar L, Tan PL, Muzina M, Zwain S, Buchanan C. Live encapsulated porcine islets from a type 1 diabetic patient 9.5 yr after xenotransplantation. Xenotransplantation 2007;14:157–61.

[145] Wang YY, Liu PF, Du J, Sun Y, Li FH, Duan YR. Targeted siRNA delivery by anti-HER2 antibody-modified nanoparticles of mPEG-chitosan diblock copolymer. J Biomater Sci—Polymer Ed 2013;24:1219–32.

[146] Watts RL, Raiser CD, Stover NP, Cornfeldt ML, Schweikert AW, Allen RC, et al. Stereotaxic intrastriatal implantation of human retinal pigment epithelial (hRPE) cells attached to gelatin microcarriers: a potential new cell therapy for Parkinson's disease. J Neural Transm Suppl 2003;:215–27.

[147] Bloch J, Bachoud-Levi AC, Deglon N, Lefaucheur JP, Winkel L, Palfi S, et al. Neuroprotective gene therapy for Huntington's disease, using polymer-encapsulated cells engineered to secrete human ciliary neurotrophic factor: results of a phase I study. Hum Gene Ther 2004;15:968–75.

[148] Aebischer P, Schluep M, Deglon N, Joseph JM, Hirt L, Heyd B, et al. Intrathecal delivery of CNTF using encapsulated genetically modified xenogeneic cells in amyotrophic lateral sclerosis patients. Nat Med 1996;2:696–9.

[149] Lim F, Sun AM. Microencapsulated islets in diabetic rats. Science 1981;213:1146.
[150] Luca G, Fallarino F, Calvitti M, Mancuso F, Nastruzzi C, Arato I, et al. Xenograft of microencapsulated Sertoli cells reverses T1DM in NOD mice by inducing neogenesis of beta cells. Transplantation 2010;90:1352–7.
[151] Gaba RC, Garcia-Roca R, Oberholzer J. Pancreatic islet cell transplantation: an update for interventional radiologists. J Vasc Interv Radiol 2012;23:583–94.
[152] Abalovich AG, Bacque MC, Grana D, Milei J. Pig pancreatic islet transplantation into spontaneously diabetic dogs. Transplant Proceed 2009;41:328–30.
[153] Ngoc PK, Phuc PV, Nhung TH, Thuy DT, Nguyen TMN. Improving the efficacy of type 1 diabetes therapy by transplantation of immunoisolated insulin-producing cells. Hum Cell 2011;24:86–95.
[154] Godfrey KJ, Mathew B, Bulman JC, Shah O, Clement S, Gallicano GI. Stem cell-based treatments for Type 1 diabetes mellitus: bone marrow, embryonic, hepatic, pancreatic and induced pluripotent stem cells. Diabetic Med 2012;29:14–23.
[155] Tuch BE, Hughes TC, Evans MDM. Encapsulated pancreatic progenitors derived from human embryonic stem cells as a therapy for insulin-dependent diabetes. Diabetes Metab Res Rev 2011;27:928–32.
[156] Basta G, Montanucci P, Luca G, Boselli C, Noya G, Barbaro B, et al. Long-term metabolic and immunological follow-up of nonimmunosuppressed patients with type 1 diabetes treated with microencapsulated islet allografts: four cases. Diabetes Care 2011;34:2406–9.
[157] Korsgren O, Nilsson B. Improving islet transplantation: a road map for a widespread application for the cure of persons with type I diabetes. Curr Opin Organ Transplant 2009;14:683–7.
[158] Kiessling FM. Science to practice: are theranostic agents with encapsulated cells the key for diabetes therapy? Radiology 2011;260:613–5.
[159] Wong A, Hwa A, Hellman D, Greenstein J. Surrogate insulin-producing cells. F1000 Med Rep 2012;4.
[160] Read TA, Sorensen DR, Mahesparan R, Enger PO, Timpl R, Olsen BR, et al. Local endostatin treatment of gliomas administered by microencapsulated producer cells. Nat Biotechnol 2001;19:29–34.
[161] Li AA, Shen F, Zhang T, Cirone P, Potter M, Chang PL. Enhancement of myoblast microencapsulation for gene therapy. J Biomed Mater Res B Appl Biomater 2006;77:296–306.
[162] Read TA, Farhadi M, Bjerkvig R, Olsen BR, Rokstad AM, Huszthy PC, et al. Intravital microscopy reveals novel antivascular and antitumor effects of endostatin delivered locally by alginate-encapsulated cells. Cancer Res 2001;61:6830–7.
[163] Hao SG, Su LP, Gu XL, Moyana T, Xiang J. A novel approach to tumor suppression using microencapsulated engineered J558/TNF-α cells. Exp Oncol 2005;27:56–60.
[164] Moran DM, Koniaris LG, Jablonski EM, Cahill PA, Halberstadt CR, McKillop IH. Microencapsulation of engineered cells to deliver sustained high circulating levels of interleukin-6 to study hepatocellular carcinoma progression. Cell Transplant 2006;15:785–98.
[165] Both GW. Recent progress in gene-directed enzyme prodrug therapy: an emerging cancer treatment. Curr Opin Mol Ther 2009;11:421–32.
[166] Karle P, Muller P, Renz R, Jesnowski R, Saller R, von Rombs K, et al. Intratumoral injection of encapsulated cells producing an oxazaphosphorine activating cytochrome P450 for targeted chemotherapy. Gene Ther Cancer 1998;451:97–106.
[167] Lohr M, Hoffmeyer A, Kroger JC, Freund M, Hain J, Holle A, et al. Microencapsulated cell-mediated treatment of inoperable pancreatic carcinoma. Lancet 2001;357:1591–2.
[168] Cirone P, Bourgeois JM, Austin RC, Chang PL. A novel approach to tumor suppression with microencapsulated recombinant cells. Hum Gene Ther 2002;13:1157–66.
[169] Goren A, Dahan N, Goren E, Baruch L, Machluf M. Encapsulated human mesenchymal stem cells: a unique hypoimmunogenic platform for long-term cellular therapy. FASEB J 2010;24:22–31.
[170] Spuch C, Antequera D, Portero A, Orive G, Hernández RM, Molina JA, et al. The effect of encapsulated VEGF-secreting cells on brain amyloid load and behavioral impairment in a mouse model of Alzheimer's disease. Biomaterials 2010;31:5608–18.
[171] Tresco PA, Winn SR, Aebischer P. Polymer encapsulated neurotransmitter secreting cells. Potential treatment for Parkinson's disease. ASAIO J 1992;38:17–23.
[172] Tresco PA, Winn SR, Tan S, Jaeger CB, Greene LA, Aebischer P. Polymer-encapsulated PC12 cells: long-term survival and associated reduction in lesion-induced rotational behavior. Cell Transplant 1992;1:255–64.
[173] Lindner M, Plone MA, Frydel B, Kaplan FA, Krueger PM, Bell WJ, et al. Intraventricular encapsulated calf adrenal chromaffin cells: viable for at least 500 days *in vivo* without detectable adverse effects on behavioral/cognitive function or host immune sensitization in rats. Restor Neurol Neurosci 1997;11:21–35.
[174] Sajadi A, Bensadoun JC, Schneider BL, Lo Bianco C, Aebischer P. Transient striatal delivery of GDNF via

encapsulated cells leads to sustained behavioral improvement in a bilateral model of Parkinson disease. Neurobiol Dis 2006;22:119–29.

[175] Grandoso L, Ponce S, Manuel I, Arrue A, Ruiz-Ortega JA, Ulibarri I, et al. Long-term survival of encapsulated GDNF secreting cells implanted within the striatum of parkinsonized rats. Int J Pharm 2007;343:69–78.

[176] Yasuhara T, Shingo T, Kobayashi K, Takeuchi A, Yano A, Muraoka K, et al. Neuroprotective effects of vascular endothelial growth factor (VEGF) upon dopaminergic neurons in a rat model of Parkinson's disease. EurJ Neurosci 2004;19:1494–504.

[177] Skinner SJ, Lin H, Geaney MS, Gorba T, Elliott RB, Tan PL. Restoration of motor control and dopaminergic activity in rats with unilateral 6-hydroxydopamine lesions. Regen Med 2011;6:319–26.

[178] Winn SR, Hammang JP, Emerich DF, Lee A, Palmiter RD, Baetge EE. Polymer-encapsulated cells genetically-modified to secrete human nerve growth-factor promote the survival of axotomized septal cholinergic neurons. Proc Natl Acad Sci USA 1994;91:2324–8.

[179] Nagahara AH, Bernot T, Moseanko R, Moseanko R, Brignolo L, Blesch A, et al. Long-term reversal of cholinergic neuronal decline in aged non-human primates by lentiviral NGF gene delivery. Exp Neurol 2009;215:153–9.

[180] Garcia P, Youssef I, Utvik JK, Florent-Bechard S, Barthelemy V, Malaplate-Armand C, et al. Ciliary neurotrophic factor cell-based delivery prevents synaptic impairment and improves memory in mouse models of Alzheimer's disease. J Neurosci 2010;30:7516–27.

[181] Klinge PM, Harmening K, Miller MC, Heile A, Wallrapp C, Geigle P, et al. Encapsulated native and glucagon-like peptide-1 transfected human mesenchymal stem cells in a transgenic mouse model of Alzheimer's disease. Neurosci Lett 2011;497:6–10.

[182] Wahlberg LU, Lind G, Almqvist PM, Kusk P, Tornøe J, Juliusson B, et al. Targeted delivery of nerve growth factor via encapsulated cell biodelivery in Alzheimer disease: a technology platform for restorative neurosurgery. J Neurosurg 2012;117:340–7.

[183] Mai G, Huy NT, Morel P, Mei J, Andres A, Bosco D, et al. Treatment of fulminant liver failure by transplantation of microencapsulated primary or immortalized xenogeneic hepatocytes. Xenotransplantation 2005;12:457–64.

[184] Liu ZC, Chang TMS. Effects of bone marrow cells on hepatocytes: when co-cultured or co-encapsulated together. Artif Cells Blood Substit Immobil Biotechnol 2000;28:365–74.

[185] Liu ZC, Chang TMS. Transdifferentiation of bioencapsulated bone marrow cells into hepatocyte-like cells in the 90% hepatectomized rat model. Liver Transplant 2006;12:566–72.

CHAPTER

11

Cell Engineering with Nanoparticles for Cell Imaging

David C. Yeo, Riddhima Minocha and Chenjie Xu

Division of Bioengineering, School of Chemical and Biomedical Engineering, Nanyang Technological University, Singapore

OUTLINE

11.1 INTRODUCTION

The cell is generally considered as the basic building block of life. Understanding its structure, spatial location, and function is critical in revealing biological and physiological phenomena such as growth, proliferation, regeneration, cancer, and death. Since the first observations made by Van Leeuwenhoek and Hooke in the seventeenth century, advances in cellular biology have become closely intertwined with that of imaging technology [1]. Whereas compound microscopes were state of the art 300 years ago, the field of imaging has vastly expanded to encompass a wide range of imaging modalities such as fluorescence, luminescence, Raman, magnetic resonance, radionuclide, (photo)-acoustic, and X-ray based imaging [2]. This multitude of imaging techniques, with their specific advantages

J.M. Karp & W. Zhao (Eds): Micro- and Nanoengineering of the Cell Surface.
DOI: http://dx.doi.org/10.1016/B978-1-45-573146-6.00011-8

and limitations, have since been utilized in biology and medicine to study both biological and physiological phenomena from singular cellular units to whole organisms.

To be visualized and distinguished from the background by an imaging modality, cells of interest have to generate a contrast signal. Although some cells may generate endogenous contrast agents (e.g., the metabolic cofactor, nicotinamide adenine dinucleotide or NADH) [3–5], such biophysical phenomena is comparatively uncommon. Instead, a more reliable method is to engineer cells to produce detectable signals. Currently, the most widely used and reproducible approach is centered upon genetic engineering of cells with fluorescent proteins [6,7]. Proximity-based imaging probes (e.g., Förster resonance energy transfer, FRET) which rely on donor and acceptor fluorescent proteins are yet another technique commonly adopted to decipher cell signaling dynamics [8–10]. The major limitation of these methods is the usage of genomic modification to express fluorescent proteins within cells. Not only are gene transfection techniques considerably inefficient (e.g., 5–10% transfection rate for human embryonic stem cells [11]), they are also unsuitable for clinical usage since cells with random mutation sites might be detrimental to patient health [12].

An alternative way to engineering cells for imaging is to utilize nanoparticles (NPs) that exhibit unique optical, magnetic, or electronic properties due to their basic structural units, grains, particle fibers or other constituent components <100 nm in one or more dimensions. Unlike genetic engineering, NP-based cell engineering does not require the alteration of cellular genotype. Following 20 years of investigation, cells of interest can be engineered with NPs through one of the following strategies: intracellular internalization [13,14], functionalization of the plasma membrane [15,16], and the modification of extracellular matrix (ECM) [14,17–19]. The most common modification strategy, intracellular particle internalization has already been adopted in the clinic (e.g., the labeling and tracking of dendritic cells) [20,21], this preference may be attributed to their ease in usage and high labeling efficiency (>97% cells are labeled using NPs with cationic surface modification [22,23]).

This chapter starts with a brief overview of major imaging modalities for cell imaging and their most relevant NP-based contrast agents. Next, the focus switches to describe relevant strategies to engineer cells with NPs, namely: intracellular internalization of NPs, NP functionalization of the plasma membrane, and ECM NP immobilization. A schematic of cell engineering methods as well as detection methods discussed in the book chapter is provided in Figure 11.1. Their usage, especially for cell tracking *in vivo* is highlighted. Finally, the challenges during the translation from bench to bedside using NP engineering will be discussed.

11.2 IMAGING MODALITIES FOR NP ENGINEERED CELLS

The most common imaging modalities for cell imaging in laboratory are fluorescence imaging, Raman imaging, and photoacoustic imaging. While these give high imaging resolution, they are less amenable for imaging in deeper tissue locations. Instead, modalities like magnetic resonance and radionuclide contrast can be applied for whole body clinical applications.

Fluorescence imaging modalities covered in this chapter include epifluorescence, confocal, and multiphoton microscopy. Commonly used NPs for fluorescence imaging encompass quantum dots (QDs), upconversion NPs, nanodiamonds, silica NPs, graphene QDs, and fluorescent-labeled NPs [24]. QDs are semiconductor NPs typically comprised of materials such as cadmium selenide (CdSe) and germanium. Upconversion is a process where sequential absorption of two or more photons

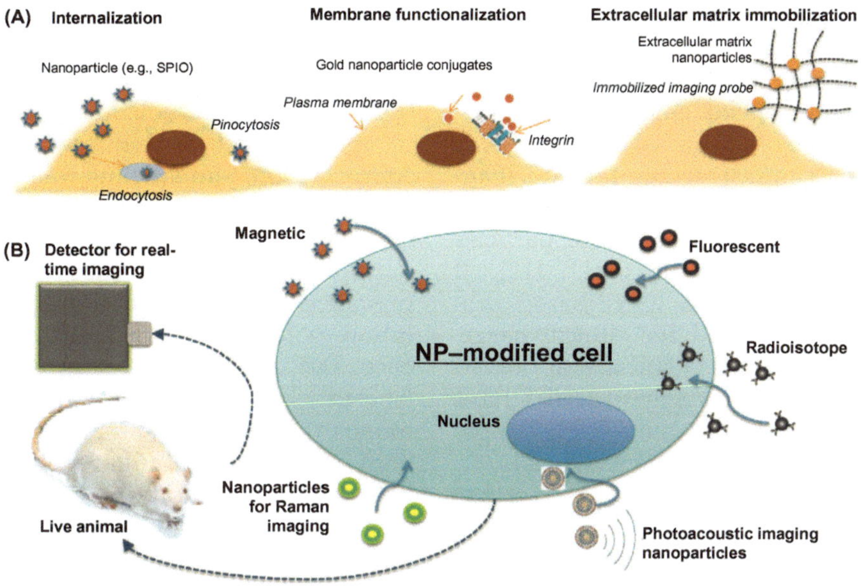

FIGURE 11.1 Engineering cells using NPs for cell imaging: (A) three methods to engineer cells with NPs: intracellular internalization, membrane functionalization, and ECM immobilization, and (B) NP engineered cells can be imaged with magnetic resonance imaging (MRI), fluorescent imaging, radioisotope-aided imaging, photoacoustic imaging, and Raman imaging.

leads to light emission at shorter wavelengths—anti-Stokes nonlinear emission behavior, aiding higher sensitivity and penetration depths. Nanodiamonds are biologically inert with low exocytosis activity and possess higher fluorescence lifetimes compared to organic fluorophores [14]. Silica, an extremely versatile material for NP fabrication, has been used as biocompatible shells for reducing cytotoxicity of many substances, such as QDs [24].

Raman imaging is based on red-shifted (Stokes) and blue-shifted (anti-Stokes) light scattering. Its advantages lie in its specificity (narrow peaks), allowing Raman signals to be easily differentiated from autofluorescence. On the other hand, Raman signals are considerably weaker, 12–14 orders of magnitude weaker than fluorescent signals. The introduction of single-walled carbon nanotubes (SWNTs), graphene, gold (Au) NPs, and silver NPs significantly enhances Raman signals [24].

Photoacoustic detection is based on optical excitation and ultrasonic detection, typically using pulsed nanosecond laser excitation and acoustic impulse response detection. Unlike other optical imaging methods, photoacoustic can achieve up to 7 cm imaging depth due to the low acoustic scattering in tissue (approximately 10^3 fold lower than optical scattering) [25]. Of NPs that enable photoacoustic detection: dielectric core Au-coated nano shells, Au nanorods, and Au nanocages have been commonly applied. Silicate matrices also afford a greater range of flexibility in particle fabrication by incorporating sensor dyes, imaging contrast agents, drugs, and cell targeting ligands [26].

MRI is the most popular modality for cellular imaging in large animals and humans due to its superior temporal and spatial contrast, absence of harmful ionizing radiation, and unlimited penetration depth. Magnetic contrast can be generated through the exchange of water protons with gadolinium (Gd^{3+}) resulting in T_1—nuclear spin–lattice relaxation time, also known as "positive contrast." On the other hand, paramagnetic agents generate T_2—spin–spin relaxation time or "negative contrast." NPs used in MRI as contrast agents include magnetic NPs (such as iron oxide NPs (IONPs)), NPs containing Gd^{3+}, and micro-sized particles containing magnetic NPs or Gd^{3+} [12]. It should be noted that seven IONP-based products have been approved for clinical usage. Despite this success, some clinically approved cell labeling products (e.g., Feridex IV) have ceased operations, impeding the reliable supply of clinical grade cell imaging agents [20,27].

11.3 STRATEGIES FOR CELL ENGINEERING WITH NPs

11.3.1 Intracellular Internalization of NPs

Cellular internalization of NPs is an endocytosis process, which is closely related with several parameters such as (i) size, (ii) surface coating, and (iii) labeling condition of particles. The usage of different particle sizes for cell labeling can affect the manner in which cells are internalized. Cellular mechanisms responsible for internationalization of micro/nanoparticles include: (i) phagocytosis—"cell eating" (particles in the range of 50 nm–4 μm [28]); (ii) macropinocytosis (particles are bigger than 200 nm [29]); and (iii) clathrin/caveolae pit endocytosis (particles are smaller than 200 nm [29])—"cell drinking" [30]. To ensure high intracellular loading capacity, NPs typically are 200 nm in size for better cellular retention [31]. Furthermore, shape is another consideration in particle design with macrophages favoring ellipsoidal shapes over spherical ones during attachment and uptake. The endocytosis capacity of cells can be further enhanced by incorporating ligand targeting/binding modifications on cell membrane receptors, as well as cationic polymers such as poly-L-lysine [22,23,31]. Other parameters to optimize cell internalization include varying parameters such as NP concentration, serum concentration, and labeling time [32].

Upon NP internalization, cells can be visualized through relevant imaging modalities. For example, using fluorescence imaging (the most prevalent imaging modality due to high resolution (single cell imaging resolution), low cost, and availability), lung stem cells were labeled with 100 nm fluorescent nanodiamonds (FNDs) to monitor their engraftment in both injured and uninjured lung alveoli following transplantation [14]. Promisingly, labeled cells were still detected 15 days after labeling. Nanoflares (Au NPs functionalized with a dense layer of fluorescent emitting aptamers) have also been utilized in live cell applications for studying cell signaling pathways [33,34]. These nanostructures consist of a gold NP functionalized with a high concentration of oligonucleotides ($\sim$8.4 pmol/cm^2) hybridized to fluorophores. These aptamer nanoflares were developed to detect physiological changes in ATP levels and readily entered cells. The detected fluorescent signal levels correlated closely with changes of the intracellular ATP levels [33].

Similar to fluorescence imaging, the burgeoning popularity of photoacoustic and Raman imaging modalities can also be improved through NP internalization [24]. For example, SWNTs have been conjugated with protamine (SWNT–PEG–PRO) to enable efficient cellular uptake and detection using Raman imaging [35]. Crucially, the addition of arginine-enriched protamine enhanced the uptake of SWNTs in stem cells. Subsequent incorporation of metallic catalysts facilitated enhanced magnetic and photoacoustic contrast [35]. This versatility of SWNTs as labeling agents in stem cells was further exploited by Gambhir et al., fabricating optical dye-doped

SWNTs that enhanced photoacoustic contrast 100-fold. This enhancement improved the sensitivity during the detection of $\alpha_V\beta_3$ positive tumors in mouse models [36].

Being the most popular imaging modality for large animals and humans [37], MRI has received plenty of interest for the design and application of contrast agents. To facilitate the potential translation from bench to bedside, most of the current cell labeling uses iron oxide-based magnetic NPs ranging from ultrasmall (35 nm) to micron dimensions. The surface of those NPs is often functionalized with targeting ligands (e.g., antibody [38,39], polycationic polymer [40,41], viral capsid [42–44]) for improved uptake. One early study used superparamagnetic iron oxide (SPIO) NPs to label dendritic cells *ex vivo*. This strategy relied upon the ability of dendritic cells to endocytose SPIO NPs. This modality was as sensitive as radionuclide ^{111}In agents with a greater tracking accuracy in a melanoma patient [45]. In another recent report, three FDA approved substances (heparin, protamine, and ferumoxytol—carbohydrate-coated SPIO) were combined to form self-assembled NPs ($\sim$200 nm) that act as clinically relevant cell tracers. The individual particle constituents were incubated within serum-free media and allowed to self-assemble and label the respective cell types. These contrast agents were considerably inert and did not affect functional behavior (viability, proliferation, freeze thaw, migration, and differentiation) to any significant degree. Crucially, as few as 10^3 cells could be detected by T_2 weighted MRI, when implanted into a rat brain [46].

Although IONPs are more commonly studied, it should be brought to the reader's attention that non-IONPs like gadolinium oxide NPs are generating great interest as labeling agents for MRI imaging. Although SPIOs primarily generate T_2 contrast, they induce hypo-T_1 intensity that can be misinterpreted as blood clots or hemorrhages. On the other hand, Gd^{3+} paramagnetic contrast agents typically generate increased T_1 signal intensity through the exchange of water protons with the Gd^{3+} containing complex which reduces this confusion. Recent advances in the implementation of T_1 contrast agents include the generation of a gadolinium hexanedione NPs with greater than three times accumulation and greater than or equal to twofold contrast in human mesenchymal stem cells (MSCs) compared to commercial gadolinium labeling agents. In addition, by conjugating folate receptor targeting ligands, the *in vitro* and *in vivo* detection of tumor cells can be achieved using T_1 contrast imaging [47]. NP internalization methods to enable cell imaging are widely used and will likely be mainstays for cell imaging in the coming years.

11.3.2 Cell Membrane Functionalization with NPs

The second strategy for engineering cells with NPs is to label the cell plasma membrane with NPs. The cell surface contains a multitude of biological moieties that can be utilized as recognition sites for modification. In one recent case, mouse embryonic stem cells (mESCs) were engineered by labeling the plasma membrane with poly-lactic-glycolic acid (PLGA) NPs conjugated with an antibody for the oligosaccharide antigen stage-specific embryonic antigen 1 (SSEA-1). PLGA NPs eluted leukemia inhibitory factor (LIF) cytokine to support long-term embryonic stem cell renewal [16]. Karp et al. also utilized the covalent attachment of fluorescent aptamer (fApt) sensors on the MSC surface for the recognition of platelet-derived growth factor (PDGF) *in vivo*. Unlike other imaging agents that emit a passive signal, this novel "cell surface" sensor facilitated the monitoring of intercellular signaling and increased expression when homing to PDGF rich sites [15]. Based on similar principles, fApts immobilized on glia cell surfaces were able to monitor extracellular adenine (Ade)—a crucial gliotransmitter. Remarkably, the choice of fApt was able to achieve rapid (10 s) response time in sensing proximal Ade transmission dynamics, a significant improvement over existing techniques, albeit not at the order of milliseconds demonstrated in physiological neurotransmission [48].

A surface energy transfer (SET) "nanoruler" principle was used to measure the distance of cell membrane receptor protein tyrosine kinase (PTK7) receptor and their ligand binding sites. This system comprised of an antibody-coupled fluorophore and an aptamer−Au NP conjugate. By employing varying sizes of Au NPs, different signal emissions were recorded, providing information of the distance between the antibody binding site and the aptamer binding sites [49]. Using similar NP engineering principles, cell surface features can be measured using Au NP-labeled surface receptors. To detect direct interactions of surface receptors on live cells, 40 nm Au NPs were conjugated using fibronectin to integrins expressed on the plasma membrane of HeLa cells. The authors suggested that plasmon coupling between individual Au NPs could be used to examine dynamic or transient interactions below the diffraction limit [50].

11.3.3 ECM Modifications

The ECM is another location with the potential for immobilizing imaging probes. An advantage of this strategy compared to internalization and plasma membrane methods is the retention of imaging probes, thus reducing the loss of probe density through cell division and exocytosis [17]. A highly successful approach utilizing the ECM to sequester probes is the co-encapsulation of L-arginine liposomes that exhibit chemical exchange saturation transfer (CEST) contrast with hepatocytes to function as a bioartificial liver (BAL) [18]. The numerous N−H bonds on the L-arginine structure facilitate exchange with secreted acidic species to detect pH changes noninvasively through MRI. This is a highly promising cell tracking method with clinical translational potential. A similar optical-based pH NP sensor strategy has been developed to be compatible for sequestration in agarose dishes, potentially useful as cheap, robust, and contactless bacteria sensors for the food industry [19]. Imaging through ECM immobilization has also been carried out to detect matrix metalloproteinase activity in two disease models—tumor xenografts and an inflammation model. This was achieved through the conjugation of the dabcyl luciferase quencher and luciferase to a GPLGVRGC peptide (a MMP-2/9 substrate) immobilized on collagen *in vivo* [17]. The addition of an inert ^{64}Cu isotope normalizes sensor concentration, making relevant to image disease progression in real time.

11.4 CHALLENGES AND OUTLOOK

Despite the promise demonstrated by the above techniques, existing limitations still remain. The authors envision the following as the most pressing of challenges: cytotoxicity of NPs; decreased signal strength caused by NP exocytosis and dilution during cell division; and the inability to report cellular function following transplantation [17].

Ideally, NPs for cell engineering should have either minimal or no influence over the original properties of the implanted cell. Evaluating cell function consists of evaluating viability, proliferation, "stemness," cytokine secretion profile, gene profile, differentiation, and homing parameters. For example, although SPIOs with cationic coatings (e.g., poly-L-lysine) displayed minimal influence on MSC viability and proliferation, they impaired the osteogenic and chondrogenic differentiation of

MSCs [51,52]. Unexpectedly, the internalization of silica NPs into MSCs led to the increased secretion of cytokines critical for cardiac regeneration [13]. The trend to develop NP imaging agents that have few detrimental effects on cellular function is the preference for incorporating FDA approved materials. In a recent example, the self-assembly of three FDA approved materials—ferumoxytol, heparin, and protamine—did not exhibit significant toxicity and damage on a variety of cells (neural stem cells, T cells, MSCs, and hematopoietic stem cells), and enabled the detection of ~1000 cells with MRI in a rat brain model [53].

The second challenge is the loss of signal due to NP dilution impairing its longitudinal imaging ability. This is especially the case for stem cells that not only exocytose NPs but also undergo frequent rounds of cell division. Moreover, magnetic contrast is less reliable over longer periods of tracking, since iron oxide degrades *in vivo*, contributing to signal attenuation [37]. A promising method to minimize these effects is to label cells with particles of larger dimensions, which can reduce exocytosis rates [54]. As demonstrated in our previous achievements, the entrapment of IONPs (<20 nm) within PLGA microspheres (1–2 μm) resulted in an increase in R_2 weighted MRI signal (fivefold) and prolonged signal duration beyond that of naked IONPs (threefold) [23]. These achievements are illustrated in Figure 11.2, which covers the experimental scheme, MRI signal enhancement using submicron particles, and *in vivo* achievements. A greater understanding of micro/nanoparticle interactions with cells may facilitate the design of imaging agents [28,30,32,55] with a sufficient period of intracellular residence to enable longitudinal tracking postimplantation.

Finally, developing NP imaging strategies that report on *in vivo* cellular function can

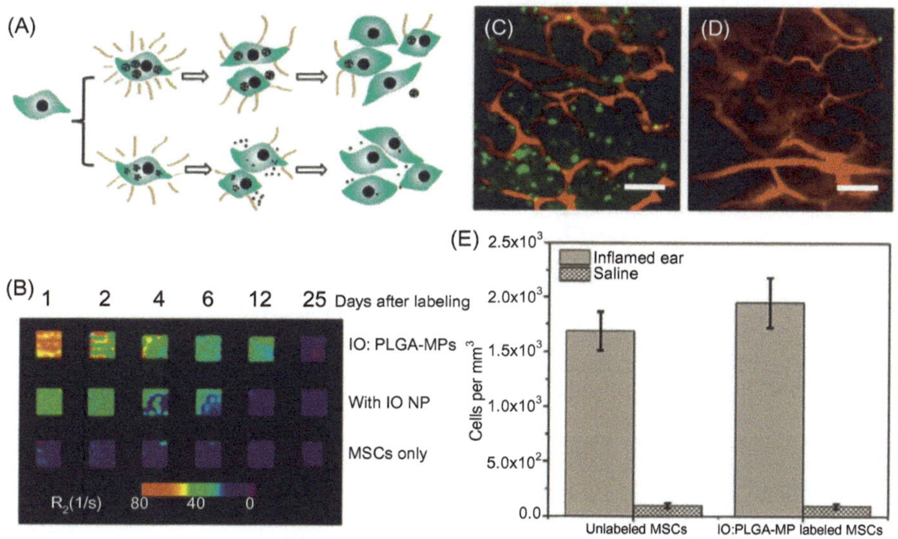

FIGURE 11.2 Micron-sized magnetic particles for MSCs engineering. (A) Experimental scheme detailing how PLGA-doped iron oxide microparticles (IO:PLGA-MPs) enhance labeling agent retention in MSCs over naked IONP. (B) R_2 weighted magnetic resonance images of agarose (3%) gels with entrapped IO:PLGA-MPs, IONP, and unmodified MSCs over a period of 25 days. Representative images of DilC$_{18}$(5) (DiD) and IO:PLGA-MPs labeled MSCs in an (C) inflamed and (D) un-inflamed ear. (E) Comparison of IO:PLGA-MPs labeling effects on MSC migration in an inflamed ear model. *Source: Reprinted (adapted) with permission from Ref. [23]. Copyright 2012 American Chemical Society.*

increase the yield of data content from a single study (i.e., clinical trial). Currently, most strategies are not able to reveal postimplantation behavior such as viability and differentiation status. For example, during islet transplantation, magnetic NPs were used to label islets and track in real time an islet graft in diabetic mice [56,57]. This strategy is limited in its ability to report the survival of transplanted islets and the level of insulin secretion, since transplant efficacy was gauged by measuring physiological blood glucose levels. Such passive cell tracking technologies are unable to determine specific insulin secretion rates or the extent of cell death. Instead, the ability to monitor post-implantation behavior can assist the optimization of islet transplantation by determining if insulin secretion is mediated by a small subset of cells or the majority. A potential solution for this hurdle lies in CEST MRI imaging. This method utilizes CEST probes (FDA approved L-arginine) immobilized in the extracellular environment of an alginate capsule. Changes in proton content (pH 5.5–7.5) led to CEST contrast increase, enabling *in vivo* detection of cell death through indirect means [18]. The experimental schematic as well as key findings of this report are shown in Figure 11.3. On the other hand, this method may have limited usage for cells that secrete high levels of lactic acid (thereby low pH microenvironment) such as embryonic stem cells. Other promising solutions may be based on technologies that contain molecular recognition elements such as Au NP-based nanoflares [33,34] and molecular beacons [58]. Their precision in detecting specific molecular sequences as well as their versatility suggests a strong potential for reporting cell function.

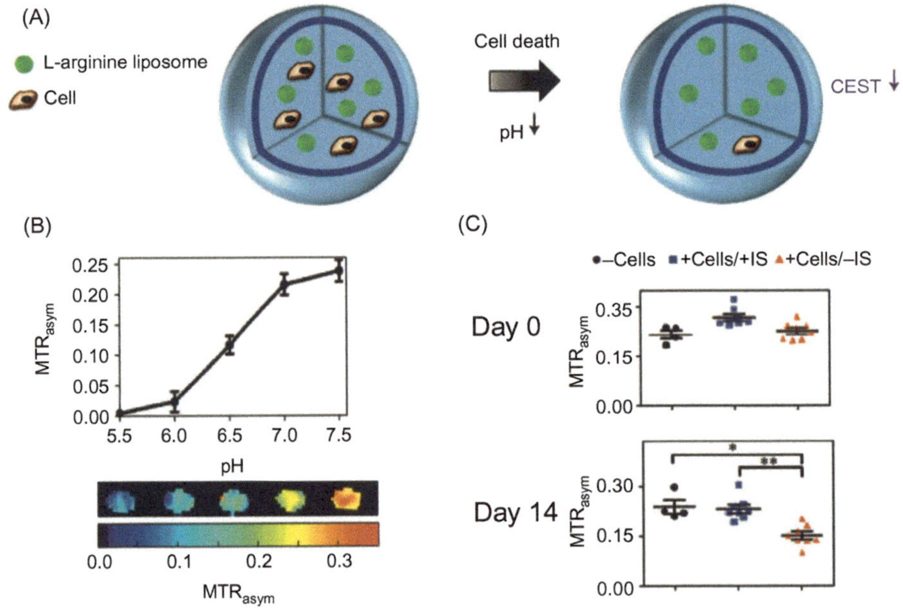

FIGURE 11.3 CEST MRI contrast to evaluate implanted cell viability. (A) ECM immobilized L-arginine MRI probes in alginate capsules (LipoCEST) to detect changes in pH as an indirect indicator of cell death. (B) Sensitivity of CEST contrast (MTR_{asym}) in LipoCEST capsules to changes in pH levels. (C) The indirect evaluation of implanted hepatocyte viability through CEST contrast using LipoCEST capsules. *Source: Reprinted by permission from Ref [18]. Copyright 2013 Macmillan Publishers Ltd.*

11.5 CONCLUSION

This chapter gives a brief overview on NP-based cell engineering strategies including intracellular internalization, plasma membrane, and ECM modifications. These strategies are universal to many imaging modalities as well as cell types. Their biggest advantage is their simplicity, convenience, high efficiency, and "inertness"—minimal change to the cell's original properties. Nevertheless, while several of these strategies are poised for clinical usage, they can be further improved by incorporating designs that: (i) provide stable signal intensity; (ii) report cell function; yet have (iii) minimal influence on cellular function. Returning to the themes discussed at the beginning of this chapter, advancements in NP engineering methods to enable clinically viable imaging tools will usher in significant advances for medicine, biology, and healthcare.

Acknowledgments

Project was supported by the Nanyang Technological University Start-Up Grant, Singapore and the Singapore MOE AcRF Tier 1 Research Fund.

References

[1] Megason SG, Fraser SE. Digitizing life at the level of the cell: high-performance laser-scanning microscopy and image analysis for in toto imaging of development. Mech Dev 2003;120:1407–20.

[2] Pysz MA, Gambhir SS, Willmann JK. Molecular imaging: current status and emerging strategies. Clin Radiol 2010;65:500–16.

[3] Wright BK, Andrews LM, Markham J, Jones MR, Stringari C, Digman MA, et al. NADH distribution in live progenitor stem cells by phasor-fluorescence lifetime image microscopy. Biophys J 2012;103:L7–9.

[4] Quinn KP, Bellas E, Fourligas N, Lee K, Kaplan DL, Georgakoudi I. Characterization of metabolic changes associated with the functional development of 3D engineered tissues by non-invasive, dynamic measurement of individual cell redox ratios. Biomaterials 2012;33:5341–8.

[5] Caorsi V, Toepfer C, Sikkel MB, Lyon AR, MacLeod K, Ferenczi MA. Non-linear optical microscopy sheds light on cardiovascular disease. PLoS ONE 2013;8:e56136.

[6] Giepmans BN, Adams SR, Ellisman MH, Tsien RY. The fluorescent toolbox for assessing protein location and function. Science 2006;312:217–24.

[7] Lippincott-Schwartz J, Patterson GH. Development and use of fluorescent protein markers in living cells. Science 2003;300:87–91.

[8] Ozawa T, Yoshimura H, Kim SB. Advances in fluorescence and bioluminescence imaging. Anal Chem 2013;85:590–609.

[9] Schroeder T. Long-term single-cell imaging of mammalian stem cells. Nat Methods 2011;8:S30–5.

[10] Snippert HJ, Clevers H. Tracking adult stem cells. EMBO Rep 2011;12:113–22.

[11] Villa-Diaz LG, Garcia-Perez JL, Krebsbach PH. Enhanced transfection efficiency of human embryonic stem cells by the incorporation of DNA liposomes in extracellular matrix. Stem Cells Dev 2010;19:1949–57.

[12] Gu E, Chen WY, Gu J, Burridge P, Wu JC. Molecular imaging of stem cells: tracking survival, biodistribution, tumorigenicity, and immunogenicity. Theranostics 2012;2:335–45.

[13] Jokerst JV, Khademi C, Gambhir SS. Intracellular aggregation of multimodal silica nanoparticles for ultrasound-guided stem cell implantation. Sci Transl Med 2013;5:177ra135.

[14] Wu TJ, Tzeng YK, Chang WW, Cheng CA, Kuo Y, Chien CH, et al. Tracking the engraftment and regenerative capabilities of transplanted lung stem cells using fluorescent nanodiamonds. Nat Nanotechnol 2013;8(9):682–9.

[15] Zhao W, Schafer S, Choi J, Yamanaka YJ, Lombardi ML, Bose S, et al. Cell-surface sensors for real-time probing of cellular environments. Nat Nanotechnol 2011;6:524–31.

[16] Corradetti B, Freile P, Pells S, Bagnaninchi P, Park J, Fahmy TM, et al. Paracrine signalling events in embryonic stem cell renewal mediated by affinity targeted nanoparticles. Biomaterials 2012;33:6634–43.

[17] Xia Z, Xing Y, Jeon J, Kim YP, Gall J, Dragulescu-Andrasi A, et al. Immobilizing reporters for molecular imaging of the extracellular microenvironment in living animals. ACS Chem Biol 2011;6:1117–26.

[18] Chan KW, Liu G, Song X, Kim H, Yu T, Arifin DR, et al. MRI-detectable pH nanosensors incorporated into hydrogels for *in vivo* sensing of transplanted-cell viability. Nat Mater 2013;12:268–75.

[19] Wang XD, Meier RJ, Wolfbeis OS. Fluorescent pH-sensitive nanoparticles in an agarose matrix for imaging of bacterial growth and metabolism. Angew Chem Int Ed Engl 2013;52:406–9.

[20] Ahrens ET, Bulte JW. Tracking immune cells *in vivo* using magnetic resonance imaging. Nat Rev Immunol 2013;13:755–63.

[21] Dwyer RM, Ryan J, Havelin RJ, Morris JC, Miller BW, Liu Z, et al. Mesenchymal stem cell-mediated delivery of the sodium iodide symporter supports radionuclide imaging and treatment of breast cancer. Stem Cells 2011;29:1149–57.

[22] Sarkar D, Ankrum JA, Teo GS, Carman CV, Karp JM. Cellular and extracellular programming of cell fate through engineered intracrine-, paracrine-, and endocrine-like mechanisms. Biomaterials 2011;32: 3053–61.

[23] Xu C, Miranda-Nieves D, Ankrum JA, Matthiesen ME, Phillips JA, Roes I, et al. Tracking mesenchymal stem cells with iron oxide nanoparticle loaded poly (lactide-co-glycolide) microparticles. Nano Lett 2012;12: 4131–9.

[24] Gao Y, Cui Y, Chan JK, Xu C. Stem cell tracking with optically active nanoparticles. Am J Nucl Med Mol Imaging 2013;3:232.

[25] Wang LV, Hu S. Photoacoustic tomography: *in vivo* imaging from organelles to organs. Science 2012; 335:1458–62.

[26] Yang X, Stein EW, Ashkenazi S, Wang LV. Nanoparticles for photoacoustic imaging. Wiley Interdiscip Rev Nanomed Nanobiotechnol 2009;1:360–8.

[27] Wang Y, Xu C, Ow H. Commercial nanoparticles for stem cell labeling and tracking. Theranostics 2013;3: 544–60.

[28] Champion JA, Walker A, Mitragotri S. Role of particle size in phagocytosis of polymeric microspheres. Pharm Res 2008;25:1815–21.

[29] Rejman J, Oberle V, Zuhorn IS, Hoekstra D. Size-dependent internalization of particles via the pathways of clathrin- and caveolae-mediated endocytosis. Biochem J 2004;377:159–69.

[30] Duncan R, Richardson SC. Endocytosis and intracellular trafficking as gateways for nanomedicine delivery: opportunities and challenges. Mol Pharm 2012; 9:2380–402.

[31] Ankrum JA, Miranda OR, Ng KS, Sarkar D, Xu C, Karp JM. Engineering cells with intracellular agent-loaded microparticles to control cell phenotype. Nat Protocols 2014;9:233–45.

[32] Gratton SEA, Ropp PA, Pohlhaus PD, Luft JC, Madden VJ, Napier ME, et al. The effect of particle design on cellular internalization pathways. Proc Natl Acad Sci USA 2008;105:11613–8.

[33] Zheng D, Seferos DS, Giljohann DA, Patel PC, Mirkin CA. Aptamer nano-flares for molecular detection in living cells. Nano Lett 2009;9:3258–61.

[34] Seferos DS, Giljohann DA, Hill HD, Prigodich AE, Mirkin CA. Nano-flares: probes for transfection and mRNA detection in living cells. J Am Chem Soc 2007;129:15477–9.

[35] Wang C, Ma X, Ye S, Cheng L, Yang K, Guo L, et al. Protamine functionalized single-walled carbon nanotubes for stem cell labeling and *in vivo* Raman/ magnetic resonance/photoacoustic triple-modal imaging. Adv Funct Mater 2012;22:2363–75.

[36] de la Zerda A, Bodapati S, Teed R, May SY, Tabakman SM, Liu Z, et al. Family of enhanced photoacoustic imaging agents for high-sensitivity and multiplexing studies in living mice. ACS Nano 2012;6: 4694–701.

[37] Cromer Berman SM, Walczak P, Bulte JW. Tracking stem cells using magnetic nanoparticles. Wiley Interdiscip Rev Nanomed Nanobiotechnol 2011;3:343–55.

[38] Remsen LG, McCormick CI, Roman-Goldstein S, Nilaver G, Weissleder R, Bogdanov A, et al. MR of carcinoma-specific monoclonal antibody conjugated to monocrystalline iron oxide nanoparticles: the potential for noninvasive diagnosis. Am J Neuroradiol 1996;17:411–8.

[39] Artemov D, Mori N, Okollie B, Bhujwalla ZM. MR molecular imaging of the Her-2/neu receptor in breast cancer cells using targeted iron oxide nanoparticles. Magn Reson Med 2003;49:403–8.

[40] Babič M, Horák D, Trchová M, Jendelová P, Glogarová Ki, Lesný P, et al. Poly(L-lysine)-modified iron oxide nanoparticles for stem cell labeling. Bioconjugate Chem 2008;19:740–50.

[41] McBain SC, Yiu HHP, El Haj A, Dobson J. Polyethyleneimine functionalized iron oxide nanoparticles as agents for DNA delivery and transfection. J Mater Chem 2007;17:2561–5.

[42] Huang X, Bronstein LM, Retrum J, Dufort C, Tsvetkova I, Aniagyei S, et al. Self-assembled virus-like particles with magnetic cores. Nano Lett 2007;7: 2407–16.

[43] Shenton W, Mann S, Cölfen H, Bacher A, Fischer M. Synthesis of nanophase iron oxide in lumazine synthase capsids. Angew Chem Int Ed 2001;40:442–5.

[44] Anderson EA, Isaacman S, Peabody DS, Wang EY, Canary JW, Kirshenbaum K. Viral nanoparticles donning a paramagnetic coat: conjugation of MRI contrast agents to the MS2 capsid. Nano Lett 2006;6:1160–4.

[45] de Vries IJ, Lesterhuis WJ, Barentsz JO, Verdijk P, van Krieken JH, Boerman OC, et al. Magnetic resonance tracking of dendritic cells in melanoma patients for monitoring of cellular therapy. Nat Biotechnol 2005;23:1407–13.

[46] Thu MS, Bryant LH, Coppola T, Jordan EK, Budde MD, Lewis BK, et al. Self-assembling nanocomplexes

by combining ferumoxytol, heparin and protamine for cell tracking by magnetic resonance imaging. Nat Med 2012;18:463–7.

[47] Cheng Z, Thorek DL, Tsourkas A. Gadolinium-conjugated dendrimer nanoclusters as a tumor-targeted T1 magnetic resonance imaging contrast agent. Angew Chem Int Ed 2010;49:346–50.

[48] Tokunaga T, Namiki S, Yamada K, Imaishi T, Nonaka H, Hirose K, et al. Cell surface-anchored fluorescent aptamer sensor enables imaging of chemical transmitter dynamics. J Am Chem Soc 2012;134: 9561–4.

[49] Chen Y, O'Donoghue MB, Huang YF, Kang H, Phillips JA, Chen X, et al. A surface energy transfer nanoruler for measuring binding site distances on live cell surfaces. J Am Chem Soc 2010;132:16559–70.

[50] Rong G, Wang H, Skewis LR, Reinhard BM. Resolving sub-diffraction limit encounters in nanoparticle tracking using live cell plasmon coupling microscopy. Nano Lett 2008;8:3386–93.

[51] Chang YK, Liu YP, Ho JH, Hsu SC, Lee OK. Amine-surface-modified superparamagnetic iron oxide nanoparticles interfere with differentiation of human mesenchymal stem cells. J Orthop Res 2012;30:1499–506.

[52] Bulte JW, Kraitchman DL, Mackay AM, Pittenger MF. Chondrogenic differentiation of mesenchymal stem cells is inhibited after magnetic labeling with ferumoxides. Blood 2004;104:3410–2 [author reply 3412–3413].

[53] Thu MS, Bryant LH, Coppola T, Jordan EK, Budde MD, Lewis BK, et al. Self-assembling nanocomplexes by combining ferumoxytol, heparin and protamine for cell tracking by magnetic resonance imaging. Nat Med 2012;18:463–7.

[54] Chithrani BD, Chan WCW. Elucidating the mechanism of cellular uptake and removal of protein-coated gold nanoparticles of different sizes and shapes. Nano Lett 2007;7:1542–50.

[55] Agarwal R, Singh V, Jurney P, Shi L, Sreenivasan SV, Roy K. Mammalian cells preferentially internalize hydrogel nanodiscs over nanorods and use shape-specific uptake mechanisms. Proc Natl Acad Sci USA 2013;110:17247–52.

[56] Xu C, Mu L, Roes I, Miranda-Nieves D, Nahrendorf M, Ankrum JA, et al. Nanoparticle-based monitoring of cell therapy. Nanotechnology 2011;22:494001.

[57] Evgenov NV, Medarova Z, Dai G, Bonner-Weir S, Moore A. *In vivo* imaging of islet transplantation. Nat Med 2006;12:144–8.

[58] Larsson HM, Lee ST, Roccio M, Velluto D, Lutolf MP, Frey P, et al. Sorting live stem cells based on *Sox2* mRNA expression. PLoS ONE 2012;7:e49874.

CHAPTER

12

Micro/Nano-Engineering of Cells for Delivery of Therapeutics

Oren Levy, Edward Han, Jessica Ngai, Priya Anandakumaran, Zhixiang Tong, Kelvin S. Ng and Jeffrey M. Karp

Division of Biomedical Engineering, Department of Medicine, Center for Regenerative Therapeutics, Brigham and Women's Hospital, Boston, MA; Harvard Medical School, Harvard University, Boston, MA; Harvard Stem Cell Institute, Cambridge, MA; Harvard-MIT Division of Health Sciences and Technology, Cambridge, MA, USA

OUTLINE

J.M. Karp & W. Zhao (Eds): Micro- and Nanoengineering of the Cell Surface.
DOI: http://dx.doi.org/10.1016/B978-1-45-573146-6.00012-X

In this chapter, we discuss micro/nano-engineering strategies to enable efficient targeting of transplanted cells to diseased/damaged tissues that should improve the efficacy of cell therapy and permit the use of cells as drug delivery vehicles. To introduce the tremendous clinical potential and the challenges of cell therapy, we begin by describing a selection of current cell therapies and the importance of cell targeting. We will then discuss bioengineering approaches to modify the cell surface to improve cell targeting to desired sites. This will be followed by in-depth discussion of targeted cell-based drug delivery and its clinical potential.

12.1 CELL THERAPY—SUCCESS AND CURRENT CHALLENGES

12.1.1 Hematopoietic Stem Cell Transplantation

Over 10,000 bone marrow transplants are performed every year in the United States to treat patients with severe blood disorders [1]. Hematopoietic stem cells (HSCs), capable of giving rise to the entire hematopoietic system [2], are the key therapeutic component of the transplanted bone marrow units and are therefore widely used to treat many blood malignancies, including myeloma, leukemia, and lymphoma, as well as immune deficiencies and other hematological disorders [3–6]. In most cases, the patient's own diseased hematopoietic system is first ablated and then reconstituted with healthy cells using HSC transplantation (HSCT). Following systemic transplantation, the first key cellular process that affects HSCT efficacy is the homing of HSCs to the bone marrow, where an optimal niche is provided for hematopoiesis [7]. Akin to leukocytes, HSCs roll on, adhere to, and migrate across the vascular endothelium at specific tissues, guided by chemotactic gradients [8,9]. Lack of homing capacity has been shown to drastically reduce HSCT efficacy [10]. After homing, engraftment is required to retain and maintain HSCs in a conductive microenvironment for long-term survival and proper function (e.g., self-renewal, differentiation). Enhancing homing and engraftment will likely improve efficacy and reduce the dosage of HSC infusion, allowing more patients to be treated with the same donor pool. This is important since HSCs are difficult to expand *in vitro* without inducing undesirable differentiation and cell death [11,12].

Because HSCT is typically allogeneic, a main challenge in HSCT is significant immune response following transplantation. Not only may the host immune response compromise the engraftment and survival of transplanted HSCs [13], patients are also at risk for graft-versus-host disease (GvHD), which can lead to death [14]. The risk of an immune response can be lessened via prophylactic immune suppression [15]. Haplotype matching based on the human leukocyte antigen (HLA) further ameliorates the risk, although this limits the number of potential donors [6]. Nowadays, HSCs can be efficiently harvested from multiple tissue sources, including cord blood, substantially increasing the number of patients who can benefit from HSCT [16,17]. Importantly, genetic modification enables the potential use of autologous HSCs: for some diseases, patients' cells can be corrected *ex vivo* and returned to the same patient, providing a promising platform when matched donors are unavailable [18].

However, significant challenges still remain for HSCT. In addition to GvHD and poor engraftment of transplanted cells, disease relapse often occurs after allogeneic (unmatched), haploidentical (matched), or autologous transplantations [19]. Although continuous efforts are made to engineer cells to improve therapy, successful implementation into the clinic is challenging due to safety issues and preservation of engineered properties post-transplantation [20]. Genetic modification entails potential risks such as abnormal splicing, inactivation of tumor suppressor genes, or accidental oncogene activation, that may result in development of leukemia [21]. Nongenetic modifications may circumvent some risks while conferring better control over cell fate post-transplantation, as we will discuss in detail later.

12.1.2 Pancreatic Islet Transplantation

Another cell therapy in growing demand is pancreatic islet transplantation for treatment of type 1 diabetes (T1D) [22]. In T1D, the pancreas does not produce sufficient levels of insulin, mostly due to autoimmune destruction of beta cells, resulting in impaired glucose metabolism [23]. Patients with T1D need to constantly monitor their blood glucose and receive insulin injections to reduce hyperglycemia when necessary. Poor control of blood glucose may lead to hypoglycemia, which can be fatal, or chronic hyperglycemia, which may result in complications such as heart disease, kidney disease, and nerve damage [24,25]. Successful islet transplantation would restore autonomous, precise control of blood glucose, freeing patients from constant monitoring, and adjustment of blood glucose. Allogeneic islets are required for T1D [25], since the patient's own islets are typically dysfunctional. Unfortunately, islets are composed of terminally differentiated cells and cannot be expanded *ex vivo*, while donor morbidity is clinically unjustifiable. Elaborate procedures for quickly obtaining fresh islets from cadavers are therefore required, especially given the fact that islets from at least two deceased donors are required to treat a single patient [26].

Like HSCT, allogeneic islet transplantation faces immune rejection and donor shortage. But unlike HSCs, islets are large multicellular aggregates containing multiple cell types and have to be administered locally instead of systemically (i.e., via the bloodstream) [27]. This greatly affects the repertoire of strategies for cell engineering. For example, genetic modification becomes more difficult since transfection efficiency will vary according to the location of the cell within the islet, whereas tissue engineered islet with defined compositional and structural features becomes instrumental for controlling cell fate and function following local transplantation [28]. However, while homing may be less relevant, engraftment and survival remain crucial for the success of islet transplantation: islets have to be vascularized quickly post-transplantation to maintain sufficient viability for long-term function [29].

Although the host immune response may be reduced by genetically downregulating foreign antigens [30–32], encapsulation inside biocompatible scaffolds such as alginate or porous pouches can sufficiently shield islets from both cellular and humoral components of the host immune system and thereby offer stronger and longer term protection [33–35]. Native islets deliver insulin directly into the hepatic portal vein; hence the gold standard has been intraportal implantation [36], which requires invasive surgery. Scaffold technologies may potentially enable minimally invasive [37] islet transplantation, such as subcutaneous implantation, given that a variety of problems will be addressed, such as capability of islets to detect changes in blood glucose, delivery of relevant amounts of insulin into the bloodstream, scaffold degradation over time, foreign body response against the scaffold, and nutrient transport to encapsulated islets [38]. Supplementing scaffold technologies with genetic strategies, such as accelerating vascularization by ectopically expressing pro-angiogenic factors [39], or promoting islet viability and functionality by ectopically expressing hormones [40] may be necessary to achieve sufficient efficacy of islet transplantation in the future [26]. An emerging paradigm is to derive pancreatic cells from precursors *in vitro* before transplantation [41,42]. Cells may then be administered systemically as single cells instead of aggregates, allowing them to efficiently home to and engraft at sites (e.g., hepatic or pancreatic capillary beds) where optimal islet physiology may be achieved.

12.1.3 Mesenchymal Stem Cell Therapy

Mesenchymal stromal cells, otherwise known as mesenchymal stem cells (MSCs), have emerged during the past two decades as promising candidates for cell therapy [43–45]. Unlike pancreatic islets, MSCs are readily isolated from bone marrow and other adult tissues without significant donor morbidity. Unlike HSCs, MSCs can be easily expanded *ex vivo* to obtain a sufficient quantity for cell transplantation [46]. Moreover, their immune-evasive and potent immunomodulatory properties following systemic infusion permit allogeneic transplantation and have prompted their use in over 380 clinical trials for treating multiple diseases, including lupus nephritis, GvHD, multiple sclerosis, and cardiovascular diseases [47]. While results from pre-clinical animal studies have been encouraging and hundreds of millions of allogeneic MSCs have been safely administered, clinical trials have produced only mixed results for efficacy endpoints and MSC translational potential has not yet been fully realized [48–51]. This could be attributed to the innate heterogeneity of MSC population, significant variability in their donor/tissue sources, manufacturing processes, as well as their diverse characterization methods [52]. One of the major challenges for MSC therapy is poor homing to diseased or damaged tissues, potentially due to minimal expression of key homing ligands on MSC surface [49,53,54]. Since the MSC secretome serves a crucial role in their therapeutic impact, increasing the number of transplanted MSCs at disease sites at the same total dosage will likely significantly improve the efficacy of MSC therapy [55].

12.1.4 Embryonic Stem Cells and Induced Pluripotent Stem Cells

Embryonic stem cells (ESCs) are perhaps the most promising yet most controversial cell type being investigated for therapy. Unlike adult stem cells, ESCs have an unlimited capacity for self-renewal, can differentiate into all three germ layers (ectoderm, endoderm, and mesoderm), and thus can generate almost any cell type [56]. However, adverse effects, such as formation of teratomas, and the many ethical issues that plague the use of ESCs, significantly hinder their translation into the clinic [57]. Numerous studies demonstrated the therapeutic potential of ESCs. For example, Min et al. [58] were able to significantly improve left ventricular function in post-infarcted rats after ESC transplantation. Brustle et al. [59] demonstrated that ESCs differentiate into oligodendrocyte and astrocyte precursors and induce axon myelination in a rat model of dysmyelinating Pelizaeus–Merzbacher disease. Besides cardiac and neural applications, ESC studies have spanned virtually every discipline of the biomedical field [60]. Better understanding of the mechanisms governing ESC differentiation and transformation may advance the development of safe and effective ESC therapies in the future.

Induced pluripotent stem cells (IPSCs) bypass ethical issues of ESCs and are also starting to be tested in clinical trials [61]. IPSCs are generated by reprogramming somatic cells to regain pluripotency [61,62], first demonstrated when Takahashi and Yamanaka [63] successfully reprogrammed adult fibroblasts into an embryonic-like state. Theoretically, this would enable a virtually limitless quantity of highly desirable ESC-like cells, while avoiding any ethical issues. One of the main questions is whether IPSCs are truly as pluripotent as ESCs [62]. Different groups have shown significant differences between IPSCs and ESCs. For example, the differentiation potential of IPSCs appears intrinsically lower than that of ESCs

[64]. In addition, IPSCs seem to have an increased propensity to differentiate back into the cells they were formed from [65]. Despite these differences, IPSCs have been used in ample studies spanning a multitude of biomedical applications. For example, Wernig et al. [66] were able to demonstrate that neurons derived from IPSCs were able to functionally integrate into the brains of rats with Parkinson's disease. Much of the potential of IPSCs remains untapped, however, making this field an exciting source of potential therapies.

Among the challenges discussed above, poor cell targeting to disease sites following transplantation seems to be a common drawback shared by all the aforementioned cell therapies. We will now discuss micro/nano-engineering approaches to control cell homing/targeting. These approaches mainly involve the cell surface, the key interface where the cell interacts with its environment during the homing process.

12.2 CELL SURFACE ENGINEERING TO IMPROVE CELL TARGETING

One of the major challenges in cell therapy is poor control over cell homing to disease sites following transplantation—to maximize therapeutic impact, delivering a sufficient amount of functional cells to target organs is crucial. Improving cell homing would also reduce the number of transplanted cells needed to achieve clinical improvement. This has enormous significance since cell expansion *ex vivo* has both biological and financial costs [48], and not every cell type can be expanded. For MSCs, which exert their immunomodulatory effects primarily via their secretome [55,67], increasing their bioavailability at the disease site is likely to significantly improve their therapeutic impact [68].

12.2.1 Affinity-Based Cell Targeting

Different bioengineering approaches have been exploited to modify the cell surface to enhance cell targeting to desired sites. A generalized concept is to trigger receptor–ligand interactions between the surface of transplanted cells and the cells at the target site. Cells can be engineered to overexpress biological or artificial receptors or antibodies to specifically target cells expressing the corresponding ligands or antigens [69] (Figure 12.1).

Whole and fragmented antibodies have been incorporated into cell surfaces to enhance cell localization to target sites. For instance, coating MSCs with whole antibodies against MAdCAM increased their homing to inflamed bowel and improved therapeutic efficacy on inflammatory bowel disease [70]. Likewise, in adoptive immunotherapy, autologous T cells may be modified with antibodies *ex vivo* and redirected to target specific cells, such as cancer cells. Whole antibodies have been engineered to create chimeric T-cell receptors (TCRs), which signal the T cell to launch a killing response upon binding to malignant cells [71,72].

Fragmented antibodies, such as single chain variable fragments (scFv) antibodies which do not contain the Fc domain but retain the high specificity and affinity of whole antibodies, have also been used to modify the cell surface [73]. Omitting the Fc domain eliminates potential binding of Fc receptors onto the cell surface and facilitates vector design for genetic engineering, but removes the signaling ability intrinsic to the Fc region. Fragmented antibodies have been engineered to create bispecific adaptors (diabodies) that link the specificity of two antibodies into a single agent and act as a bridge between two cell types. Cells have been genetically modified to secrete diabodies which bind tumor cells at one end and bind/activate T cells on the other, directing T-cell cytotoxic activity specifically toward tumors and significantly inhibiting tumor growth [74–76]. Fragmented

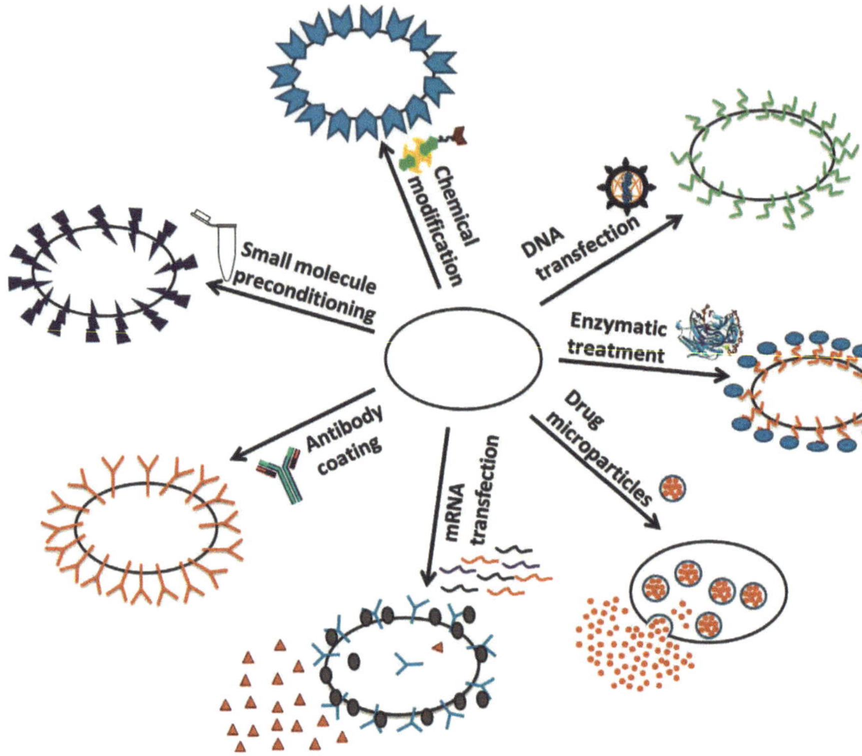

FIGURE 12.1 Bioengineering strategies to improve cell targeting to disease sites.

antibodies, instead of whole antibodies, are preferred for this approach since they penetrate tissues better due to their reduced size [77].

These two approaches represent how both membrane-anchored and secreted factors may enhance cell targeting via the cell surface. The same concept extends to receptors beyond antibodies. We will next discuss how such factors may be imparted to cells that natively underexpress them.

12.2.2 Bioengineering Approaches to Accomplish Affinity-Based Cell Targeting

12.2.2.1 Genetic Modification

Genetic modification has been employed to introduce new genes or to knock out specific genes to confer added functionalities to cells. The stability and specificity of genetic modifications have made it an attractive technique to achieve enhanced cell homing through cell surface modifications. The use of viral vectors such as adenoviruses and lentiviruses, as well as nonviral reagents such as lipid and polymeric micro/nanoparticles, has been widely explored to deliver DNA/RNA to cells to overexpress specific receptors to achieve improved homing to target sites.

12.2.2.1.1 VIRAL VECTORS

Viral vectors have been utilized as gene delivery vehicles to overexpress receptors on cell surfaces due to their high transfection efficiencies. Yu et al. [78] found that transducing MSCs with lentiviral vectors encoding the gene for the chemokine receptor CXCR4 resulted in enhanced engraftment and neuroprotection in rats subjected to middle cerebral artery occlusion. This was facilitated via interactions between CXCR4 and, its ligand, SDF-1, which is expressed along the ischemic boundary of the brain, and was shown to promote the migration of CXCR4-MSCs toward the ischemic zone [78]. In another study, increased homing of MSCs to the brain in a malignant glioma model was induced by transducing cells with retroviruses encoding EGFR [79]. Ectopic expression of EGFR on the surface of MSCs resulted in increased targeting toward its ligands EGF, HB-EGF, and TGF-alpha, all of which are present on glioma cells [79]. Retrovirally induced CXCR4 overexpression was also demonstrated to improve MSC homing to infarcted myocardium, also resulting in improved cardiac function following cell treatment [80]. In another study, retroviral delivery to increase CCR1 expression on MSC surface prior to transplantation increased cell viability, migration, and engraftment to infarcted myocardium [81]. Furthermore, viral-induced expression of integrin alpha 4 on MSC surface was shown to target MSCs to bone in immunocompetent mice, where MSCs exhibited differentiation into osteoblasts and osteocytes in the growth plate of recipient mouse limb bones [82]. Other studies have used genetically modified affinity-based targeting to increase delivery of systemically administered MSCs to tumor sites. For example, MSCs have been genetically modified with adenoviral vectors to express artificial receptors that can bind to the tumor cell marker erbB2 [69]. These erbB2-receptor-modified MSCs were retained in high concentrations in the lungs of erbB2-expressing mice, and they were targeted toward ovarian tumors in ovarian xenograft tumor models [69]. In another study, monocytes were adenovirally transduced to express a chimeric CD64 receptor to target CEA-expressing tumor cells, resulting in significantly reduced *in vivo* tumor growth rates in xenograft studies [83].

T cells which have been genetically modified with viral vectors to express chimeric TCRs that recognize tumor markers have also been evaluated. For instance, peripheral blood-derived human T cells were transduced with retroviruses encoding chimeric TCR specific to CD30 and were found to specifically target $CD30^+$ lymphoma cells. Following engraftment with $CD30^+$ lymphoma cells, the TCR-modified T cells secreted high levels of IFN-γ to induce killing of specific tumor cells [84]. Furthermore, human lymphocytes which have been genetically engineered to express tumor-specific chimeric receptors have been translated to phase I clinical trials [85] due to their high potency against tumors in animal models, which has resulted in partial or complete remission of cancer [72,86]. Despite their high transfection efficiency, viral vectors are challenged by several safety issues such as immunogenicity and potential oncogene activation, resulting in the search for and development of nonviral vectors to genetically engineer cells [87].

12.2.2.1.2 NONVIRAL VECTORS

A primary challenge in nonviral transfection is overcoming the electrostatic repulsion between negatively charged nucleic acids and the negatively charged cell membrane to facilitate uptake of DNA/RNA by cells. Common transfection methods include reagent-based methods such as cationic lipids or polymers, and device-based methods such as electroporation. Despite the lower transfection efficiency of nonviral vectors in comparison to viral vectors, the reduced immune response and toxicity of nonviral vectors have made them attractive gene delivery vehicles [88–94].

Kim et al. demonstrated that MSCs transfected with plasmids encoding the gene for CXCR1 through electroporation resulted in superior migration of MSCs toward gliomas. The increase in migration was attributed to the binding interaction between CXCR1 and IL-8, which is released by glioma cells [95]. MSCs have also been genetically modified through electroporation to express scFv antibodies on their surface to target EGFRvIII, which is highly expressed in many glioblastoma (GB) tumors. The modified MSCs were found to bind to U87-EGFRvIII cells at high levels *in vitro*, and they experienced a sevenfold increase in retention in human U87-EGFRvIII expressing tumors *in vivo* [96].

Finally, peripheral blood human T cells, which were transfected via electroporation with chimeric TCRs specific to CD20, were found to preferentially migrate towards $CD20^+$ lymphoma cells. Following coculture with $CD20^+$ lymphoma cells, the TCR-modified T cells exhibited cytotoxic characteristics through an increased secretion of IFN-γ [97].

Recently, we have demonstrated the use of lipofectamine to deliver *in vitro* transcribed mRNA to control MSC phenotype following transplantation [98]. Unlike delivery of DNA, which could result in stable transfection but risks integration of foreign DNA into the genome, mRNA transfection elicits a transient, yet functionally significant, expression of genes of interest. Via transfection with pseudouridine-modified mRNA (to increase translation efficiency), we have expressed sialyl Lewis X (SLeX) and PSGL-1 on MSC surface, resulting in improved homing of systemically transplanted MSCs to healthy and irradiated bone marrow as well as to distant sites of local inflammation [98]. The transient effect of mRNA transfection may not be desired for all clinical applications. However, this may be partially addressed by administering multiple infusions to patients and further research is needed to fully elucidate the translatability of this approach into the clinic.

Genetic engineering has the potential to stably and specifically modify cell surfaces to achieve increased cell targeting to different disease sites. However, this approach raises potential safety risks, mostly due to potential integration of transfected DNA into the genome of the host cell and the use of viruses for delivery. An alternative approach, which is safer and may result in more transient cell surface alteration, is chemical or enzymatic modification of the cell surface prior to transplantation.

12.2.2.2 Chemical Modification

Another technique to modify the cell surface is to chemically conjugate molecules or nanoparticles (NPs) onto existing functional groups on cell surface proteins, polysaccharides, or lipids. It is also possible to metabolically or genetically introduce new, reactive functional groups onto the surface of cells to act as anchors for other molecules to remodel the cell surface [99]. However, chemical conjugation is a simpler and more transient method to modify the cell surface.

12.2.2.2.1 SURFACE FUNCTIONAL GROUPS

Functional groups such as thiols and amines are naturally present on the cell surface and are readily available for covalent conjugation. Other functional groups such as aldehydes and ketones have been used as anchors for surface-modifying molecules; however, they must first be exposed through chemical or enzymatic treatment of cell surface carbohydrates. Primary amine groups have been utilized as sites for covalent conjugation through biotinylation, in which the amine groups are reacted with amine-reactive biotin, such as *N*-hydroxy-succinimide biotin derivatives. The cell surface can then be functionalized with other biotinylated molecules or NPs through a streptavidin bridge [100,101]. We previously found that covalently coupling SLeX to the surface of MSCs through a streptavidin–biotin bridge resulted in a robust cell rolling on P-selectin-coated substrates *in vitro*, as well as an increased homing efficiency to inflamed tissue *in vivo*, without compromising the intrinsic properties of MSCs [102,103].

12.2.2.2.2 LIPID INSERTION

Another approach involving the chemical conjugation of molecules to the cell surface takes advantage of the hydrophobic portion of the cell membrane to immobilize antibodies onto the cell surface. This two-step method consists of first coating cells with palmitate-conjugated proteins which are stabilized onto the cell membrane through the insertion of the hydrophobic palmitate moiety into the phospholipid bilayer. The palmitated proteins, which have high binding affinity for the Fc region of antibodies, allow cell surfaces to be functionalized with a variety of antibodies. These antibodies then act as artificial receptors for soluble or cell surface antigens [104]. Kim et al. [104] showed that cells coated with palmitate-conjugated protein A followed by rabbit anti-mouse IgG demonstrated enhanced adhesiveness to surface-bound IgG-positive mouse B cells *in vitro*. Others investigated MSCs coated with palmitated protein G (PPG) followed by antibodies against ICAM-1 and found an increased binding of anti-ICAM-1 MSCs to activated endothelial cells in flow chamber studies [105]. Similarly, Lo et al. [106] recently attached PSGL-1 (via PPG) to MSC surface, resulting in MSC tethering and rolling on stimulated endotheilal cell monolayers. Finally, Dennis et al. [107] focused on enhancing the engraftment of pre-chondrocytes to cartilage defects by coating the cell surface with lipidated protein G and antibodies which are specific to cartilage matrix antigens. Painted cells which were injected into rabbit cartilage explants with a partial thickness defect preferentially bound to the exposed cartilage material within the defect.

Another example of cell painting to improve cell homing involves binding antibodies utilizing the interaction between glycosylphosphatidylinositol (GPI) and cell surface proteins [108]. Sheep erythrocytes surface loaded with GPI-anchored scFv antibodies against CD-40 were found to specifically target and form rosettes with CD20-positive cancer cells [109]. Finally, Kean et al. [110] improved cell homing of MSCs to sites of myocardial infarction (MI) by coating MSCs with MI-specific peptides which have palmitic acid tails to facilitate integration into the cell membrane. Peptide-coated MSCs which were injected into mice with MI were localized at the heart in higher numbers than native MSCs [110].

12.2.2.3 *Enzymatic Modification*

It is also possible to directly modify existing cell surface molecules through enzymatic transformations. This method was employed to improve HSC homing and engraftment by *ex vivo* fucosylation of cord blood in order to achieve an increase in HSC tethering to activated endothelial cells expressing P-selectin and E-selectin [111]. This concept was further applied to human MSCs, where enzymatically modifying native CD44 glycoforms into hematopoietic cell E-selectin/L-selectin ligand resulted in targeting of modified MSCs to bone marrow in a murine model [112,113].

12.2.2.4 *Physical Targeting*

Physical targeting strategies utilize appropriate devices or procedures to enhance cell targeting to a specific site and may involve cell surface modifications. Although the concept of using devices or procedures to enhance cell targeting appears simple, a number of issues associated with gaining physical and direct access to organs severely complicate this method. While guiding devices exist, such as catheters, which can directly introduce cells into the brain [114,115], heart [116], and into wounds [117] with positive therapeutic outcomes, other physical methods exist like loading cells with magnetic compounds to redirect cell localization via application of a magnetic field to the target site. The initial purpose of using magnetic compounds was to monitor cell migration *in vivo* with magnetic resonance imaging; however, it has since evolved to be used to redirect cell localization both *in vitro* and *in vivo*. Nakashima et al. [118] showed that magnetically labeled NK cells targeted human osteosarcoma cells under a magnetic field *in vitro*. Other groups managed to magnetically target MSCs in animal models. For example, iron oxide-labeled MSCs were shown to specifically concentrate in the liver when external magnets were placed over it [119]. The same group showed that bone marrow stromal cells injected into the cerebrospinal fluid were able to aggregate on the surface of the spinal cord nearest to the magnetic field [119]. Another study focused on delivering magnetic NP-loaded endothelial cells to the steel surfaces of intra-arterial stents in a rat carotid artery stenting model [120]. Loading cells with nano/microparticles can be used beyond cell localization, especially for drug delivery applications.

12.3 CELL-BASED DRUG DELIVERY

The ability of cells to target specific tissues presents new opportunities for drug delivery. Harnessing cells as vehicles for drug delivery potentially overcomes challenges faced by current treatments. Local administration of therapeutics is undesirable when the disease is systemic or requires invasive access to target sites. Meanwhile, challenges of systemic administration include host toxicity, uncontrolled drug levels at desired sites, premature inactivation of the drug [121], and fast clearance of the agent from the blood, consequently requiring frequent repeat dosing. The usage of drug carriers is critical in the field of drug delivery and has been researched intensively [121]. Encapsulation of therapeutic agents in micro/nanoparticles increases safety by reducing toxicity and increases efficacy by protecting the drug from early degradation, permitting sustained therapeutic effects of systemically administered drugs. However, targeted delivery, which maximizes efficacy by releasing drugs only at disease sites, remains challenging. Particles passively target tissues where the vasculature is leaky (e.g., liver, kidney, and tumor). Although immobilizing ligands on the particle surface can actively target particles to specific tissues and counter passive targeting, excessive active targeting limits the depth of tissue penetration, and may also compromise the drug release kinetics [122]. In contrast, cells have homing and migratory properties; they can target and penetrate specific tissues even without modification. Therefore, cells with known or controllable tissue targets following transplantation emerge as a promising platform for targeted drug delivery. Like a Trojan horse, a cellular carrier protects the drug, infiltrates into the target tissue, and locally releases therapeutic cargo. Sustained drug release may be achieved by loading cells with drug-containing particles [123]. Not only does this arm the cell with additional therapeutic capabilities, its fate and phenotype may also be controlled by the released drug, like in the case of MSCs [55,98]. Herein, we will discuss several nanosystems, cell types, delivery methods, and disease models that utilize this strategy (Figure 12.2).

12.3.1 Nanosystem Types and Key Characteristics

The largest group of particulate carriers is NPs which encompass nanospheres, nanocapsules, micelles, dendrimers, nanocrystals, and nanogolds. NPs are chosen for their small size, high surface-to-volume ratio, and surface functionality. The most commonly used NP materials are polymers and lipids, which typically encapsulate hydrophobic drugs. Polymer-based NPs are made from synthetic polymers such as poly(lactic acid) (PLA) and poly(lactide-*co*-glycolide) (PLGA) or natural polymers such as gelatin, alginate, collagen, and chitosan [124–126]. PLA and PLGA NPs are especially

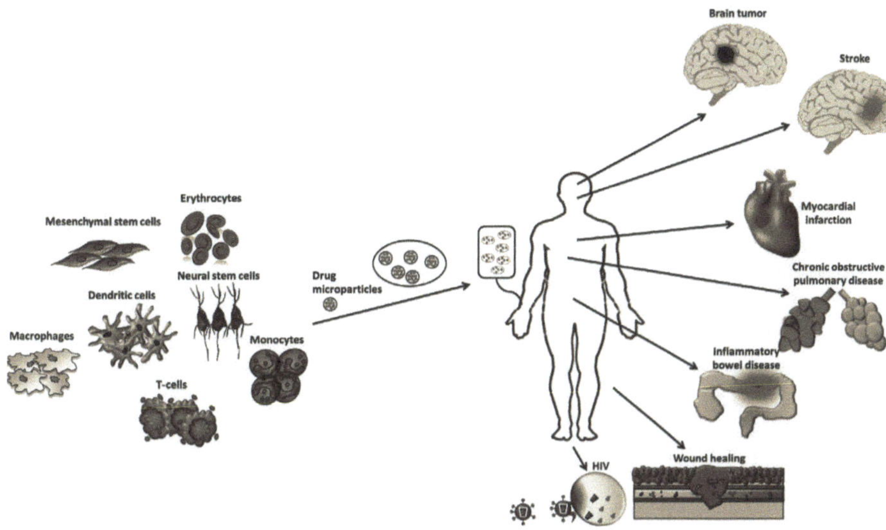

FIGURE 12.2 Multiple cell types can be used as drug delivery vehicles to treat a wide array of tragic diseases.

attractive since their degradation time can be tuned from days to months based on copolymer configuration and molecular weight [127]. However, without special treatments to render their surfaces hydrophilic, they are unstable in the aqueous biological environment and are unsuitable for systemic delivery. Lipid-based NPs typically have an amphiphilic coating to stabilize their hydrophobic cores in an aqueous environment. There are two common types of lipid-based NPs: solid lipid NPs [128] which consist of a solid lipid core and lipid nanocapsules [129] which have an oily (liquid) center [130]. Although lipid-based NPs may be more biologically stable, tuning their biodegradation *in vivo* is challenging. Other lipid-based NPs include micelles and liposomes, with liposomes being particularly attractive for their ability to encapsulate hydrophilic molecules [131].

Nucleic acids represent an important class of therapeutic hydrophilic molecules. Safety concerns over viral vectors have shifted research focus toward nonviral vectors [132]. Cationic liposomes and cationic polymer-based vectors interact with negatively charged molecules such as DNA while maintaining an overall positive charge on the surface, enabling interaction with the negatively charged cell surface and subsequent endocytosis, thus preventing DNA from systemic enzymatic degradation. Cargo is released into the cytoplasm via several mechanisms of endosomal escape such as membrane disruption/reassembly [133]. Similarly, cationized polysaccharides such as spermine-dextran [134] and spermine-pullulan [135] can be recognized by cell surface receptors to promote internalization, with biodegradability serving as another advantage. Although recent *in vivo* results have shown toxicity and low transfection efficiency, charged polysaccharides still demonstrated remarkable potential for use in tissue regeneration and cancer target therapy [91,136,137].

Another promising nanosystem is carbon nanotubes (CNTs) [138,139], which can immobilize and intracellularly deliver small molecules, nucleic acids, and proteins as long as the cargo can adsorb onto the CNT surface, typically by pi–pi stacking interactions [140,141]. Unfortunately, studies have also associated CNTs with cell toxicity [142], reducing its potential for clinical translation.

Protein adsorption on the surface of particles was originally viewed as undesirable as it promotes clearance by phagocytes. Efforts have since shifted toward using cellular binding and internalization properties to their advantage, such as to target specific cell types or to develop cellular drug carriers [143]. Internalization efficiency of particles by cells depends on cell type, particle size, surface chemistry, surface topography, shape, and mechanical properties [89]. Additional ligands can be conjugated to the particle surface to promote internalization via specific endocytic pathways [144].

12.3.2 Candidate Therapeutic Agents and Cell Types for Cell-Based Delivery

The types of therapeutic agents to be encapsulated in particles or cells depend not only on the disease of interest but also their stability during encapsulation and release kinetics after encapsulation. Encapsulation in cells warrants extra attention because the released drug may affect the cell and vice versa. For example, cell-based delivery of cytotoxic chemotherapeutics requires careful tuning of release kinetics, since excessive burst release may kill the cellular carrier before it reaches the target tissue [145]. An appealing strategy is to deliver prodrugs—drugs that are administered as inactive molecules, but are converted into their active form after specific chemical or enzymatic activity intracellularly or extracellularly. For instance, one can use the resident enzymes present in cells to convert prodrug into diffusible drug that can be released into the circulation [146]. Many therapeutic cargos, including nucleoside/nucleotide analogs [147], glucocorticoid analogs [148–151], enzymes [152–154], toxins [155,156], mRNA [98], peptides [157–167], and antisense peptide nucleic acids [168–171], are available as prodrugs.

An ideal cellular carrier must not only withstand its therapeutic cargo but also home to target tissues for efficient drug delivery. The most obvious, yet valuable, approach is to harness the innate homing properties of specific cells. Examples include leukocytes and stem cells that are known to home to tumors and sites of inflammation [172–175]. However, the innate homing properties may be suboptimal; hence cell surface engineering can be applied to maximize cell delivery to specific sites, as discussed in the previous section. The next section reviews a selection of cell types and their tropism. Discussion of cancer applications will be deferred to the following section, where we provide a case study on how multiple cell studies have been used as Trojan horses to target tumors.

12.3.2.1 Erythrocytes

One of the deficiencies of using pure drugs or drug delivery nanosystems is their short life spans—they are quickly cleared by immune cells, liver, and kidneys. Frequent injections of the drug or nanosystem are required to resupply the effective dosage. Known for their long circulating life spans (up to 120 days), erythrocytes or red blood cells (RBCs) have been harnessed for sustained drug delivery [176]. Drugs or drug-loaded particles may be conjugated as "backpacks" onto RBC surfaces where immune-evasive molecules reside [177] or encapsulated within RBC membranes [178]. Both strategies "camouflage" the drugs from clearance pathways. RBCs also have other advantages such as abundance ($\sim$5.4 million cells/mm^3 blood), size and shape uniformity [176], as well as long life span in circulation [146].

Particularly, RBC modification can be used in blood disorders. Fibrinolysis, the process of breaking down blood clots, is used to treat conditions due to hypercoagulation such as stroke and brain ischemia. Coupling fibrinolytic agents such as tissue-type plasminogen activator (tPA), streptokinase, and urokinase onto RBCs have been shown to sustain fibrinolytic activity without affecting RBC circulation [179–181]. Furthermore, when an antibody against collagen was attached on the RBC membrane with streptokinase, RBCs were able to attach to immobilized collagen and lyse the overlaying fibrin clots [182]. This approach may potentially be used to target sites where integrity of the vascular endothelium is compromised, since coagulation can be triggered by exposure of basement membrane proteins (e.g., collagen) to blood.

Dexamethasone is a potent glucocorticoid drug that has anti-inflammatory and immunosuppressant properties. Unfortunately, dexamethasone treatment has severe adverse effects, including redistribution of fat, muscle wasting, acne, bruising, thinning of skin, osteoporosis, exacerbation of diabetes mellitus, suppression of growth in children, and cataract. In light of these systemic effects, Pierige et al. developed a dexamethasone prodrug that utilizes enzymes in RBCs for its activation. The prodrug is encapsulated into autologous RBCs. When applied as a therapy, it is dephosphorylated by RBC enzymes and releases active dexamethasone. Owing to the slow, yet sustained, release of effective doses of dexamethasone, patients fighting chronic obstructive pulmonary disease [149], cystic fibrosis [150], and bowel inflammatory disease [183–186] were able to see benefits from treatments without adverse effects [147].

Although utilizing RBCs as drug carriers may achieve sustained systemic delivery, circulation times of "camouflaged" drugs were not as long as expected, typically lasting less than a week in stark contrast with the 120-day life span of unmodified RBCs [178]. Biomolecules on the RBC membranes that are responsible for prolonged circulation may be irreversibly damaged during RBC modification. In particular, RBCs lack tissue penetration properties and are unsuitable for delivering drugs deep into solid tissues. Consequently, targeted cell-based drug delivery strategies favor immune and stem cells, mostly for their innate tropism toward inflamed, hypoxic, and cancerous tissues [68].

12.3.2.2 Stem Cells

12.3.2.2.1 NEURAL STEM CELLS

Neural stem cells (NSCs) exhibit tropism toward intracranial gliomas and medullablastomas in animal models [187–189]. However, NSCs are only found in the brain [190] which poses significant donor morbidity. The challenge of harvesting enough autologous NSCs for clinical applications is a downfall of using NSCs. Other cell types—fetal brain, adult allogeneic brain, and ESCs—are being investigated as potential alternatives for autologous NSCs [191].

12.3.2.2.2 MESENCHYMAL STEM CELLS

MSCs exhibit tropism toward sites of injury [192] and tumors [174,175], and their homing efficiency can be enhanced by genetic, chemical, enzymatic, and other modifications as previously described. Unlike NSCs and other stem cells, MSCs can be feasibly obtained at sufficient numbers for clinical applications since they can be isolated from many adult tissues [193–202] at significant yields using minimally invasive methods and, more importantly, can be readily expanded *ex vivo* in large scales. Recently, our group has explored the potential of using MSC as cellular carriers via several bioengineering approaches. For instance, MSCs carrying intracellular dexamethasone-loaded PLGA microparticles could program the phenotypes of the MSC carrier as well as neighboring cells [103]. We have also demonstrated the use of mRNA modification on MSCs to suppress local inflammation via targeted delivery of the anti-inflammatory cytokine interleukin-10 following systemic MSC administration [98]. Moreover, MSCs can also be genetically modified to promote wound repair [203,204].

12.3.2.3 Leukocytes

Leukocytes, also known as white blood cells, possess natural homing properties toward tumors as well as other disease sites due to their major role in the immune response. Isolation of leukocytes is minimally invasive (via blood) and relatively straightforward. Consequently, multiple types of leukocytes have been implicated for Trojan horse delivery of therapeutics [205].

12.3.2.3.1 MONOCYTES

Monocytes are one of the first cells recruited by damaged, infected, or diseased tissues. After arriving at the site, monocytes differentiate into macrophages that phagocytose pathogens and recruit additional immune cells to the disease site [206]. At tumor sites, monocytes may become tumor-associated macrophages (TAMs), which constitute up to 70% of the tumor mass in breast cancer [206]. As drug carriers, monocytes may infiltrate into the hypoxic core of tumors and force surrounding TAMs to prevent proliferation [200,207–210], tumor neoangiogenesis [158–160,201], invasion [161,164,167,207], and metastasis [207,211–218] of malignant epithelial cells. An advantage of using monocytes or macrophages as transport vehicles is their innate phagocytic capability which provides high efficiency in particle loading. The use of macrophages as cellular carriers for NPs arose from the observation that systemically administered of NPs for imaging were ingested by endogenous macrophages that migrated and accumulated in and around tumors [219–221]. Another advantage of macrophages as cellular carriers is their ability to cross the blood brain barrier (BBB). For instance, macrophages preincubated with NPs containing a retroviral drug and administered in a murine model of HIV-1 encephalitis were observed to cross the BBB, increase local drug concentration for 14 days, and thereby suppress HIV-1 replication [222,223].

12.3.2.3.2 T CELLS

T cells play a key role in adaptive immunity, for instance by mediating cellular responses against infected or cancerous cells [224]. T cells have potential as chaperones of surface-attached therapeutic cargo, supported by their previous use in clinical trials for adoptive T-cell therapy in cancer, ease of harvest, and well-established protocols for their genetic modifications and expansion [225]. They have also shown promising infiltration into tumor lesions by crossing the barriers imposed by endothelial and stromal tissues [226].

12.3.2.3.3 DENDRITIC CELLS

Dendritic cells (DCs) are antigen-presenting cells, mostly found in the lymph nodes. After a local infection, DCs process and present antigens on their surfaces in order to activate naïve T cells and B cells [227]. Cell–cell interactions between T cells and DCs occur through direct contact [228] and paracrine signaling. There have only been a limited number of studies employing DCs as cellular vehicles. For example, DCs have been employed to systemically deliver oncolytic reovirus for killing melanoma cells and shown effective protection for the virus against the preexisting antiviral immunity [229]. Interestingly, the principle of "antigen presentation" on DCs was mimicked by immobilizing receptors and co-stimulatory ligands onto synthetic particles to activate adaptive immunity [230].

12.3.3 Case Study: "Trojan Horse" Cell Therapy for Cancers

Tumors are characterized by chaotic vasculature, high interstitial pressure, and dense extracellular matrices, presenting major obstacles for treatment via systemically administered drugs or drug-loaded particles [126]. Often, drugs and particles are unable to penetrate into the tumor and, as a result, are rapidly cleared. An example is GBs, which are extremely malignant brain tumors, known for their rapid growth and vascularization [231]. Current treatments revolve around surgery, radiation, and chemotherapy with DNA alkylating agents such as temozolomide (TMZ). However, there are several limitations with current treatments such as the poor effectiveness of TMZ for TMZ-sensitive GB patients as well as GB resistance against chemotherapy and radiotherapy. In addition, glioblastoma stem cells (GSCs) possess infiltrative, proliferative, and progressive characteristics, rendering them more resistant to hypoxic and acidotic tumor microenvironments [232]. Although GBs are highly vascularized, penetrating the BBB is yet another challenge toward effectively delivering therapeutic agents to the tumor. In fact, the BBB prevents uptake of large molecules and of over 98% of small-molecule drugs available [233]. Likewise, NP transport efficiency across the BBB is poor. To deliver antitumor agents across the BBB and deep into GBs, cellular carriers—namely monocytes and stem cells—are attractive candidates due to their ability to infiltrate into gliomas [234–238].

Known for their tropism toward intracranial tumors, NSCs have been used to transport toxic molecules [239–244], proliferation inhibitors [245], anti-angiogenic agents [246,247], and cytokines with toxic molecules [240]. Additionally, NSCs have been genetically modified to yield cytokines such as IL-18, IL-2 [248–250], IFN-γ [251], and tumor necrosis factor-related apoptosis-inducing ligand (TRAIL) [252–254] for modifying increased immune response at gliomas. NSCs have also been engineered to secrete cytosine deaminase (CD) that activates the prodrug 5-fluorocytosine (5-FC), so that cytotoxic effects of the systemically administered drug are localized within the tumor [189]. Off-target homing of antitumor NSCs is potentially dangerous as antitumor agents can harm healthy cells. To reduce off-target actions, overexpression of a pH-sensitive viral fusogen on the NSC surface has been used to fuse NSCs specifically with tumor cells, since tissue pH is lower at tumor sites [255]. These engineered NSCs carried a suicide gene, allowing selective killing of any syncytium formed by the NSCs and residual, unfused NSCs.

Monocytes or macrophages can also cross the BBB, but unlike NSCs, monocytes are isolated with much less invasive procedures. Choi et al. envisioned loading monocytes with gold nanoshells such that they home to the tumor site, differentiate into TAMs, migrate and deliver the nanoshells into hypoxic regions of the tumor. Once in place, near-infrared illumination would be used to heat the nanoshells and destroy the TAMs and adjacent tumor cells [205]. Nanoshell-based photothermal ablation therapy has shown promising results toward reducing tumor in mice, with over 90% remission rate [256]. Combining these concepts, monocyte carriers of gold nanoshells were successfully used to ablate human breast carcinoma tumor spheroids [172,205,219]. In another study employing a murine glioma model, cyclodextrin-based NPs were internalized by TAMs. Due to their tumor-tropic properties, the TAMs migrated within and around the tumor, demonstrating their capability to distribute therapeutic agents across the tumor [257]. T cells have also been used for nanoshell-based photothermal ablation [258], demonstrating the general effectiveness of leukocytes as tumor-homing cellular carriers.

Another use of T cells as a cellular vehicle followed from observations that viral particles can "hitchhike" on the T-cell surface, be taken up by tumor cells *in vivo*, and be released at the tumor site [259]. Extending the success of this strategy, numerous groups have used tumor antigen-specific T cells to carry oncolytic viruses to tumor deposits [260–262]. The same strategy can be applied to deliver NPs for cancer diagnostics and therapeutics, where the use of viruses is undesirable. Such agents include small-molecule drugs, antibody–drug conjugates, aptamers, and magnetic imaging agents [263,264]. The Irvine group used thioreactive maleimide groups to crosslink nanocarriers, surface proteins, or lipids to the T-cell surface for up to a week. Their *in vivo* results showed successful biodistribution of

T-cell-coupled NPs in tumor models, while systemic injection without cellular carriers resulted in fast clearance of the particles in liver and spleen [265].

MSCs are yet another cell type that exhibits tumor-tropic properties. Systemically administered MSCs carrying intracellular PLA and lipid nanocapsules demonstrated successful delivery of nanocapsules to gliomas [175]. NPs have also been conjugated onto the MSC surface using biotin–streptavidin bridges without attenuating MSC tropism toward tumor spheroids [266]. In multiple studies, MSCs were genetically engineered using viral and nonviral vectors to secrete immunostimulators such as IL-2, IL-12, IFN-beta, IL-18, and IL-7 to suppress tumor growth in glioma and other cancer models [203,250,251,267–276]. MSCs are also being used to deliver oncolytic adenoviruses to tumor sites [277,278]. More complex systems combining gene modification and prodrugs such as the CD/5-FC system (also applied on NSCs) have been tested on MSCs [279,280]. Another system uses herpes simplex virus thymidine kinase (HSVtk), which activates the chemotherapeutic prodrug ganciclovir [281]. Retrovirally transfected MSCs carried HSVtk to the tumor site, such that systemically administered ganciclovir was preferentially activated in the tumor, resulting in tumor suppression in GB patients [281]. The same system appeared to be less efficacious when NSCs are used instead [241,282–284]. Yet another combination involves rabbit carboxylesterase and CPT-11, in which genetically modified adipose-derived MSCs were used to deliver carboxylesterase that enzymatically cleaves the CPT-11 prodrug to release the active drug SN-38 to combat brainstem gliomas [285]. Despite these advances, thorough investigation is required to fully elucidate the interaction between MSCs and tumor environment, especially when MSCs have been suggested to promote tumor growth [286], and clinical trials have not shown long-term engraftment of MSCs in tumors [287].

While most cellular drug delivery vehicles are still in early stages of research, one cell type has been successfully translated into FDA-approved therapy. DCs have become a part of FDA-approved therapeutic cancer vaccine (against prostate cancer), in which antigen-loaded autologous DCs are injected into patients [288]. In addition, another group has proposed a method to track cells that have been loaded with NPs. Noh et al. encapsulated near-infrared fluorophores and iron oxide NPs together with model antigens such that cells carrying the NPs could be tracked in real time. Injection of those DCs resulted in tumor suppression [289]. DCs also show tropism toward lymph nodes which enables the use of DCs to carry immunosuppressive drugs to lymphoid tissues in order to suppress T-cell proliferation [290].

Apart from solid tumors, blood cancers may also be treated with cellular carrier. In a pilot clinical trial, autologous RBCs were loaded with asparaginase so they could remove asparagine—a nonessential amino acid—from blood. Asparagine has been shown to increase lymphoblastic proliferation which is crucial for the progression of lymphoblastic leukemia [291].

12.4 CONCLUDING REMARKS

Cell therapy provides hope for treating many tragic diseases. Harnessing the innate therapeutic properties of certain cell types, such as MSCs, and using cells as vehicles for targeted delivery of therapeutics are promising strategies to further advance cell-based therapy. However, for cell therapy to fully realize its clinical potential, better control over cell fate following transplantation must be achieved. Bioengineering approaches, such as genetic, enzymatic, or chemical surface modifications, may enable specific delivery of systemically infused cells to desired organs, although this has yet to be tested in humans.

Cells loaded with micro/nanoparticles containing specific drugs can then be directed to diseased organs, generating high local levels of therapeutics in the desired location, minimizing systemic toxicity, and significantly improving treatment efficacy. Tailoring drug release from cells has potential to provide a highly regulated and sustained therapeutic impact, and multiple cell infusions can further maximize the effect. Furthermore, mRNA engineering can be utilized to simultaneously but transiently control multiple cell properties, enabling improved cell homing and delivery of soluble biologics to distant diseased sites, while avoiding potential risks associated with DNA transfection. Targeted delivery of biologics is a major challenge that potentially could be overcome by use of cell-based approaches.

To circumnavigate current challenges in cell therapy, a concerted effort from clinicians, cell processing professionals, cell biologists, and bioprocess engineers is required [292]. Developing large quantities of cells that are immune evasive and that can be readily engineered to accurately deliver therapeutics including biologics to distant desired sites would have immense clinical potential.

References

[1] BeTheMatch.org. Homepage. Available at: <http://bethematch.org/>.

[2] Osawa M, Hanada K, Hamada H, Nakauchi H. Long-term lymphohematopoietic reconstitution by a single CD34-low/negative hematopoietic stem cell. Science 1996;273(5272):242–5.

[3] Guimaraes FA, Oliveira-Cardoso EA, Mastropietro AP, Voltarelli JC, Santos MA. Impact of autologous hematopoetic stem cell transplantation on the quality of life of patients with multiple sclerosis. Arq Neuropsiquiatr 2010;68(4):522–7.

[4] Buckley RH, Schiff SE, Schiff RI, et al. Hematopoietic stem-cell transplantation for the treatment of severe combined immunodeficiency. N Engl J Med 1999;340 (7):508–16.

[5] Pfender N, Saccardi R, Martin R. Autologous hematopoietic stem cell transplantation as a treatment option for aggressive multiple sclerosis. Curr Treat Options Neurol 2013;15(3):270–80.

[6] Van Zant G, Liang Y. Concise review: hematopoietic stem cell aging, life span, and transplantation. Stem Cells Transl Med 2012;1(9):651–7.

[7] Lymperi S, Ferraro F, Scadden DT. The HSC niche concept has turned 31. Has our knowledge matured? Ann N Y Acad Sci 2010;1192:12–8.

[8] Lapidot T, Dar A, Kollet O. How do stem cells find their way home? Blood 2005;106(6):1901–10.

[9] Kavanagh DP, Kalia N. Hematopoietic stem cell homing to injured tissues. Stem Cell Rev 2011;7(3): 672–82.

[10] Frenette PS, Subbarao S, Mazo IB, von Andrian UH, Wagner DD. Endothelial selectins and vascular cell adhesion molecule-1 promote hematopoietic progenitor homing to bone marrow. Proc Natl Acad Sci USA 1998;95(24):14423–8.

[11] Magnusson M, Sierra MI, Sasidharan R, et al. Expansion on stromal cells preserves the undifferentiated state of human hematopoietic stem cells despite compromised reconstitution ability. PLoS ONE 2013;8 (1):e53912.

[12] Dmitrieva RI, Anisimov SV. Optional protocols of hematopoietic stem cell expansion *in vitro*. Tsitologiia 2013;55(1):11–5.

[13] Huang Y, Rezzoug F, Chilton PM, Grimes HL, Cramer DE, Ildstad ST. Matching at the MHC class I K locus is essential for long-term engraftment of purified hematopoietic stem cells: a role for host NK cells in regulating HSC engraftment. Blood 2004;104(3): 873–80.

[14] Ratanatharathorn V, Ayash L, Lazarus HM, Fu J, Uberti JP. Chronic graft-versus-host disease: clinical manifestation and therapy. Bone Marrow Transplant 2001;28(2):121–9.

[15] Marmont AM, Horowitz MM, Gale RP, et al. T-cell depletion of HLA-identical transplants in leukemia. Blood 1991;78(8):2120–30.

[16] Xu F, Deeg HJ. Current status of allogeneic hematopoietic cell transplantation for MDS. Curr Pharm Des 2012;18(22):3215–21.

[17] Paun O, Lazarus HM. Allogeneic hematopoietic cell transplantation for acute myeloid leukemia in first complete remission: have the indications changed? Curr Opin Hematol 2012;19(2):95–101.

[18] Aiuti A, Biasco L, Scaramuzza S, et al. Lentiviral hematopoietic stem cell gene therapy in patients with Wiskott–Aldrich syndrome. Science 2013;341(6148): 1233151.

[19] Phillips GL. Allogeneic hematopoietic stem cell transplantation (HSCT) for high-risk acute myeloid leukemia (AML)/myelodysplastic syndrome (MDS): how

can we improve outcomes in the near future? Leukemia Res 2012;36(12):1490–5.

[20] Riviere I, Dunbar CE, Sadelain M. Hematopoietic stem cell engineering at a crossroads. Blood 2012;119 (5):1107–16.

[21] Watts KL, Adair J, Kiem HP. Hematopoietic stem cell expansion and gene therapy. Cytotherapy 2011;13 (10):1164–71.

[22] Hesse UJ, Weyer J, Meyer G, Isselhard W, Pichlmaier H. Long-term results after porcine islet transplantation. Transplant Proc 1989;21(1 Pt 3):2763–4.

[23] Acharjee S, Ghosh B, Al-Dhubiab BE, Nair AB. Understanding type 1 diabetes: etiology and models. Can J Diab 2013;37(4):269–76.

[24] Sachedina S, Toth C. Association of comorbidities with increasing severity of peripheral neuropathy in diabetes mellitus. World J Diab 2013;4(4):135–44.

[25] Schneider DA, Kretowicz AM, von Herrath MG. Emerging immune therapies in type 1 diabetes and pancreatic islet transplantation. Diab Obes Metab 2013;15(7):581–92.

[26] van der Windt DJ, Bottino R, Kumar G, et al. Clinical islet xenotransplantation: how close are we? Diabetes 2012;61(12):3046–55.

[27] Merani S, Toso C, Emamaullee J, Shapiro AM. Optimal implantation site for pancreatic islet transplantation. Br J Surg 2008;95(12):1449–61.

[28] Kodama S, Kojima K, Furuta S, Chambers M, Paz AC, Vacanti CA. Engineering functional islets from cultured cells. Tissue Eng Part A 2009;15(11):3321–9.

[29] Korsgren O, Lundgren T, Felldin M, et al. Optimising islet engraftment is critical for successful clinical islet transplantation. Diabetologia 2008;51(2):227–32.

[30] O'Connell PJ, Cowan PJ, Hawthorne WJ, Yi S, Lew AM. Transplantation of xenogeneic islets: are we there yet? Curr Diab Rep 2013;13(5):687–94.

[31] Komoda H, Miyagawa S, Kubo T, et al. A study of the xenoantigenicity of adult pig islets cells. Xenotransplantation 2004;11(3):237–46.

[32] Thompson P, Badell IR, Lowe M, et al. Islet xenotransplantation using gal-deficient neonatal donors improves engraftment and function. Am J Transplant 2011;11 (12):2593–602.

[33] Jeong JH, Yook S, Lee H, Ahn CH, Lee DY, Byun Y. Effects of surface camouflaged islet transplantation on pathophysiological progression in a db/db type 2 diabetic mouse model. Biochem Biophys Res Commun 2013;433(4):513–8.

[34] Elliott RB, Escobar L, Calafiore R, et al. Transplantation of micro- and macroencapsulated piglet islets into mice and monkeys. Transplant Proc 2005;37(1):466–9.

[35] Kumagai-Braesch M, Jacobson S, Mori H, et al. The TheraCyte device protects against islet allograft rejection in immunized hosts. Cell Transplant 2013;22(7):1137–46.

[36] Shapiro AM, Ricordi C, Hering BJ, et al. International trial of the Edmonton protocol for islet transplantation. N Engl J Med 2006;355(13):1318–30.

[37] O'Cearbhaill ED, Ng KS, Karp JM. Emerging medical devices for minimally invasive cell therapy. Mayo Clinic Proceed 2014;89(2):259–73.

[38] Vaithilingam V, Fung C, Ratnapala S, et al. Characterisation of the xenogeneic immune response to microencapsulated fetal pig islet-like cell clusters transplanted into immunocompetent C57BL/6 mice. PLoS ONE 2013;8(3):e59120.

[39] Narang AS, Cheng K, Henry J, et al. Vascular endothelial growth factor gene delivery for revascularization in transplanted human islets. Pharm Res 2004;21(1):15–25.

[40] Park JB, Jeong JH, Lee M, Lee DY, Byun Y. Xenotransplantation of exendin-4 gene transduced pancreatic islets using multi-component (alginate, poly-L-lysine, and polyethylene glycol) microcapsules for the treatment of type 1 diabetes mellitus. J Biomater Sci Polym Ed 2013;24(18):2045–57.

[41] Lumelsky N, Blondel O, Laeng P, Velasco I, Ravin R, McKay R. Differentiation of embryonic stem cells to insulin-secreting structures similar to pancreatic islets. Science 2001;292(5520):1389–94.

[42] Bonner-Weir S, Weir GC. New sources of pancreatic beta-cells. Nat Biotechnol 2005;23(7):857–61.

[43] Singer NG, Caplan AI. Mesenchymal stem cells: mechanisms of inflammation. Annu Rev Pathol 2011;6:457–78.

[44] Hoogduijn MJ, Popp F, Verbeek R, et al. The immunomodulatory properties of mesenchymal stem cells and their use for immunotherapy. Int Immunopharmacol 2010;10(12):1496–500.

[45] Liang J, Zhang H, Hua B, et al. Allogeneic mesenchymal stem cells transplantation in treatment of multiple sclerosis. Mult Scler 2009;15(5):644–6.

[46] Dominici M, Le Blanc K, Mueller I, et al. Minimal criteria for defining multipotent mesenchymal stromal cells. The international society for cellular therapy position statement. Cytotherapy 2006;8(4):315–7.

[47] Ankrum JA, Ong JF, Karp JM. Mesenchymal stem cells: immune evasive, not immune privileged. Nat Biotechnol. 2014;32(3):252–60.

[48] Francois M, Galipeau J. New insights on translational development of mesenchymal stromal cells for suppressor therapy. J Cell Physiol 2012;227(11):3535–8.

[49] Ankrum J, Karp JM. Mesenchymal stem cell therapy: two steps forward, one step back. Trends Mol Med 2010;16(5):203–9.

[50] Bernardo ME, Ball LM, Cometa AM, et al. Co-infusion of *ex vivo*-expanded, parental MSCs prevents life-threatening acute GVHD, but does not reduce the risk of graft failure in pediatric patients undergoing allogeneic umbilical cord blood transplantation. Bone Marrow Transplant 2011;46(2):200–7.
[51] Kuzmina LA, Petinati NA, Parovichnikova EN, et al. Multipotent mesenchymal stromal cells for the prophylaxis of acute graft-versus-host disease—a phase II study. Stem Cells Int 2012;2012:968213.
[52] Mendicino M, Bailey AM, Wonnacott K, Puri RK, Bauer SR. MSC-based product characterization for clinical trials: an FDA perspective. Cell Stem Cell 2014;14(2):141–5.
[53] Sarkar D, Spencer JA, Phillips JA, et al. Engineered cell homing. Blood 2011;118(25):e184–91.
[54] Rombouts WJ, Ploemacher RE. Primary murine MSC show highly efficient homing to the bone marrow but lose homing ability following culture. Leukemia 2003;17(1):160–70.
[55] Ranganath SH, Levy O, Inamdar MS, Karp JM. Harnessing the mesenchymal stem cell secretome for the treatment of cardiovascular disease. Cell Stem Cell 2012;10(3):244–58.
[56] Wobus AM, Boheler KR. Embryonic stem cells: prospects for developmental biology and cell therapy. Physiol Rev 2005;85(2):635–78.
[57] Knoepfler PS. Deconstructing stem cell tumorigenicity: a roadmap to safe regenerative medicine. Stem cells (Dayton, OH) 2009;27(5):1050–6.
[58] Min JY, Yang Y, Converso KL, et al. Transplantation of embryonic stem cells improves cardiac function in postinfarcted rats. J Appl Physiol 2002;92(1): 288–96.
[59] Brustle O, Jones KN, Learish RD, et al. Embryonic stem cell-derived glial precursors: a source of myelinating transplants. Science 1999;285(5428):754–6.
[60] Brignier AC, Gewirtz AM. Embryonic and adult stem cell therapy. J Allergy Clin Immunol 2010;125(Suppl. 2):S336–44.
[61] Yamanaka S. Induced pluripotent stem cells: past, present, and future. Cell Stem Cell 2012;10(6):678–84.
[62] Robinton DA, Daley GQ. The promise of induced pluripotent stem cells in research and therapy. Nature 2012;481(7381):295–305.
[63] Takahashi K, Yamanaka S. Induction of pluripotent stem cells from mouse embryonic and adult fibroblast cultures by defined factors. Cell 2006;126(4):663–76.
[64] Feng Q, Lu SJ, Klimanskaya I, et al. Hemangioblastic derivatives from human induced pluripotent stem cells exhibit limited expansion and early senescence. Stem Cells 2010;28(4):704–12.
[65] Hu Q, Friedrich AM, Johnson LV, Clegg DO. Memory in induced pluripotent stem cells: reprogrammed human retinal-pigmented epithelial cells show tendency for spontaneous redifferentiation. Stem Cells 2010;28(11):1981–91.
[66] Wernig M, Zhao J-P, Pruszak J, et al. Neurons derived from reprogrammed fibroblasts functionally integrate into the fetal brain and improve symptoms of rats with Parkinson's disease. Proc Nat Acad Sci USA 2008;105(15):5856–61.
[67] Lai RC, Arslan F, Lee MM, et al. Exosome secreted by MSC reduces myocardial ischemia/reperfusion injury. Stem Cell Res 2010;4(3):214–22.
[68] Karp JM, Leng Teo GS. Mesenchymal stem cell homing: the devil is in the details. Cell Stem Cell 2009;4(3):206–16.
[69] Komarova S, Roth J, Alvarez R, Curiel DT, Pereboeva L. Targeting of mesenchymal stem cells to ovarian tumors via an artificial receptor. J Ovarian Res 2010;3:12.
[70] Ko IK, Kim BG, Awadallah A, et al. Targeting improves MSC treatment of inflammatory bowel disease. Mol Ther 2010;18(7):1365–72.
[71] Willemsen RA, Debets R, Chames P, Bolhuis RL. Genetic engineering of T cell specificity for immunotherapy of cancer. Hum Immunol 2003;64(1):56–68.
[72] Brentjens RJ, Santos E, Nikhamin Y, et al. Genetically targeted T cells eradicate systemic acute lymphoblastic leukemia xenografts. Clin Cancer Res 2007;13(18 Pt 1):5426–35.
[73] Hudson PJ, Souriau C. Engineered antibodies. Nat Med 2003;9(1):129–34.
[74] Blanco B, Holliger P, Vile RG, Alvarez-Vallina L. Induction of human T lymphocyte cytotoxicity and inhibition of tumor growth by tumor-specific diabody-based molecules secreted from gene-modified bystander cells. J Immunol 2003;171(2):1070–7.
[75] Compte M, Blanco B, Serrano F, et al. Inhibition of tumor growth *in vivo* by in situ secretion of bispecific anti-CEA x anti-CD3 diabodies from lentivirally transduced human lymphocytes. Cancer Gene Ther 2007;14 (4):380–8.
[76] Chan JK, Hamilton CA, Cheung MK, et al. Enhanced killing of primary ovarian cancer by retargeting autologous cytokine-induced killer cells with bispecific antibodies: a preclinical study. Clin Cancer Res 2006;12(6):1859–67.
[77] Chames P, Van Regenmortel M, Weiss E, Baty D. Therapeutic antibodies: successes, limitations and hopes for the future. Br J Pharmacol 2009;157(2):220–33.
[78] Yu X, Chen D, Zhang Y, et al. Overexpression of CXCR4 in mesenchymal stem cells promotes migration, neuroprotection and angiogenesis in a rat model of stroke. J Neurol Sci 2012;316(1–2):141–9.

[79] Kuwashima N, Nishimura F, Eguchi J, et al. Delivery of dendritic cells engineered to secrete IFN-alpha into central nervous system tumors enhances the efficacy of peripheral tumor cell vaccines: dependence on apoptotic pathways. J Immunol 2005;175(4):2730–40.

[80] Cheng Z, Ou L, Zhou X, et al. Targeted migration of mesenchymal stem cells modified with CXCR4 gene to infarcted myocardium improves cardiac performance. Mol Ther 2008;16(3):571–9.

[81] Huang J, Zhang Z, Guo J, et al. Genetic modification of mesenchymal stem cells overexpressing CCR1 increases cell viability, migration, engraftment, and capillary density in the injured myocardium. Circ Res 2010;106(11):1753–62.

[82] Kumar S, Ponnazhagan S. Bone homing of mesenchymal stem cells by ectopic alpha 4 integrin expression. FASEB J 2007;21(14):3917–27.

[83] Biglari A, Southgate TD, Fairbairn LJ, Gilham DE. Human monocytes expressing a CEA-specific chimeric CD64 receptor specifically target CEA-expressing tumour cells *in vitro* and *in vivo*. Gene Ther 2006;13 (7):602–10.

[84] Hombach A, Muche JM, Gerken M, et al. T cells engrafted with a recombinant anti-CD30$^+$ receptor target autologous CD30$^+$ cutaneous lymphoma cells. Gene Ther 2001;8(11):891–5.

[85] Carroll RG, June CH. Programming the next generation of dendritic cells. Mol Ther 2007;15(5):846–8.

[86] Gade TP, Hassen W, Santos E, et al. Targeted elimination of prostate cancer by genetically directed human T lymphocytes. Cancer Res 2005;65(19):9080–8.

[87] Walther W, Stein U. Viral vectors for gene transfer: a review of their use in the treatment of human diseases. Drugs 2000;60(2):249–71.

[88] De Smedt SC, Demeester J, Hennink WE. Cationic polymer based gene delivery systems. Pharm Res 2000;17(2):113–26.

[89] Green JJ, Zhou BY, Mitalipova MM, et al. Nanoparticles for gene transfer to human embryonic stem cell colonies. Nano Lett 2008;8(10): 3126–30.

[90] McBain SC, Yiu HH, Dobson J. Magnetic nanoparticles for gene and drug delivery. Int J Nanomed 2008;3(2):169–80.

[91] Okazaki A, Jo J-I, Tabata Y. A reverse transfection technology to genetically engineer adult stem cells. Tissue Eng 2007;13(2):245–51.

[92] Prabha S, Labhasetwar V. Critical determinants in PLGA/PLA nanoparticle-mediated gene expression. Pharm Res 2004;21(2):354–64.

[93] Salem AK, Searson PC, Leong KW. Multifunctional nanorods for gene delivery. Nat Mater 2003;2 (10):668–71.

[94] Vonarbourg A, Passirani C, Desigaux L, et al. The encapsulation of DNA molecules within biomimetic lipid nanocapsules. Biomaterials 2009;30(18):3197–204.

[95] Cho NH, Cheong TC, Min JH, et al. A multifunctional core–shell nanoparticle for dendritic cell-based cancer immunotherapy. Nat Nanotechnol 2011;6(10):675–82.

[96] Balyasnikova IV, Franco-Gou R, Mathis JM, Lesniak MS. Genetic modification of mesenchymal stem cells to express a single-chain antibody against EGFRvIII on the cell surface. J Tissue Eng Regen Med 2010;4(4):247–58.

[97] Jensen MC, Cooper LJ, Wu AM, Forman SJ, Raubitschek A. Engineered CD20-specific primary human cytotoxic T lymphocytes for targeting B-cell malignancy. Cytotherapy 2003;5(2):131–8.

[98] Levy O, Zhao W, Mortensen LJ, et al. mRNA-engineered mesenchymal stem cells for targeted delivery of interleukin-10 to sites of inflammation. Blood 2013;122(14):e23–32.

[99] Kayser H, Geilen CC, Paul C, Zeitler R, Reutter W. Incorporation of *N*-acyl-2-amino-2-deoxy-hexoses into glycosphingolipids of the pheochromocytoma cell line PC 12. FEBS Lett 1992;301(2):137–40.

[100] Chakraborty K, Bose A, Pal S, et al. Neem leaf glycoprotein restores the impaired chemotactic activity of peripheral blood mononuclear cells from head and neck squamous cell carcinoma patients by maintaining CXCR3/CXCL10 balance. Int Immunopharmacol 2008;8(2):330–40.

[101] Krishnamachari Y, Pearce ME, Salem AK. Self-assembly of cell–microparticle hybrids. Adv Mater 2008;20(5):989–93.

[102] Sarkar D, Vemula PK, Teo GS, et al. Chemical engineering of mesenchymal stem cells to induce a cell rolling response. Bioconjug Chem 2008;19(11): 2105–9.

[103] Sarkar D, Ankrum JA, Teo GS, Carman CV, Karp JM. Cellular and extracellular programming of cell fate through engineered intracrine-, paracrine-, and endocrine-like mechanisms. Biomaterials 2011;32 (11):3053–61.

[104] Kim SA, Peacock JS. The use of palmitate-conjugated protein a for coating cells with artificial receptors which facilitate intercellular interactions. J Immunol Methods 1993;158(1):57–65.

[105] Ko IK, Kean TJ, Dennis JE. Targeting mesenchymal stem cells to activated endothelial cells. Biomaterials 2009;30(22):3702–10.

[106] Lo CY, Antonopoulos A, Dell A, Haslam SM, Lee T, Neelamegham S. The use of surface immobilization of P-selectin glycoprotein ligand-1 on mesenchymal stem cells to facilitate selectin mediated cell tethering and rolling. Biomaterials 2013;34(33):8213–22.

[107] Dennis JE, Cohen N, Goldberg VM, Caplan AI. Targeted delivery of progenitor cells for cartilage repair. J Orthop Res 2004;22(4):735–41.

[108] Englund PT. The structure and biosynthesis of glycosyl phosphatidylinositol protein anchors. Annu Rev Biochem 1993;62:121–38.

[109] Hamdy N, Goustin AS, Desaulniers JP, Li M, Chow CS, Al-Katib A. Sheep red blood cells armed with anti-CD20 single-chain variable fragments (scFvs) fused to a glycosylphosphatidylinositol (GPI) anchor: a strategy to target CD20-positive tumor cells. J Immunol Methods 2005;297(1–2):109–24.

[110] Kean TJ, Duesler L, Young RG, et al. Development of a peptide-targeted, myocardial ischemia-homing, mesenchymal stem cell. J Drug Target 2012;20(1):23–32.

[111] Xia L, McDaniel JM, Yago T, Doeden A, McEver RP. Surface fucosylation of human cord blood cells augments binding to P-selectin and E-selectin and enhances engraftment in bone marrow. Blood 2004;104(10):3091–6.

[112] Sackstein R, Merzaban JS, Cain DW, et al. *Ex vivo* glycan engineering of CD44 programs human multipotent mesenchymal stromal cell trafficking to bone. Nat Med 2008;14(2):181–7.

[113] Sackstein R. Directing stem cell trafficking via GPS. Methods Enzymol 2010;479:93–105.

[114] Kawabata K, Migita M, Mochizuki H, et al. *Ex vivo* cell-mediated gene therapy for metachromatic leukodystrophy using neurospheres. Brain Res 2006;1094 (1):13–23.

[115] Parr AM, Kulbatski I, Tator CH. Transplantation of adult rat spinal cord stem/progenitor cells for spinal cord injury. J Neurotrauma 2007;24(5):835–45.

[116] Poh KK, Sperry E, Young RG, Freyman T, Barringhaus KG, Thompson CA. Repeated direct endomyocardial transplantation of allogeneic mesenchymal stem cells: safety of a high dose, "off-the-shelf", cellular cardiomyoplasty strategy. Int J Cardiol 2007;117(3):360–4.

[117] Mi Q, Riviere B, Clermont G, Steed DL, Vodovotz Y. Agent-based model of inflammation and wound healing: insights into diabetic foot ulcer pathology and the role of transforming growth factor-beta1. Wound Repair Regen 2007;15(5):671–82.

[118] Nakashima Y, Deie M, Yanada S, Sharman P, Ochi M. Magnetically labeled human natural killer cells, accumulated *in vitro* by an external magnetic force, are effective against HOS osteosarcoma cells. Int J Oncol 2005;27(4):965–71.

[119] Arbab AS, Jordan EK, Wilson LB, Yocum GT, Lewis BK, Frank JA. *In vivo* trafficking and targeted delivery of magnetically labeled stem cells. Hum Gene Ther 2004;15(4):351–60.

[120] Polyak B, Fishbein I, Chorny M, et al. High field gradient targeting of magnetic nanoparticle-loaded endothelial cells to the surfaces of steel stents. Proc Natl Acad Sci USA 2008;105(2):698–703.

[121] Roth JC, Curiel DT, Pereboeva L. Cell vehicle targeting strategies. Gene Therapy 2008;15(10):716–29.

[122] Peer D, Karp JM, Hong S, Farokhzad OC, Margalit R, Langer R. Nanocarriers as an emerging platform for cancer therapy. Nat Nanotechnol 2007;2(12):751–60.

[123] Ferreira L, Karp JM, Nobre L, Langer R. New opportunities: the use of nanotechnologies to manipulate and track stem cells. Cell Stem Cell 2008;3(2): 136–46.

[124] Jain RA. The manufacturing techniques of various drug loaded biodegradable poly(lactide-*co*-glycolide) (PLGA) devices. Biomaterials 2000;21(23):2475–90.

[125] Jain RA, Rhodes CT, Railkar AM, Malick AW, Shah NH. Comparison of various injectable protein-loaded biodegradable poly(lactide-*co*-glycolide) (PLGA) devices: *in situ*-formed implant versus *in situ*-formed microspheres versus isolated microspheres. Pharm Dev Technol 2000;5(2):201–7.

[126] Jain RK, Stylianopoulos T. Delivering nanomedicine to solid tumors. Nat Rev Clin Oncol 2010;7(11):653–64.

[127] Lin SY, Chen KS, Teng HH, Li MJ. *In vitro* degradation and dissolution behaviours of microspheres prepared by three low molecular weight polyesters. J Microencapsul 2000;17(5):577–86.

[128] Wissing SA, Kayser O, Muller RH. Solid lipid nanoparticles for parenteral drug delivery. Adv Drug Deliv Rev 2004;56(9):1257–72.

[129] Heurtault B, Saulnier P, Pech B, Proust JE, Benoit JP. A novel phase inversion-based process for the preparation of lipid nanocarriers. Pharm Res 2002;19 (6):875–80.

[130] Beduneau A, Saulnier P, Anton N, et al. Pegylated nanocapsules produced by an organic solvent-free method: evaluation of their stealth properties. Pharm Res 2006;23(9):2190–9.

[131] Collnot EM, Ali H, Lehr CM. Nano- and microparticulate drug carriers for targeting of the inflamed intestinal mucosa. J Control Release 2012;161(2):235–46.

[132] Lechardeur D, Verkman AS, Lukacs GL. Intracellular routing of plasmid DNA during non-viral gene transfer. Adv Drug Deliv Rev 2005;57(5):755–67.

[133] Liang W, Lam J. Endosomal escape pathways for non-viral nucleic acid delivery systems. Molecular regulation of endocytosis; Dr. Brian Ceresa (editor), InTech (Rijeka, Croatia); 2012: [Chapter 17].

[134] Chen Y-Z, Yao X-L, Tabata Y, Nakagawa S, Gao J-Q. Gene carriers and transfection systems used in the recombination of dendritic cells for effective cancer immunotherapy. Clin Dev Immunol 2010;2010:565643

[135] Nagane K, Jo J-i, Tabata Y. Promoted adipogenesis of rat mesenchymal stem cells by transfection of small Interfering RNA complexed with a cationized dextran. Tissue Eng Part A 2010;16(1):21–31.
[136] Jo J-I, Nagaya N, Miyahara Y, et al. Transplantation of genetically engineered mesenchymal stem cells improves cardiac function in rats with myocardial infarction: benefit of a novel nonviral vector, cationized dextran. Tissue Eng 2007;13(2):313–22.
[137] Thakor DK, Teng YD, Obata H, Nagane K, Saito S, Tabata Y. Nontoxic genetic engineering of mesenchymal stem cells using serum-compatible pullulan-spermine/DNA anioplexes. Tissue Eng C Methods 2011;17(2):131–44.
[138] Zhu L, Chang DW, Dai L, Hong Y. DNA damage induced by multiwalled carbon nanotubes in mouse embryonic stem cells. Nano Lett 2007;7(12):3592–7.
[139] Mooney E, Dockery P, Greiser U, Murphy M, Barron V. Carbon nanotubes and mesenchymal stem cells: biocompatibility, proliferation and differentiation. Nano Lett 2008;8(8):2137–43.
[140] Iijima S. Helical microtubules of graphitic carbon. Nature 1991;354(6348):56–8.
[141] Zhang W, Zhang Z, Zhang Y. The application of carbon nanotubes in target drug delivery systems for cancer therapies. Nanoscale Res Lett 2011;6:555.
[142] Magrez A, Kasas S, Salicio V, et al. Cellular toxicity of carbon-based nanomaterials. Nano Lett 2006;6 (6):1121–5.
[143] Kocbek P, Obermajer N, Cegnar M, Kos J, Kristl J. Targeting cancer cells using PLGA nanoparticles surface modified with monoclonal antibody. J Control Release 2007;120(1–2):18–26.
[144] Bareford LM, Swaan PW. Endocytic mechanisms for targeted drug delivery. Adv Drug Deliv Rev 2007;59 (8):748–58.
[145] Gao Z, Zhang L, Hu J, Sun Y. Mesenchymal stem cells: a potential targeted-delivery vehicle for anti-cancer drug, loaded nanoparticles. Nanomedicine 2013;9(2):174–84.
[146] Biagiotti S, Paoletti MF, Fraternale A, Rossi L, Magnani M. Drug delivery by red blood cells. IUBMB Life 2011;63(8):621–31.
[147] Pierigè F, Serafini S, Rossi L, Magnani M. Cell-based drug delivery. Adv Drug Deliv Rev 2008;60 (2):286–95.
[148] Crinelli R, Antonelli A, Bianchi M, Gentilini L, Scaramucci S, Magnani M. Selective inhibition of NF-kB activation and TNF-alpha production in macrophages by red blood cell-mediated delivery of dexamethasone. Blood Cells Mol Dis 2000;26(3):211–22.
[149] Rossi L, Serafini S, Cenerini L, et al. Erythrocyte-mediated delivery of dexamethasone in patients with chronic obstructive pulmonary disease. Biotechnol Appl Biochem 2001;33(Pt 2):85–9.
[150] Rossi L, Castro M, D'Orio F, et al. Low doses of dexamethasone constantly delivered by autologous erythrocytes slow the progression of lung disease in cystic fibrosis patients. Blood Cells Mol Dis 2004;33 (1):57–63.
[151] Sprandel U, Way JL. International society for the use of resealed erythrocytes. Meeting Irsee G. erythrocytes as drug carriers in medicine. New York, NY: Plenum Press; 1997.
[152] Magnani M, Fazi A, Mangani F, Rossi L, Mancini U. Methanol detoxification by enzyme-loaded erythrocytes. Biotechnol Appl Biochem 1993;18:217–26:Pt 3.
[153] Magnani M, Laguerre M, Rossi L, et al. Acetaldehyde dehydrogenase-loaded erythrocytes as bioreactors for the removal of blood acetaldehyde. Alcohol Clin Exp Res 1989;13(6):849.
[154] Rossi L, Bianchi M, Magnani M. Increased glucose metabolism by enzyme-loaded erythrocytes *in vitro* and *in vivo* normalization of hyperglycemia in diabetic mice. Biotechnol Appl Biochem 1992;15(2):207.
[155] de Chastellier C, Lang T, Thilo L. Phagocytic processing of the macrophage endoparasite, *Mycobacterium avium*, in comparison to phagosomes which contain *Bacillus subtilis* or latex beads. Eur J Cell Biol 1995;68 (2):167–82.
[156] Rossi L, Brandi G, Malatesta M, et al. Effect of listeriolysin O-loaded erythrocytes on *Mycobacterium avium* replication within macrophages. J Antimicrob Chemother 2004;53(5):863–6.
[157] Antonelli A, Crinelli R, Bianchi M, et al. Efficient inhibition of macrophage TNF-alpha production upon targeted delivery of K48R ubiquitin. Br J Haematol 1999;104(3):475–81.
[158] Fraternale A, Casabianca A, Orlandi C, et al. Macrophage protection by addition of glutathione (GSH)-loaded erythrocytes to AZT and DDI in a murine AIDS model. Antiviral Res 2002;56(3):263–72.
[159] Fraternale A, Casabianca A, Rossi L, et al. Erythrocytes as carriers of reduced glutathione (GSH) in the treatment of retroviral infections. J Antimicrob Chemother 2003;52(4):551–4.
[160] Palamara AT, Perno CF, Aquaro S, Bue MC, Dini L, Garaci E. Glutathione inhibits HIV replication by acting at late stages of the virus life cycle. AIDS Res Hum Retroviruses 1996;12(16):1537–41.
[161] Buhl R, Jaffe HA, Holroyd KJ, et al. Systemic glutathione deficiency in symptom-free HIV-seropositive individuals. Lancet 1989;2(8675):1294–8.
[162] Garaci E, Palamara AT, Di Francesco P, Favalli C, Ciriolo MR, Rotilio G. Glutathione inhibits replication and expression of viral proteins in cultured cells

infected with Sendai virus. Biochem Biophys Res Commun 1992;188(3):1090–6.

[163] Magnani M, Fraternale A, Casabianca A, et al. Antiretroviral effect of combined zidovudine and reduced glutathione therapy in murine AIDS. AIDS Res Hum Retroviruses 1997;13(13):1093–9.

[164] Mihm S, Galter D, Droge W. Modulation of transcription factor NF kappa B activity by intracellular glutathione levels and by variations of the extracellular cysteine supply. FASEB J 1995;9(2):246–52.

[165] Murata Y, Shimamura T, Hamuro J. The polarization of T(h)1/T(h)2 balance is dependent on the intracellular thiol redox status of macrophages due to the distinctive cytokine production. Int Immunol 2002;14 (2):201–12.

[166] Palamara AT, Perno CF, Ciriolo MR, et al. Evidence for antiviral activity of glutathione: *in vitro* inhibition of herpes simplex virus type 1 replication. Antiviral Res 1995;27(3):237–53.

[167] Staal FJ, Roederer M, Israelski DM, et al. Intracellular glutathione levels in T cell subsets decrease in HIV-infected individuals. AIDS Res Hum Retroviruses 1992;8(2):305–11.

[168] Arima H, Sakamoto T, Aramaki Y, Ishidate K, Tsuchiya S. Specific inhibition of nitric oxide production in macrophages by phosphorothioate antisense oligonucleotides. J Pharm Sci 1997;86(10):1079–84.

[169] Chiarantini L, Cerasi A, Fraternale A, et al. Inhibition of macrophage iNOS by selective targeting of antisense PNA. Biochemistry 2002;41(26):8471–7.

[170] Egholm M, Buchardt O, Christensen L, et al. PNA hybridizes to complementary oligonucleotides obeying the Watson–Crick hydrogen-bonding rules. Nature 1993;365(6446):566–8.

[171] Wittung P, Kajanus J, Edwards K, Nielsen P, Norden B, Malmstrom BG. Phospholipid membrane permeability of peptide nucleic acid. FEBS Lett 1995;365 (1):27–9.

[172] Choi M-R, Stanton-Maxey KJ, Stanley JK, et al. A cellular Trojan horse for delivery of therapeutic nanoparticles into tumors. Nano Lett 2007;7(12):3759–65.

[173] Chen X, Lin X, Zhao J, et al. A tumor-selective biotherapy with prolonged impact on established metastases based on cytokine gene-engineered MSCs. Mol Ther 2008;16(4):749–56.

[174] Brennen WN, Chen S, Denmeade SR, Isaacs JT. Quantification of mesenchymal stem cells (MSCs) at sites of human prostate cancer. Oncotarget 2013;4 (1):106–17.

[175] Roger M, Clavreul A, Venier-Julienne M-C, et al. Mesenchymal stem cells as cellular vehicles for delivery of nanoparticles to brain tumors. Biomaterials 2010;31(32):8393–401.

[176] Hamidi M, Tajerzadeh H. Carrier erythrocytes: an overview. Drug Deliv 2003;10(1):9–20.

[177] Beck Z, Brown BK, Wieczorek L, et al. Human erythrocytes selectively bind and enrich infectious HIV-1 virions. PLoS ONE 2009;4(12):e8297.

[178] Hu CM, Zhang L, Aryal S, Cheung C, Fang RH. Erythrocyte membrane-camouflaged polymeric nanoparticles as a biomimetic delivery platform. Proc Natl Acad Sci USA 2011;108(27):10980–5.

[179] Murciano JC, Medinilla S, Eslin D, Atochina E, Cines DB, Muzykantov VR. Prophylactic fibrinolysis through selective dissolution of nascent clots by tPA-carrying erythrocytes. Nat Biotechnol 2003;21 (8):891–6.

[180] Ganguly K, Krasik T, Medinilla S, et al. Blood clearance and activity of erythrocyte-coupled fibrinolytics. J Pharmacol Exp Ther 2005;312(3):1106–13.

[181] Armstead WM, Ganguly K, Riley J, et al. RBC-coupled tPA prevents whereas tPA aggravates JNK MAPK-mediated impairment of ATP- and Ca-sensitive K channel-mediated cerebrovasodilation after cerebral photothrombosis. Transl Stroke Res 2012;3(1):114–21.

[182] Muzykantov VR, Sakharov DV, Smirnov MD, Samokhin GP, Smirnov VN. Immunotargeting of erythrocyte-bound streptokinase provides local lysis of a fibrin clot. Biochim Biophys Acta 1986;884 (2):355–62.

[183] Castro M, Rossi L, Papadatou B, et al. Long-term treatment with autologous red blood cells loaded with dexamethasone 21-phosphate in pediatric patients affected by steroid-dependent Crohn disease. J Pediatr Gastroenterol Nutr 2007;44(4): 423–6.

[184] Annese V, Latiano A, Rossi L, et al. Erythrocytes-mediated delivery of dexamethasone in steroid-dependent IBD patients—a pilot uncontrolled study. Am J Gastroenterol 2005;100(6):1370–5.

[185] Annese V, Latiano A, Rossi L, et al. The polymorphism of multi-drug resistance 1 gene (MDR1) does not influence the pharmacokinetics of dexamethasone loaded into autologous erythrocytes of patients with inflammatory bowel disease. Eur Rev Med Pharmacol Sci 2006;10(1):27–31.

[186] Castro M, Knafelz D, Rossi L, et al. Periodic treatment with autologous erythrocytes loaded with dexamethasone 21-phosphate for fistulizing pediatric Crohn's disease: case report. J Pediatr Gastroenterol Nutr 2006;42(3):313–5.

[187] Aboody KS, Brown A, Rainov NG, et al. Neural stem cells display extensive tropism for pathology in adult brain: evidence from intracranial gliomas. Proc Natl Acad Sci USA 2000;97(23):12846–51.

[188] Benedetti S, Pirola B, Pollo B, et al. Gene therapy of experimental brain tumors using neural progenitor cells. Nat Med 2000;6(4):447–50.
[189] Kim SK, Kim SU, Park IH, et al. Human neural stem cells target experimental intracranial medulloblastoma and deliver a therapeutic gene leading to tumor regression. Clin Cancer Res 2006;12(18):5550–6.
[190] Ming GL, Song H. Adult neurogenesis in the mammalian central nervous system. Annu Rev Neurosci 2005;28:223–50.
[191] Singh G. Sources of neuronal material for implantation. Neuropathology 2001;21(2):110–4.
[192] Chen L, Tredget EE, Wu PYG, Wu Y. Paracrine factors of mesenchymal stem cells recruit macrophages and endothelial lineage cells and enhance wound healing. PLoS ONE 2008;3(4):e1886.
[193] Friedenstein AJ, Piatetzky II S, Petrakova KV. Osteogenesis in transplants of bone marrow cells. J Embryol Exp Morphol 1966;16(3):381–90.
[194] Pittenger MF, Mackay AM, Beck SC, et al. Multilineage potential of adult human mesenchymal stem cells. Science 1999;284(5411):143–7.
[195] Zuk PA, Zhu M, Mizuno H, et al. Multilineage cells from human adipose tissue: implications for cell-based therapies. Tissue Eng 2001;7(2):211–28.
[196] Bieback K, Kern S, Kluter H, Eichler H. Critical parameters for the isolation of mesenchymal stem cells from umbilical cord blood. Stem Cells 2004;22(4):625–34.
[197] Erices A, Conget P, Minguell JJ. Mesenchymal progenitor cells in human umbilical cord blood. Br J Haematol 2000;109(1):235–42.
[198] Goodwin HS, Bicknese AR, Chien SN, Bogucki BD, Quinn CO, Wall DA. Multilineage differentiation activity by cells isolated from umbilical cord blood: expression of bone, fat, and neural markers. Biol Blood Marrow Transplant 2001;7(11):581–8.
[199] Kogler G, Sensken S, Airey JA, et al. A new human somatic stem cell from placental cord blood with intrinsic pluripotent differentiation potential. J Exp Med 2004;200(2):123–35.
[200] van Netten JP, George EJ, Ashmead BJ, Fletcher C, Thornton IG, Coy P. Macrophage–tumour cell associations in breast cancer. Lancet 1993;342(8875):872–3.
[201] Lewis CE, Pollard JW. Distinct role of macrophages in different tumor microenvironments. Cancer Res 2006;66(2):605–12.
[202] Jiang Y, Vaessen B, Lenvik T, Blackstad M, Reyes M, Verfaillie CM. Multipotent progenitor cells can be isolated from postnatal murine bone marrow, muscle, and brain. Exp Hematol 2002;30(8):896–904.
[203] Peng LH, Tsang SY, Tabata Y, Gao JQ. Genetically-manipulated adult stem cells as therapeutic agents and gene delivery vehicle for wound repair and regeneration. J Control Release 2012;157(3):321–30.
[204] Schäffler A, Büchler C. Concise review: adipose tissue-derived stromal cells—basic and clinical implications for novel cell-based therapies. Stem Cells 2007;25(4):818–27.
[205] Choi MR, Stanton-Maxey KJ, Stanley JK, et al. A cellular Trojan horse for delivery of therapeutic nanoparticles into tumors. Nano Lett 2007;7(12):3759–65.
[206] Kelly PM, Davison RS, Bliss E, McGee JO. Macrophages in human breast disease: a quantitative immunohistochemical study. Br J Cancer 1988;57(2):174–7.
[207] Goswami S, Sahai E, Wyckoff JB, et al. Macrophages promote the invasion of breast carcinoma cells via a colony-stimulating factor-1/epidermal growth factor paracrine loop. Cancer Res 2005;65(12):5278–83.
[208] Lewis C, Murdoch C. Macrophage responses to hypoxia: implications for tumor progression and anti-cancer therapies. Am J Pathol 2005;167(3):627–35.
[209] O'Sullivan C, Lewis CE, Harris AL, McGee JO. Secretion of epidermal growth factor by macrophages associated with breast carcinoma. Lancet 1993;342(8864):148–9.
[210] Tsutsui S, Yasuda K, Suzuki K, Tahara K, Higashi H, Era S. Macrophage infiltration and its prognostic implications in breast cancer: the relationship with VEGF expression and microvessel density. Oncol Rep 2005;14(2):425–31.
[211] al-Shukri S, Korneev IA. The prognostic factors in bladder cancer patients. Urol Nefrol (Mosk) 1996;6:49–53.
[212] Hagemann T, Robinson SC, Schulz M, Trumper L, Balkwill FR, Binder C. Enhanced invasiveness of breast cancer cell lines upon co-cultivation with macrophages is due to TNF-alpha dependent up-regulation of matrix metalloproteases. Carcinogenesis 2004;25(8):1543–9.
[213] Hanada T, Nakagawa M, Emoto A, Nomura T, Nasu N, Nomura Y. Prognostic value of tumor-associated macrophage count in human bladder cancer. Int J Urol 2000;7(7):263–9.
[214] Leek RD, Lewis CE, Whitehouse R, Greenall M, Clarke J, Harris AL. Association of macrophage infiltration with angiogenesis and prognosis in invasive breast carcinoma. Cancer Res 1996;56(20):4625–9.
[215] Mantovani A, Sozzani S, Locati M, Allavena P, Sica A. Macrophage polarization: tumor-associated macrophages as a paradigm for polarized M2 mononuclear phagocytes. Trends Immunol 2002;23(11):549–55.
[216] Ohno S, Ohno Y, Suzuki N, et al. Correlation of histological localization of tumor-associated macrophages

with clinicopathological features in endometrial cancer. Anticancer Res 2004;24(5C):3335–42.
[217] Oosterling SJ, van der Bij GJ, Meijer GA, et al. Macrophages direct tumour histology and clinical outcome in a colon cancer model. J Pathol 2005;207 (2):147–55.
[218] Wyckoff J, Wang W, Lin EY, et al. A paracrine loop between tumor cells and macrophages is required for tumor cell migration in mammary tumors. Cancer Res 2004;64(19):7022–9.
[219] Kah JC, Wong KY, Neoh KG, et al. Critical parameters in the pegylation of gold nanoshells for biomedical applications: an *in vitro* macrophage study. J Drug Target 2009;17(3):181–93.
[220] Metz S, Bonaterra G, Rudelius M, Settles M, Rummeny EJ, Daldrup-Link HE. Capacity of human monocytes to phagocytose approved iron oxide MR contrast agents *in vitro*. Eur Radiol 2004;14(10):1851–8.
[221] Oude Engberink RD, Blezer EL, Hoff EI, et al. MRI of monocyte infiltration in an animal model of neuroinflammation using SPIO-labeled monocytes or free USPIO. J Cereb Blood Flow Metab 2008;28(4):841–51.
[222] Dou H, Destache CJ, Morehead JR, et al. Development of a macrophage-based nanoparticle platform for antiretroviral drug delivery. Blood 2006;108(8):2827–35.
[223] Dou H, Grotepas CB, McMillan JM, et al. Macrophage delivery of nanoformulated antiretroviral drug to the brain in a murine model of neuroAIDS. J Immunol 2009;183(1):661–9.
[224] Janeway CA Jr. Frontiers of the immune system. Nature 1988;333(6176):804–6.
[225] June CH. Principles of adoptive T cell cancer therapy. J Clin Invest 2007;117(5):1204–12.
[226] Harlin H, Meng Y, Peterson AC, et al. Chemokine expression in melanoma metastases associated with $CD8^+$ T-cell recruitment. Cancer Res 2009;69 (7):3077–85.
[227] Randolph GJ, Angeli V, Swartz MA. Dendritic-cell trafficking to lymph nodes through lymphatic vessels. Nat Rev Immunol 2005;5(8):617–28.
[228] Huppa JB, Davis MM. T-cell-antigen recognition and the immunological synapse. Nat Rev Immunol 2003;3(12):973–83.
[229] Ilett EJ, Barcena M, Errington-Mais F, et al. Internalization of oncolytic reovirus by human dendritic cell carriers protects the virus from neutralization. Clin Cancer Res 2011;17(9):2767–76.
[230] Uto T, Akagi T, Yoshinaga K, Toyama M, Akashi M, Baba M. The induction of innate and adaptive immunity by biodegradable poly(gamma-glutamic acid) nanoparticles via a TLR4 and MyD88 signaling pathway. Biomaterials 2011;32(22):5206–12.
[231] Vajkoczy P, Menger MD. Vascular microenvironment in gliomas. J Neurooncol 2000;50(1–2):99–108.
[232] Ischenko I, Seeliger H, Schaffer M, Jauch KW, Bruns CJ. Cancer stem cells: how can we target them? Curr Med Chem 2008;15(30):3171–84.
[233] Pardridge WM. Blood-brain barrier drug targeting: the future of brain drug development. Mol Interv 2003;3(2):90–105,151.
[234] Anderson SA, Glod J, Arbab AS, et al. Noninvasive MR imaging of magnetically labeled stem cells to directly identify neovasculature in a glioma model. Blood 2005;105(1):420–5.
[235] Brekke C, Williams SC, Price J, Thorsen F, Modo M. Cellular multiparametric MRI of neural stem cell therapy in a rat glioma model. Neuroimage 2007;37 (3):769–82.
[236] Bulte JW, Wu C, Brechbiel MW, et al. Dysprosium-DOTA-PAMAM dendrimers as macromolecular T2 contrast agents. Preparation and relaxometry. Invest Radiol 1998;33(11):841–5.
[237] Bulte JW, Kraitchman DL. Monitoring cell therapy using iron oxide MR contrast agents. Curr Pharm Biotechnol 2004;5(6):567–84.
[238] Wu X, Hu J, Zhou L, et al. *In vivo* tracking of superparamagnetic iron oxide nanoparticle-labeled mesenchymal stem cell tropism to malignant gliomas using magnetic resonance imaging. Laboratory investigation. J Neurosurg 2008;108(2):320–9.
[239] Barresi V, Belluardo N, Sipione S, Mudo G, Cattaneo E, Condorelli DF. Transplantation of prodrug-converting neural progenitor cells for brain tumor therapy. Cancer Gene Ther 2003;10(5):396–402.
[240] Ito S, Natsume A, Shimato S, et al. Human neural stem cells transduced with IFN-beta and cytosine deaminase genes intensify bystander effect in experimental glioma. Cancer Gene Ther 2010;17(5):299–306.
[241] Li S, Gao Y, Tokuyama T, et al. Genetically engineered neural stem cells migrate and suppress glioma cell growth at distant intracranial sites. Cancer Lett 2007;251(2):220–7.
[242] Li S, Tokuyama T, Yamamoto J, Koide M, Yokota N, Namba H. Potent bystander effect in suicide gene therapy using neural stem cells transduced with herpes simplex virus thymidine kinase gene. Oncology 2005;69(6):503–8.
[243] Li S, Tokuyama T, Yamamoto J, Koide M, Yokota N, Namba H. Bystander effect-mediated gene therapy of gliomas using genetically engineered neural stem cells. Cancer Gene Ther 2005;12(7):600–7.
[244] Zhao M, Liang C, Li A, et al. Magnetic paclitaxel nanoparticles inhibit glioma growth and improve the survival of rats bearing glioma xenografts. Anticancer Res 2010;30(6):2217–23.

[245] Kim SK, Cargioli TG, Machluf M, et al. PEX-producing human neural stem cells inhibit tumor growth in a mouse glioma model. Clin Cancer Res 2005;11(16):5965–70.

[246] Lorico A, Mercapide J, Solodushko V, Alexeyev M, Fodstad O, Rappa G. Primary neural stem/progenitor cells expressing endostatin or cytochrome P450 for gene therapy of glioblastoma. Cancer Gene Ther 2008;15(9):605–15.

[247] van Eekelen M, Sasportas LS, Kasmieh R, et al. Human stem cells expressing novel TSP-1 variant have anti-angiogenic effect on brain tumors. Oncogene 2010;29(22):3185–95.

[248] Ehtesham M, Kabos P, Kabosova A, Neuman T, Black KL, Yu JS. The use of interleukin 12-secreting neural stem cells for the treatment of intracranial glioma. Cancer Res 2002;62(20):5657–63.

[249] Stagg J, Lejeune L, Paquin A, Galipeau J. Marrow stromal cells for interleukin-2 delivery in cancer immunotherapy. Hum Gene Ther 2004;15 (6):597–608.

[250] Xu G, Jiang XD, Xu Y, et al. Adenoviral-mediated interleukin-18 expression in mesenchymal stem cells effectively suppresses the growth of glioma in rats. Cell Biol Int 2009;33(4):466–74.

[251] Gunnarsson S, Bexell D, Svensson A, Siesjo P, Darabi A, Bengzon J. Intratumoral IL-7 delivery by mesenchymal stromal cells potentiates IFNγ-transduced tumor cell immunotherapy of experimental glioma. J Neuroimmunol 2010;218(1–2):140–4.

[252] Corsten MF, Miranda R, Kasmieh R, Krichevsky AM, Weissleder R, Shah K. MicroRNA-21 knockdown disrupts glioma growth *in vivo* and displays synergistic cytotoxicity with neural precursor cell delivered S-TRAIL in human gliomas. Cancer Res 2007;67 (19):8994–9000.

[253] Ehtesham M, Kabos P, Gutierrez MA, et al. Induction of glioblastoma apoptosis using neural stem cell-mediated delivery of tumor necrosis factor-related apoptosis-inducing ligand. Cancer Res 2002;62(24):7170–4.

[254] Shah K, Tung CH, Yang K, Weissleder R, Breakefield XO. Inducible release of TRAIL fusion proteins from a proapoptotic form for tumor therapy. Cancer Res 2004;64(9):3236–42.

[255] Zhu D, Lam DH, Purwanti YI, et al. Systemic delivery of fusogenic membrane glycoprotein-expressing neural stem cells to selectively kill tumor cells. Mol Ther 2013;21(8):1621–30.

[256] O'Neal DP, Hirsch LR, Halas NJ, Payne JD, West JL. Photo-thermal tumor ablation in mice using near infrared-absorbing nanoparticles. Cancer Lett 2004;209(2):171–6.

[257] Alizadeh D, Zhang L, Hwang J, Schluep T, Badie B. Tumor-associated macrophages are predominant carriers of cyclodextrin-based nanoparticles into gliomas. Nanomedicine 2010;6(2):382–90.

[258] Kennedy LC, Bear AS, Young JK, et al. T cells enhance gold nanoparticle delivery to tumors *in vivo*. Nanoscale Res Lett 2011;6(1):283.

[259] Cole C, Qiao J, Kottke T, et al. Tumor-targeted, systemic delivery of therapeutic viral vectors using hitchhiking on antigen-specific T cells. Nat Med 2005;11(10):1073–81.

[260] Ilett EJ, Prestwich RJ, Kottke T, et al. Dendritic cells and T cells deliver oncolytic reovirus for tumour killing despite pre-existing anti-viral immunity. Gene Ther 2009;16(5):689–99.

[261] Power AT, Bell JC. Taming the Trojan horse: optimizing dynamic carrier cell/oncolytic virus systems for cancer biotherapy. Gene Ther 2008;15(10):772–9.

[262] Russell SJ, Peng KW. The utility of cells as vehicles for oncolytic virus therapies. Curr Opin Mol Ther 2008;10(4):380–6.

[263] Alley SC, Okeley NM, Senter PD. Antibody–drug conjugates: targeted drug delivery for cancer. Curr Opin Chem Biol 2010;14(4):529–37.

[264] Banerjee R, Katsenovich Y, Lagos L, McIintosh M, Zhang X, Li CZ. Nanomedicine: magnetic nanoparticles and their biomedical applications. Curr Med Chem 2010;17(27):3120–41.

[265] Stephan MT, Moon JJ, Um SH, Bershteyn A, Irvine DJ. Therapeutic cell engineering with surface-conjugated synthetic nanoparticles. Nat Med 2010;16(9):1035–41.

[266] Cheng H, Kastrup CJ, Ramanathan R, et al. Nanoparticulate cellular patches for cell-mediated tumoritropic delivery. ACS Nano 2010;4(2):625–31.

[267] Nakamura K, Ito Y, Kawano Y, et al. Antitumor effect of genetically engineered mesenchymal stem cells in a rat glioma model. Gene Ther 2004;11 (14):1155–64.

[268] Hong X, Miller C, Savant-Bhonsale S, Kalkanis SN. Antitumor treatment using interleukin-12-secreting marrow stromal cells in an invasive glioma model. Neurosurgery 2009;64(6):1139–46, [discussion 1146–1137].

[269] Nakamizo A, Marini F, Amano T, et al. Human bone marrow-derived mesenchymal stem cells in the treatment of gliomas. Cancer Res 2005;65(8):3307–18.

[270] Kim SM, Lim JY, Park SI, et al. Gene therapy using TRAIL-secreting human umbilical cord blood-derived mesenchymal stem cells against intracranial glioma. Cancer Res 2008;68(23):9614–23.

[271] Menon LG, Kelly K, Yang HW, Kim SK, Black PM, Carroll RS. Human bone marrow-derived mesenchymal stromal cells expressing S-TRAIL as a cellular

delivery vehicle for human glioma therapy. Stem Cells 2009;27(9):2320–30.

[272] Sasportas LS, Kasmieh R, Wakimoto H, et al. Assessment of therapeutic efficacy and fate of engineered human mesenchymal stem cells for cancer therapy. Proc Natl Acad Sci USA 2009;106 (12):4822–7.

[273] Miletic H, Fischer Y, Litwak S, et al. Bystander killing of malignant glioma by bone marrow-derived tumor-infiltrating progenitor cells expressing a suicide gene. Mol Ther 2007;15(7):1373–81.

[274] Kinoshita Y, Kamitani H, Mamun MH, et al. A gene delivery system with a human artificial chromosome vector based on migration of mesenchymal stem cells towards human glioblastoma HTB14 cells. Neurol Res 2010;32(4):429–37.

[275] Uchibori R, Okada T, Ito T, et al. Retroviral vector-producing mesenchymal stem cells for targeted suicide cancer gene therapy. J Gene Med 2009;11 (5):373–81.

[276] Gu C, Li S, Tokuyama T, Yokota N, Namba H. Therapeutic effect of genetically engineered mesenchymal stem cells in rat experimental leptomeningeal glioma model. Cancer Lett 2010;291(2):256–62.

[277] Sonabend AM, Ulasov IV, Tyler MA, Rivera AA, Mathis JM, Lesniak MS. Mesenchymal stem cells effectively deliver an oncolytic adenovirus to intracranial glioma. Stem Cells 2008;26(3):831–41.

[278] Josiah DT, Zhu D, Dreher F, Olson J, McFadden G, Caldas H. Adipose-derived stem cells as therapeutic delivery vehicles of an oncolytic virus for glioblastoma. Mol Ther 2010;18(2):377–85.

[279] Negroni L, Samson M, Guigonis JM, Rossi B, Pierrefite-Carle V, Baudoin C. Treatment of colon cancer cells using the cytosine deaminase/5-fluorocytosine suicide system induces apoptosis, modulation of the proteome, and Hsp90beta phosphorylation. Mol Cancer Ther 2007;6(10):2747–56.

[280] Altanerova V, Cihova M, Babic M, et al. Human adipose tissue-derived mesenchymal stem cells expressing yeast cytosinedeaminase:uracil phosphoribosyltransferase inhibit intracerebral rat glioblastoma. Int J Cancer 2012;130(10):2455–63.

[281] Pulkkanen KJ, Yla-Herttuala S. Gene therapy for malignant glioma: current clinical status. Mol Ther 2005;12(4):585–98.

[282] Amano S, Li S, Gu C, et al. Use of genetically engineered bone marrow-derived mesenchymal stem cells for glioma gene therapy. Int J Oncol 2009;35 (6):1265–70.

[283] Song C, Xiang J, Tang J, et al. Thymidine kinase gene modified bone marrow mesenchymal stem cells as vehicles for antitumor therapy. Hum Gene Ther 2011;22(4):439–49.

[284] Mori K, Iwata J, Miyazaki M, et al. Bystander killing effect of tymidine kinase gene-transduced adult bone marrow stromal cells with ganciclovir on malignant glioma cells. Neurol Med Chir (Tokyo) 2010;50 (7):545–53.

[285] Choi SA, Lee JY, Wang KC, et al. Human adipose tissue-derived mesenchymal stem cells: characteristics and therapeutic potential as cellular vehicles for prodrug gene therapy against brainstem gliomas. Eur J Cancer 2012;48(1):129–37.

[286] Djouad F, Plence P, Bony C, et al. Immunosuppressive effect of mesenchymal stem cells favors tumor growth in allogeneic animals. Blood 2003;102(10):3837–44.

[287] von Bahr L, Batsis I, Moll G, et al. Analysis of tissues following mesenchymal stromal cell therapy in humans indicates limited long-term engraftment and no ectopic tissue formation. Stem Cells 2012;30(7):1575–8.

[288] DeFrancesco L. Landmark approval for Dendreon's cancer vaccine. Nat Biotechnol 2010;28(6):531–2.

[289] Noh YW, Jang YS, Ahn KJ, Lim YT, Chung BH. Simultaneous *in vivo* tracking of dendritic cells and priming of an antigen-specific immune response. Biomaterials 2011;32(26):6254–63.

[290] Azzi J, Tang L, Moore R, et al. Polylactide-cyclosporin a nanoparticles for targeted immunosuppression. FASEB J 2010;24(10):3927–38.

[291] Kravtzoff R, Colombat PH, Desbois I, et al. Tolerance evaluation of L-asparaginase loaded in red blood cells. Eur J Clin Pharmacol 1996;51(3–4):221–5.

[292] Brandenberger R, Burger S, Campbell A, Fong T, Lapinskas E, Rowley JA. Cell therapy bioprocessing. BioProcess Int 2011;9(s1):30–7.

CHAPTER

13

Molecular Engineering of Cell and Tissue Surfaces with Polymer Thin Films

John T. Wilson[1] *and Elliot L. Chaikof*[2]

[1]Department of Chemical and Biomolecular Engineering, Vanderbilt University, Nashville, TN, USA
[2]Department of Surgery, Beth Israel Deaconess Medical Center, Boston, MA; Harvard Medical School, Boston, MA; Wyss Institute for Biological Inspired Engineering of Harvard University, Cambridge, MA, USA

OUTLINE

13.1 INTRODUCTION

From household paints to semiconductors to implantable medical devices, engineered polymer thin films are at the core of many of the influential and emerging technologies that have and continue to play important roles in society and in our daily lives. Polymer films

J.M. Karp & W. Zhao (Eds): Micro- and Nanoengineering of the Cell Surface.
DOI: http://dx.doi.org/10.1016/B978-1-45-573146-6.00013-1

take on many different facets with respect to their physicochemical properties, including thickness, wettability, reactivity, mechanical properties, conductivity, and optical properties, to name only a few. Polymeric thin films can be assembled, deposited or grown on surfaces using a variety of mechanisms including solvent casting, spraying, coating, covalent conjugation, surface initiated polymerization, or self-assembly (i.e., electrostatic interactions, hydrogen bonding). They have been created using an astonishing number of different polymers, including both synthetic and naturally occurring polymers of diverse architecture and properties. The exquisite ability of polymer thin films to control diverse physical and biochemical phenomena at interfaces combined with their enormous utility as surface engineering tools has sparked considerable excitement towards translating polymer thin film technologies for use in cell surface engineering applications. However, unlike conventional inanimate substrates (e.g., glass, ceramic, metallic, or polymeric supports), which are largely passive bystanders of polymer film deposition or growth, cell surfaces present complex and dynamic interfaces capable of chemically and physically restructuring in response to film constituents. Cellular membranes comprise lipids, proteins, and carbohydrates that are essential for maintaining native cell function and phenotype, and, as such, re-engineering these highly dynamic and diverse chemical landscapes with polymer thin films imposes a number of unique challenges and design restrictions which may not normally be encountered in more traditional thin film engineering applications. This chapter will introduce design key principles for engineering living cell and tissue surfaces with polymer thin films, will describe the main classes of cell surface supported polymer films, and will provide notable examples of how such films are being used in biomedical applications.

13.2 GENERAL DESIGN PRINCIPLES AND CONSIDERATIONS

As in medicine, "Do no harm," should be the most important rule in the design of polymer thin films for cell surface engineering applications. Film constituents, solvents, and film deposition methods and conditions must be minimally toxic to cells and should not interfere with cellular functions that are essential to the desired therapeutic application. Therefore, film fabrication methods that use organic solvents, high temperatures, vacuum, or the use of initiators, monomers, catalysts, and polymers with high cellular toxicity are not suitable for cell surface engineering applications. Rather, film formation techniques must be achieved using cell-compatible constituents and film assembly procedures that can be performed under physiological or near-physiological conditions defined by narrow ranges of pH, ionic strength, temperature (e.g., temperature 25–37°C, pH ~7–7.5, osmolarity ~300 mOsM), and mechanical forces. Additionally, to limit toxicity and promote surface deposition, film constituents ideally should not freely pass through the plasma membrane; this is not generally a problem for most hydrophilic macromolecules but can be an issue when using polycations or low molecular weight polymer precursors such as monomers and initiators. Indeed, many initiators and monomers commonly employed in free radical polymerizations exert significant cytotoxicity. Additionally, thin film deposition techniques such as spin coating, spray coating, or chemical vapor deposition are not suitable for cell surface engineering or would require significant modifications from existing standards. For example, chemical vapor deposition (CVD), a popular and widely used approach to generate polymer thin films, is not currently useful for depositing films on cell surfaces due, in part, to the use of vacuum during the film formation. These stringent requirements dramatically reduce the available canon of film constituents and fabrication techniques available to the cell surface engineer. This has spawned a number of highly innovative thin film engineering approaches to circumvent these limitations, many of which will be described in this chapter.

A second major challenge, which, like toxicity, is a common challenge throughout the field of cell surface engineering is the considerable chemical heterogeneity intrinsic to the cell surface. The cell surface is a complex and nonuniform nanostructured material composed of diverse types of lipids, proteins, and carbohydrates, which can vary in relative composition as function of both time and space along the cell surface in a highly cell-type-specific manner. This considerable complexity presents a major challenge for producing uniform films as well as controlling interactions between film components and specific biomacromolecules on the cell surface; interactions that have important implications for regulation of cell function and behavior. The dynamic nature of the cell surface also imposes a major challenge in generating stable polymer thin films. The plasma membrane is a constant state of flux, with lipid and protein components continuously being internalized via endocytosis, degraded, and replaced with fresh molecules [1]. This is a tightly regulated process in which turnover rates, which can vary from hours to weeks, are very specific to a given molecular species [2]. Hence, polymer films formed on cell surfaces are highly prone to intracellular internalization and degradation as a consequence of this process, rendering control of stability, both during and after film formation, a major consideration in film engineering.

Finally, the ultimate objective of most applications of cell surface engineering is application in clinical cell therapy therefore, polymer thin films are subject to an onslaught of mechanical and biochemical forces in the *in vivo* environment. Polymer films may be exposed to considerable forces as cells are introduced into the bloodstream or migrate through tissues. Depending on the application and cell type, this requires care in generating films that are neither too weak to withstand such forces nor so robust as to hinder cell functions that are dependent on specific mechanical properties of the cell membrane. Additionally, the biochemical environments experienced by cells *in vivo*, particularly in disease states, can have a dramatic impact on the integrity of the film. Likewise, polymer films can be recognized by the host as a foreign material, triggering inflammation, complement activation thrombosis, and even immunity that can be deleterious to both the engineered cell and the health of the host [3–5]. There exist no universal solutions to these challenges. Films must be engineered considering the intended application and the underlying biology or physiology, the type of cell being used, and the environment in which it will be introduced. The next section of this chapter describes some of the important types of thin films that have successfully met these challenges for a number of important biomedical challenges.

13.3 CELL SURFACE ENGINEERING WITH POLYMER THIN FILMS

13.3.1 Covalent Conjugation of Polymer Monolayer Films to Cell Surfaces

The cell surface presents a multitude of reactive groups, including amines, carboxylic acids, and thiols, that are abundant within the natural canon of proteins and carbohydrates present on the cell surface [6,7]. Moreover, through enzymatic, chemical, metabolic, or genetic approaches, additional reactive moieties such as aldehydes, ketones, and azido groups may be introduced to the cell surface milieu [7,8]. As described elsewhere in this book, the development of bioorthogonal and chemoselective conjugation chemistries has revolutionized our ability to reengineer cell surfaces with an enormous diversity of molecules. Accordingly, these reactive sites can serve as anchors for covalent conjugation of appropriately functionalized polymers, representing the most common strategy through which polymer films have been generated on cell surfaces (Figure 13.1A). Among the covalent conjugation chemistries, derivatization of cell surface amines has been the most extensively explored, primarily using polymers functionalized with *N*-hydroxysuccinimide (NHS) activated esters [9–12], and to a lesser extent benzotriazole carbonate (BTC) [13], cyanuric chloride [14], isocyanate derivatives [15], and aldehydes [16].

The covalent conjugation of poly(ethylene glycol) (PEG) chains to the cell surface, most commonly to amino groups, represents far and away the most common example of a covalently anchored polymer thin film. PEGylation of cell and tissue surfaces has been utilized for a number of applications, some of which will be described later in this chapter. Accordingly, covalent conjugation of PEG provides an excellent case study for illustrating some of the key challenges and design parameters associated with fabrication of covalently grafted polymer monolayers. In general, hydrophilic and neutral polymers, such as PEG, associate poorly with the cell surface [17]; the same is also true for negatively charged polymers (polyanions), which are electrostatically repelled by the negatively charged cell surface. Consequently, a large stoichiometric excess of reactive PEG is commonly used to increase the surface density of polymer chains immobilized to the cell surface [17]. The need for large molar excess is further exacerbated when using coupling chemistries that involve nucleophilic attack by cell surface amines (e.g., NHS–esters), which are subject to competing ester hydrolysis under aqueous conditions. Although reactive groups (e.g., maleimide) for other cell surface nucleophiles (e.g., thiols) may have slower rates of hydrolysis, the number of reactive groups on the cell surface is usually considerably lower than the number of amines. Not only is the considerable excess of reactive polymer needed is not only a highly inefficient use of reactive polymer, but such high polymer concentrations can exert cytotoxicity. Two approaches to increasing the surface density of chains while minimizing bulk polymer concentration is to

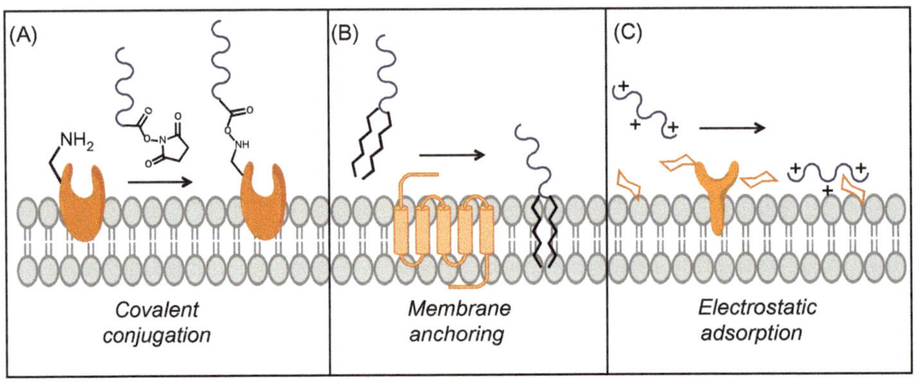

FIGURE 13.1 General approaches to engineering of cell surfaces with polymer monolayers include covalent conjugation of polymers to the cell surface (A), spontaneous insertion of amphiphilic polymers into cell membranes (B), and electrostatic adsorption of polycations to negatively charged cell surfaces (C).

perform multiple sequential reactions [18,19] or to utilize reactions that target multiple anchor groups [20,21]. Increased reaction efficiency has also been described using 2-iminothiolane (Traut's reagent) to convert surface amines to thiol groups to which maleimide-functionalized polymers can subsequently be attached [22]. This approach not only exploits the slower hydrolysis rates of maleimides, conjugation efficiency is further improved by introducing a carbon spacer that renders the resultant thiol more accessible to bulky polymer chains [22]. Recently, a technique termed diffusion modulated macromolecular cell derivatization (DMMCD) has been used to increase the conjugation efficiency of macromolecular species to cell surfaces (Figure 13.2) [17]. By mixing the reactive PEG–NHS with an excess of a nonreactive macromolecule additive such as dextran or hyperbranched polyglycerol (HPG), the surface PEGylation could be increased up to 10-fold. The authors postulate that the synergistic effects of two phenomena can explain this increase in efficiency. First, the diffusion of one type of macromolecule can be enhanced by the presence of a second macromolecule in higher concentration, a process known as enhanced molecular transport. This more rapid diffusion of reactive PEG to the cell surface causes an apparent increase in polymer concentration at the cellular interface, resulting in increased grafting. This enhancement in bulk phase diffusion was postulated to operate synergistically with an increase in the ability of PEG chains to penetrate the glycocalyx, an effect mediated by interactions between the additive macromolecule and the cell surface. Hence, while covalent conjugation protocols must be empirically optimized with respect to polymer type, conjugation chemistry, cell type, and desired application, employing techniques such as DMMCD can widen the window of reaction conditions that may be explored and represents an important tool for improving covalent film deposition on cell surfaces.

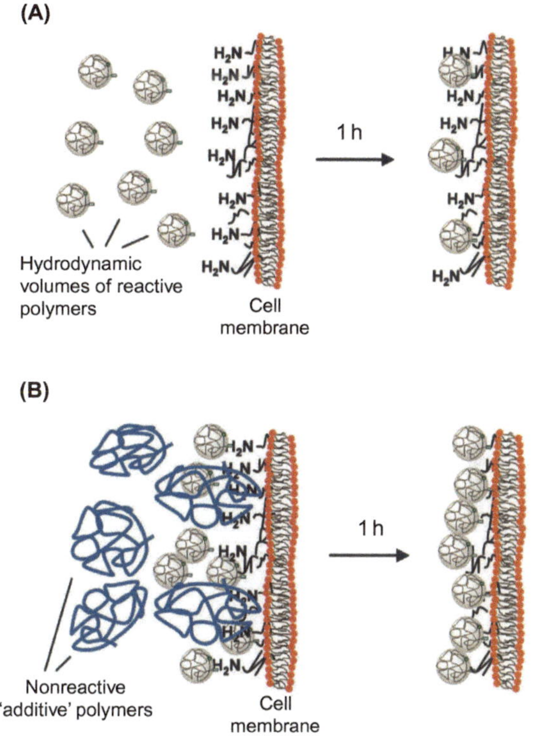

FIGURE 13.2 Diffusion modulated macromolecular cell derivatization (DMMCD) can be used to enhance covalent conjugation of reactive macromolecules to the cell surface. (A) Representation of conventional approach to cell surface modification using an amine-reactive polymer. (B) Cell-compatible surface grafting process in the presence of a nonreactive polymer additive (e.g., dextran or HPG). *Source: Adapted with permission from Ref. [17]. Copyright 2010 American Chemical Society.*

Another important consideration for covalent grafting of polymer films to cell surfaces is the molecular specificity, or often lack thereof, of the conjugation. As most covalently anchored films exploit native amine or thiol groups, a diverse assortment of cell surface proteins may be modified indiscriminately and to various degrees. This has important implications for not only film properties and performance, but also the resultant cellular behavior and function. For example, conjugation of polymer chains to cell surface proteins involved in cell adhesion or migration (e.g., integrins) may dramatically influence the ability of cells to adhere to substrates or migrate. By contrast, failure to conjugate PEG to key epitopes on red blood cells (RBCs) can result in inefficient masking of antigens from recognition by cognate antibodies [23]. When increased molecular specificity is necessary, metabolic or genetic introduction of nonnaturally occurring reactive groups (e.g., azides) [8] or the use of enzymatic ligation strategies (e.g., biotin ligase) [24] may be harnessed to link polymer chains to specific molecules on the cells' surface. While such approaches offer exquisite specificity, they are not suitable for all cell types or surface molecules. Improved specificity may also be conferred through control of the molecular weight and architecture of the reactive polymer. Smaller chains or chains that adopt a more compact conformation may more readily access certain cell surface proteins or buried residues. For example, conjugation of 5 kD PEG chains to RBCs has been shown to block antibody binding to D antigens more effectively than 20kD PEG chains, presumably due to the increased access of smaller chains to this particular surface antigen [22]. The molecular specificity of the conjugation can also have important implications for film stability as the turnover rate for surface proteins can vary dramatically [7] and may also be influenced by polymer conjugation [25]. Mass spectroscopy has recently been used to identify the specific cell surface proteins that were linked to polymeric nanoparticles via a thioether bond formed between thiols on the surface of T cells and maleimide-functionalized nanoparticles [26]. Interestingly, the authors found that many of the proteins involving key T cell functions, such as migration and antigen recognition, were not modified, whereas modified proteins were those that tended to accumulate at the immunological synapse. This helped to explain their finding that nanoparticles conjugated to T cells tended to concentrate at the immunological synapse yet did not interfere with migration or antigen recognition. The use of techniques and analyses such as this are critical to understanding patterns of conjugation that have significant impact on both cell behavior and the performance characteristics of the polymer film.

13.3.2 Membrane-Anchored Polymer Thin Films

Covalent chemical manipulation of native cell surface molecules may negatively affect their biological activity and perturb fundamental cell–cell and cell–matrix interactions and signaling processes critical to cell phenotype and survival [23,27,28]. This has spawned interest in noncovalent cell surface modification strategies, the most common of which harnesses passive insertion of a hydrophobic anchor into the phospholipid bilayer of the plasma membrane (Figure 13.1B). Amphiphilic polymers, which contain both hydrophilic and hydrophobic domains, can exploit this entropically favorable process to introduce polymer chains to the cell surface [29]. Amphiphilic polymers of a variety of compositions and architectures, including diblock [28,30], branched [31], and graft copolymers [11,25], can be used to modify cell surfaces, which offers an enormous degree of flexibility in film design. Polymer–phospholipid conjugates represent the most common type of polymer for noncovalent cell surface engineering due to the highly favorable interactions between diacyl lipids and native lipids in the plasma membrane. A key design variable in engineering this class of polymer thin film is the selection of the hydrophobic anchor(s), which determines the efficiency of membrane insertion. In comparing the cell surface density of polysaccharide chains linked to different hydrophobic anchors, a phospholipid (1,2-dipalmitoyl-glycero-3-phosphoethanolamino), a pyrene group, and single octyldodecyl chain, it was found that the degree of polymer incorporation was greatest when using the phospholipid, followed next by the pyrene moiety with the octyldodecyl anchor providing the least incorporation [28]. Liu et al. [30] reported similar findings in demonstrating that the surface density of an oligonucleotide was highest when using a diacyllipid anchor, followed by cholesterol, and, finally, an octyldodecyl anchor. This work also illustrated how chemistry of the hydrophilic segment can also influence membrane insertion efficiency. Introduction of a PEG spacer between the diacyllipid anchor and the oligonucleotide dramatically reduced the degree of insertion, owing to steric repulsion of PEG from the cell surface [30]. Another key consideration in the design of amphiphilic polymers for cell surface engineering applications is to ensure membrane insertion rather than membrane lysis. Many polymer-based surfactants can lyse cell membranes (e.g., Triton X-100), and therefore careful consideration and optimization of polymer design and film deposition conditions (e.g., concentration, time) is critical.

Film stability is perhaps the biggest limitation of polymer films assembled using membrane-anchors, as lipids and integral membrane proteins are in a constant state of flux, continuously being internalized, degraded, and replaced [1,32]. However, conjugation of polymers to lipid anchors may prolong surface residence time of lipids by interfering with natural mechanisms of endocytosis and recycling. For example, the lipid 1,2-distearoyl-sn-glycero-3-phosphatidylethanolamine (DSPE) is rapidly internalized into T lymphoblasts within an hour, whereas a PEG–DSPE conjugate could be observed on the surface 3–24 h later and appeared to be released directly into the surrounding culture media as opposed to first being endocytosed and subsequently exocytosed [25]. Such differences in not only in the membrane-bound half-life of polymer films, but also the mechanism of film instability can be an important consideration for some applications. For example, polymer films used for facilitating drug delivery from a cell surface to the surrounding environment (discussed below) could leverage a release mechanism from the cell surface; by contrast, internalization of the film may result in drug degradation in endo/lysosomal compartments [33] or be leveraged to achieve controlled intracellular drug delivery from the surface. Alkyl chain length of polymer–phospholipid conjugates can also influence stability. In comparing PEG–lipid conjugates with diacyl chain lengths of 12, 14, and 16 methylene units, Inui et al. [1,25] found that the stability of lipids on the cell surface increased with increasing alkyl chain length; longer chains, however, were not explored.

An alternative to using polymers with a single lipid anchor is to engineer copolymers bearing multiple hydrophobic moieties to provide multipoint membrane attachment. Graft copolymers comprised of a hydrophilic poly(vinyl alcohol) (PVA) backbone and grafted 14-carbon alkyl chains have recently been used as a facile tool for cell surface engineering [2,11,25,34,35]. By incorporating additional functional groups along the polymer backbone, such as thiols or biotins, such films provide a convenient approach for controlling the chemical composition of the cell surface [3–5,34,36]. Compared to a PEG–lipid conjugate containing the same number of methylene units in the alkyl chains, the PVA–alkyl copolymer remained on the cell surface longer, potentially due to the multiple attachment points [6,7,25]. Though dependent on cell type, coupling chemistry, and polymer properties, polymer thin films assembled through membrane insertion can be comparable to that of covalent modification, as illustrated by a comparative study in which similar stability was demonstrated between surface modification with an NHS–PEG reagent and noncovalent approaches utilizing PEG–lipid and PVA–alkyl polymers. Regardless, most polymers were excluded from the cell surface within 24 h. The short half-life of membrane-anchored polymers is a key consideration in their applicability. Strategies that incorporate crosslinkable sites into the membrane-anchor, such as diacetylene, which upon insertion with cell membranes can be UV polymerized *in situ* [7,8,37,38] may provide a mechanism for enhancing film stability.

13.3.3 Electrostatic Adsorption of Polymers

The rich abundance of carbohydrates and glycoproteins renders cell surfaces highly negatively charged, a property that can be harnessed for noncovalently engineering cell surfaces through the use of polycations that interact electrostatically with this negatively charged interface (Figure 13.1C). While monovalent electrostatic interactions with cells are weak due to the high ionic strength of physiological solutions, the multiple positive charges of polycations can promote stable binding via multivalent interactions. However, a significant challenge associated with this facile approach to cell surface engineering has been, and remains, the high cytotoxicity associated with many polycations [9–12,39–42]. Indeed, many polycations commonly used for fabrication of polymer thin films on inanimate surfaces (e.g., silicon wafers and glass slides) can elicit damage to cells by generating pores in the cell membrane resulting in unregulated efflux of molecules across the plasma membrane and ultimately cell death [13,43–45]. Moreover, notwithstanding the toxicity of most polycations, this propensity to rapidly translocate across the membrane in a non-energy-dependent manner, rather than adsorb to the extracellular surface, renders many polycations largely ineffective for cell surface engineering applications. This concept is demonstrated in Figure 13.3A, which shows a confocal micrograph of pancreatic islets incubated with a fluorescently labeled polycation, poly(L-lysine) (PLL). It is obvious in evaluating the images at sufficient magnification that PLL is localized intracellularly, colocalized with the nucleus and distributed broadly within the intracellular space of individual cells within the multi-cellular cluster. Not surprisingly, many such polycations have found utility as carriers in gene delivery applications, due in part to their ability to enhance intracellular uptake [14,33]. By contrast, a poly(L-lysine)-*graft*-poly(ethylene glycol) (PLL-*g*-PEG) copolymer specifically designed for cell surface

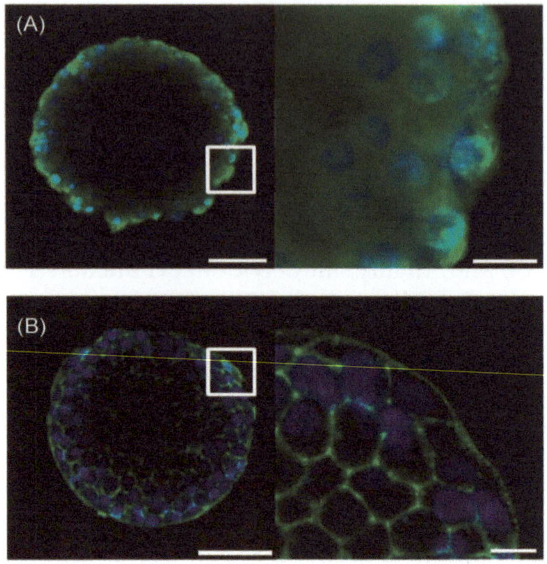

FIGURE 13.3 Polycations used for cell surface engineering must adsorb to cell surfaces with minimal intracellular accumulation. Fluorescent micrographs of pancreatic islets incubated with fluorescently labeled PLL (A) and PLL-*g*-PEG (B) demonstrate distinct differences in the spatial distribution of the polycations. It is evident that PLL, which is cytotoxic, is broadly distributed through the cytosol and vesicular compartments of individual cells within the islet, whereas PLL-*g*-PEG, which displays reduced toxicity, can be observed on the surface of individual cells within the islet. This contrast in behavior highlights the need for careful selection or design of polycations for cell surface engineering applications. *Source: Adapted with permission from Ref. [27]. Copyright 2010 American Chemical Society.*

applications [15,27,46], is observed predominantly on the extracellular surface (Figure 13.3B). Such a distinction may not be clear with lower magnification imaging and indeed in some published reports it is not; this has been and continues to be a weakness in the field that has confounded interpretation of data. Therefore, careful analysis of cellular localization of film constituents is a vital element of all cell surface engineering endeavors but is perhaps most critical when using polycations which have a propensity for perturbing the integrity of the cell membrane. The toxicity and membrane disruptive properties of polycations is dependent on a number of variables, including the type of positive charge (e.g., primary, secondary, tertiary amine), degree of ionization, concentration, contact time, deposition temperature, molecular weight, charge density, hydrophobicity, and chain conformation [16,39,40,42,45]. Additionally, polycation toxicity can vary widely depending on the cell type studied and the method through which viability is assessed [17,47]. Consequently, there is a considerable amount of disagreement and uncertainty in the literature regarding the use of polycations in cell surface engineering applications. For example, PAH has been reported to be both toxic and nontoxic to primary human pancreatic islets [17,48,49]; similar disparity exists for a number of other polycations.

The cytotoxicity of polycations therefore requires careful selection of not only the appropriate polycation properties, but also deposition conditions. Low concentrations, short contact times, and low temperatures can be utilized to allow polycations to be deposited on cell surfaces while minimizing toxic effects. Additionally, naturally derived polycations such as chitosan and protamine tend to exhibit lower toxicity profiles and, therefore, have been used in a number of cell surface engineering applications [16,18,19,50,51]. Another approach has been to re-engineer polycation properties to render them less toxic and membrane destabilizing. While not explicit in their seminal work, Hubbell and coworkers first described that the grafting of PEG to a polycation, specifically PLL, can mitigate cytotoxicity and enable adsorption of polymer chains on the cell surface through electrostatic interactions [20,21,52]. Through adsorption of PLL-*g*-PEG copolymers, they were able to noncovalently introduce PEG chains to the surface of RBCs to prevent lectin-induced hemagglutination. This concept has recently been extended with the goal of further understanding the relationship between the degree of PEG grafting and PEG chain length and polycation cytotoxicity [22,27,46]. In this work, a library of 24 PLL_{MW}-*g*[*D*]-PEG_n ($P_{MW}Pn[D]$) graft copolymers of variable PLL backbone molecular weights (P_{MW}) was synthesized, where MW is the PLL molecular weight in kDa, P*n* is the PEG graft length, where *n* is the number of monomeric repeats, and *D* is the degree of grafting, defined as the percentage of backbone lysine groups grafted to PEG chains. These polymers were incubated with pancreatic islets, multicellular aggregates that were focus of these investigations, and cell viability was subsequently assessed. It was found that polycation toxicity decreased as charge density was reduced but, interestingly, at a fixed molar concentration and degree of grafting, increasing the PEG graft length decreased cytotoxicity, suggesting synergism between the degree of grafting and PEG chain length in reducing toxicity. Significantly, for each length of PEG chain used, a critical degree of grafting, D_c, could be identified where copolymers did not exert significant toxicity under the conditions tested, and D_c was found to decrease exponentially with increasing PEG chain length. Confocal microscopy of pancreatic islets incubated with fluorescently labeled PLL and $P_{MW}Pn[D_c]$ revealed very distinct contrasts in behavior (Figure 13.3). PLL, which is highly toxic, translocated across the plasma membrane and was localized within the cytoplasm of individual cells, whereas labeled $P_{MW}Pn[D_c]$ copolymers adsorbed to the apical surface of individual cells within pancreatic islets, indicating maintenance of cell membrane integrity and minimal endocytosis of copolymers. Such contrasting behavior suggests that conjugation of PEG chains to PLL inhibits the capacity of PLL to cross the cell membrane, most likely through inhibition of membrane pore formation, consistent with observed reductions in toxicity.

PLL-*g*-PEG copolymers engineered for enhanced cytocompatibility through control of structural variables can also be used as "cell surface active" polymeric carriers for ligands and reactive groups [22,27]. This technology, termed PAINTS (Polycations Assembled at Interfaces for Noncovalent Tissue Surfacing), utilizes copolymers bearing terminally functionalized PEG grafts to molecularly paint cell surfaces with a diversity of chemical functional groups. PEG grafts were terminally functionalized with biotin, hydrazide, and azide moieties, to selectively capture streptavidin-, aldehyde-, and cyclooctyne-labeled molecules, (e.g. carbohydrates and peptides) respectively, on cell surfaces [17,27,53]. The introduction of hydrazide groups, which are a useful handle for immobilizing oligosaccharides and glycoconjugates, is particularly notable since covalent generation of hydrazide groups on cell surfaces is not feasible using the commonly employed NHS–ester reagents. Additionally, this approach facilitates presentation of cell surface azido groups within minutes, providing a facile and rapid alternative to metabolic engineering approaches for chemically targeting cell surfaces via copper-free click chemistry, albeit with reduced molecular specificity. For example, a "one-pot" modification of cell surfaces with bioactive peptides could be achieved through concomitant adsorption of an azide-functionalized copolymer and chemoselective ligation of a cyclooctyne-modified peptide. Co-adsorption of differentially functionalized PLL-*g*-PEG copolymers was also demonstrated for simultaneous display of multiple functional groups on cell surfaces, and in principle, a library of copolymers bearing a diverse array of small molecules (e.g., peptides, oligosaccharides, other reactive groups) could be used combinatorially to achieve exquisite control over the composition of cell surfaces in a single step. Cell surface engineering using PLL-*g*-PEG copolymers also offers a number of advantages over conventional covalent approaches, particularly NHS–ester coupling. In contrast to NHS–ester conjugates, PLL-*g*-PEG copolymers allow cells to be functionalized in aqueous solvents containing primary amines (e.g., cell culture media) and may be dissolved well in advance of application without the hydrolysis associated with NHS–esters. Using a biotin-functionalized polymer as a model, adsorbed PLL-*g*-PEG copolymers generated surface densities of functional groups similar to that obtained by treatment with an NHS–PEG (biotin) reagent. However, pancreatic islets treated with the NHS–ester reagent adopted an altered morphology, possibly a result of impaired cell–cell adhesion due to covalent modification of surface proteins. However, the relative stability of this approach in relation to covalent strategies has not yet been evaluated.

13.3.4 Multilayer Polymer Thin Films

Over the past two decades, layer-by-layer (LbL) assembly of polymer thin films has emerged as one of the most versatile and facile bottom-up approaches for engineering surfaces of defined biological and physicochemical properties [23,54–60]. The premise behind the assembly of multilayer thin films is quite simple (Figure 13.4): polymer A binds to a surface via a number of possible interactions (electrostatic, hydrogen bonding, covalent conjugation), unbound polymer is then washed away,

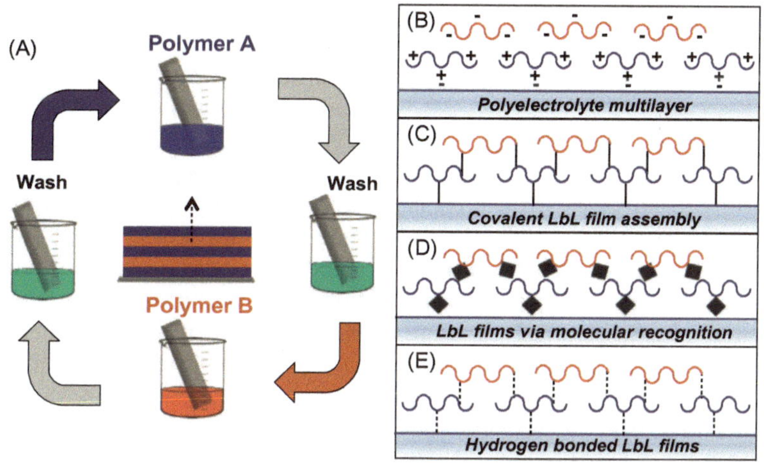

FIGURE 13.4 Assembly of LbL thin films via alternating deposition of polymers with complementary binding properties. (A) General approach to film deposition involves incubation of a substrate into a solution of polymer A, rinsing unbound polymer from the surface, incubating the substrate in solution B, which has an affinity for binding the adsorbed species A, followed by another wash step. This process is repeated until the desired number of layers is generated. Films can be assembled via a number of mechanisms including formation of PEM films via alternating deposition of polycations and polyanions (B), covalent LbL assembly via deposition of polymers functionalized with orthogonally reactive groups (C), use of molecular recognition to drive film assembly (D), and harnessing hydrogen-bonding interactions between layers to assemble the multilayer film (E).

and the surface is then incubated with a second polymer, polymer B, that has an affinity for the bound polymer A. This completes the assembly of a polymer bilayer, and this process can be repeated indefinitely to form more complex multilayered assemblies. Through incorporation of enzymes and other proteins, nucleic acids, liposomes, biologically active nanoparticles, polymers functionalized with bioactive motifs, and guest–host supramolecular complexes, LbL films provide an immensely versatile platform for controlling the biochemical properties of surfaces. Moreover, film thickness, permeability, mechanical properties, and surface chemistry may be tailored through film design, providing additional mechanisms for manipulating biophysical phenomenon at interfaces. Relative to polymer monolayers, LbL films offer considerably more control over film properties, enabling a wider range of possible applications. For example, through control of film thickness and/or permeability, LbL films can be used to regulate the diffusion of molecules to and from the cell surface. Additionally, LbL films provide a three-dimensional scaffold for the presentation or controlled release of bioactive molecules and/or drug carriers and facilitate significantly higher drug loading than polymer monolayers. Moreover, LbL films may provide a route for engineering significantly more stable cell surface modifications since LbL films are anchored multivalently to the cell surface, and become much thicker than the lipid bilayer as successive layers are deposited and effectively "glued" together. This creates a cell–material interface with a very high aspect ratio from the perspective of the cell, which has been shown to inhibit endocytosis [8,61,62]. Furthermore, the mechanical properties of the film may be modulated [24,46] to render it difficult for cells to endocytose the film and its associated cargo [22,63]. Therefore, multilayer polymer thin films offer an enormous opportunity to expand the molecular repertoire of available cell surface modifications beyond what is currently possible with genetic and metabolic approaches or covalent and noncovalent chemistries.

However, until recently, this powerful surface engineering technique has been largely inaccessible to living cell surfaces, due largely to the cytotoxicity inherent to conventional LbL film constituents and/or architectures. Polyelectrolyte multilayer (PEM) films (Figure 13.4B), which are assembled through alternating deposition of polycations and polyanions [7,54], represent the most common and versatile LbL film architecture. Here, a polycation is first deposited onto the negatively charged cell surface, followed by a polyanion, until the desired number of layers is obtained. Cells are washed after each polymer deposition step to eliminate unbound polyelectrolyte. While many (though not all) polyanions are highly cytocompatible, the well-documented toxicity elicited by most polycations [25,39–42] (see above) in direct contact with the cell membrane poses a significant hurdle in employing PEMs in cell surface engineering. Nonetheless, through careful selection of polycations and assembly routines, a number of reports have described the assembly of PEM films on mammalian cells or cellular aggregates without significantly compromising cell viability. Owing to its reduced cytotoxicity relative to synthetic polycations, chitosan, a naturally derived polysaccharide, has been commonly employed as the polycation component in the assembly of PEM films. For example, Rajagopalan et al. [26,50] reported use of chitosan in the assembly of PEMs on hepatocytes, Pickup and colleagues [23,27,28,51] have described the assembly of chitosan/alginate films on MIN6 cells, and Tabrizian and coworkers [29,64] have utilized phosphorylcholine-modified chitosan for coating RBCs with PEM films. Hence, chitosan may be an attractive choice of polycation for generating PEMs on cell surfaces, though it should be noted that dissolution of chitosan requires slightly acidic conditions (~pH 6.0–6.5), which may not be suitable for some cell types. Another approach to minimize polycation cytotoxicity has been to deposit an initial basement layer onto the cell surface to serve as a molecular spacer between polycations and the cell surface. For example, a cationic lipid conjugate was first inserted into cell membranes to enable deposition of a polyanionic alginate spacer layer prior to polycation deposition [28,30,65]. While such approaches may reduce direct contact between the polycation and the cell membrane, the capacity of polycations to diffuse within PEM films [31,66] may still permit direct interaction with the cell surface.

Another approach to designing PEM films for cell surface engineering has recently been reported by the team of Wilson and Chaikof [11,25,46] (Figure 13.5A–C). To circumvent the molecular hurdle of polycation cytotoxicity, they exploited PEG-dependent conformational changes in polycation (PLL) structure to design cytocompatible polycations for direct cell surface assembly of PEM films with tunable physicochemical and biochemical properties. A major challenge in designing films using PLL-*g*-PEG copolymers is a complex interplay between copolymer structural variables that influence cell viability and PEM growth in opposing manners. For example, decreasing polycation charge density generally attenuates cytotoxicity [28,39], while insufficiently charged species may be incapable of participating in film growth [30,67,68]. Similarly, PLL-*g*-PEG copolymers bearing long PEG chains can create steric barriers to protein adsorption and molecular recognition and similarly may hinder electrostatic interactions necessary to drive film assembly [30,69]. By mapping the relationships between copolymer properties and both cell viability and PEM film growth (using alginate as the polyanion), they were able to unveil a narrow window, where films of diverse and unique composition, thickness, and mechanical properties could be generated without

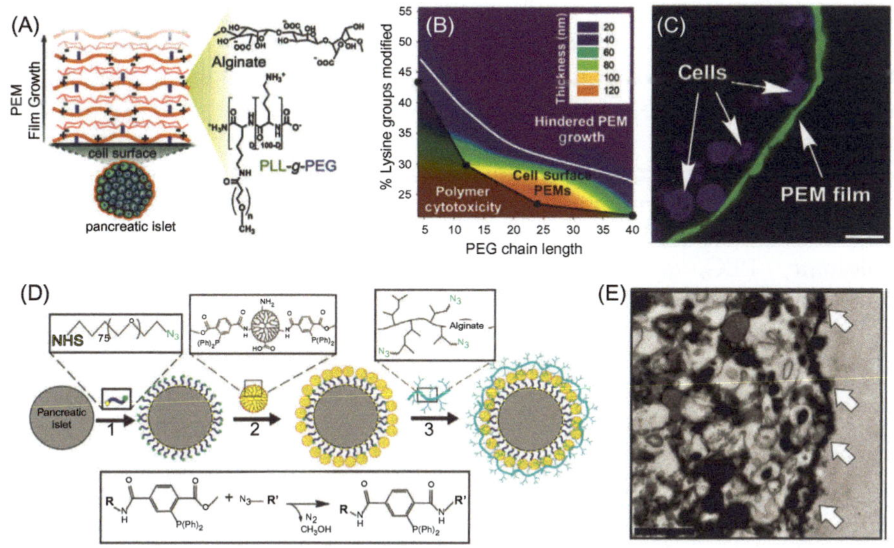

FIGURE 13.5 Examples of LbL thin films assembled on cell and tissue surfaces. PEM films could be assembled on the surface of pancreatic islets via LbL deposition of appropriately designed PLL-*g*-PEG copolymers and alginate (A). A library of PLL-*g*-PEG polymers composed of different degrees of lysine modification and different lengths of grafted PEG chains was synthesized and relationships between polymer structural properties, cytotoxicity, and film growth were established (B). The gray overlay delineates the region where polymers were found to have measurable cytotoxicity and the white line approximates the boundary of another region where film growth was limited. Between these regions a window in polymer structure could be found where PEM film growth could be achieved using cytocompatible polycations. Confocal microscopy demonstrated the deposition of fluorescently labeled alginate on the extracellular surface of cells (C). LbL films can also be deposited on islets via covalent Staudinger ligation using alternating deposition of PAMAM dendrimers functionalized with methyl-2-(diphenylphosphino)terephthalate moieties and alginate bearing azide groups to drive film formation (D). Film deposition on the islet surface can be visualized via TEM and estimated to be 70–100 nm thick (E). *Source: A–C: Adapted with permission from Ref. [46]. Copyright 2011 American Chemical Society. D and E: Adapted with permission from Ref. [71]. Copyright 2013 American Chemical Society.*

compromising cell viability or function (Figure 13.5B). Additionally, through use of functionalized polyelectrolytes, a diversity of reactive groups could be incorporated into films, providing a strategy for further tailoring film properties and integrating bioactive molecules. By elucidating the effect of key film design variables (e.g., polycation structure, layer number) on both film properties and cellular viability/function, this work serves as a model for the rational design of cell surface supported multilayer thin films.

While PEM films represent arguably the most versatile architecture, the polycation toxicity challenge has also spawned efforts to design new architectures that are less dependent on electrostatic interactions and instead seek to utilize covalent bonds, molecular recognition, and hydrogen bonding to drive film growth. Covalently assembled films (Figure 13.4C) utilize the toolbox of bioorthogonal and chemoselective chemistries described briefly previously in this chapter and elsewhere in this book to link polymer layers both to the cell surface and to each other. Iwata and coworkers [1,32,36] described the assembly of a multilayer PVA film via covalent disulfide exchange reactions. To generate the initial PVA layer, the cell surface was modified noncovalently with maleimide groups using a maleimide-functionalized phospholipid and thiol-modified PVA was subsequently bound covalently via thiol-ene reaction. Through alternating deposition of thiol-modified and pyridyl disulfide-functionalized PVA, covalently crosslinked thin films of PVA could be assembled. Similarly, covalent films could be generated through alternating deposition of thiol- and maleimide-functionalized polymers (e.g., alginate, PEG, and lipid-based micelles). Significantly, these films could be grown to be approximately 30 μm thick, substantially thicker than most LbL assemblies and were shown to be stable on the surface of pancreatic islets for at least 30 days, representing a very stable surface modification [25,70]. The design of covalently assembled films is also exemplified by the recent work of Stabler and coworkers [33,71] who generated covalently crosslinked LbL films on the surface of viable and functional pancreatic islets using the chemoselective Staudinger ligation to link polymers functionalized with azide and phosphine groups (Figure 13.5D–E). Specifically, they utilized alternating deposition of a hyperbranched alginate bearing azide groups and poly(amidoamine) (PAMAM) dendrimers functionalized with methyl-2-(diphenylphosphino)terephthalate moieties. This not only provided a mechanism for orthogonal covalent crosslinking but also a strategy through which to integrate azide- or phosphine-functionalized molecules into the film. Interestingly, since PAMAM and alginate are polycationic and polyanionic, respectively, this allowed for generation of hybrid PEM–covalent films as well as comparison of films assembled via both mechanisms. Similar to the approach taken by Wilson and Chaikof [46], the degree of positive charge of phosphine-modified PAMAM dendrimers was attenuated via reaction of residual primary amines with glutaric anhydride (GA). Indeed, decreasing the PAMAM positive charge reduced toxicity and enabled deposition of an initial phosphine-rich layer on the cell surface, which served as the foundation for covalent immobilization of azide-functionalized hyperbranched alginate. Through repeated alternating deposition of azide–alginate and PAMAM–phosphine polymers, multilayer thin films could thereby be grown. Deposition of six bilayers generated films approximately 70–100 nm thick on the cell surface as measured by TEM (Figure 13.5E); this represents one of the more rigorous attempts to characterize film properties on the cell surface, a significant technical challenge. However, the amount of alginate deposited on the surface and the uniformity of the coating tended to decrease as the PAMAM cationic charge was decreased, prompting the investigators to fabricate films assembled completely through covalent interactions. This could be achieved by first introducing azide groups to the cell surface using a heterobifunctional NHS–PEG–azide, followed by covalent LbL assembly of a net neutral phosphine-modified PAMAM dendrimer and azide-modified hyperbranched alginate. Their results demonstrate that covalent cross-linking of LbL films could significantly enhance film stability, though films assembled purely via covalent interactions demonstrated a higher degree of surface inhomogeneity. This was attributed to incomplete and nonuniform coverage of the initial PEG-azide layer via NHS coupling relative to the more complete coverage achieved via electrostatic deposition of the PAMAM dendrimer. This highlights the need to balance the surface density and uniformity of the basement layer with film stability and toxicity, and also suggests that hybrid films assembled via multiple mechanisms may provide an ideal combination of these two important variables.

Cytocompatible LbL films can also be generated on cell surfaces using molecular recognition between two cognate-binding partners to drive film assembly (Figure 13.4D). Film growth is initiated using polymers functionalized with motifs that bind endogenous surface molecules or by first introducing an exogenous group (e.g., biotin) to the cell surface. Through alternating deposition of polymers or biomacromolecules that specifically bind to one another film growth can be achieved. This is perhaps best exemplified through the use of the biotin–streptavidin interaction, which represents one of the strongest noncovalent interactions known in nature. Wilson et al. [48] utilized electrostatic adsorption of cytocompatible poly(L-lysine)-*g*-poly(ethylene glycol)(biotin) (PPB) copolymers to create a biotin-rich cellular interface. This enabled binding of streptavidin; since streptavidin has four binding sites for biotin, a fraction remains unoccupied, which facilitated the binding of another layer of PPB and the regeneration of the biotin-rich interface. Therefore, through alternating deposition of streptavidin and PPB, PEG-rich multilayer films could be assembled on cell surface without compromising cell viability. Films can also be fabricated using natural biopolymers and/or biomacromolecules with complementary binding affinities. An elegant example of this concept was described by Akashi and coworkers [72] who generated thin films comprised of fibronectin and gelatin on the surface of adherent fibroblasts. Although both film components are negatively charged, they can be used to generate LbL films through binding of gelatin to fibronectin via a collagen-binding domain.

LbL assembly driven by hydrogen bonds has emerged as a powerful technique for generating films under physiologic conditions using polymers that carry no charge and, consequently, have recently been explored as an alternative to PEM films for cell surface engineering (Figure 13.4E). Originally developed for surface engineering and nanoencapsulation of yeast cells [73], hydrogen-bonded LbL films assembled through alternating deposition of tannic acid (TA) and poly(*N*-vinylpyrrolidone) (PVPON) have recently been translated for mammalian cell surface engineering [74]. Film growth is initiated through deposition of PVPON which bears pyrrolidone rings that contain a proton-accepting carbonyl group to facilitate hydrogen bonding with carbonyl, amide, and hydroxyl groups on the cell surface. The TA layer can subsequently be deposited through hydrogen bonding between PVPON carbonyl groups and hydroxyl groups on TA. This deposition cycle can be repeated to generate films entirely stabilized by hydrogen-bonding interactions. Moreover, TA possesses antioxidant properties and is capable of scavenging free radicals. Hence, these films possess intrinsically anti-inflammatory properties, as evidenced by reduced pro-inflammatory cytokine production when incubated with activated macrophages.

Despite the enormous potential offered by LbL films for cell surface engineering, a number of challenges and design considerations must be addressed when creating such films. As with any surface modification, film stability is a key consideration. While LbL films offer potential for generating more stable films, particularly the covalently assembled films described above, they are nonetheless susceptible to endocytosis and chemical and/or enzymatic degradation [51,71,74]. Moreover, film stability must also be accounted for *during* the film assembly process, as deposited layers may be endocytosed prior to addition of subsequent layers. Minimizing polymer incubation times and assembling films at reduced temperatures, where endocytosis is inhibited, can help ensure complete film assembly prior to returning cells to physiological temperatures. This, of course, must be balanced with the sensitivity of cells to lower temperatures over the time of the coating process. The coating process itself also represents a significant practical challenge in LbL film assembly, as the numerous deposition and wash steps are both time consuming and laborious. Most approaches have utilized centrifugation to separate cells from coating and wash solutions, a process that can mechanically damage cells, is time consuming and subject to human variability.

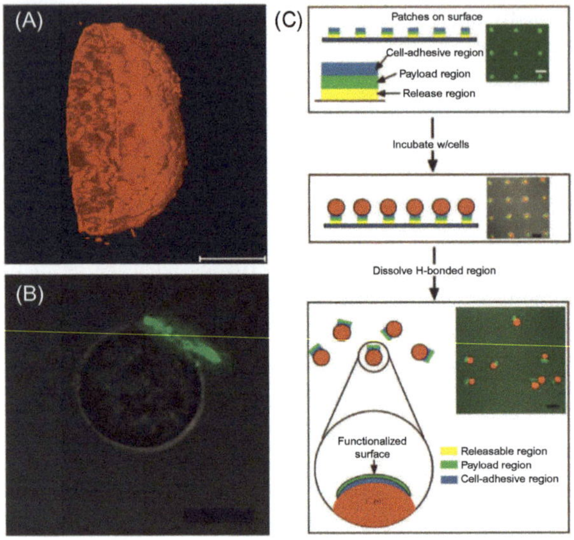

FIGURE 13.6 LbL thin films can be assembled directly on cell and tissue surfaces to create conformal coatings (A) or be attached to only a part of a cell, creating a cell "backpack" for anisotropic surface modification (B). Cell backpacks can be created by first assembling LbL films containing three regions on a micropatterned substrate. The region proximal to the substrate is designed for triggered released from the surface, followed by a payload region for carrying drug cargo, and a terminal cell adhesion region through which to tether cells to film surface. Cells can bind surface-bound patches of film via receptor–ligand interactions, and upon release of the film from the substrate become functionalized with the LbL film backpack (C). *Source: A: Adapted with permission from Ref. [48]. Copyright 2008 American Chemical Society. B and C: Adapted with permission from Ref. [76]. Copyright 2008 American Chemical Society.*

Methods for automating assembly of LbL films on cells and tissues have been explored and are currently in development [75]. Automated methods will not only save time and increase the efficiency of the coating process, but also allow for high throughput optimization of film design and process variables.

The direct assembly of LbL films on cell and tissue surfaces results in the formation of a thin coating that conforms to the geometry of the surface (Figure 13.6A). While this is favorable in some applications, such as generating barriers to diffusion of molecules to the cell surface, it can be disadvantageous in certain applications if the film interferes with cell–cell and cell–matrix interactions that are essential for desired cellular function. The team of Irvine and Rubner [76] has devised an elegant solution to this challenge that enables cell surface engineering with LbL films that does not completely occlude the cell surface. Rather than using the cell surface as the foundation upon which to generate films, they have focused on generation of LbL films that can be attached to cells postassembly, generating a so-called cell backpack (Figure 13.6B,C). In this approach, films are assembled onto a photolithographic patterned surface to create distinct patches of LbL film. The film is comprised of a temperature and pH-responsive basement region that allows for triggered release of the film from the substrate, a central payload region, and an outer cell-adhesive region that contains hyaluronic acid (HA), a ligand recognized by CD44 receptors on the surface of lymphocytes. Cells are then seeded onto the array of patches where they can bind to the patches via receptor–ligand interactions. Upon a change in temperature or pH, the PEM patches, now connected to cells, are released from the underlying substrate yielding a suspension of cells modified with a "backpack" that may contain diverse functional payloads. Significantly, cells maintained their native ability to migrate, a phenomenon dependent on molecular interactions between the cell surface and underlying substrate. This approach may prove particularly advantageous in applications where anisotropic cell surface modification is desired and also allows films to be assembled under nonphysiologic conditions using constituents that may be toxic in soluble form but are not when deposited on a surface. This approach greatly expands the possible options available for using LbL films in cell surface engineering applications where conformal cell coating is not required.

13.4 BIOMEDICAL APPLICATIONS

13.4.1 Molecular Camouflage and Nano-Encapsulation

Cell- and tissue-based therapies continue to emerge as promising strategies for disease treatment. Early examples include blood transfusion and organ transplantation, now mainstays of modern medicine. Substantial advancement in areas such as adoptive T cell transfer [77], engineered induced pluripotent stem (iPS) cells [78], and mesenchymal stem cells (MSCs) [79] therapies has led to numerous promising clinical trials that continue to ignite excitement for the burgeoning field of cell-based therapy. However, in many instances, organs, cells, and tissues are derived from a nonautologous source, resulting in recognition and destruction of graft by the host immune repertoire that target epitopes present on the foreign cell surface and vice versa (i.e., destruction of host tissue by foreign immune cells). Therefore, cell and organ transplantation often necessitates the use of immunosuppressive drugs to prevent the destruction and rejection of the transplanted tissue by the host immune system. While considerable advancements have been made over the past two decades in the development of immunosuppressive drug therapy, systemic immunosuppression requires life-long treatment and continues to impose a number of adverse side effects. This challenge fueled one of the earliest applications of cell surface engineering: the use of polymer thin films to effectively hide or "camouflage" transplanted cells from detection by the immune system. To accomplish this, polymer thin films on the cell surface must provide a physical barrier between the cellular and molecular components of the immune system, including T cells, antibodies, and complement proteins, and their targets on the transplanted cell surface [80] (Figure 13.7). At the same time, polymer films must allow the diffusion of small molecules such as oxygen, glucose, and ions as well as the transport and binding of other molecules essential to cell viability and desired function. Therefore, in designing such films, it is essential to consider what cell functionalities are essential for a given application and which may

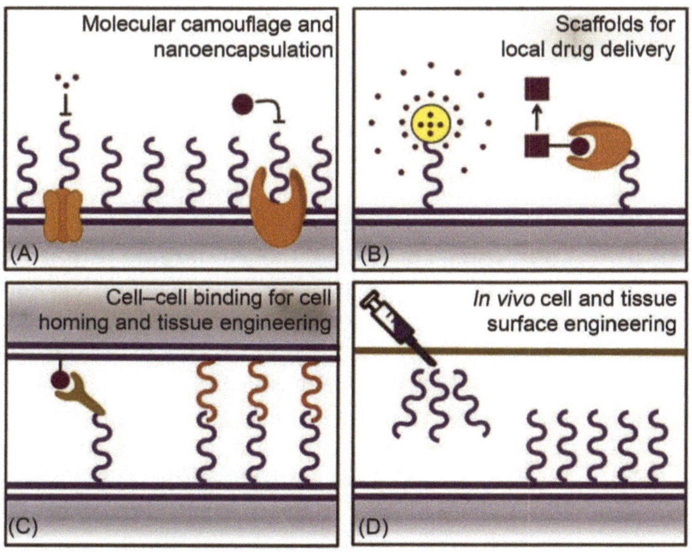

FIGURE 13.7 Biomedical application of cell surface engineering with polymer thin films include enabling molecular camouflage and nanoencapsulation to mitigate inflammatory and immune responses that occur during cell transplantation (A) (ii) serving as linkers and scaffolds for the presentation of immobilized enzymes and nanoparticles for controlled drug delivery (B), (iii) mediating cell–cell binding interactions for improving cell homing or driving self-assembly of multicellular structures for tissue engineering applications (C), and (iv) modifying of cell and tissue surfaces *in vivo* for improved integration of implantable devices and enhanced performance of therapeutics (D).

be compromised at the benefit of gaining increased immune protection. RBC transfusion and islet transplantation are two prominent examples where the immune protection conferred by polymeric barriers can outweigh any deleterious effects on cell function.

There are two general classes of barrier: steric barriers, which prevent or inhibit molecular interactions as a result of the spatial structure of the film; and permselective barriers, which exclude diffusion of molecules on the basis of molecular size, charge, and/or physical configuration. The most common approach utilized in RBC transfusion and islet encapsulation has been to generate steric barriers via immobilization of PEG chains to the cell or tissue surface to prevent molecular recognition between cell surface receptors and soluble ligands. PEG is a hydrated, flexible polymer chain due to repeating, highly mobile, ether units, which creates a large hydrodynamic volume that allows the polymer chain to act as a polymer exclusion layer on the cell surface. PEG can be synthesized with a range of molecular weights and architectures (linear, branched, and star). These films have mostly been formed through covalent coupling of PEG to amine groups on the cell surface or by direct insertion of PEG–lipid conjugates into the cell membrane, although other approaches have been used including PEG-based copolymers designed to adsorb to cell surfaces [48,52,81,82].

The most common use of PEG in cell surface engineering has been in the pursuit of the "universal" RBC for blood transfusion. Although the discovery of ABO blood group antigens on RBCs has revolutionized the categorization and matched transfusion of blood, patients that receive multiple transfusions develop alloimmunity against minor antigens on RBCs that consequently prohibits identification of appropriate blood donors [83]. Moreover, limitations in donor blood supply, particularly for group O type, continue to drive interest in strategies to generate universal RBCs [23]. To prevent immune rejection of RBCs, surface-bound PEG chains must efficiently mask blood group antigens (e.g., A, B, Rh) from recognition by cognate antibodies in circulation; even a small amount of residual antibody-binding activity can have deleterious effects on RBC survival [23]. Additionally, surface PEGylation must not dramatically hinder the functionality or circulation time of RBCs nor interfere with natural splenic RBC clearance mechanisms [23,83]. Toward addressing this considerable challenge, investigations have focused on inhibiting antibody recognition of RBCs by optimizing the mechanism of PEG immobilization (i.e., conjugation chemistry), surface grafting density, molecular weight, and polymer architecture. Due to the relative lack of molecular specificity of PEGylation reagents and the diversity of antigens present on the RBC surface, there exists a complex and incompletely understood interplay between these variables in achieving complete masking of blood antigens without compromising *in vivo* survival and circulation properties. This is exemplified by studies demonstrating that only RBCs PEGylated at and below grafting concentrations 0.4 mM of 5 kDa PEG and 2 mM of 20 kDa PEG survived in mice, despite having normal *in vitro* cell structure, function, and viability [84].

The surface density of PEG and the efficiency of antigen masking does not necessarily correlate with the solution concentration of reactive PEG, which is confined by an upper limit where RBC morphology and function are compromised [23,85]. Hence, largely empirical approaches are commonly utilized to determine optimal conjugation conditions. For example, for a 5 kD cyanuric chloride-functionalized PEG, it was found that a maximal 15 mg/ml PEG concentration, pH of 8.7, temperature of 14°C, and reaction time of 30 min were most suitable for RBCs [85]. Likewise, the dependence on PEG molecular weight and architecture in masking RBC antigens is not entirely intuitive. Several groups have demonstrated that PEG chains of different molecular weight are more effective at masking certain blood group antigens; for example, 20 kD PEG has been shown to mask A antigens better than 5 kD PEG, whereas D antigens were more efficiently blocked by 5 kD PEG chains [22]. Similarly, in comparing linear PEG chains to highly branched polyglycerol (HPG), PEG chains formed larger exclusion zones and, consequently, masked major blood antigens better than HPGs, though the more compact HPG structure was found to be more effective than PEG at protecting certain minor antigens [84]. As the inhibition of antibody binding is dependent on steric blockade of cognate antigen epitopes, this result may reflect differences in the accessibility of amino groups within specific blood antigens to different sized PEG chains. Therefore, using combinations of molecular weight PEG can lead to more effective antigen masking. Additionally, while amine-reactive chemistries have been most commonly explored, some evidence suggests that the maleimide linkage may be more stable under certain environments encountered during transfusion, although further study is necessary to more thoroughly characterize the properties of RBCs modified via different chemistries *in vivo* [86].

Optimization of PEG conjugation chemistries has yielded a number of preclinical studies in animal models that illustrate both the promise and challenges of RBC PEGylation. Scott et al. [14] were among the first to evaluate PEGylated RBCs *in vivo*, using cyanuric chloride coupling to conjugate methoxy poly (ethylene glycol) (mPEG) to RBC surface amines. PEGylated RBCs lost their ABO group reactivity and were more resistant to phagocytosis by monocytes *in vitro* [14], but, significantly, appeared to be functionally normal *in vivo* [87]. However, the concentration of reactive mPEG used to achieve normal *in vivo* RBC survival (0.4 mM) was not found to efficiently block antigen binding *in vitro*; conversely, the high concentrations used to mask antigens *in vitro* lead to poor *in vivo* survival. The group also demonstrated that PEGylated sheep RBCs exhibited significantly prolonged survival when transfused into mice, with ~90% less antibody produced in response to mPEG-coated RBCs, illustrating the promise of using PEG coatings to improve transfusion across species. In a study comparing the conjugation of 2, 5, and 20 kD mPEG coupled using cyanuric chloride, BTC, and NHS ester amine-reactive chemistries, they found that the 20 kD mPEG chains to provide better immunocamouflage and, using a 20 kD BTC-functionalized mPEG, reported normal survival of RBCs in mice [13]. Recently, Chen et al. have also demonstrated no ill effects of RBC PEGylation on the recipient microcirculation, as indicated by no changes in hematocrit, hemoglobin concentration, blood vessel diameter, blood flow velocities, and the interstitial partial oxygen pressure (pO_2) before, during, and after the injections of PEG–RBCs coated with a 20 kD mPEG [12]. While RBCs have been the major target, PEGylation of other blood cell types has also been explored. mPEG modification of peripheral blood mononuclear cells effectively decreased antibody recognition and inhibited T cell activation [88]. Similarly, PEGylated donor T cells prevented allorecognition and initiation of graft-versus-host disease *in vivo* [89], and PEGylation of allografts prolonged survival in a bone marrow transplantation model [90]. Platelet concentrates have also been modified by PEG to prevent bacteria binding and biofilm formation [91]. Attempts have even been made to perfuse entire hearts with reactive PEG *ex vivo* to attenuate hyperacute xenograft rejection upon transplantation [92].

Despite promising *in vivo* reports throughout the last two decades, generation of universal RBCs through surface PEGylation has yet to achieve clinical realization. In addition to the continued challenge of obtaining complete antigen masking, the discovery of anti-PEG antibodies that can shorten the survival and circulation time of PEGylated RBCs poses another significant barrier to the use of PEG in the engineering of "universal" RBCs [23]. This has prompted recent investigation into other surface conjugated polymers. Kizhakkedathu and colleagues [84,93,94] have recently explored NHS activated HPG as an alternative to PEG. Like PEG, HPG is highly hydrated and biocompatible but is more compact and contains multiple hydroxyl groups for linking therapeutic molecules. Through appropriate control of HBG molecular weight and grafting concentrations, HBG-functionalized RBCs demonstrated normal *in vivo* circulation times and splenic clearance mechanisms [93]. Though PEG grafting provided better shielding and protection of ABO antigens from antibody recognition than HPG polymers [84], HBGs provided superior protection against some minor antigens [84,93], which is important in chronic blood transfusion where alloantibodies to minor antigens develop. Importantly, HPG-grafted RBCs do not appear to be immunogenic, as a similar RBC circulation profile was observed upon repeated administration in mice over several months. Covalent conjugation of polyethyloxazoline has also been explored as a PEG alternative [95], and though polyethyloxazoline grafting better preserved RBC morphology, mPEG was superior for masking blood group antigens.

As covalent grafting of polymer chains can damage RBCs through modification of integral membrane proteins, resulting in sequestration of RBCs by the immune system, noncovalent approaches have been explored as an alternative to covalent grafting of PEG and other polymer chains. Elbert and Hubbell [52] were among the first to utilize a noncovalent strategy, using PLL-*g*-PEG copolymers that adsorbed electrostatically to RBC surfaces. By controlling the length of the PLL backbone and the PEG grafting ratio, they created an adsorbed film that was able to prevent lectin-induced hemagglutination; however, camouflage of specific RBC surface antigens was not explored. More recently a conceptually similar approach has been explored by Grandfils and coworkers [96] using poly(2-dimethylamino ethylmethacrylate) (PDMAEMA) and PDMAEMA–PEG copolymers, though a low efficiency of blood group antigen masking was observed for all polymers evaluated.

Tabrizian and colleagues [64,97] have recently taken a different polymer film engineering approach to generating a universal RBC. Instead relying on a single polymer layer to form a steric barrier between RBC surface antigens and antibodies, they instead assembled a PEM thin film on the surface of RBCs. Key to their design was selection of a polyelectrolyte pair, particularly the polycation, and a film architecture that did not result in RBC lysis or loss of function while also providing a barrier to antibody binding. Moreover, an optimal balance between solution cell density and polyelectrolyte concentration had to be determined to avoid cell aggregation during the film buildup process due to cross-bridge formation. To achieve this, the group assembled a film consisting of an inner protective shell made of four bilayers of chitosan-*graft*-phosphorylcholine (CH-PC)/alginate (AL) to prevent RBC lysis and an outer shell composed of two bilayers of alginate/PLL-*g*-PEG terminated with a negatively charged alginate layer to repel proteins from the RBC surface. This film could be assembled without RBC lysis and did not comprise RBC oxygen uptake. Significantly, a multilayered shell composed of 13 biocompatible polyelectrolyte layers of $(AL/CH\text{-}PC)_4-(AL/PLL\text{-}PEG)_2\text{-}AL$ was found to dramatically inhibit the binding of antibodies against A, B, D, and Rh RBC surface antigens, as measured in an agglutination assay. While LbL films had previously been deposited on fixed RBC templates [98], this work represents the first to develop an LbL film that can be assembled on functional RBCs. It remains to be seen how LbL-coated RBCs will function *in vivo*, particularly given the possibility that other key RBC properties may have been altered as a result of film assembly, for example, RBC mechanical properties or accessibility of the essential RBC membrane protein CD47 [84]. While covalent conjugation of PEG is still the gold standard in RBC camouflage technology, there remains a significant need for the development of new polymer thin-film-based approaches to enhance RBC transfusion.

The replacement of compromised insulin production through transplantation of pancreatic islets has long been conceived as a promising therapy for the treatment of diabetes [99]. However, immune recognition and destruction of transplanted cells remains a major limitation to the widespread clinical realization of islet transplantation [80]. Host reaction to transplanted islets can be generally categorized into two intertwined phases. The first phase, which happens immediately upon transplantation, results in activation of the coagulation cascade and induction of an inflammatory response initiated by islet-surface-derived factors [99]. Recruitment of immune cells and production of proinflammatory cytokines leads to nonspecific destruction of a large fraction of injected cell mass within days of transplantation. The second phase of the response is mediated by adaptive immunity and the generation of antibodies and T cells that recognize foreign antigens on the islet surface, resulting in destruction of the graft [100]. Both phases of the response are largely driven by molecular recognition of islet-surface molecules by cells (e.g., macrophages, cytotoxic T cells) and molecules (e.g., complement, antibodies) of the host immune system, and, hence, molecular camouflage of these cell surface mediators using polymer thin films has emerged as a promising strategy for improving the outcome of islet transplantation.

Similar to surface modification of RBCs, optimized chemistries and conditions have been developed that allow PEG chains to be displayed on the surface of islets without compromising viability or function. These strategies have primarily utilized amine-reactive PEG [10,15,18,19,101–104] (e.g., NHS) or passive membrane insertion of PEG–lipid conjugates [16,105]. The efficacy of islet PEGylation in attenuating immune responses and improving islet function has been evaluated in a number of animals of both allo- and xenotransplantation. Byun and colleagues [10] have demonstrated that covalent conjugation of PEG (5 kD)–propionic acid to islets extended islet survival from 5 to 11 days in a rat allograft model. Histological evaluation demonstrated the presence of the PEG on islets 30 days after transplantation, a significant finding considering the rapid turnover of many cell surface molecules. While PEGylation prevented graft infiltration by host immune cells, it did not prevent host cell recruitment to the graft site [10,18]. Though the PEG film appeared to impede recognition of the islet by immune cells, it did not appear to inhibit the release of antigens and immunostimulatory compounds (e.g., chemokines, cytokines), resulting in lymphocyte recruitment and localized production of toxic mediators. Hence, it was postulated that islet PEGylation could synergize with systemic administration of immunosuppressive drugs, enabling prolonged islet survival while reducing the dose, and side effects, of immunosupressive drug required. Indeed, in a model of allotransplantation, the combination of islet PEGylation with low-dose cyclosporin A extended islet survival time of PEGylated islets from 12 to more than 100 days, the lifetime of the rats used in this study [10]. More recently, the group has developed an innovative approach through the use of 6- and 8-arm PEG stars terminally functionalized with catechol groups, which react with islet-surface amines [101]. Here, in a more challenging xenograft model of islet transplantation, conjugation of the PEG stars did not significantly enhance islet survival relative to unmodified islets (~11 days), but doubled survival (21 days) when administered in combination with the immunosupressive drug Tacrolimus. Survival could be extended further (>50 days) using a combination of PEGylation, Tacrolimus, and anti-CD154 mAb. Similarly, Dong et al. [102] demonstrated that islet PEGylation resulted in long-term (>100 days) normoglycemia in 30% of recipients, whereas no long-term normoglycemia was observed for mice receiving uncoated islets. This effect could be extended to 57% by tethering nanoparticles loaded with the immunomodulatory drug leukemia inhibitory factor (LIF) to PEG chains. Currently, islet PEGylation is not able to completely protect islets from the host immune system, a consequence of both the relatively short stability of the modification compared to the life span of the graft as well as the complex immunological mechanisms underlying islet destruction [80]. Nonetheless, the synergy between PEGylation and low-dose immunosuppression holds considerable promise for improving the outcome of islet transplantation while reducing the potentially severe side effects of immunosuppressive drugs. In this regard, cell surface PEGylation should be considered to be part of the therapeutic arsenal when designing combination immunosuppressive drug regimens.

While the relatively short residence time of most monolayer polymer thin films on cell surfaces limits their efficacy in completely preventing immune rejection of islets over the desired life span of the graft, the surface residence time of such thin films may be more appropriate to camouflage *early* cellular and molecular recognition events that lead to deleterious inflammatory responses [99]. For example, Contreras and coworkers [103,104] have used islet-surface PEGylation in a xenogenic model of intraportal islet transplantation, where xenoreactive natural antibodies and complement act rapidly to destroy the islet graft. Within the immediate posttransplant period (2 weeks), animals that received islets treated with PEG presented significantly better control of glucose than animals receiving nonmodified islets. PEG with a molecular weight of 5 kD performed slightly better than 2 kD PEG, and capping surface-grafted PEG with albumin proved most efficacious, which was attributed to the enhanced capacity of thicker films and/or larger steric barriers. More recently, Teramura and Iwata [105] have utilized a noncovalent approach to islet-surface PEGylation using lipid-modified PEG in both allograft and autograft [106] models of islet transplantation. In both reports, surface engineering of islets with PEG resulted in improved islet engraftment, extending the duration of graft survival and permitting infusion of a fewer number of islets to achieve normoglycemia, mostly likely through inhibition of early inflammatory responses. However, as is the case in adaptive immunity to transplanted islets, the presence of a molecular blockade may not be sufficient to completely abrogate responses, which may also be mediated by the release of chemoattractants and proinflammatory compounds from the islet [99]. Therefore, molecular camouflage has been combined with surface immobilization of active regulators of inflammatory and thrombotic responses. For example, PEGylation has been combined with surface immobilization of heparin, thrombomodulin (TM), and urokinase, all active regulators of the coagulation cascade [9,16,34,107].

While islet PEGylation has clearly demonstrated some encouraging results, it remains unclear whether or not surface-grafted PEG will ultimately remain stable enough to provide protection for the anticipated lifetime of a human islet graft. Moreover, data suggests that the efficacy of PEGylation is at least partially limited by the lack of a defined pore structure and dependence on a steric exclusion effect in order to provide an immunoprotective barrier. Inspired largely by the concept of islet microencapsulation, which provides a higher level of immunoprotection than PEGylation but is limited by high transplant volumes and diffusional limitations [80], several groups have recently began to explore the possibility of constructing nano-thin films of controlled permeability and surface chemistry directly on the surface of pancreatic islets via LbL polymer self-assembly (discussed above), effectively creating conformal, permselective membranes of nanoscale thickness. Wilson and Chaikof [46,48] were the first to describe *in vivo* investigations into the ability of LbL films to protect islets from host inflammatory and immune responses. In their most recent report [46], islets coated with an eight bilayer PEM film comprised of PLL-*g*-PEG and alginate showed a trend toward improved engraftment *in vivo* in a murine allograft model but failed to improve rates of conversion to normoglycemia to a statistically significant level. While this study marked the first of its kind in that viable cells coated with PEM films were shown to function in an *in vivo* environment, the data suggest that the films provided an insufficient barrier for protection against early mediators of islet destruction. To date, no other *in vivo* investigations into the use of LbL-coated islets, or other cell types, have been reported. Instead, other investigators have focused on evaluating the barrier capacity of the film *in vitro* by assessing the ability of the film to prevent binding of fluorescently labeled ligands for islet-surface targets. For example, Pickup and colleagues [51] demonstrated reduced penetration of an antibody against class I major histocompatibility complex upon coating MIN6 cell spheroids with chitosan/alginate PEM films. The assembly of LbL films on cell surfaces is a nascent area of research, with most efforts focusing on developing new film types and deposition strategies to circumvent the attendant fabrication and toxicity challenges; however, tools for evaluating cell surface film permeability and barrier capacity, among other properties, are in their infancy.

13.4.2 Polymer Thin Films for Enhancing Cell and Drug Delivery

Cell-based therapies have recently found utility in the treatment of numerous pathologies, including heart and vascular disease, stroke and spinal injury, musculoskeletal disorders, cancer, and diabetes. The field of cell surface engineering has enabled modification of cell surfaces with an enormous diversity of biologic drugs, as well as drug delivery vehicles, that bestow the cell with new biochemical functionalities to improve therapeutic performance (Figure 13.7). For example, cell surfaces can be engineered with specific targeting moieties to facilitate cell homing to an injured area [31,108,109], enzymes that can enhance localized production of therapeutic products [9,20,107], and nanoparticle carriers for sustained release of drugs that can improve cell function and survival [26,102,110]. In some instances, the local production or release of drugs from the cell surface has been shown to be superior to parental drug administration [110], which may be limited by systemic toxicity, inadequate circulation time, and lack of targeting. In many such cell surface engineering applications, polymer thin films play an essential role in enabling immobilization of exogenous bioactive species.

As in many other applications described herein, surface immobilized PEG serves as the most common and notable example of a polymer thin film that enables the introduction of exogenous bioactive molecules to the cell surface. The high biocompatible and water solubility of PEG render it an attractive choice as a linker and spacer between the cell surface and the tethered species. Moreover, the increasing commercial availability of heterobifunctional PEG molecules with diverse reactive groups has increased the accessibility of cell surface engineering with PEG-based linkers. Work by Chaikof and coworkers serves as a good illustration for the use of functionalized PEG films for immobilization of biological active biomacromolecules. Their team developed a two-step process for covalently conjugating recombinant TM, an important inhibitor of thrombosis and inflammation, to the surface of pancreatic islets in a chemo- and bio-orthogonal manner [9]. The group first introduced triphenylphosphine moieties to the cell surface through use of a heterobifunctional NHS–PEG–triphenylphosphine linker. This enabled subsequent chemoselective Staudinger ligation of a recombinant TM construct engineered with a C-terminal azido (N_3) group. This process could be achieved without compromising the viability or function of the islet. By increasing the surface density of TM on the islet surface, the thrombin-mediated conversion of protein C to activated protein C (APC) was dramatically enhanced. Similarly, Iwata and colleagues have utilized lipid–PEG conjugates with a terminal maleimide group for covalent anchoring of thiol-modified TM, urokinase, a thrombolytic enzyme, and complement receptor 1 [107,111]. The team has also utilized graft copolymers comprised of a hydrophilic PVA backbone functionalized with both pendant 14-carbon alkyl chains and maleimide groups and used this carrier for conjugation of similar biomacromolecules [34]. Given the deleterious effects of thrombosis and inflammation on islet survival in the immediate posttransplant period, arming islet surfaces with an arsenal of anticoagulant and anti-inflammatory molecules offers a promising strategy for improving clinical outcomes of islet transplantation.

Heterobifunctional PEG has also been an essential tool for engineering cell surfaces with nanoparticles for controlled drug release. Recently, Dong et al. [102] utilized an NHS–PEG–biotin compound to first introduce biotin groups on islet surfaces, allowing subsequent immobilization of ~100 nm PLGA nanoparticles coated with avidin molecules. The PLGA particles were designed to enable efficient loading and controlled release of LIF, a cytokine that can improve islet engraftment through reduction of inflammation, induction of immunological tolerance, and promotion of islet beta cell survival and growth. They postulated that islet viability, insulin secretion, and *in vivo* function could be enhanced through local and sustained of LIF to both islet cells and the surrounding milieu. Indeed, they demonstrated that while islet PEGylation resulted in long-term (>100 days) normoglycemia in 30% of recipients, this effect could be extended to 57% by tethering LIF-loaded nanoparticles to the cell surface. This work is predated by the seminal work of Irvine and colleagues, who tethered drug-loaded nanoparticles to the surface of T cells and hematopoietic stem cells to improve *in vivo* survival. In this study, however, nanoparticles were tethered directly to cell surface thiols without the use of a polymer film as a linker.

Holden and colleagues have recently demonstrated another manner by which cell surface polymer films can potentially be used for drug delivery from cell surfaces [112]. In a proof-of-concept study, the group covalently tethered PEGylated poly(amido amine) (PAMAM) dendrimers, a common polymeric drug carrier, to the surface of macrophages. To do so, they first generated cell surface aldehydes via mild sodium periodate treatment. Resultant aldehydes were then coupled to amine groups on the dendrimer via Schiff base formation, a bond that was subsequently reduced to improve stability using sodium cyanoborohydride. However, as conceded in their report, surface modification of highly endocytotic cells such as macrophages and dendritic cells poses a significant challenge for stable surface modification. Toward addressing this challenge, the team of Rubner and Mitragotri utilized the "cell backpack approach" described earlier in this chapter to functionalize macrophage surfaces with patches of LbL polymer films that were largely resistant to phagocytosis [63]. Because macrophages struggle to phagocytose very flat and disk-shaped particles [61,62], such as the LbL patches used, stable cell surface modification can be achieved. To exemplify the potential utility of polymer films for drug delivery, PEM patches were loaded with a layer containing bovine serum albumin as a model biologic payload, which could be released from the film over several hours. Owing to the versatility of such PEM films in loading diverse therapeutic agents, this strategy offers great potential for cell surface engineering of highly phagocytic cells, such as dendritic cells currently used in cancer immunotherapy [113]; however, the ability of therapeutically relevant cells modified in this manner to function *in vivo* remains to be demonstrated. Though both of the aforementioned studies stopped short of loading the polymer film with therapeutically relevant drugs, by harnessing the ability of macrophages to migrate to infected sites, damaged tissue, and tumors, these reports establish an important foundation for using drug-loaded polymer films to empower cell-mediated delivery of therapeutics.

While some cell-based therapies can be directly injected into the target tissue (e.g., pancreatic islet transplantation) or have the intrinsic capacity to locate pathologic tissue (e.g., adoptive T cell transfer), many therapeutic cell products inefficiently home to the desired site because the infused cells lack key receptors required for homing. This results in only a small fraction of the infused cells, as well as any associated drug cargo, arriving at the site of interest; hence, engineering cell surfaces with molecules to improve the efficiency and specificity of targeting is an area of active investigation [31,108,109]. In this application, polymers have also played an important role (Figure 13.7). Sarkar et al. introduced the selectin ligand sialyl Lewis^x (sLe^x) to the surface of MSCs via a three-step process, first covalently conjugating biotin to MSCs via NHS coupling, followed by immobilization with streptavidin, and lastly displaying a biotinylated tetravalent sLe^x construct with a poly-(acrylamide) polymer linker [108]. By maximizing the surface density of sLe^x, the MSC homing to a site of inflammation *in vivo* via selectin-mediated rolling as well as subsequent extravasation through the endothelium was significantly enhanced. Here, not only did the poly(acrylamide) linker provide a facile strategy for immobilization of sLe^x, the authors also suggest that the polymer helped space the sLe^x groups at a distance from the cell surface comparable to that of native ligands, providing a sufficient length to bypass the steric and electrostatic repulsion provided by the endothelial glycocalyx. More recently, Jeong et al. utilized an HPG functionalized with both octadecyl chains and a peptide sequence that targeted vascular endothelial adhesion molecule (VCAM) overexpressed by inflamed blood vessels [31]. This multimodal polymer efficiently incorporated with the surface of MSCs through insertion of the octadecyl chains into the lipid bilayer, resulting in a surface-bound HPG polymer monolayer displaying the targeting peptides on the surface in a single step and without any covalent modification. By engineering the cell surface with this novel "cell-guidance" molecule, MSCs more efficiently adhered to inflamed endothelial cells in an *in vitro* flow camber model. While the *in vivo* homing capabilities of this technology have not been assessed, these works serve as excellent illustrations of how multifunctional polymers, specifically engineered to interact with the cell surface, and carry biologically relevant molecule, can provide a versatile and facile approach for re-engineering the molecular landscape of cell and tissue surfaces.

13.4.3 Tissue Engineering

The ability to organize individual cells in a spatially controlled manner, particularly in three dimensions, offers great potential for *de novo* engineering of tissues, for creating *in vitro* models of pathology, and for modulating the cellular composition of a target tissue. Polymer thin films have played an important role toward this end (Figure 13.7). A relatively simple but important application has been to use cells modified with polymeric thin films to generate 3D cell aggregates [114–119]. In these applications, the polymer film serves as "glue" that helps hold the cells together, a synthetic extracellular matrix of sorts that drives cellular assembly, and provides initial structural support. The use of simple polymer films to assemble cell aggregates is exemplified by the work of Yu and coworkers who have utilized covalent linkages between hydrazide-functionalized PEI [114] or cationic dendrimers [115] and cells engineered with surface aldehydes through mild periodate treatment. Here, the polycation serves to bring the cells together through electrostatic crossbridge formation, allowing hydrazone bonds to subsequently be formed between the cells. Of note, the polymer concentrations and contact times were optimized in these studies to minimize potential toxicity. This linkage was reported only to last for about 5–7 days after which time the aggregate structure was maintained by natural cell–cell and cell–matrix interactions. This strategy has been used to dramatically accelerate the formation of spheroids comprised of carcinoma cell line for use as a 3D cell culture model for *in vitro* drug penetration studies [116]. More recently, the group has utilized an entirely noncovalent approach using oleyl–PEG conjugated to a polypropylenimine hexadecaamine dendrimer [117]. Here, the positively charged dendrimer serves to concentrate the polymer on the surface such that when cells are brought into close contact with each other through mechanical means (e.g., centrifugation, micromanipulation), they may be linked together through insertion of the hydrophobic oleyl groups into adjacent cell membranes. The ability to link cells into complex patterns was demonstrated using optical tweezers to spatially organize engineered cells; though perhaps not a scalable strategy for tissue engineering applications, it serves to demonstrate the enormous potential of this approach in controlling tissue architecture and may prove useful in understanding cell–cell interactions in hierarchical cellular architectures.

Gartner and Bertozzi [118] have also described a versatile and elegant approach to generating 3D cellular assemblies. By covalently modifying cell surfaces with oligonucleotide sequences through Staudinger ligation or copper-free click chemistry, cells engineered with complementary DNA strands self-assembled in a highly specific manner to create heterotopic 3D microtissues of defined composition and connectivity. Cell connectivity via DNA hybridization could be reversed through melting of DNA sequences at elevated temperatures or through addition of DNase, providing simple and convenient mechanisms for eliminating extracellular DNA after tissue assembly. As a simple but telling demonstration of tissue engineering using this approach, a murine B-cell line that is dependent on the cytokine IL-3 for survival was linked via DNA hybridization to CHO cells engineered to express IL-3. Only when the B cells were linked to CHO cells were they able to survive, demonstrating the formation of a paracrine signaling network in three dimensions. In contrast to approaches utilizing specific receptor–ligand interactions (e.g., biotin–avidin) or orthogonally reactive chemistries, this approach to tissue self-assembly is particularly powerful since the number of complementary DNA sequences that may be employed to drive assembly is practically unlimited. This provides a combinatorial strategy that, in principle, could allow precise control of the arrangement of a diversity of modified cell types in a single step. Moreover, the group has also demonstrated the ability to use this approach for templating cells on substrates in a controlled manner [120,121], offering further potential for this technology in engineering tissues from scaffolds of more complex geometries.

The formation of cellular aggregates has also been driven through the use of patches of multilayer thin films assembled using the previously described "cell backpack" approach. To utilize the LbL patches for this application, Swiston et al. [119] mixed cells with a suspension of PEM patches that had been released from the underlying substrate. The cell surface linked to the patches through receptor–ligand interactions, here, CD44 on B cell surfaces and HA deposited on the surface of the PEM patch. In this scenario, backpacks, displaying multiple ligands on their surface, are free to associate with multiple cell membranes simultaneously, thereby linking multiple cells together. By controlling the dimensions of the backpack and the ratio of cells to backpacks, the size of cell aggregates could be tailored. This approach to generating self-assembling cell aggregates may be particularly advantageous over strategies that use a polymer monolayer in that the PEM patches provide a versatile method for incorporating a diversity of bioactive molecules that can be released from the film after aggregate assembly, delivering key chemical cues to direct the fate of the aggregate.

LbL films have also recently emerged as a strategy for creating layered assemblies of cells for tissue engineering applications. Rajagopalan et al. [50] assembled cell-adhesive chitosan/DNA PEM films on adherent cell monolayers in order to promote the adhesion and growth of a second layer of cells, the film effectively serving as a glue between layers in a multilayered cellular construct. This approach is limited, however, by the use of chitosan as a PEM film constituent, owing to its cytotoxicity and low solubility in neutral buffers. Akashi and coworkers [122] later reported a conceptually similar approach to generate 3D cellular multilayers but employed LbL films comprised of the extracellular matrix proteins fibronectin and gelatin, which interact with each other through biorecognition rather than electrostatic interactions. Using this approach, they were able to generate assemblies comprised of four cell layers, both homo- and heterotopic, that could be subsequently removed from the underlying substrate to which cell were initially adhered, providing a strategy for engineering cell sheets. Given the conformal nature of films assembled through LbL assembly, these approaches could reasonably be applied to generating cellular multilayers on more geometrically complex substrates as well, which might further expand the potential of this approach in tissue engineering applications.

In addition to *de novo* assembly of cellular assemblies, investigators have also used thin-film-based approaches to integrate exogenous cell types with naturally occurring cell aggregates, most notably pancreatic islets. This has been exemplified by the work of Iwata and colleagues who have utilized a number of different polymer thin-film-based approaches to encapsulate pancreatic islets with a layer of living cells. This not only creates both a physical barrier between the islet and the surrounding environment but also a biochemical barrier generated by factors expressed or secreted by the immobilized cell. The group has utilized phospholipid–PEG(biotin) to introduce biotin groups to the cell surface, facilitating subsequent immobilization of streptavidin-coated HEK293 cells [35]. They have also utilized an approach similar to Gartner and Bertozzi wherein polyDNA–PEG–lipid conjugates were used to noncovalently engineer cell surfaces with specific DNA sequences. By incubating poly(adenine) functionalized HEK293 cells with poly(thymine) functionalized islets, HEK293 cells could be specifically immobilized on the islet surface through DNA hybridization [123].

13.4.4 *In vivo* Modification of Cells and Tissues with Polymer Thin Films

While cell-based therapy and cellularized tissue engineered constructs offer incredible potential for treating a number of diseases, challenges such as cell sourcing, cell and tissue preservation and banking, and notably, high cost remain potential barriers to the clinical translation of these technologies. Hence, *in situ* fabrication of thin films in living organisms using specifically engineered polymers or film deposition strategies may provide a solution to some of these translational challenges as well as open new opportunities in drug delivery and tissue engineering (Figure 13.7). Elbert and Hubbell [124] were the first to demonstrate *in vivo* surface remodeling of tissue surfaces with polymer thin films. Building upon their previous work using PLL-*g*-PEG copolymers to noncovalently modify cell and tissue surfaces with PEG brushes *in vitro*, they postulated that self-assembly of polymers with a tissue-binding domain (PLL) and a tissue repelling domain (PEG) could be used to prevent formation of surgical adhesions in a rat model. In this model, an ischemic injury is made to the uterine horn that untreated results in fibrous adhesions to other organs in the peritoneal cavity. By applying a solution of PLL-*g*-PEG to both the injury site and to the organs in the vicinity of the uterine horn, the incidence of surgical adhesion could be reduced by 88% owing to disruption of cell–tissue adhesive interactions as well as inhibition of fibrin formation. This work highlights a number of potential advantages of using such a noncovalent approach for *in vivo* tissue remodeling, most notably that the modification was achieved nearly instantaneously, no leaving groups or other reaction byproducts were generated, and the modification could be accomplished in a complex biochemical milieu comprised of proteins and carbohydrates that might otherwise compete with covalent methods to modify the cell surface.

In vivo cell surface engineering has also been used to improve the efficacy of therapeutics. Irvine and coworkers [30] have recently described a noncovalent approach for *in situ* engineering of tissues with oligonucleotides, specifically CpG oligodeoxynucleotide, a single-stranded DNA stimulator of the immune system that is being explored as an adjuvant for vaccination and as a cancer immunotherapeutic. Toward this end, they conjugated CpG to a diacyl lipid and postulated that the amphiphilic construct would enable spontaneous *in vivo* cell surface engineering via membrane insertion of the diacyl chains. They found that CpG–diacyllipid conjugate was retained at an intratumoral injection site significantly longer than unmodified CpG, a relatively small and hydrophilic macromolecule that rapidly diffused from the injection site. The authors hypothesized that cell surface engineering with CpG provided a mechanism for prolonging residence time in the tissue, creating a depot that enhanced localized immune attack of the tumor. While the insertion of the conjugate into cell membranes *in vivo* was not explicitly demonstrated, and it is conceivable that micellization of CpG–lipid conjugates might also effect biodistribution kinetics. This study serves as an excellent example of the potential benefit that can be achieved through *in vivo* engineering of cells and tissues with therapeutic agents.

Integration of implantable and injectable biomaterials and tissue engineered scaffolds with the surrounding host tissue remains a major challenge to the field and an active area of research. To address this challenge, several groups have developed implantable polymeric materials that covalently react with surrounding cells and tissue surfaces. These covalent strategies must utilize chemistries that do not rely on toxic catalysts or generate harmful by-products and reactive groups must be sufficiently stable to allow for aqueous solubilization prior to injection. Elisseeff and coworkers [125,126] have described the synthesis of a multifunctional condrotin sulfate (CS) bearing both acrylate and aldehyde groups for use in cartilage repair applications. A solution of the polymer can be applied to a defect in cartilage, where aldehyde groups on the polymer bind to proteins on cells and extracellular matrix through Schiff base formation. The excess polymer is then washed away, generating a CS thin film functionalized with acrylate groups on the tissue surface. Acrylate groups at the tissue surface could then be covalently integrated with an injectable acrylate-based hydrogel formed via *in situ* photoinitiated polymerization. Thus, the polymer thin film serves as a bridge between the injured tissue surface and the injectable biomaterial scaffold, promoting tissue integration at the interface. Similarly, Brubaker et al. [127] recently utilized a four-arm PEG bearing terminal catechol groups to effectively glue pancreatic islets to tissue surfaces. Oxidation of the catechol-functionalized polymer solution with sodium periodate induces rapid hydrogel formation with o-quinone groups generated during the process capable of covalently coupling with primary amines at cell surfaces. The group leveraged this strategy to apply a thin polymer film that covered pancreatic islets deposited on the epididymal fat pad and external surfaces of the liver. By mixing the sodium periodate and polymer solutions immediately before deposition, the polymer gelled *in situ* while covalently coupling with both islets and tissue surfaces, creating a stable interface between two. This allowed islets to be immobilized in anatomic positions not possible without the use of such an adhesive. Significantly, the film elicited minimal acute or chronic inflammatory responses and permitted normoglycemic recovery and islet revascularization.

13.5 CONCLUSION

Over the past century, engineered polymer thin films have had an enormous technological impact, and their use as functional coatings has enabled the realization of a plethora of products and devices that now are mainstays of daily life. However, most polymer thin film technologies, even those intended for coating of medical devices, were not developed for cell surface engineering applications, where the health and survival of the cellular support is of utmost importance. Recent advances in polymer and organic chemistry, combined with an improving understanding of cell–polymer interactions, are now allowing the rich surface engineering toolbox afforded by polymer thin films to be translated to living cellular interfaces. This has opened up a number of new and exciting areas in the field of cell surface engineering with significant potential for clinical impact. Notably, the recent advances in engineering of cell surfaces with LbL thin films is ushering in a new wave of cell encapsulation and molecular camouflage research. For the past several decades, the field has largely been dominated at two ends of the length scale: large microcapsules, which provided maximal protection but imposed mass transfer restrictions, and surface immobilized polymer chains, which did not add to the transplant volume, but lacked permselectivity. Many of the LbL films currently in the early phases of development may provide the solution to this conundrum. Moreover, LbL films, particularly covalently crosslinked LbL films, are also well poised to circumvent the stability challenge associated with many cell surface engineering approaches. Rapid growth and development in the field of cell-based therapeutics has also identified a need for the delivery of drugs to support the survival and function of transplanted cells. This merger of cellular therapy and drug delivery has spawned highly innovative research focused on engineering of cell surfaces with nanoparticle vehicles and thin films for local delivery and controlled release. While polymer films have thus far taken only a modest role in this area, with some of the more prominent reports described here, the inevitable need to increase drug loading and control drug release profiles is likely to invoke the use of polymer films as linkers or scaffolds in the near future. For some applications, LbL films, such as the cell backpack, may be uniquely suited to deliver large amounts of cargo in a highly controlled manner. For others, it may be necessary to use a diversity of linker chemistries and immobilization methods to maximize immobilization of enzymes or nanoparticle delivery vehicles. Finally, while most cell surface engineering endeavors are performed *ex vivo* and under controlled conditions, the emerging field of regenerative medicine and tissue engineering has led to the development of injectable polymers that can integrate, both covalently and noncovalently, with the surface of cells and tissue. *In vivo* cell surface engineering strategies are also beginning to merge with the drug delivery paradigms of controlled release depots and targeted drug delivery, where immobilization of drugs or drug carriers on cell surfaces can enhance local retention time, improve uptake at a desired site, or modulate the pharmacokinetics of the drug. In all of these *in vivo* applications, polymers will likely play an important role in mediating the interaction with the target tissue or otherwise improving the therapeutic efficacy. Looking forward, engineered polymer thin films will undoubtedly continue to play a major role in the development, improvement, and optimization technologies that will define this century. The innovation and research over the past several decades now allows cell surface engineering to be added to this list of technologies.

References

[1] Lingwood D, Simons K. Lipid rafts as a membrane-organizing principle. Science 2010;327:46–50.

[2] Vellai T, Takács-Vellai K. Regulation of protein turnover by longevity pathways. Adv Exp Med Biol 2010;694:69–80.

[3] Love RJ, Jones KS. The recognition of biomaterials: pattern recognition of medical polymers and their adsorbed biomolecules. J Biomed Mater Res A 2013;101:2740–52.

[4] Tang L, Hu W. Molecular determinants of biocompatibility. Expert Rev Med Devices 2005;.

[5] Luttikhuizen DT, Harmsen MC, Van Luyn MJA. Cellular and molecular dynamics in the foreign body reaction. Tissue Eng 2006;12:1955–70.

[6] Kellam B, De Bank PA, Shakesheff KM. Chemical modification of mammalian cell surfaces. Chem Soc Rev 2003;32:327–37.

[7] Stephan MT, Irvine DJ. Enhancing cell therapies from the outside in: cell surface engineering using synthetic nanomaterials. Nano Today 2011;6:309–25.

[8] Mahal LK, Bertozzi CR. Engineered cell surfaces: fertile ground for molecular landscaping. Chem Biol 1997;4:415–22.

[9] Stabler CL, Sun X-L, Cui W, Wilson JT, Haller CA, Chaikof EL. Surface re-engineering of pancreatic islets with recombinant azido-thrombomodulin. Bioconjugate Chem 2007;18:1713–5.

[10] Yun Lee D, Hee Nam J, Byun Y. Functional and histological evaluation of transplanted pancreatic islets immunoprotected by PEGylation and cyclosporine for 1 year. Biomaterials 2007;28:1957–66.

[11] Teramura Y, Kaneda Y, Totani T, Iwata H. Behavior of synthetic polymers immobilized on a cell membrane. Biomaterials 2008;29:1345–55.

[12] Chen PCY, Huang W, Stassinopoulos A, Cheung ATW. Effects of pegylated hamster red blood cells on microcirculation. Artif Cells Blood Substit Immobil Biotechnol 2008;36:295–309.

[13] Bradley AJ, Murad KL, Regan KL, Scott MD. Biophysical consequences of linker chemistry and polymer size on stealth erythrocytes: size does matter. Biochim Biophys Acta 2002;1561:147–58.

[14] Scott MD, Murad KL, Koumpouras F, Talbot M, Eaton JW. Chemical camouflage of antigenic determinants: stealth erythrocytes. Proc Natl Acad Sci USA 1997;94:7566–71.

[15] Panza JL, Wagner WR, Rilo HL, Rao RH, Beckman EJ, Russell AJ. Treatment of rat pancreatic islets with reactive PEG. Biomaterials 2000;21:1155–64.

[16] Teramura Y, Iwata H. Islets surface modification prevents blood-mediated inflammatory responses. Bioconjugate Chem 2008;19:1389–95.

[17] Rossi NAA, Constantinescu I, Brooks DE, Scott MD, Kizhakkedathu JN. Enhanced cell surface polymer grafting in concentrated and nonreactive aqueous polymer solutions. J Am Chem Soc 2010;132:3423–30.

[18] Lee DY, Park SJ, Lee S, Nam JH, Byun Y. Highly poly (ethylene) glycolylated islets improve long-term islet allograft survival without immunosuppressive medication. Tissue Eng 2007;13:2133–41.

[19] Lee DY, Yang K, Lee S, Chae SY, Kim K-W, Lee MK, et al. Optimization of monomethoxy-polyethylene glycol grafting on the pancreatic islet capsules. J Biomed Mater Res 2002;62:372–7.
[20] Wilson JT, Haller CA, Qu Z, Cui W, Urlam MK, Chaikof EL. Biomolecular surface engineering of pancreatic islets with thrombomodulin. Acta Biomater 2010;6:1895–903.
[21] Nacharaju P, Manjula BN, Acharya SA. Thiolation mediated pegylation platform to generate functional universal red blood cells. Artif Cells Blood Substit Immobil Biotechnol 2007;35:107–18.
[22] Nacharaju P, Boctor FN, Manjula BN, Acharya SA. Surface decoration of red blood cells with maleimidophenyl–polyethylene glycol facilitated by thiolation with iminothiolane: an approach to mask A, B, and D antigens to generate universal red blood cells. Transfusion 2005;45:374–83.
[23] Garratty G. Modulating the red cell membrane to produce universal/stealth donor red cells suitable for transfusion. Vox Sang 2008;94:87–95.
[24] Howarth M, Takao K, Hayashi Y, Ting AY. Targeting quantum dots to surface proteins in living cells with biotin ligase. Proc Natl Acad Sci USA 2005;102:7583–8.
[25] Inui O, Teramura Y, Iwata H. Retention dynamics of amphiphilic polymers PEG–lipids and PVAalkyl on the cell surface. ACS Appl Mater Interfaces 2010;2:1514–20.
[26] Stephan MT, Stephan SB, Bak P, Chen J, Irvine DJ. Synapse-directed delivery of immunomodulators using T-cell-conjugated nanoparticles. Biomaterials 2012;33:5776–87.
[27] Wilson JT, Krishnamurthy VR, Cui W, Qu Z, Chaikof EL. Noncovalent cell surface engineering with cationic graft copolymers. J Am Chem Soc 2009;131:18228–9.
[28] Rabuka D, Forstner MB, Groves JT, Bertozzi CR. Noncovalent cell surface engineering: incorporation of bioactive synthetic glycopolymers into cellular membranes. J Am Chem Soc 2008;130:5947–53.
[29] White SH. Membrane protein insertion: the biology–physics nexus. J Gen Physiol 2007;129:363–9.
[30] Liu H, Kwong B, Irvine DJ. Membrane anchored immunostimulatory oligonucleotides for *in vivo* cell modification and localized immunotherapy. Angew Chem Int Ed Engl 2011;50:7052–5.
[31] Jeong JH, Schmidt JJ, Kohman RE, Zill AT, DeVolder RJ, Smith CE, et al. Leukocyte-mimicking stem cell delivery via *in situ* coating of cells with a bioactive hyperbranched polyglycerol. J Am Chem Soc 2013;135:8770–3.
[32] Kumari S, Mg S, Mayor S. Endocytosis unplugged: multiple ways to enter the cell. Cell Res 2010;20:256–75.
[33] Pack DW, Hoffman AS, Pun S, Stayton PS. Design and development of polymers for gene delivery. Nat Rev Drug Discov 2005;4:581–93.
[34] Totani T, Teramura Y, Iwata H. Immobilization of urokinase on the islet surface by amphiphilic poly (vinyl alcohol) that carries alkyl side chains. Biomaterials 2008;29:2878–83.
[35] Teramura Y, Iwata H. Islet encapsulation with living cells for improvement of biocompatibility. Biomaterials 2009;30:2270–5.
[36] Teramura Y, Kaneda Y, Iwata H. Islet-encapsulation in ultra-thin layer-by-layer membranes of poly(vinyl alcohol) anchored to poly(ethylene glycol)-lipids in the cell membrane. Biomaterials 2007;28:4818–25.
[37] Mech A, Orynbayeva Z, Irgebayev K, Kolusheva S, Jelinek R. Screening membrane interactions of pesticides by cells decorated with chromatic polymer nanopatches. Chem Res Toxicol 2009;22:90–6.
[38] Orynbayeva Z, Kolusheva S, Livneh E, Lichtenshtein A, Nathan I, Jelinek R. Visualization of membrane processes in living cells by surface-attached chromatic polymer patches. Angew Chem Int Ed Engl 2005;44:1092–6.
[39] Fischer D, Li Y, Ahlemeyer B, Krieglstein J, Kissel T. *In vitro* cytotoxicity testing of polycations: influence of polymer structure on cell viability and hemolysis. Biomaterials 2003;24:1121–31.
[40] Hunter AC. Molecular hurdles in polyfectin design and mechanistic background to polycation induced cytotoxicity. Adv Drug Delivery Rev 2006;58:1523–31.
[41] Parhamifar L, Larsen AK, Hunter AC, Andresen TL, Moghimi SM. Polycation cytotoxicity: a delicate matter for nucleic acid therapy—focus on polyethylenimine. Soft Matter 2010;6:4001.
[42] Chanana M, Gliozzi A, Diaspro A, Chodnevskaja I, Huewel S, Moskalenko V, et al. Interaction of polyelectrolytes and their composites with living cells. Nano Lett 2005;5:2605–12.
[43] Hong S, Leroueil PR, Janus EK, Peters JL, Kober M-M, Islam MT, et al. Interaction of polycationic polymers with supported lipid bilayers and cells: nanoscale hole formation and enhanced membrane permeability. Bioconjugate Chem 2006;17:728–34.
[44] Menger FM, Seredyuk VA, Kitaeva MV, Yaroslavov AA, Melik-Nubarov NS. Migration of poly-L-lysine through a lipid bilayer. J Am Chem Soc 2003;125:2846–7.
[45] Leroueil PR, Hong S, Mecke A, Baker JR, Orr BG, Banaszak Holl MM. Nanoparticle interaction with biological membranes: does nanotechnology present a Janus face? Acc Chem Res 2007;40:335–42.
[46] Wilson JT, Cui W, Kozlovskaya V, Kharlampieva E, Pan D, Qu Z, et al. Cell surface engineering with polyelectrolyte multilayer thin films. J Am Chem Soc 2011;133:7054–64.

[47] Fakhrullin RF, Zamaleeva AI, Minullina RT, Konnova SA, Paunov VN. Cyborg cells: functionalisation of living cells with polymers and nanomaterials. Chem Soc Rev 2012;41:4189–206.
[48] Wilson JT, Cui W, Chaikof EL. Layer-by-layer assembly of a conformal nanothin PEG coating for intraportal islet transplantation. Nano Lett 2008;8:1940–8.
[49] Krol S, del Guerra S, Grupillo M, Diaspro A, Gliozzi A, Marchetti P. Multilayer nanoencapsulation. New approach for immune protection of human pancreatic islets. Nano Lett 2006;6:1933–9.
[50] Rajagopalan P, Shen CJ, Berthiaume F, Tilles AW, Toner M, Yarmush ML. Polyelectrolyte nano-scaffolds for the design of layered cellular architectures. Tissue Eng 2006;12:1553–63.
[51] Zhi Z-L, Liu B, Jones PM, Pickup JC. Polysaccharide multilayer nanoencapsulation of insulin-producing β-cells grown as pseudoislets for potential cellular delivery of insulin. Biomacromolecules 2010;11:610–6.
[52] Elbert DL, Hubbell JA. Self-assembly and steric stabilization at heterogeneous, biological surfaces using adsorbing block copolymers. Chem Biol 1998;5:177–83.
[53] Krishnamurthy VR, Wilson JT, Cui W, Song X, Lasanajak Y, Cummings RD, et al. Chemoselective immobilization of peptides on abiotic and cell surfaces at controlled densities. Langmuir 2010;26:7675–8.
[54] Decher G. Fuzzy nanoassemblies: toward layered polymeric multicomposites. Science 1997;277:1232–7.
[55] Hammond PT. Building biomedical materials layer-by-layer. Mater Today 2012;15(5):196–206.
[56] Jiang C, Tsukruk VV. Freestanding nanostructures via layer-by-layer assembly. Adv Mater 2006;18 (7):829–40.
[57] Tang Z, Wang Y, Podsiadlo P, Kotov NA. Biomedical applications of layer-by-layer assembly: from biomimetics to tissue engineering. Adv Mater 2006;18:3203–24.
[58] Becker AL, Johnston APR, Caruso F. Layer-by-layer-assembled capsules and films for therapeutic delivery. Small 2010;6:1836–52.
[59] Jewell CM, Lynn DM. Multilayered polyelectrolyte assemblies as platforms for the delivery of DNA and other nucleic acid-based therapeutics. Adv Drug Delivery Rev 2008;60:979–99.
[60] Boudou T, Crouzier T, Ren K, Blin G, Picart C. Multiple functionalities of polyelectrolyte multilayer films: new biomedical applications. Adv Mater Weinheim 2010;22:441–67.
[61] Champion JA, Mitragotri S. Role of target geometry in phagocytosis. Proc Natl Acad Sci USA 2006;103: 4930–4.
[62] Champion JA, Katare YK, Mitragotri S. Particle shape: a new design parameter for micro-and nanoscale drug delivery carriers. J Control Release 2007;121:3–9.
[63] Doshi N, Swiston AJ, Gilbert JB, Alcaraz ML, Cohen RE, Rubner MF, et al. Cell-based drug delivery devices using phagocytosis-resistant backpacks. Adv Mater 2011;23:H105–9.
[64] Mansouri S, Merhi Y, Winnik FM, Tabrizian M. Investigation of layer-by-layer assembly of polyelectrolytes on fully functional human red blood cells in suspension for attenuated immune response. Biomacromolecules 2011;12:585–92.
[65] Miura S, Teramura Y, Iwata H. Encapsulation of islets with ultra-thin polyion complex membrane through poly(ethylene glycol)-phospholipids anchored to cell membrane. Biomaterials 2006;27:5828–35.
[66] Picart C, Mutterer J, Richert L, Luo Y, Prestwich GD, Schaaf P, et al. Molecular basis for the explanation of the exponential growth of polyelectrolyte multilayers. Proc Natl Acad Sci USA 2002;99:12531–5.
[67] Glinel K, Moussa A, Jonas AM, Laschewsky A. Influence of polyelectrolyte charge density on the formation of multilayers of strong polyelectrolytes at low ionic strength. Langmuir 2002;18:1408–12.
[68] Schoeler B, Sharpe S, Hatton TA, Caruso F. Polyelectrolyte multilayer films of different charge density copolymers with synergistic nonelectrostatic interactions prepared by the layer-by-layer technique. Langmuir 2004;20:2730–8.
[69] Huang N-P, Michel R, Vörös J, Textor M, Hofer R, Rossi A, et al. Poly(L-lysine)-*g*-poly(ethylene glycol) layers on metal oxide surfaces: surface-analytical characterization and resistance to serum and fibrinogen adsorption. Langmuir 2001;17:489–98.
[70] Teramura Y, Oommen OP, Olerud J, Hilborn J, Nilsson B. Microencapsulation of cells, including islets, within stable ultra-thin membranes of maleimide-conjugated PEG–lipid with multifunctional crosslinkers. Biomaterials 2013;34:2683–93.
[71] Gattás-Asfura KM, Stabler CL. Bioorthogonal layer-by-layer encapsulation of pancreatic islets via hyperbranched polymers. ACS Appl Mater Interfaces 2013;5:9964–74.
[72] Matsusaki M, Kadowaki K, Nakahara Y, Akashi M. Fabrication of cellular multilayers with nanometer-sized extracellular matrix films. Angew Chem Int Ed 2007;46:4689–92.
[73] Kozlovskaya V, Harbaugh S, Drachuk I, Shchepelina O, Kelley-Loughnane N, Stone M, et al. Hydrogen-bonded LbL shells for living cell surface engineering. Soft Matter 2011;7:2364–72.
[74] Kozlovskaya V, Zavgorodnya O, Chen Y, Ellis K, Tse HM, Cui W, et al. Ultrathin polymeric coatings based on hydrogen-bonded polyphenol for protection of pancreatic islet cells. Adv Funct Mater 2012;22:3389–98.

[75] Mets JM, Wilson JT, Cui W, Chaikof EL. An automated process for layer-by-layer assembly of polyelectrolyte multilayer thin films on viable cell aggregates. Adv Healthc Mater 2013;2(2):266–70.
[76] Swiston AJ, Cheng C, Um SH, Irvine DJ, Cohen RE, Rubner MF. Surface functionalization of living cells with multilayer patches. Nano Lett 2008;8:4446–53.
[77] Restifo NP, Dudley ME, Rosenberg SA. Adoptive immunotherapy for cancer: harnessing the T cell response. Nat Rev Immunol 2012;12:269–81.
[78] Bellin M, Marchetto MC, Gage FH, Mummery CL. Induced pluripotent stem cells: the new patient? Nat Rev Mol Cell Bio 2012;13:713–26.
[79] Mizuno H, Tobita M, Uysal AC. Concise review: adipose-derived stem cells as a novel tool for future regenerative medicine. Stem Cells 2012;30:804–10.
[80] Wilson JT, Chaikof EL. Challenges and emerging technologies in the immunoisolation of cells and tissues. Adv Drug Delivery Rev 2008;60:124–45.
[81] Winblade ND, Schmökel H, Baumann M, Hoffman AS, Hubbell JA. Sterically blocking adhesion of cells to biological surfaces with a surface-active copolymer containing poly(ethylene glycol) and phenylboronic acid. J Biomed Mater Res 2002;59:618–31.
[82] Riquelme BD, Dumas D, Fontana A, Delannoy M, Valverde JR, Sondag D, et al. Hemocompatibility and biofunctionality of two poly(2-(dimethylamino)ethyl methacrylate-co-poly(ethyleneglycol) copolymers. J Biomed Mater Res A 2011;99:445–54.
[83] Scott MD, Murad KL. Cellular camouflage: fooling the immune system with polymers. Curr Pharm Des 1998;4:423–38.
[84] Chapanian R, Constantinescu I, Rossi NAA, Medvedev N, Brooks DE, Scott MD, et al. Influence of polymer architecture on antigens camouflage, CD47 protection and complement mediated lysis of surface grafted red blood cells. Biomaterials 2012;33:7871–83.
[85] Hashemi-Najafabadi S, Vasheghani-Farahani E, Shojaosadati SA, Rasaee MJ, Armstrong JK, Moin M, et al. A method to optimize PEG-coating of red blood cells. Bioconjugate Chem 2006;17:1288–93.
[86] Moore MS, Okelberry E, Cordingley K, Drake A, Robinett Z. DePEGylation studies: PEG-RBC stability in conditions consistent with massive transfusion. Clin Lab Sci 2011;24:227–32.
[87] Murad KL, Mahany KL, Brugnara C, Kuypers FA, Eaton JW, Scott MD. Structural and functional consequences of antigenic modulation of red blood cells with methoxypoly(ethylene glycol). Blood 1999;93:2121–7.
[88] Murad KL, Gosselin EJ, Eaton JW, Scott MD. Stealth cells: prevention of major histocompatibility complex class II-mediated T-cell activation by cell surface modification. Blood 1999;94:2135–41.
[89] Chen AM, Scott MD. Immunocamouflage: prevention of transfusion-induced graft-versus-host disease via polymer grafting of donor cells. J Biomed Mater Res A 2003;67:626–36.
[90] Huang Y, Feng S, Tang R, Du B, Xu K, Pan X. Efficacy of pretreatment of allografts with methoxypolyethylene glycol–succinimidyl–propionic acid ester in combination with an anti-OX40L monoclonal antibody in relieving graft-versus-host disease in mice. Int J Hematol 2010;92:609–16.
[91] Greco CA, Maurer-Spurej E, Scott MD, Kalab M, Nakane N, Ramírez-Arcos SM. PEGylation prevents bacteria-induced platelet activation and biofilm formation in platelet concentrates. Vox Sang 2011;100:336–9.
[92] Stuhlmeier KM, Lin Y. Camouflaging endothelial cells: does it prolong graft survival? Biochim Biophys Acta 1999;1428:177–90.
[93] Chapanian R, Constantinescu I, Medvedev N, Scott MD, Brooks DE, Kizhakkedathu JN. Therapeutic cells via functional modification: influence of molecular properties of polymer grafts on *in vivo* circulation, clearance, immunogenicity, and antigen protection. Biomacromolecules 2013;14:2052–62.
[94] Chapanian R, Constantinescu I, Brooks DE, Scott MD, Kizhakkedathu JN. *In vivo* circulation, clearance, and biodistribution of polyglycerol grafted functional red blood cells. Biomaterials 2012;33:3047–57.
[95] Kyluik-Price DL, Li L, Scott MD. Comparative efficacy of blood cell immunocamouflage by membrane grafting of methoxypoly(ethylene glycol) and polyethyloxazoline. Biomaterials 2014;35:412–22.
[96] Cerda-Cristerna BI, Cottin S, Flebus L, Pozos-Guillén A, Flores H, Heinen E, et al. Poly(2-dimethylamino ethylmethacrylate)-based polymers to camouflage red blood cell antigens. Biomacromolecules 2012; 13:1172–80.
[97] Mansouri S, Fatisson J, Miao Z, Merhi Y, Winnik FM, Tabrizian M. Silencing red blood cell recognition toward anti-A antibody by means of polyelectrolyte layer-by-layer assembly in a two-dimensional model system. Langmuir 2009;25:14071–8.
[98] Neu B, Voigt A, Mitlöhner R, Leporatti S, Gao CY, Donath E, et al. Biological cells as templates for hollow microcapsules. J Microencapsul 2001;18:385–95.
[99] Wilson JT, Chaikof EL. Thrombosis and inflammation in intraportal islet transplantation: a review of pathophysiology and emerging therapeutics. J Diabetes Sci Technol 2008;2:746–59.
[100] Weir GC, Bonner-Weir S. Scientific and political impediments to successful islet transplantation. Diabetes 1997;46:1247–56.

[101] Jeong J-H, Hong SW, Hong S, Yook S, Jung Y, Park J-B, et al. Surface camouflage of pancreatic islets using 6-arm-PEG-catechol in combined therapy with tacrolimus and anti-CD154 monoclonal antibody for xenotransplantation. Biomaterials 2011;32:7961–70.
[102] Dong H, Fahmy TM, Metcalfe SM, Morton SL, Dong X, Inverardi L, et al. Immuno-isolation of pancreatic islet allografts using pegylated nanotherapy leads to long-term normoglycemia in full MHC mismatch recipient mice. PLoS ONE 2012;7:e50265.
[103] Contreras JL, Xie D, Mays J, Smyth CA, Eckstein C, Rahemtulla FG, et al. A novel approach to xenotransplantation combining surface engineering and genetic modification of isolated adult porcine islets. Surgery 2004;136:537–47.
[104] Xie D, Smyth CA, Eckstein C, Bilbao G, Mays J, Eckhoff DE, et al. Cytoprotection of PEG-modified adult porcine pancreatic islets for improved xenotransplantation. Biomaterials 2005;26:403–12.
[105] Teramura Y, Iwata H. Surface modification of islets with PEG–lipid for improvement of graft survival in intraportal transplantation. Transplantation 2009;88:624–30.
[106] Teramura Y, Iwata H. Improvement of graft survival by surface modification with poly(ethylene glycol)-lipid and urokinase in intraportal islet transplantation. Transplantation 2011;91:271–8.
[107] Chen H, Teramura Y, Iwata H. Co-immobilization of urokinase and thrombomodulin on islet surfaces by poly(ethylene glycol)-conjugated phospholipid. J Control Release 2011;150:229–34.
[108] Sarkar D, Spencer JA, Phillips JA, Zhao W, Schafer S, Spelke DP, et al. Engineered cell homing Blood. 2011;118(25):e184–91.
[109] Ko IK, Kean TJ, Dennis JE. Targeting mesenchymal stem cells to activated endothelial cells. Biomaterials 2009;30:3702–10.
[110] Stephan MT, Moon JJ, Um SH, Bershteyn A, Irvine DJ. Therapeutic cell engineering with surface-conjugated synthetic nanoparticles. Nat Med 2010;16:1031–7.
[111] Luan NM, Teramura Y, Iwata H. Immobilization of soluble complement receptor 1 on islets. Biomaterials 2011;32:4539–45.
[112] Holden CA, Yuan Q, Yeudall WA, Lebman DA, Yang H. Surface engineering of macrophages with nanoparticles to generate a cell-nanoparticle hybrid vehicle for hypoxia-targeted drug delivery. Int J Nanomed 2010;5:25–36.
[113] Vacchelli E, Vitale I, Eggermont A, Fridman WH, Fučíková J, Cremer I, et al. Trial watch: dendritic cell-based interventions for cancer therapy. Oncoimmunology 2013;2:e25771.
[114] Ong S-M, He L, Thuy Linh NT, Tee Y-H, Arooz T, Tang G, et al. Transient inter-cellular polymeric linker. Biomaterials 2007;28:3656–67.
[115] Zhao D, Ong S-M, Yue Z, Jiang Z, Toh Y-C, Khan M, et al. Dendrimer hydrazides as multivalent transient inter-cellular linkers. Biomaterials 2008;29:3693–702.
[116] Ong S-M, Zhao Z, Arooz T, Zhao D, Zhang S, Du T, et al. Engineering a scaffold-free 3D tumor model for in vitro drug penetration studies. Biomaterials. 2010;31(6):1180–90.
[117] Mo X, Li Q, Lui LWY, Zheng B, Kang CH, Nugraha B, et al. Rapid construction of mechanically-confined multi-cellular structures using dendrimeric intercellular linker. Biomaterials 2010;31(29):7455–67.
[118] Gartner ZJ, Bertozzi CR. Programmed assembly of 3-dimensional microtissues with defined cellular connectivity. Proc Natl Acad Sci USA 2009;106:4606–10.
[119] Swiston AJ, Gilbert JB, Irvine DJ, Cohen RE, Rubner MF. Freely suspended cellular "backpacks" lead to cell aggregate self-assembly. Biomacromolecules 2010;11:1826–32.
[120] Onoe H, Hsiao SC, Douglas ES, Gartner ZJ, Bertozzi CR, Francis MB, et al. Cellular microfabrication: observing intercellular interactions using lithographically-defined DNA capture sequences. Langmuir 2012;28:8120–6.
[121] Hsiao SC, Shum BJ, Onoe H, Douglas ES, Gartner ZJ, Mathies RA, et al. Direct cell surface modification with DNA for the capture of primary cells and the investigation of myotube formation on defined patterns. Langmuir 2009;25:6985–91.
[122] Kadowaki K, Matsusaki M, Akashi M. Control of cell surface and functions by layer-by-layer nanofilms. Langmuir 2010;26:5670–8.
[123] Teramura Y, Minh LN, Kawamoto T, Iwata H. Microencapsulation of islets with living cells using PolyDNA–PEG–Lipid conjugate. Bioconjugate Chem 2010;21:792–6.
[124] Elbert DL, Hubbell JA. Reduction of fibrous adhesion formation by a copolymer possessing an affinity for anionic surfaces. J Biomed Mater Res 1998;42:55–65.
[125] Wang DA, Williams CG, Yang F, Elisseeff JH. Enhancing the tissue–biomaterial interface: tissue-initiated integration of biomaterials. Adv Funct Mater 2004;14:1152–9.
[126] Wang D-A, Varghese S, Sharma B, Strehin I, Fermanian S, Gorham J, et al. Multifunctional chondroitin sulphate for cartilage tissue-biomaterial integration. Nat Mater 2007;6:385–92.
[127] Brubaker CE, Kissler H, Wang L-J, Kaufman DB, Messersmith PB. Biological performance of mussel-inspired adhesive in extrahepatic islet transplantation. Biomaterials. 2010;31(3):420–7.

CHAPTER

14

Biofunctionalization of Hydrogels for Engineering the Cellular Microenvironment

Maniraj Bhagawati and Sanjay Kumar

Department of Bioengineering, University of California, Berkeley, CA, USA

OUTLINE

J. M. Karp & W. Zhao (Eds): Micro- and Nanoengineering of the Cell Surface.
DOI: http://dx.doi.org/10.1016/B978-1-45-573146-6.00014-3

14.1 THE 3D EXTRACELLULAR MILIEU

Tissues are composed of cells and the extracellular matrix (ECM), which provides not only physical support in the form of a scaffold but also a variety of biochemical and biomechanical signals essential for cellular function and homeostasis. The ECM is an intricate network of biopolymers that is secreted and subsequently modified, degraded, and organized by the cells adherent to it. Therefore, cells and their ECM form a closed, bidirectional loop of information transfer in which the ECM provides biochemical and mechanical signals that are conveyed from the cell membrane through the cytoskeleton to the nucleus to direct cell behavior, with cells in turn remodeling the ECM under the regulation of ECM-derived inputs [1]. Although all ECMs are nominally composed of proteins and polysaccharides, each tissue has an ECM that is distinct not only in its biochemical composition but also with respect to its physical properties, including stiffness, microstructure, and porosity. A large and still-growing body of work has demonstrated the functional importance of the ECM in regulating morphogenesis and organogenesis and mediating disease processes that occur as a result of abnormal or deregulated ECM [2–5]. Thus, the ability to fabricate *in vitro* cellular environments that can mimic the complex properties of the ECM holds profound value for a wide variety of applications ranging from basic cell biological studies to regenerative medicine and drug delivery.

The biochemical components of the ECM have been extensively described in several previous reviews [6–8] (Figure 14.1), and a variety of ECM proteins have been incorporated into *in vitro* systems for cell culture applications [9–12]. These include fibrous proteins such as collagen and fibrin, which play important roles in providing mechanical support in tissue. More specialized and tissue-specific ECM proteins, such as fibronectin (FN) and laminin, have also been included in these culture systems [13,14]. Importantly, these proteins are not merely inert scaffolds but also specifically bind cell surface adhesion receptors. The most notable among these are the integrins, which are heterodimeric transmembrane proteins that physically link the ECM and the cellular cytoskeleton [15–20]. Domain-mapping analyses of various ECM proteins have yielded the identity of several short peptide sequences that define integrin-binding specificity [21]. The best-known such sequence, and the one that has most often been included in synthetic matrices for enabling cell adhesion, is the arginine–glycine–aspartate (RGD) sequence, which was originally derived from FN repeat III10 [22]. Since then this sequence has also been found and shown to mediate cell adhesion in several other ECM proteins such as vitronectin, fibrinogen, laminin, and tenascin [21].

A second category of ECM macromolecules that has also been extensively used for fabrication of *in vitro* cell culture matrices is glycosaminoglycans (GAGs). GAGs are large linear polysaccharides consisting of disaccharide units containing an amino sugar (either GlcNAc or GalNAc) and an uronic acid (either glucuronic acid and/or iduronic acid).

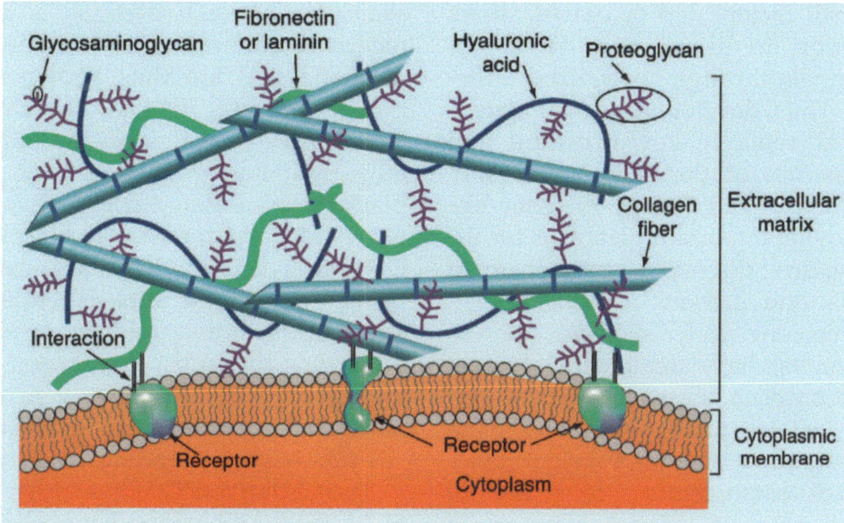

FIGURE 14.1 A highly simplified representation of some of the major components of the ECM and their interactions with cells. *Source: Reproduced with permission from Ref. [44].*

The amino sugar is very often sulfated, and thus GAGs are highly negatively charged polymers. GAGs, which are defined by the constituent sugar monomers and the bonds linking them together, include hyaluronans, heparan sulfates (HSs), chondroitin sulfate, dermatan sulfate, and keratan sulfate. Except for hyaluronan, all GAGs can be found covalently conjugated with proteins to form proteoglycans. *In vivo*, GAGs and proteoglycans perform multiple structural and mechanical roles and can trigger cell signaling by engaging cell surface receptors [23]. Another major function of GAGs, especially HS, is to locally sequester and concentrate a diverse set of growth factors and cytokines [24]. This property has inspired the use of HS (and the structurally similar molecule heparin) for enabling the entrapment of GFs in several synthetic matrices [25,26].

As alluded to earlier, the composition of the ECM is not static but is instead subject to active cell-driven remodeling, which is an important mechanism for dynamically regulating various aspects of both morphogenesis and tissue homeostasis and disease [27,28]. The major mechanism for this remodeling is degradation of the various components of the ECM by the action of cell-secreted enzymes. ECM proteins including collagens, FN, and laminin are degraded by enzymes such as matrix metalloproteinases (MMPs) [29,30] and plasmin [31]. Proteoglycans can be degraded by ADAMTS proteinases [32,33], and enzymes like hyaluronidases [34] and heparanase are specialized to degrade specific GAGs. In the context of synthetic cell culture scaffolds, cell-mediated matrix degradation can both create room for tissue growth and angiogenesis and provide a convenient mechanism for clearing the material following *in vivo* implantation. For these reasons, as we discuss later, there has been a significant effort to develop synthetic ECMs that are amenable to cell-mediated degradation.

14.2 MIMICKING THE ECM

To summarize, the ECM represents a complex, cell-interactive meshwork of proteins, proteoglycans, and GAGs that sequesters and organizes growth factors and cytokines, offers structural support to tissue, and transduces biochemical signals through specialized adhesion receptors. Thus, developing *in vitro* materials systems that replicate key structural and functional properties of the ECM could both facilitate the study of cell biology in tissue-like settings and provide important platforms for tissue engineering, drug discovery, and delivery of therapeutic cells and proteins *in vivo* [35–41]. Hydrogels, which are water-swollen, 3D networks of polymers, have demonstrated great potential to accomplish these goals. Hydrogel networks can be structurally stabilized by a number of mechanisms such as covalent crosslinking, physical entanglement, formation of crystallites, and assembly through a variety of noncovalent interactions including hydrogen bonding and electrostatic forces [42,43]. The highly water-swollen nature of hydrogels confers important mechanical properties and facilitates the encapsulation of cells. Hydrogel materials often exhibit strong biocompatibility and high permeability for oxygen, nutrients, and other water-soluble metabolites [44–49].

Hydrogels have been fabricated from both natural and synthetic polymers. Natural polymers are typically biocompatible, inherently biodegradable, and often contain motifs that direct specific biological functions. ECM-derived proteins such as elastin, fibrin, collagen, and the collagen derivative gelatin have been traditionally used to form hydrogels [50–53]. Despite their advantages, ECM proteins derived from tissue can suffer from concerns over immunogenicity, retention of infectious agents, poor scalability, and high batch-to-batch variability. Furthermore, hydrogels based on natural polymers offer limited opportunity for bottom-up design of desired physicochemical properties or biological functionalities. For these reasons, hydrogels based on recombinantly expressed proteins have recently emerged as a complementary class of biological matrices. One of the most well-known examples of such systems is inspired from recurring amino acid sequences found in tropoelastin and are thus known as elastin-like polypeptides (ELPs) [54]. ELP-based hydrogels have proved promising for both experimentally modeling and serving as a therapeutic platform for a wide variety of tissues including nerve, skeletal muscle, and cartilage [55–59]. Peptide hydrogels have also been fabricated using amphiphilic diblock copeptides [60,61] and peptides that fold to form amphiphilic β-hairpins that subsequently self-assemble into a hydrogel network [62–64]. Advances in protein- and peptide-based hydrogels have been covered in several recent reviews [65,66].

Tissue-derived GAGs and a variety of non-ECM polysaccharides have also been extensively used for fabricating hydrogels. These polysaccharides can be modified with diverse functional groups such as acrylates, thiols, and amines to allow crosslinking. Hyaluronic acid (HA), which is ubiquitous in the ECM of many tissues, has been extensively used to fabricate hydrogels for application in a wide range of study systems [67,68]. Since HA levels have been shown to be elevated in the stroma of cancers [69,70], hydrogels based on HA have been used for modeling the tumor microenvironment in breast, colon, ovarian, and lung cancers [71–73]. HA also forms a large fraction of the ECM in brain, and so HA-based hydrogels have also been used to model brain ECM *in vitro*. Such hydrogels have found particular use in the study of the brain tumor glioblastoma multiforme [74,75]. Other polysaccharides that have been utilized for fabricating hydrogels include heparin, chitosan, dextran, and alginate [76–80].

Synthetic polymer hydrogel systems have a long history of use in medical devices, implants, and other materials, and remain of significant interest for use as ECM scaffolds. In general, synthetic polymers offer greater batch-to-batch consistency and significantly more physicochemical tailorability than native biological materials. For example, poly(ethyleneglycol)

(PEG) is perhaps the most widely used synthetic polymer hydrogel system. PEG has several unique properties such as high solubility in water and various organic solvents, low intrinsic toxicity and immunogenicity, and comparatively low nonspecific protein adsorption [81–85]. In addition, the hydroxyl end groups of PEG are easily modified with other chemical functionalities that can be used for crosslinking as well as for functionalization with a variety of cell-reactive moieties. While a detailed discussion of all synthetic polymer systems currently under use for hydrogel fabrication is beyond the scope of this chapter, several extensive reviews have been written on this subject [44,86,87].

While native, recombinant, and synthetic hydrogels have their relative advantages and disadvantages, many of these systems share the need for chemical "postprocessing" to achieve maximal biological function. For example, whereas native protein hydrogel systems such as those based on collagen and fibrin encode a rich palette of biological instruction, they may require chemical crosslinking to achieve the desired mechanical properties for a given application or conjugation of growth factors to promote specific phenotypes. Conversely, whereas a PEG hydrogel may have ideal physical properties for a given biological application, its bio-inert nature requires all adhesive and biodegradation properties to be built into the system, typically by attachment of proteins and peptides to the polymer backbone. We now discuss these conjugation strategies (Figure 14.2) in greater detail.

14.2.1 Covalent Immobilization of ECM-Derived Proteins and Growth Factors Through Reaction of Amino Acid Side Chains

The side chains of several amino acids offer chemically reactive handles for bioconjugation. These include the amine group of lysine, the thiol group of cysteine, and the carboxylic acid groups of aspartic and glutamic acid. With the development of orthogonal genetic codes and synthetic tRNA systems that enable the expression of recombinant proteins containing unnatural amino acids, the reactive functionalities available on proteins have expanded even further [88–90]. Therefore, covalent coupling of these functionalities with reactive groups on hydrogels or hydrogel precursors is one of the most heavily utilized strategies for immobilization of proteins and peptides on hydrogels. Full-length FN, laminin, and collagen, as well as adhesive peptide motifs derived from these and other proteins have been incorporated into hydrogels using this strategy [91–93]. Growth factors and derived peptide motifs have also been immobilized on a variety of matrices using covalent reaction with lysine and cysteine residues (Figure 14.2) [94,95].

Lysines can be directly conjugated to scaffold carboxylic acids or their *N*-hydroxysuccinimide (NHS) derivatives leading to the formation of stable amide bonds. In this strategy, the carboxylic acid has to be incorporated into the hydrogel prior to protein immobilization. If the polymer does not already contain carboxylate functionality, it can be introduced either by copolymerizing carboxylic acid containing monomers with the polymer backbone or by post hoc carboxylate functionalization of the assembled hydrogel. The first strategy has proved valuable for modifying acrylamide-based hydrogels copolymerized with NHS-ester moieties with ECM proteins such as FN, vitronectin, and collagen [91]. Similarly, copolymerization of acrylic acid with PEG-diacrylate (PEGDA) has facilitated the formation of hydrogels which could be efficiently functionalized with several short peptide motifs derived from FN, laminin, and collagen [92]. The carboxylic acid groups already present in HA have also been used to immobilize rat tail collagen I [96]. As examples of the second strategy, poly(vinyl-alcohol) (PVA) hydrogels have been carboxylate functionalized with 11-bromoundecanoicacid, activated to yield NHS-esters, and subsequently

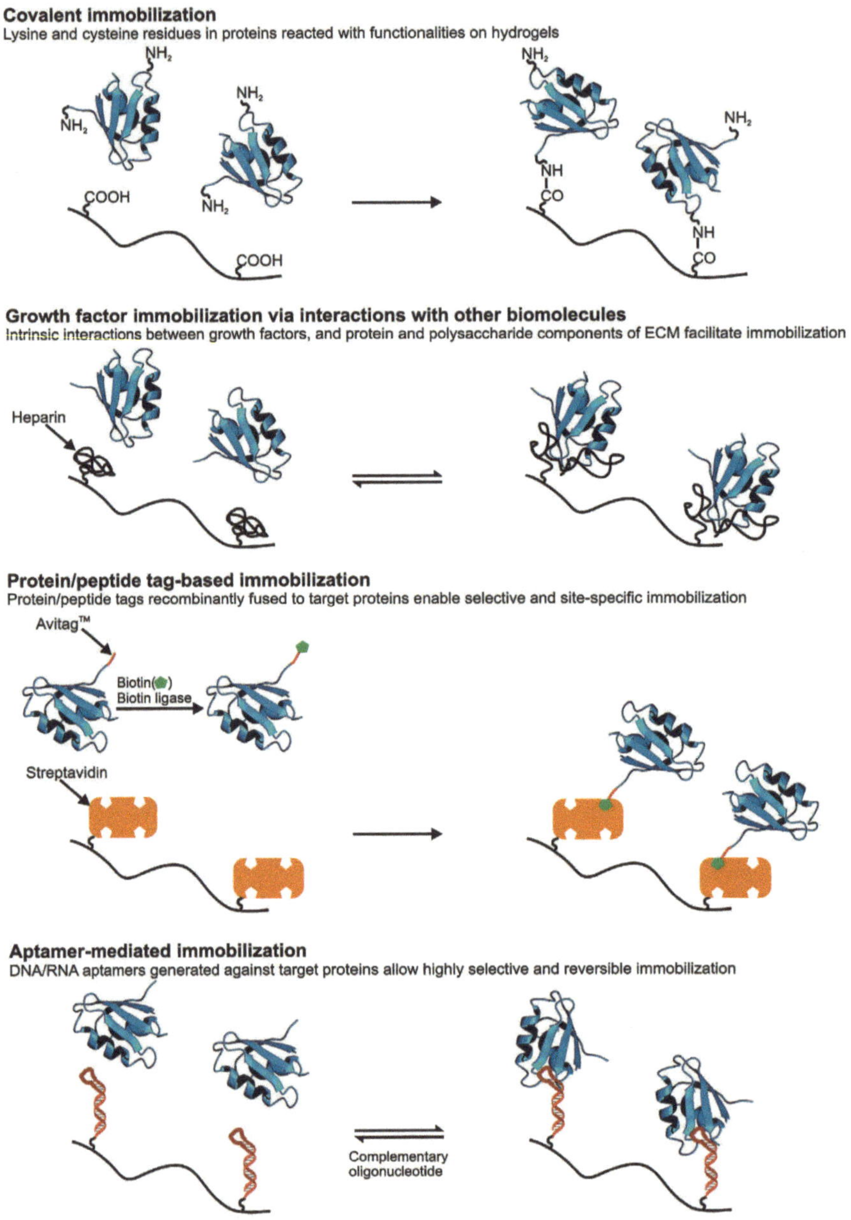

FIGURE 14.2 Strategies for capturing exogenous and cell-derived proteins on hydrogel matrices.

conjugated with FN [97]. The amine-reactive photo-linkable crosslinker sulfo-SANPAH has also been widely used to activate hydrogels, especially defined stiffness polyacrylamide hydrogels, for immobilization of ECM proteins through lysine residues [96,98]. Sulfo-SANPAH has also been used to conjugate laminin to methylcellulose hydrogels [99].

As an alternative to directly reacting the amine groups of the target protein to the polymer backbone, these amines can instead be conjugated to a small molecule that can react/interact with chemical functionalities available on the polymer backbone. For example, proteins have frequently been modified with acrylate groups, which allow these proteins to be co-photopolymerized with synthetic polymers containing acrylate groups to create bioactive hydrogels. This strategy has been used to conjugate FN to hydrogels made from PEG [100] and HA [101], thus demonstrating the flexibility of its application. Acryl-PEG-NHS has also been used to modify integrin-binding peptides such as RGD and YIGSR [102,103], as well as a variety of growth factors including fibroblast growth factor-2 (FGF-2, also known as basic FGF) [94] and platelet-derived growth factor-BB (PDGF-BB) [104], in order to facilitate the incorporation of these bioactive species into PEGDA hydrogels. Similarly, lysine residues in epidermal growth factor (EGF) have been functionalized with iodoacetamide-NHS to enable conjugation of this growth factor on HA hydrogels carrying thiol groups [105]. Another secondary functionality that enables high-affinity noncovalent conjugation is biotin, which can be used to link the modified biomolecule to avidin-functionalized hydrogels [106–108].

Although the abundance of lysines in proteins makes lysine an attractive residue to use for immobilization, it is challenging if not impossible to precisely control which lysine (or lysines) in a given protein reacts with the hydrogel matrix. This leads to an inhomogeneous distribution of protein orientation, which could lead to some fraction of the immobilized protein molecules to be denatured or sterically shielded from their interaction partners on the cell surface. Furthermore, conjugation of functionally critical lysine residues could compromise or eliminate the bioactivity of the protein. Therefore, cysteine residues, which are significantly rarer in proteins, can be used for more selective immobilization. Using recombinant DNA technology, cysteines can also be introduced into target proteins at precisely specified positions along the polypeptide chain, which enables a high degree of control over the orientation of immobilization. The glutathione S-transferase fusion tag which is often used for affinity purification of recombinantly expressed proteins also contains cysteine residues that have been used for immobilization on thiol-reactive matrices [109].

Thiols from cysteine residues can take part in several reactions that have been demonstrated to be very useful for biofunctionalization of hydrogels. One of the most widely used strategies is the Michael addition reaction in which a thiol reacts with an alkene, often in the context of an acrylate or methacrylate moiety. This chemistry has been used to attach FN to PEGDA, which was then photo-crosslinked to yield a hydrogel network [109]. A similar scheme has also been used to first generate FN-functionalized PEGDA and then crosslink thiolated-HA into the PEG network to create a three-component hydrogel [93]. A similar strategy has also been used to fabricate injectable HA hydrogels decorated with RGD peptide for the *in vivo* delivery of encapsulated cells [110]. In this study, fibroblasts were added to a mixture of thiolated-HA and RGD-modified PEGDA. The resulting mixture was then subcutaneously injected into the flanks of nude mice, leading to the formation of hydrogels *in vivo*. In a strategy employing the same chemistry but in an opposite topology, HA has been methacrylated using methacrylic anhydride and subsequently reacted with RGD peptides containing a cysteine residue [75,111]. PEGDA has also been conjugated with a thiolated transforming growth factor-β (TGF-β), which enabled the fabrication of TGF-β-functionalized PEG hydrogels [112]. Vinylsulfone moieties, which react more selectively and rapidly with thiols [113], have also been used for functionalization of hydrogels. PEG macromers carrying vinylsulfone terminal groups have been reacted with cysteine-terminated RGD and other peptides to introduce integrin-binding sites in hydrogels [114]. The same reaction has also been utilized to incorporate vascular endothelial growth factor (VEGF) carrying a genetically engineered cysteine residue into PEG hydrogels [115]. Thiol groups can also react with maleimides in a variation of the Michael addition reaction which proceeds with significantly faster kinetics [116]. This reaction has been extensively used for immobilizing peptides derived from ECM proteins onto PEG-based hydrogels [117]. Maleimide-terminated multiarm PEGs (often referred to as star-PEGs) have been modified with RGD and FN through cysteine residues. PEG arms with unreacted maleimide groups could be subsequently reacted with thiol modified PEGs thus crosslinking the polymers into a hydrogel [118]. Heparin functionalized with maleimide groups could also be incorporated into such hydrogels to enhance the bioactivity of the hydrogel system and endow it with growth-factor-binding properties [24]. PEG-maleimide has also been reacted with VEGF in order to allow its incorporation in PEG hydrogels [95]. Another fast and thiol-specific reaction that may be carried out in the presence of living cells is the photoactivated thiolene reaction [119], which has been utilized for immobilization of FN- and laminin-derived peptides on PEG hydrogels [120]. A major advantage of this technique over the earlier strategies is its photosensitive nature, which makes it amenable for spatial patterning of peptides and proteins on hydrogels. Indeed, 3D micropatterning of RGD peptides on PEG-based hydrogels has been achieved using this chemistry [121]. In an elegant approach utilizing click chemistry for hydrogel crosslinking and the thiolene reaction for peptide immobilization, PEG hydrogels with independently controllable mechanical and biochemical parameters could be fabricated [122].

Use of reactive amine and thiol groups already present in the protein sequence is a very attractive approach due to its flexibility and ease of application. However, in addition to the limitations described earlier, a shortcoming of these strategies is their lack of selectivity against cell- and medium-derived proteins, which might themselves couple to the polymer backbone. In addition to compromising yield by consuming reactive sites, this could confound studies that seek to delineate the effect of specific ECM proteins or derived peptides upon cellular processes. Therefore, several methods have been established that enable the selective immobilization of target proteins of interest onto hydrogel platforms, some of which are covered below.

14.2.2 Immobilization of Growth Factors Through Interactions with Other Biomolecules

Growth factors interact with several components of the ECM. In this section we describe ECM proteins and GAGs that exhibit such interactions and hydrogel systems that have taken advantage of some of these interactions to entrap cell-derived factors (Figure 14.2).

As described earlier, the interactions between GAGs and growth factors have been extensively characterized and leveraged for biomaterial design. GAG–protein interactions are largely based on electrostatic attraction between cationic residues on the proteins and anionic sulfate and carboxyl groups on the GAGs, although hydrogen bonding and van der Waals interaction are also believed to contribute [123,124]. Through these interactions GAGs are thought to sequester growth factors, providing a depot effect that regulates these proteins' local concentration, diffusion, and signaling intensity [125]. Moreover, the association of growth factors with heparin and HS has been shown to increase growth factor lifetime by protecting these proteins from enzymatic degradation [126,127].

Among the GAGs, heparin and HS are probably the most extensively studied

TABLE 14.1 Interactions Between Growth Factors and ECM Components

ECM Component	Interacting GF
Heparin/heparan sulfate	FGFs [24,128], TGF-βs [283], VEGF [284,285], PDGFs [286,287], HGF [136,288], BMPs [289–291]
Collagens	PDGFs [151], HGF [152], TGF-β1 [153]
Vitronectin	IGF-II [154], TGF-β1 and 2, EGF, VEGF, FGF-2 [156], HGF [141]
Fibronectin	HGF [141], TGF-β1 [140], VEGF [137,144,145], CTGF [142], FGFs [137,146], PDGFs [137,143], PIGF-2, 3, BMP-2, 7, NGF, BDNF [137]
Fibrinogen	VEGF [138,149], FGFs [138,148], PIGF-2,3, PDGFs, BMP-2,2/7, TGF-β1,2, BDNF [138]
Tenascin-C	PDGFs, VEGF, PIGF-2,3, FGFs, TGF-β1,2, BMP-2, BDNF, HGF [150]

molecules based on their interaction with diverse growth factors such as FGFs, TGF-β, VEGF, and PDGF-A and -B [24] (Table 14.1). FGFs are among the best-studied heparin-binding proteins, and structural studies have shown that heparin and HS interact with both FGF and its receptor, while simultaneously promoting receptor dimerization [128,129]. Hydrogels based on heparin and HS have therefore found extensive use as platforms for sustained release of growth factors. A heparin-crosslinked PEG-based hydrogel loaded with VEGF demonstrated gradual release of active VEGF over a period of 3 weeks [130]. Subcutaneous implantation of these hydrogels in mice was shown to induce angiogenesis around the implantation area. Taking this a step further, a similar strategy was used to enable the simultaneous sustained delivery of VEGF and FGF-2, which demonstrated superior pro-angiogenic activity over administration of single growth factors in both *in vitro* and *in vivo* models [131]. PEG-heparin hybrid hydrogels loaded with hepatocyte growth factor (HGF) were shown to not only support the culture of primary rat hepatocytes but also promote urea and albumin synthesis, which has motivated the application of this platform for *in vitro* differentiation of hepatocytes and as vehicles for transplantation [132]. Heparin has also been combined with natural polymers in order to synthesize sustained growth factor-releasing hydrogels. In a related study, heparin was incorporated into chemically crosslinked alginate hydrogels and used to entrap FGF-2 [133]. Here, regenerated axons from rat sciatic nerves grew much faster on heparin-alginate hydrogels than on alginate-only hydrogels, suggesting that heparin promoted retention and delivery of the growth factor. In another study, heparin was incorporated into photo-crosslinked alginate hydrogels and the hydrogels were loaded with bone morphogenetic protein-2 (BMP-2). It was demonstrated that BMP-2-loaded photo-crosslinked heparin-alginate hydrogels induced significantly more osteogenesis than unmodified alginate hydrogels loaded with BMP-2 [25]. Hydrogels formed from HA have also been functionalized with heparin in order to achieve controlled release of BMP-2 [26]. GAGs other than heparin are known to interact with specific growth factors [134–136], but these materials have not yet been as extensively explored as matrices for controlled release.

Growth factors have also been shown to interact with ECM protein components. Recently, several ECM proteins such as FN, fibrinogen, and tenascin-C were shown to bind a large number of diverse growth factors (Table 14.1) [137,138]. The regions of these proteins responsible for binding growth factors were localized to their heparin-binding motifs, although binding surprisingly does not require the presence of heparin. Interestingly, mutation of the positively charged residues in these motifs into serines abolishes growth factor binding indicating that the interactions may be driven by electrostatic forces.

FN is an important component of the ECM in wounds and its interaction with growth factors is believed to be important for wound healing [139]. FN is known to interact with TGF-β1, HGF, connective tissue growth factor (CTGF), PDGF-A, and VEGF with high affinity [140–144]. Using recombinant FN domains, the C-terminal heparin-binding Hep-II domain of FN (consisting of type III repeats 13 to 14) was identified as the key site for binding VEGF and FGF-2 [145,146]. Interestingly, peptide fragments containing both the $\alpha_5\beta_1$ integrin-binding domain (FN type III repeats 9 to 10) and VEGF-binding Hep-II domain significantly enhanced VEGF-induced cell responses, suggesting strong crosstalk between bound integrins and the VEGF receptor mediated by FN and possibly by other ECM-derived proteins [145]. In a high-throughput study, it was observed that the heparin-binding FN III12-14 domain binds most of the growth factors from the PDGF, VEGF, and FGF families, and some growth factors from the TGF-β and neurotrophin families with nanomolar affinity and without inhibiting growth factor activity [137]. A total of 25 previously unknown interactions with growth factors were identified in this report. Growth factor binding was enhanced in the presence of heparin, but the extent of enhancement was variable.

The therapeutic implications of this discovery were clearly demonstrated in a study in which the multiple interactions of this domain were utilized to sequester a variety of growth factors in a fibrin-based hydrogel [147]. In this study, a fusion peptide consisting of the integrin-binding FN III9-10 domain and the growth factor-binding FN III12-14 domain previously shown to have a synergistic effect on enhancing VEGF-induced cell signaling [145] was used to functionalize fibrin hydrogels. These hydrogels were shown to sequester VEGF-A165, PDGF-BB, and BMP-2, and enhanced the morphogenetic effects of these growth factors *in vitro*. Significantly, it was further demonstrated that the FN III9-10/12-14 peptide greatly enhanced the regenerative effects of the growth factors *in vivo* in a diabetic mouse model of chronic wounds and in a rat model of critical-size bone defects at doses where the growth factors delivered within fibrin only had no significant effect (Figure 14.3).

Another protein system that has been extensively studied for its interactions with growth factors is fibrinogen and its enzymatic degradation product fibrin. Fibrinogen is found in blood plasma and plays an important function in clotting by acting as a scaffold for platelet adhesion and by promoting endothelial cell mitogenesis [139]. FGF-2, which is released by disrupted blood vessels during wound formation, can bind both fibrin and fibrinogen [148], and this sequestration is necessary for endothelial cell migration, mitogenesis, and ultimately angiogenesis during wound repair. VEGF, another growth factor involved in wound healing, also interacts with both fibrinogen and fibrin [149]. A recent high-throughput study identified 15 unique interactions between the ECM proteins fibrin and fibrinogen and growth factors from the PDGF, VEGF, FGF, TGF-β, and neurotrophin families [138]. This study then demonstrated that similar binding characteristics are shown by a dimer of the heparin-binding domain of fibrinogen. Significantly, binding with this domain did not alter the ability of growth factors to trigger cell proliferation. The researchers then incorporated this domain into a synthetic PEG-based hydrogel and showed that the domain effectively sequestered FGF-2 and placental growth factor-2 (PlGF-2) (Figure 14.4). Furthermore, this synthetic bioactive hydrogel could fully recapitulate the

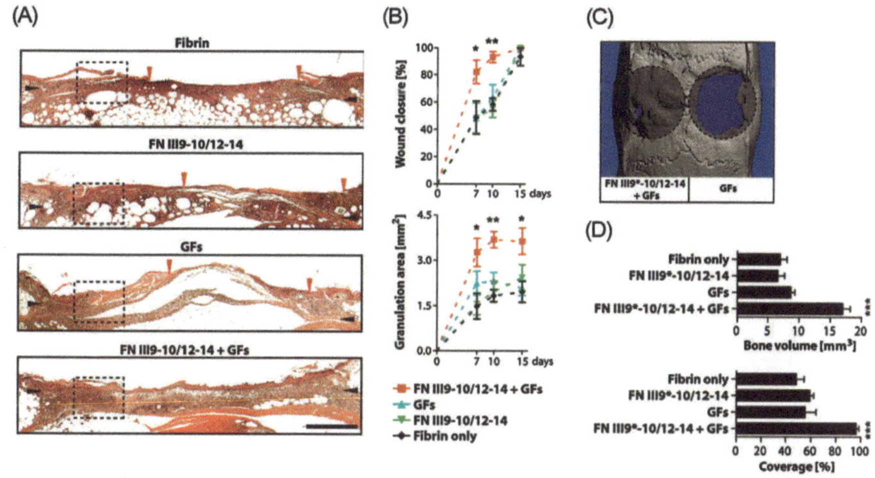

FIGURE 14.3 Enhancement of skin wound healing in diabetic mice and bone regeneration in rats by incorporation of growth factor-binding recombinant FN domains into fibrin hydrogels. (A) Histological examination of wound tissue in diabetic mice 10 days after implantation of the specified fibrin matrices. Much higher wound closure is evident for the fibrin matrices functionalized with FN III9-10/12-14 containing growth factors VEGF-A165 and PDGF-BB. Scale bar is 1 mm. (B) Quantitative analysis of the wound closure and granulation area from histological examination. (C) Representative micro-computed tomography (μCT) images of the skull of rats with critical-size calvarial defects 4 weeks after implantation of the specified fibrin matrices. The growth factors used in this study were BMP-2 and PDGF-BB. FN III9*-10/12-14 contains a point mutation in the 9th repeat which makes it specific to the $\alpha_5\beta_1$ integrin, the engagement of which is known to enhance osteoblastic differentiation of MSCs. (D) Quantitative analysis of the μCT data for implantation of four different fibrin matrices. *Source: From Ref. [147]. Reprinted with permission from AAAS.*

efficacy of a fibrin matrix in promoting wound healing in a mouse model.

Similar promiscuous growth factor-binding behavior was recently demonstrated for the heparin-binding domain of Tenascin-C (TNC) [150]. In this study, recombinant peptide fragments of TNC representing the first five TNCIII repeats (TNCIII1–5) were used to study interactions with various growth factors. Multiple growth factors of the PDGF, FGF, and TGF-β families were found to interact with this region of TNC, specifically to TNCIII5.

Collagens I–VI have been shown to bind PDGF-AA, BB, and AB with high affinity without affecting the activity of these growth factors [151]. Collagens can also bind HGF with nanomolar affinity, and HGF bound to collagen has been shown to retain its pro-motility and proliferative functions in culture [152]. Similarly, collagen was also shown to bind TGF-β [153].

Finally, insulin-like growth factor-II (IGF-II) is known to interact with vitronectin [154]. It was shown that IGF-II bound to vitronectin significantly enhanced migration of breast cancer cells in transwell assays, while unbound IGF-II did not [155]. Vitronectin is also known to interact with a variety of other growth factors such as TGF-β, EGF, FGF-2, VEGF, and HGF [141,156].

Thus, intrinsic interactions between growth factors and various components of the ECM can be utilized to entrap these factors in hydrogels based on both natural and synthetic polymers, with retention of biological activity. Furthermore, the typically reversible nature of the scaffold–factor interaction makes such strategies very attractive for therapeutic applications requiring sustained *in vivo* delivery of growth factors. A major shortcoming of such approaches is that they are not universally generalizable but

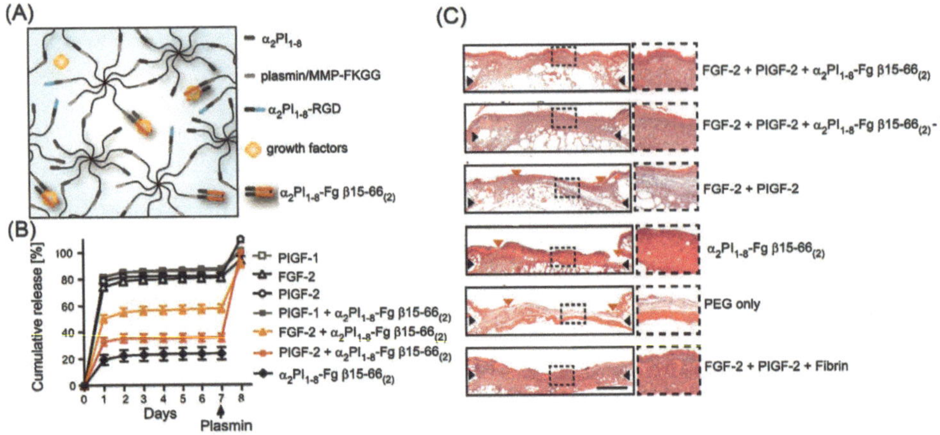

FIGURE 14.4 Entrapment of growth factors and enhancement of wound healing in diabetic mice by a fibrin-mimetic PEG matrix. (A) Schematic representation of the components for the fibrin-mimetic PEG hydrogel. The matrix contains eight-arm PEG molecules conjugated to two peptide sequences: (i) α_2PI_{1-8} peptide, and (ii) a protease-sensitive sequence fused to the FKGG peptide, which act as substrates for the transglutaminase FXIIIa. RGD fused to α_2PI_{1-8}, the growth factor binding fibrinogen domain fused to α_2PI_{1-8}, and growth factors are added to the PEG formulation prior to crosslinking. The transglutamination reaction catalyzed by FXIIIa forms isopeptide bonds between a glutamine residue in α_2PI_{1-8} and the lysine in FKGG, thus crosslinking the PEG macromers and also conjugating the RGD peptide and the fibrinogen domain to the matrix. (B) Retention of fibrin-binding growth factors in fibrin-mimetic PEG matrix functionalized with the growth-factor-binding domain of fibrinogen. PlGF-2 and FGF-2, which interact with this domain, are retained in the hydrogel only in the presence of the domain. (C) Representative histological specimens of the wound area 10 days after implantation of the specified matrices in diabetic mice. Delivering growth factors within the PEG matrix functionalized with α_2PI_{1-8}-Fg $\beta15-66_{(2)}$ enhances wound healing to an extent comparable with a fibrin matrix loaded with the growth factors. Scale bar is 1 mm. *Source: Adapted with permission from Ref. [138].*

are instead limited to specific growth factors with high affinity for a given scaffold. The following sections describe some strategies that can be generically applied for the immobilization of any target protein on hydrogels.

14.2.3 Protein/Peptide Tags for Immobilization

Recombinant DNA technology has been used to tag growth factors and ECM-derived proteins with peptides that enable their specific immobilization onto hydrogels (Figure 14.2). In this way, amino acids in the native sequence that may be critical to function are spared from chemical functionalization. Furthermore, placement of the tag in a known position relative to the bioactive portion(s) of the protein increases the likelihood that protein orientation is homogeneous. Such strategies are also more generic in that most proteins can be recombinantly fused to such tags without affecting their native function as well as expressed in material-scale quantities using bacterial, yeast, and insect cell expression systems.

14.2.3.1 Collagen-Binding Domain-Mediated Immobilization

One of the most common families of peptide tags that have been used for targeted protein immobilization on hydrogels are derived from the collagen-binding domains (CBDs) of proteins. Such domains have been obtained from four sources—the von Willebrand factor (vWF), mammalian collagenase, bacterial collagenase, and FN [157].

The CBDs derived from vWF and human collagenase are ten and seven residues long, respectively. Due to their short lengths, these peptides are preferred fusion tags for growth factors in order to impart them with collagen-binding properties. The CBD derived from vWF has been fused to several growth factors such as TGF-β [158,159], EGF [160], FGF-2 [161], BMP-2 [162,163], and BMP-3 [164]. *In vitro* experiments demonstrated that collagen matrices functionalized with the TGF-β-CBD fusion protein, but not commercial TGF-β, could promote the survival, proliferation, differentiation, and colony mineralization of osteogenic precursor cells present in rat bone marrow [159]. In an *in vivo* rat ectopic bone formation assay, implantation of collagen hydrogels functionalized with BMP-2 fused to this CBD induced the production of new bone tissue inside the implants [163]. These tissues displayed the hallmarks of mature bone such as trabecular morphology and medullary cavities. CBD-BMP-3 fusion proteins affixed to collagen hydrogels also stimulated ingrowth of new bone in a rat cranial defect model, underlining the high therapeutic potential of such growth factor fusion constructs [164]. The CBD derived from human collagenase has also been fused to a large number of growth factors in order to endow them with collagen-binding properties. Both *in vitro* and *in vivo* experiments using collagen scaffolds functionalized with growth factors such as PDGF-BB [165,166], nerve growth factor-β (NGF-β) [167], neurotrophin-3 [168], VEGF [169,170], EGF [171], and FGF-2 [172] fused to this CBD have demonstrated the great potential of this strategy for tissue engineering and medical applications. A comparison of the fusion tags derived from vWF and human collagenase showed that the latter enhanced collagen affinity. This also translated to *in vivo* activity, as collagen scaffolds functionalized with FGF-2 fused to the collagenase-derived peptide were able to stimulate greater vascularization upon subcutaneous implantation in rats [172]. The CBD of bacterial collagenase and FN are comparatively larger proteins (24 and 27 kDa respectively) and can thus affect the functionality of growth factors fused to them. Despite this, the bacterial collagenase-derived domain has been fused with FGF-1 [173,174], FGF-2 [175], EGF [175], and VEGF [176], while the CBD of FN has been combined with HGF [177], BMP-4 [178], EGF [179], and VEGF [180] to endow these growth factors with collagen-binding ability. The BMP-4 fusion protein was strongly osteoinductive in 3D cultures of human bone-marrow-derived mesenchymal stem cells in a hybrid scaffold based on collagen and poly(lactic-co-glycolic acid) [178]. *In vivo* implantation of such scaffolds demonstrated that the immobilized CBD-BMP4 retained its osteoinductive activity, as evidenced by its ability to upregulate osteogenic gene expression and biomineralization.

14.2.3.2 Factor XIIIA Transglutaminase Catalyzed Immobilization

During wound healing, fibrin clots are stabilized by the transglutaminase factor XIIIA (FXIIIa), which crosslinks glutamine residues on one fibrin molecule to lysine residues on another fibrin molecule through isopeptide bonds. This reaction is also used in the *in vitro* fabrication of fibrin hydrogels and has been extended to the conjugation of target proteins and peptides to such hydrogels [181–183]. This strategy is based on a peptide derived from the 12 N-terminal residues of α2-plasmin inhibitor (α2-PI), which was found to be a good substrate for FXIIIa and could be readily crosslinked with fibrin via the transglutamination reaction catalyzed by FXIIIa [184]. Thus, proteins or peptides fused to this FXIIIA substrate peptide could be covalently conjugated onto fibrin matrices as demonstrated for two integrin-binding peptides FN-derived RGD and collagen-derived DGEA [185]. This α2-PI-derived peptide domain was also fused with heparin biding peptides, which were then immobilized on fibrin scaffolds in order to enable the incorporation of heparin into the hydrogels [181]. This strategy was then utilized to sequester the heparin interacting growth factor, FGF-2. It was demonstrated that FGF-2 immobilized within heparin-modified fibrin enhanced neurite extension by up to 100% relative to unmodified fibrin [182]. The promiscuous growth factor-binding domain of FN was also immobilized onto fibrin hydrogels using this reaction [137,147].

Growth factors can also be recombinantly fused to the α2-PI-derived peptide sequence to allow immobilization onto fibrin matrices (Table 14.2). This was first demonstrated for NGF-β [186]. A peptide sequence that could be cleaved by the cell-secreted protease plasmin was included between the α2-PI-derived peptide and NGF-β to ensure that the growth factor, which was irreversibly immobilized onto the matrix, would be released and be able to stimulate signaling only in the vicinity of plasmin-secreting invading cells. This NGF fusion protein immobilized onto a fibrin matrix enhanced neurite extension from embryonic chick dorsal root ganglia by 50% relative to soluble NGF. Other growth factors that have been recombinantly fused to the α2-PI-derived peptide in order to enable incorporation into fibrin hydrogels include VEGF [183,187], BMP-2 [188], and IGF-1 [189]. For growth factors not amenable to recombinant expression, the peptide sequence can also be covalently reacted with surface lysines as has been done for keratinocyte growth factor [190].

TABLE 14.2 Protein and Peptide Tags That Can Be Used for Site-Specific Immobilization of Target Fusion Proteins/Peptides

Scaffold Material	Source and Size of Tag	Fused Protein/Peptide
Collagen	vWF, 10 residues	TGF-β [158,159], EGF [160], FGF-2 [161,172], BMP-2 [162,163], BMP-3 [164]
	Human collagenase, 7 residues	PDGF-BB [165,166], NGF-β [167], Neurotrophin-3 [168], VEGF [169,170], EGF [171], FGF-2 [172]
	Bacterial collagenase, 24 kDa	FGF-1 [173,174], FGF-2 [175], EGF [175], VEGF [176]
	Fibronectin, 27 kDa/40 kDa	HGF [177], BMP-4 [178], EGF [179], VEGF [180]
Fibrin, PEG conjugated to FXIIIA substrate peptide	α2-plasmin inhibitor, 12 residues	Cell-adhesive peptides [185,193], heparin-binding peptides [181], GF-binding FN domains [137,147], GF-binding fibrinogen domain [138], NGF-β [186], VEGF [115,183,187,193], BMP-2 [188], IGF-1 [189]
HA-MC functionalized with SH3-binding peptides	SH3 domains, 60–70 residues	FGF-2 [195]
Agarose, chitosan, HA-MC. All functionalized with avidin/streptavidin	Screen of peptide libraries, 15 residues (Avitag™)	CNTF [199], interferon-γ [200], PDGF-A [201]
Agarose functionalized with barnase	Barstar, 89 residues	SHH [199]

Notes: (I) Not all listed proteins with CBD fusion have been immobilized on hydrogels. Relevant proteins that have been successfully expressed and purified in bioactive collagen-binding forms have been included and are expected to bind to collagen scaffolds.
(II) The matrix material is not limiting for the other tags as long as they are modified with the interacting peptide/protein tag.

Interestingly, this enzymatic immobilization strategy has been now extended to enable functionalization of synthetic PEG-based matrices as well (Table 14.2). In this approach, multiarm PEG molecules are functionalized with either the α2-PI-derived peptide (acceptor peptide) or with the sequence FKGG (donor peptide), which was previously optimized as an efficient substrate for the transglutamination reaction catalyzed by FXIIIa [191]. Crosslinking of these two PEG formulations by FXIIIa yields a hydrogel, with inclusion of peptides or proteins fused to the acceptor peptide sequence during the crosslinking reaction incorporating these sequences into the hydrogel. Using this strategy, peptides containing the RGD sequence were immobilized onto PEG hydrogels which allowed mammalian cells to adhere and proliferate on the hydrogels [192]. The heparin-binding domain of fibrinogen, which is known to bind several growth factors, was also incorporated into PEG hydrogels using this technique in order to get fibrin-mimetic synthetic hydrogels which efficiently retained growth factors that showed interactions with the fibrinogen domain [138] (Figure 14.4). Such hydrogels could be efficiently loaded with the pro-angiogenic growth factors, FGF-2 and PlGF-2, and could fully mimic the effect of fibrin in promoting wound repair in a diabetic mouse model of impaired wound healing. Recombinant VEGF was also quantitatively incorporated into PEG-based hydrogels using this strategy [193].

14.2.3.3 *Src Homology 3 Domain-Mediated Immobilization*

Src homology 3 (SH3) domains are independently folding 60–70-residue peptide domains that bind polyproline sequences in proteins [194]. This interaction was used to immobilize FGF-2 fused to an SH3 domain on hyaluronic acid-methyl cellulose (HA-MC) composite hydrogels functionalized with SH3-binding polyproline peptide sequences [195]. Notably, this study also showed that hydrogels functionalized with two different polyproline peptide sequences having distinct affinities to the SH3 domain showed markedly different release profiles of FGF-2, which demonstrated the utility of this strategy to enable tunable release of growth factors and other protein drugs for therapeutic applications.

14.2.3.4 *Enzymatic Biotinylation of Proteins for Immobilization*

The *Escherichia coli* biotin ligase BirA catalyzes the transfer of biotin to the amino group of a lysine residue of the biotin carboxyl carrier protein subunit of acetyl-CoA carboxylase. Screening of peptide libraries has led to the discovery of a 14-residue peptide sequence that can be biotinylated by BirA at rates similar to the native substrate [196]. This sequence has since been modified by the addition of a C-terminal glutamate residue and termed the Avitag™ system. Proteins recombinantly fused to this peptide can be site-specifically biotinylated and then modified with streptavidin. This strategy has been used for conjugating recombinant proteins with fluorescent probes [197] and nanoparticles [198]. Recently, this enzymatic reaction was utilized for immobilization of recombinantly expressed growth factors and cytokines on hydrogels as well. In one report, thiolated-agarose hydrogels were first reacted with maleimide-functionalized streptavidin, which was subsequently used to capture ciliary neurotrophic factor (CNTF) fused to the AviTag that was biotinylated prior to immobilization [199]. In another paper, the pro-neuronal differentiation factor interferon-γ was recombinantly fused to the AviTag™ and immobilized on streptavidin-modified chitosan hydrogels [200]. Neural stem/progenitor cells (NSPCs) cultured on these matrices showed preferential differentiation into neurons confirming the retention of bioactivity of the cytokine even after immobilization. In order to promote differentiation of NSPCs into oligodendrocytes, the

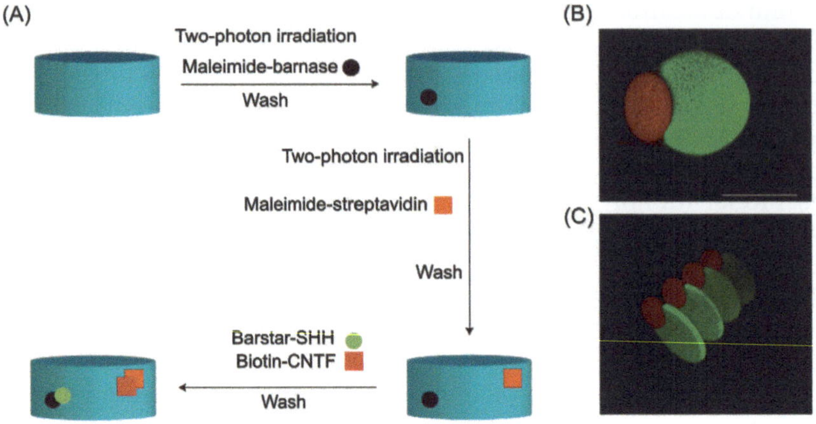

FIGURE 14.5 3D patterning of SHH and CNTF in agarose hydrogels. (A) Scheme for sequential patterning of maleimide-barnase and maleimide-streptavidin on agarose hydrogels. (B) A confocal micrograph demonstrating the localization of barstar-SHH-488 (green) and biotin-CNTF-633 (red) to the volumes patterned. (C) 3D projection of a reconstructed stack demonstrating 3D patterning capability. Scale bar is 100 μm. *Source: Adapted from Ref [199] with permission from Macmillan Publishers Ltd.*

oligodendrocyte-differentiating factor PDGF-A was biotinylated using this reaction and then immobilized on hybrid HA-MC hydrogels that were previously modified with streptavidin [201]. As expected, rat NSPCs differentiated into significantly more oligodendrocytes when cultured in such hydrogels.

14.2.3.5 *Barstar-Mediated Immobilization*

The bacterial ribonuclease barnase and its inhibitor barstar bind each other with extremely high affinity [202,203]. This strong binding has been utilized to immobilize the stem cell differentiation factor sonic hedgehog (SHH) recombinantly fused to barstar on barnase-functionalized hydrogels [199]. In this work, photosensitive caging of thiol groups on agarose hydrogels was used to pattern SHH and CNTF with a very high spatial resolution. Here, two-photon irradiation was used to uncage thiol groups in selected regions of the hydrogel followed by immobilization of barnase modified with maleimide. Next, thiol groups in another region were uncaged and reacted with streptavidin conjugated to maleimide groups. Upon immersing the hydrogel in a solution of SHH-barstar and biotin-CNTF, the proteins were independently immobilized into the target regions (Figure 14.5). The high binding affinity of both interactions ensured that binding was almost irreversible. Mouse retinal precursor cells cultured in these matrices expressed downstream effectors of both SHH and CNTF signaling pathways, indicating that the immobilized proteins retained their activity.

14.2.4 Aptamer-Mediated Immobilization

Aptamers are single stranded DNA or RNA sequences that bind with target proteins with a specificity and affinity comparable to that of antibodies [204]. These sequences are most often selected *in vitro* from a pool of random oligonucleotides, and selection conditions can be manipulated to obtain aptamers with different specificities and affinities [205]. Aptamers have been identified that bind a variety of proteins including ECM components [206–209] and growth factors [210–225] (Table 14.3).

Taking advantage of the enormous flexibility in designing and synthesizing aptamers, researchers have functionalized synthetic hydrogels with aptamers having different affinities to PDGF-BB to enable sustained release of the growth factor with a high degree of control over the release kinetics (Figure 14.2) [226,227]. Another special feature of aptamer-based protein conjugation is that binding may be readily

TABLE 14.3 Growth Factor, Cytokine, and ECM Protein Targets of Aptamers

Target	Affinity (nM)
VEGF [216]	0.14
PDGF-AB, PDGF-BB [210]	0.1
FGF-2 [212]	0.13
IGF-II [217]	2
TGF-β1 [218]	90
TGF-β2 [219]	5
KGF [214]	0.0003
Angiopoietin-1 [220]	2.8
Angiopoietin-2 [221]	2.2
Erythropoietin-α [225]	33
TNF-α [223]	7
Interferon-γ [224]	1.8
Interleukin-17A [225]	0.05
Tenascin-C [208]	5
Osteopontin [209]	18
Fibronectin [206]	Not reported

reversed by introducing the oligonucleotide sequence complementary to the aptamer. This feature was elegantly utilized to achieve specific and triggered release of multiple growth factors from a microparticle-hydrogel hybrid system [228]. Here, polystyrene microparticles were functionalized with aptamers directed against VEGF and PDGF-BB. The microparticles were then loaded with the growth factors and incorporated into hydrogels. Addition of the specific complementary sequences led to the independent release of the target growth factor without any significant effect on the release of the other growth factor. The high degree of control coupled with the low immunogenicity and toxicity of oligonucleotides makes aptamer-functionalized hydrogels a very attractive targeted drug delivery system.

Aptamers can also be selected to engage specific cell types [229]. This is based on cell-specific variations in surface proteins, including receptors and other transmembrane proteins. For example, aptamers that specifically bind to leukemia cells with sub-nanomolar affinity were recently generated and conjugated to PEG hydrogels to facilitate separation and capture of leukemic cells (Figure 14.6) [230]. Adherent cells could also be selectively released under biocompatible conditions by the addition of an endonuclease that cleaved the aptamers [231] or of an oligonucleotide sequence complementary to the aptamer [232].

14.3 ENGINEERING DEGRADABILITY INTO HYDROGELS

Degradation of hydrogels is essential for many purposes. In tissue engineering applications, degradation is required to create space in the matrix for cell migration and proliferation as well as to allow infiltration of blood vessels. When hydrogels are used for *in vivo* delivery of drugs, degradation is important for both drug release and removal of the hydrogel scaffold from the body. Importantly, the optimal degradation rate may vary significantly from one application to another, creating the need for tunable degradation kinetics. Moreover, as previously mentioned, the ECM is continuously remodeled by the action of cell-secreted enzymes, so for applications involving tissue regeneration, hydrogels must be amenable to this cell-directed degradation, with the rate of degradation ideally mirroring the rate of new tissue outgrowth. For applications requiring controlled release of therapeutic molecules, the kinetics of degradation should be adequate to maintain effective dosages of the bioactive molecule. A major consideration for hydrogels intended for *in vivo* implantation is that the degradation products should be

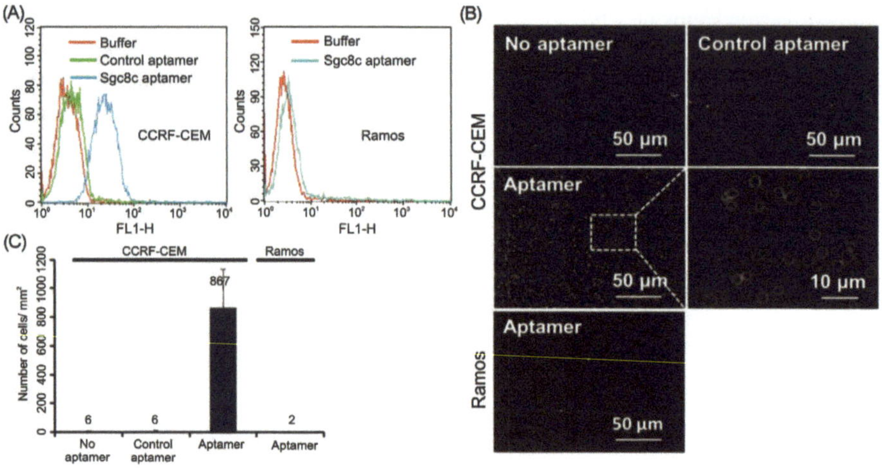

FIGURE 14.6 Cell type-specific adhesion on PEG hydrogels using aptamers. (A) Flow cytometry analysis demonstrating selective binding of the aptamer Sgc8c (fluorescently labeled) to the human leukemic cell line CCRF-CEM. In contrast, the aptamer does not bind to the nonleukemic cell line Ramos. (B) Microscopy images of PEG hydrogels functionalized with either Sgc8c or a control aptamer establishes that only CCRF-CEM cells adhere to the hydrogels. (C) A quantitative analysis of the images in (B). *Source: Reprinted from Ref. [230], with permission from Elsevier.*

nontoxic and amenable to clearance from the body. Major mechanisms of hydrogel degradation that are currently used include hydrolysis and enzymatic degradation.

14.3.1 Hydrolytic Degradation of Hydrogels

Hydrolysis causes homogenous degradation of hydrogels carrying hydrolytically sensitive macromers and crosslinkers. This leads to changes in the overall network properties and can significantly influence the diffusivity of entrapped drugs and proteins [233]. Hydrolytic degradation is spontaneous and is often preferred over enzymatic degradation in drug delivery applications, where patient-to-patient variability in the levels of circulating enzymes can lead to unpredictability in the release of drugs from enzymatically degradable hydrogels [234–236].

Hydrolytically degradable hydrogels most commonly utilize the hydrolysis of ester linkages. This mechanism has been applied for fabrication of both natural and synthetic polymer-based matrices. HA functionalized with hydrolysis-sensitive poly(lactic acid)-methacrylate (PLA-MA) has been crosslinked using the polymerization reaction of methacrylate to fabricate hydrolytically degradable hydrogels that allowed controlled release of VEGF [237]. HA has also been crosslinked with other hydrolyzable crosslinkers such as glycidyl methacrylate [238] and PEGDA [96] in order to make hydrogels susceptible to hydrolysis. Hydrogels based on synthetic polymers such as PEG have also been made sensitive to hydrolysis by incorporating hydrolyzable polymers like poly(lactic acid) (PLA) and poly (glycolic acid) into the scaffolds. The most notable example of such networks has been made using triblock polymers of the form PLA–PEG–PLA with acrylate end groups for crosslinking [81]. The ratio of PEG to PLA segments can be used to manipulate the degradation rate as well as permeability [81,239]. Degradable PEG hydrogels have also been fabricated by reacting an ester containing, amine-reactive PEG derivative with a branched PEG amine [240]. Proteins could be covalently immobilized onto this completely hydrophilic hydrogel via reaction of lysine residues with the amine-reactive groups. Multiarm PEGs terminated with hydrolyzable acrylate moieties have also been crosslinked with a PEG dithiol using the Michael addition reaction [241]. Multiarm PEGs terminated with vinylsulfone have also been often used to form hydrogels [242]. Unlike the acrylate group used previously, the vinylsulfone moiety is not hydrolyzable, and so dithiol crosslinkers containing ester groups have been used in order to render these hydrogels susceptible to hydrolytic degradation. It has been further demonstrated that the hydrolytic degradation rates of such PEG-based synthetic hydrogels can be independently modulated without affecting other physical properties. This was first accomplished by using acrylate-terminated PEG macromers incorporated with ester linkages with variable alkyl chain length (acetyl, propionyl, and butyryl esters) [243]. While the mechanical properties of the three formulations were very similar, the degradation rates were significantly higher for shorter alkyl chain lengths. Hydrolysis rates of PEG-based hydrogels have also been controlled by varying the charge of the cross-linker [244]. In this study, acrylate-terminated multiarm PEG was crosslinked using peptides containing either arginine (positively charged) or aspartate (negatively charged) residues. The arginine-based gels underwent hydrolysis 12 times faster than the aspartate-based gels.

14.3.2 Enzymatic Degradation of Hydrogels

Degradation of hydrogels by cell-secreted enzymes is often desirable in tissue engineering systems where there is an interest in locally restricting degradation to regions of cell invasion and coupling the rate of degradation with the rate of tissue formation.

Hydrogels based on natural polymers are sensitive to several cell-secreted enzymes; for example, collagen is degraded by collagenases MMP-1 and MMP-8 [245]. Therefore, hydrogels combining collagen with synthetic polymers such as PEG are also susceptible to cell-mediated degradation [246]. Fibrin hydrogels can be degraded by the action of cell-secreted MMPs and plasmin [247]. HA can be degraded by hyaluronidases [248] and the degradation rate of HA hydrogels can be controlled by covalently modifying the carboxylic groups of the macromer prior to gelation [249]. Hydrogels based on heparin can be degraded by the action of heparanase [250].

Hydrogels based on synthetic polymers can be made susceptible to enzymatic degradation as well. This has been accomplished by both functionalizing constituent macromers with enzymatic substrates or using crosslinkers consisting of peptide sequences that contain substrates for enzymes such as MMPs. These strategies have been successfully applied for creating degradable hydrogels using PEG molecules incorporating enzymatic substrates at both termini [251]. These hydrogels were crosslinked through photopolymerization of acrylate moieties at the ends of the enzyme substrate peptides, including substrates for the collagenase MMP-1 and plasmin. In another study, inclusion of an elastase-sensitive peptide in a similar model made the hydrogel sensitive to this enzyme [252]. The bioactivity of the hydrogels was further enhanced by including PEG macromers containing the cell-binding peptide KQAGDV into the polymerization reaction. This led to the formation of hydrogels with cell adhesive function and enzymatic degradability. Smooth muscle cells (SMCs), which are known to secrete elastase during migration, could be viably encapsulated in the hydrogels during polymerization. It was demonstrated that the SMCs produced more collagen in the degradable hydrogels than in nondegradable hydrogels. Similar methodologies have also been applied to multiarm PEGs, where the PEG arms were functionalized with collagenase substrates prior to crosslinking [253]. Preadipocytes cultured in these degradable hydrogels formed extensive networks resembling adipose tissue while only isolated cells could be seen in the control nondegradable hydrogels. Furthermore, the intracellular lipid droplets in degradable hydrogels were considerably enlarged, resulting in many unilocular cells, a typical feature of mature adipocytes. Multiarm PEGs with vinylsulfone functionalities have also been crosslinked with MMP-substrate peptides with terminal cysteine residues to fabricate degradable hydrogels [254]. This system allowed precise control over the degradability, matrix elasticity, and the concentration of matrix-conjugated RGD cell adhesive peptides.

MMP-1 substrate peptides have also been combined with the FXIIIa-mediated crosslinking strategy described earlier in order to fabricate MMP-1 and plasmin-sensitive PEG-based hydrogels under highly biocompatible conditions [193,255]. In this strategy, the enzymatically degradable peptide is combined with the FXIIIA substrate peptide (containing the donor FKGG sequence) in order to incorporate degradable motifs into the crosslinks. The ability to independently control biofunctionality, proteolytic degradability and crosslinking density in such hydrogels has allowed the study of the role of matrix stiffness and remodeling on 3D cell migration and formation of cellular networks (Figure 14.7) [255]. This study showed that preosteoblastic cells migrated equally efficiently in soft matrices irrespective of their degradability. However, in stiffer matrices, proteolytic degradation was necessary to facilitate migration. A similar trend was also observed in the ability of the cells to form networks in the hydrogels.

Interestingly, natural polymers have also been crosslinked using peptide sequences containing protease substrates in order to make the hydrogels susceptible to different enzymes. HA has been crosslinked using peptides containing MMP-2 substrate sequences, which rendered the hydrogels degradable by this protease [256]. For this, HA was first

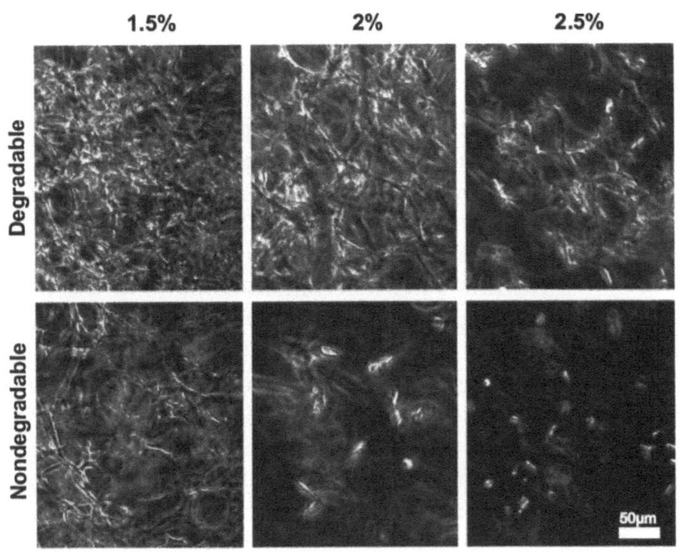

FIGURE 14.7 Matrix stiffness and MMP-sensitivity influence the formation of cellular networks. Single dispersed mouse preosteoblastic cells form dense homogeneous networks within 3 weeks in soft MMP-degradable PEG hydrogels. The networks are sparser and less uniform as matrix stiffness increases. Network formation is substantially reduced in soft nondegradable hydrogels and completely absent in stiff ones. *Source: Reprinted from Ref. [255], with permission from Elsevier.*

modified with both maleimide and methacrylate groups. The maleimides were subsequently reacted with a mixture of RGD peptides with one cysteine residue and MMP-substrate peptides with cysteine residues at each end to enable crosslinking. The methacrylate groups could also be photo-crosslinked to make the hydrogel matrix less sensitive to MMP-mediated degradation. This strategy of having two different crosslinkers allowed independent control over matrix mechanics and degradability. It was demonstrated that human mesenchymal stem cells (hMSCs) which do not secrete hyaluronidase could efficiently degrade the hydrogel by cleaving the MMP-sensitive peptides. Interestingly, the differentiation of these cells was directed by the degradability of the hydrogels independent of the matrix stiffness, with more osteogenic differentiation in the degradable hydrogels. A similar MMP-2 substrate peptide sequence has also been used to crosslink dextran to fabricate cell-degradable hydrogels which are otherwise nonhydrolyzable [257].

14.4 HYDROGEL NANOPARTICLES

Hydrogels can thus be modified with a variety of biomolecules in order to fabricate a biologically instructive microenvironment for cells. Such bioactive hydrogels are used for tissue engineering applications as well as for encapsulating cells for *in vivo* delivery. Another major area of application of hydrogels has been in the field of controlled delivery of protein drugs such as antibodies and growth factors. Hydrophilic hydrogels provide a biocompatible environment for proteins that can protect them from denaturation, enzymatic degradation, and increase their *in vivo* circulation time. These properties make hydrogels an attractive medium for engineering the surface of therapeutic cells with exogenous proteins. Synthetic nanoparticles loaded with cytokines have been used to engineer the surface of therapeutic cells in order to increase their *in vivo* viability and function [258]. Drug-loaded synthetic nanoparticles have also been used for functionalization of cells for targeted delivery using the natural homing behavior of immune cells such as T cells and macrophages [259,260]. Synthetic nanoparticles have also been independently used as delivery vehicles for a variety of therapeutic proteins [261–263]. The following section will describe a few approaches that have been used for the fabrication of hydrogel nanoparticles or nanogels.

14.4.1 Water-in-Oil Heterogeneous Emulsion

Water-in-oil (W/O or inverse emulsion) methods involve the formation of aqueous droplets (micelles) of hydrophilic polymers in a continuous oil phase by using oil-soluble surfactants, followed by crosslinking of the polymers using water-soluble crosslinkers (Figure 14.8). By using a relatively high concentration of the surfactant, micelles in the submicron range can be obtained, which then form nanogels after crosslinking [264]. Inverse emulsion methods have been used to fabricate chitosan nanogels loaded with the antitumor drug doxorubicin coupled with dextran [265]. In this study, a nanogel diameter of ~100 nm was discovered to favor tissue delivery via the enhanced permeability and retention (EPR) effect commonly exploited to deliver contrast agents and drugs to solid tumors. A similar strategy has also been applied for the preparation of HA-based nanogels [266]. Specifically, thiolated-HA was crosslinked in reverse micelles containing small interfering RNA (siRNA) to form nanogels with a diameter of 198 nm in order to enable intracellular delivery of these molecules for gene silencing. Inverse emulsion techniques have been extensively applied for

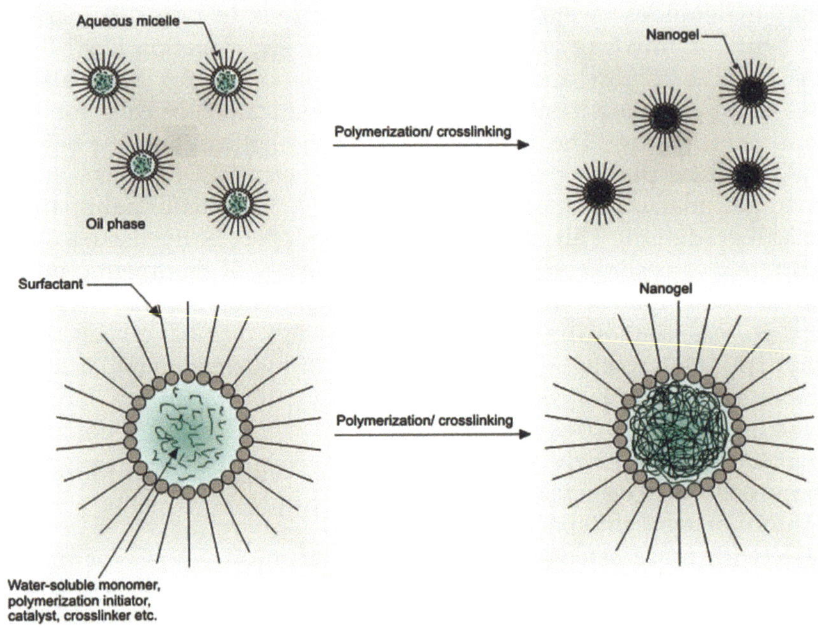

FIGURE 14.8 A schematic representation of inverse emulsion methods used for fabrication of nanogels. Aqueous micelles containing monomers, initiators, crosslinkers, etc., are stabilized in a continuous oil phase using surfactants. Polymerization and crosslinking in these submicron-sized micelles results in the formation of nanogels.

the production of nanogels based on synthetic polymers such as poly(*N*-isopropylacrylamide) [267,268], polyacrylamide (PAAm) [269,270], poly(dimethylacrylamide) [271], poly(vinylpyrrolidone) [272,273], and the amphiphilic polymer Pluronic® [274]. Inverse emulsion has also been combined with controlled radical polymerization strategies such as atom transfer radical polymerization (ATRP) and reversible addition-fragmentation chain transfer (RAFT) polymerization to produce nanogels with well-controlled size distributions. This approach was used to fabricate stable nanogels of well-controlled water-soluble poly(oligo(ethylene glycol) monomethyl ether methacrylate) (POEOMA) with a very narrow size distribution [275,276]. Similar strategies have been used to synthesize nanogels based on PAAm [277] and PEO-*b*-PHEMA block copolymers [278].

14.4.2 Dispersion/Precipitation Polymerization

In dispersion/precipitation polymerization, the monomer, the polymerization-mediating species and the initiator are initially soluble in water to form a homogeneous solution. As the polymerization proceeds to a critical point following initiation, the generated polymers become insoluble in water and thus form a separate phase. RAFT precipitation/dispersion polymerization of *N*-isopropylacrylamide using poly(*N,N'*-dimethylacrylamide) has been used to synthesize core–shell thermoresponsive nanogels [279], which could then be selectively functionalized in the core and the shell using orthogonal chemistry [280]. Aqueous dispersion RAFT polymerization has been used for the synthesis of biocompatible, antifouling, and thermosensitive core–shell nanogels based on copolymerization of oligo (ethylene glycol) methacrylates of different side chain length [281]. A similar strategy based on RAFT-controlled radical crosslinking copolymerization of *N,N*-diethylacrylamide (DEAAm) and *N,N'*-methylene bisacrylamide in aqueous dispersion polymerization was also used to synthesize PEGylated thermoresponsive core–shell nanogels [282].

14.5 CONCLUSION

Hydrogels, with their water-filled network structure and biocompatible nature, are excellent *in vitro* structural mimics of the cellular microenvironment. However, they often lack the enormous biochemical diversity that is present in the native ECM. In this chapter, we have described traditional and emerging bioconjugation strategies that have been developed to functionalize hydrogels with a variety of biomolecules in order to convert them from inert scaffolds to complex microenvironments that can not only interact with cells but can also be modified by them. This has in turn fueled many successful efforts to integrate growth factors and ECM-derived proteins and peptides to confer important bioactivity to otherwise inert materials. More recently, the ability of specific ECM components (whether native or synthetic) to sequester growth factors and cytokines has been exploited to concentrate cell-derived factors and enhance autocrine and paracrine signaling to promote tissue function. All of these strategies may be aided in the future by new peptide, protein, and aptameric tags that can be fused to target proteins and hydrogels in order to enable site-specific and selective immobilization of factors onto matrices. Moreover, the newest generation of scaffolds has begun to accommodate and direct biodegradability in highly specific ways, most notably by incorporation of crosslinks that may be enzymatically degraded to permit tissue growth and vascularization. Finally, the fabrication of hydrogel nanoparticles with "intelligent" material properties represents an important new frontier in drug delivery systems. A key recurring theme in these newest approaches is the desire to design material properties that optimally interface with biological systems. Whereas traditional biomaterial design may have traditionally focused on choosing a material that "does the least harm" in a given biological system, we are rapidly entering an era in which materials can be designed to both facilitate desirable biological processes and actively direct them.

LIST OF ABBREVIATIONS

α2-PI	α2-plasmin inhibitor
ADAMTS	A disintegrin and metalloproteinase with thrombospondin motifs
BMP	Bone morphogenetic protein
CBD	Collagen-binding domain
CTGF	Connective tissue growth factor
ECM	Extracellular matrix
FGF	Fibroblast growth factor
FN	Fibronectin
GAG	Glycosaminoglycan
HA	Hyaluronic acid
HGF	Hepatocyte growth factor
IGF	Insulin-like growth factor
MC	Methyl cellulose
MMP	Matrix metalloproteinase
NGF	Nerve growth factor
NHS	*N*-hydroxysuccinimide
PDGF	Platelet-derived growth factor
PEG	Poly(ethyleneglycol)
PEGDA	Poly(ethyleneglycol)-diacrylate
PLA	Poly(lactic acid)
PlGF	Placental growth factor
PVA	Poly(vinylalcohol)
SHH	Sonic hedgehog
SMC	Smooth muscle cell
TGF	Transforming growth factor
TNC	Tenascin-C
VEGF	Vascular endothelial growth factor
vWF	von Willebrand factor

References

[1] Gjorevski N, Nelson CM. Bidirectional extracellular matrix signaling during tissue morphogenesis. Cytokine Growth Factor Rev 2009;20(5–6):459–65.

[2] Tsang KY, Cheung MC, Chan D, Cheah KS. The developmental roles of the extracellular matrix: beyond structure to regulation. Cell Tissue Res 2010; 339(1):93–110.

[3] Jarvelainen H, Sainio A, Koulu M, Wight TN, Penttinen R. Extracellular matrix molecules: potential targets in pharmacotherapy. Pharmacol Rev 2009;61(2): 198–223.

[4] Denys H, Braems G, Lambein K, Pauwels P, Hendrix A, De Boeck A, et al. The extracellular matrix regulates cancer progression and therapy response: implications for prognosis and treatment. Curr Pharm Des 2009; 15(12):1373–84.

[5] Lu P, Weaver VM, Werb Z. The extracellular matrix: a dynamic niche in cancer progression. J Cell Biol 2012;196(4):395–406.

[6] Tanzer ML. Current concepts of extracellular matrix. J Orthop Sci 2006;11(3):326–31.

[7] Bosman FT, Stamenkovic I. Functional structure and composition of the extracellular matrix. J Pathol 2003;200(4):423–8.

[8] Frantz C, Stewart KM, Weaver VM. The extracellular matrix at a glance. J Cell Sci 2010;123(24):4195–200.

[9] Lu Q, Ganesan K, Simionescu DT, Vyavahare NR. Novel porous aortic elastin and collagen scaffolds for tissue engineering. Biomaterials 2004;25(22):5227–37.

[10] Simionescu DT, Lu Q, Song Y, Lee JS, Rosenbalm TN, Kelley C, et al. Biocompatibility and remodeling potential of pure arterial elastin and collagen scaffolds. Biomaterials 2006;27(5):702–13.

[11] Buijtenhuijs P, Buttafoco L, Poot AA, Daamen WF, van Kuppevelt TH, Dijkstra PJ, et al. Tissue engineering of blood vessels: characterization of smooth-muscle cells for culturing on collagen-and-elastin-based scaffolds. Biotechnol Appl Biochem 2004;39(Pt 2):141–9.

[12] Chan-Park MB, Shen JY, Cao Y, Xiong Y, Liu Y, Rayatpisheh S, et al. Biomimetic control of vascular smooth muscle cell morphology and phenotype for functional tissue-engineered small-diameter blood vessels. J Biomed Mater Res A 2009;88A(4):1104–21.

[13] Plow EF, Haas TA, Zhang L, Loftus J, Smith JW. Ligand binding to integrins. J Biol Chem 2000;275(29): 21785–8.

[14] Colognato H, Yurchenco PD. Form and function: the laminin family of heterotrimers. Dev Dyn 2000;218(2): 213–34.

[15] Leitinger B, Hohenester E. Mammalian collagen receptors. Matrix Biol 2007;26(3):146–55.

[16] Harburger DS, Calderwood DA. Integrin signalling at a glance. J Cell Sci 2009;122(2):159–63.

[17] Rodgers UR, Weiss AS. Cellular interactions with elastin. Pathol Biol 2005;53(7):390–8.

[18] Rodgers UR, Weiss AS. Integrin $\alpha v\beta 3$ binds a unique non-RGD site near the C-terminus of human tropoelastin. Biochimie 2004;86(3):173–8.

[19] Bax DV, Rodgers UR, Bilek MM, Weiss AS. Cell adhesion to tropoelastin is mediated via the C-terminal GRKRK motif and integrin $\alpha_V\beta_3$. J Biol Chem 2009; 284(42):28616–23.

[20] Hynes RO. Integrins: bidirectional, allosteric signaling machines. Cell 2002;110(6):673–87.

[21] Ruoslahti E. RGD and other recognition sequences for integrins. Annu Rev Cell Dev Biol 1996;12:697–715.

[22] Pierschbacher MD, Ruoslahti E. Cell attachment activity of fibronectin can be duplicated by small synthetic fragments of the molecule. Nature 1984;309(5963):30–3.

[23] Esko JD, Kimata K, Lindahl U. Proteoglycans and sulfated glycosaminoglycans. In: Varki A, Cummings RD, Esko JD, editors. Essentials of glycobiology. 2nd ed. New York, NY: Cold Spring Harbor Laboratory Press; 2009.

[24] Taipale J, Keski-Oja J. Growth factors in the extracellular matrix. FASEB J 1997;11(1):51–9.

[25] Jeon O, Powell C, Solorio LD, Krebs MD, Alsberg E. Affinity-based growth factor delivery using biodegradable, photocrosslinked heparin–alginate hydrogels. J Control Release 2011;154(3):258–66.

[26] Xu X, Jha AK, Duncan RL, Jia X. Heparin-decorated, hyaluronic acid-based hydrogel particles for the controlled release of bone morphogenetic protein 2. Acta Biomater 2011;7(8):3050–9.

[27] Lu P, Takai K, Weaver VM, Werb Z. Extracellular matrix degradation and remodeling in development and disease. Cold Spring Harb Perspect Biol 2011;3:12.

[28] Daley WP, Peters SB, Larsen M. Extracellular matrix dynamics in development and regenerative medicine. J Cell Sci 2008;121(3):255–64.

[29] Stamenkovic I. Extracellular matrix remodelling: the role of matrix metalloproteinases. J Pathol 2003; 200(4):448–64.

[30] Page-McCaw A, Ewald AJ, Werb Z. Matrix metalloproteinases and the regulation of tissue remodelling. Nat Rev Mol Cell Biol 2007;8(3):221–33.

[31] Smith HW, Marshall CJ. Regulation of cell signalling by uPAR. Nat Rev Mol Cell Biol 2010;11(1):23–36.

[32] Apte SS. A disintegrin-like and metalloprotease (reprolysin type) with thrombospondin type 1 motifs: the ADAMTS family. Int J Biochem Cell Biol 2004; 36(6):981–5.

[33] Tang BL. ADAMTS: a novel family of extracellular matrix proteases. Int J Biochem Cell Biol 2001;33(1): 33–44.

[34] Stern R, Jedrzejas MJ. Hyaluronidases: their genomics, structures, and mechanisms of action. Chem Rev 2006;106(3):818–39.

[35] Chan BP, Leong KW. Scaffolding in tissue engineering: general approaches and tissue-specific considerations. Eur Spine J 2008;17(4):467–79.

[36] Zhu J. Bioactive modification of poly(ethylene glycol) hydrogels for tissue engineering. Biomaterials 2010;31 (17):4639–56.

[37] Prestwich GD. Evaluating drug efficacy and toxicology in three dimensions: using synthetic extracellular matrices in drug discovery. Acc Chem Res 2007; 41(1):139–48.

[38] Prestwich GD. Hyaluronic acid-based clinical biomaterials derived for cell and molecule delivery in regenerative medicine. J Control Release 2011;155(2):193–9.

[39] Perale G, Rossi F, Sundstrom E, Bacchiega S, Masi M, Forloni G, et al. Hydrogels in spinal cord injury repair strategies. ACS Chem Neurosci 2011;2(7):336–45.

[40] Soler-Botija C, Bagó JR, Bayes-Genis A. A bird's-eye view of cell therapy and tissue engineering for cardiac regeneration. Ann NY Acad Sci 2012;1254(1):57–65.

[41] Johnson TD, Christman KL. Injectable hydrogel therapies and their delivery strategies for treating myocardial infarction. Expert Opin Drug Deliv 2013;10(1):59–72.

[42] Peppas NA, Hilt JZ, Khademhosseini A, Langer R. Hydrogels in biology and medicine: from molecular principles to bionanotechnology. Adv Mater 2006;18 (11):1345–60.

[43] Hoare TR, Kohane DS. Hydrogels in drug delivery: progress and challenges. Polymer (Guildf) 2008;49(8): 1993–2007.

[44] Zhu J, Marchant RE. Design properties of hydrogel tissue-engineering scaffolds. Expert Rev Med Devices 2011;8(5):607–26.

[45] Seliktar D. Designing cell-compatible hydrogels for biomedical applications. Science 2012;336(6085):1124–8.

[46] Geckil H, Xu F, Zhang X, Moon S, Demirci U. Engineering hydrogels as extracellular matrix mimics. Nanomedicine 2010;5(3):469–84.

[47] Slaughter BV, Khurshid SS, Fisher OZ, Khademhosseini A, Peppas NA. Hydrogels in regenerative medicine. Adv Mater 2009;21(32–33):3307–29.

[48] Cushing MC, Anseth KS. Hydrogel cell cultures. Science 2007;316(5828):1133–4.

[49] Brandl F, Sommer F, Goepferich A. Rational design of hydrogels for tissue engineering: impact of physical factors on cell behavior. Biomaterials 2007;28(2):134–46.

[50] Glowacki J, Mizuno S. Collagen scaffolds for tissue engineering. Biopolymers 2008;89(5):338–44.

[51] Sakai S, Hirose K, Taguchi K, Ogushi Y, Kawakami K. An injectable, *in situ* enzymatically gellable, gelatin derivative for drug delivery and tissue engineering. Biomaterials 2009;30(20):3371–7.

[52] Daamen WF, Veerkamp JH, van Hest JCM, van Kuppevelt TH. Elastin as a biomaterial for tissue engineering. Biomaterials 2007;28(30):4378–98.

[53] Mol A, van Lieshout MI, Dam-de Veen CG, Neuenschwander S, Hoerstrup SP, Baaijens FPT, et al. Fibrin as a cell carrier in cardiovascular tissue engineering applications. Biomaterials 2005;26(16):3113–21.

[54] MacEwan SR, Chilkoti A. Elastin-like polypeptides: biomedical applications of tunable biopolymers. Pept Sci 2010;94(1):60–77.

[55] Lampe KJ, Antaris AL, Heilshorn SC. Design of three-dimensional engineered protein hydrogels for tailored control of neurite growth. Acta Biomater 2013;9(3): 5590–9.

[56] Benitez PL, Sweet JA, Fink H, Chennazhi KP, Nair SV, Enejder A, et al. Sequence-specific crosslinking of electrospun, elastin-like protein preserves bioactivity and native-like mechanics. Adv Healthcare Mater 2013;2(1):114–8.

[57] Sengupta D, Gilbert PM, Johnson KJ, Blau HM, Heilshorn SC. Protein-engineered biomaterials to generate human skeletal muscle mimics. Adv Healthcare Mater 2012;1(6):785–9.

[58] McHale MK, Setton LA, Chilkoti A. Synthesis and *in vitro* evaluation of enzymatically cross-linked elastin-like polypeptide gels for cartilaginous tissue repair. Tissue Eng 2005;11(11–12):1768–79.

[59] Lim DW, Nettles DL, Setton LA, Chilkoti A. Rapid cross-linking of elastin-like polypeptides with (hydroxymethyl)phosphines in aqueous solution. Biomacromolecules 2007;8(5):1463–70.

[60] Nowak AP, Breedveld V, Pakstis L, Ozbas B, Pine DJ, Pochan D, et al. Rapidly recovering hydrogel scaffolds from self-assembling diblock copolypeptide amphiphiles. Nature 2002;417(6887):424–8.

[61] Pakstis LM, Ozbas B, Hales KD, Nowak AP, Deming TJ, Pochan D. Effect of chemistry and morphology on the biofunctionality of self-assembling diblock copolypeptide hydrogels. Biomacromolecules 2003;5(2):312–8.

[62] Pochan DJ, Schneider JP, Kretsinger J, Ozbas B, Rajagopal K, Haines L. Thermally reversible hydrogels via intramolecular folding and consequent self-assembly of a *de novo* designed peptide. J Am Chem Soc 2003;125(39):11802–3.

[63] Ozbas B, Kretsinger J, Rajagopal K, Schneider JP, Pochan DJ. Salt-triggered peptide folding and consequent self-assembly into hydrogels with tunable modulus. Macromolecules 2004;37(19):7331–7.

[64] Schneider JP, Pochan DJ, Ozbas B, Rajagopal K, Pakstis L, Kretsinger J. Responsive hydrogels from the intramolecular folding and self-assembly of a

designed peptide. J Am Chem Soc 2002;124(50): 15030–7.

[65] Altunbas A, Pochan DJ. Peptide-based and polypeptide-based hydrogels for drug delivery and tissue engineering. Top Curr Chem 2012;310:135–67.

[66] DiMarco RL, Heilshorn SC. Multifunctional materials through modular protein engineering. Adv Mater 2012;24(29):3923–40.

[67] Xu X, Jha AK, Harrington DA, Farach-Carson MC, Jia X. Hyaluronic acid-based hydrogels: from a natural polysaccharide to complex networks. Soft Matter 2012;8(12):3280–94.

[68] Burdick JA, Prestwich GD. Hyaluronic acid hydrogels for biomedical applications. Adv Mater 2011;23(12): H41–56.

[69] Toole BP, Wight TN, Tammi MI. Hyaluronan–cell interactions in cancer and vascular disease. J Biol Chem 2002;277(7):4593–6.

[70] Toole BP, Hascall VC. Hyaluronan and tumor growth. Am J Pathol 2002;161(3):745–7.

[71] Liu Y, Shu XZ, Prestwich GD. Tumor engineering: orthotopic cancer models in mice using cell-loaded, injectable, cross-linked hyaluronan-derived hydrogels. Tissue Eng 2007;13(5):1091–101.

[72] Xu X, Prestwich GD. Inhibition of tumor growth and angiogenesis by a lysophosphatidic acid antagonist in an engineered three-dimensional lung cancer xenograft model. Cancer 2010;116(7):1739–50.

[73] Serban MA, Scott A, Prestwich GD. Use of hyaluronan-derived hydrogels for three-dimensional cell culture and tumor xenografts. Curr Protoc Cell Biol 2008; Chapter 10: Unit 10 14.

[74] Jin SG, Jeong YI, Jung S, Ryu HH, Jin YH, Kim IY. The effect of hyaluronic acid on the invasiveness of malignant glioma cells: comparison of invasion potential at hyaluronic acid hydrogel and matrigel. J Korean Neurosurg Soc 2009;46(5):472–8.

[75] Ananthanarayanan B, Kim Y, Kumar S. Elucidating the mechanobiology of malignant brain tumors using a brain matrix-mimetic hyaluronic acid hydrogel platform. Biomaterials 2011;32(31):7913–23.

[76] Kirker KR, Prestwich GD. Physical properties of glycosaminoglycan hydrogels. J Polym Sci B Polym Phys 2004;42(23):4344–56.

[77] Cheng Y, Luo X, Payne GF, Rubloff GW. Biofabrication: programmable assembly of polysaccharide hydrogels in microfluidics as biocompatible scaffolds. J Mater Chem 2012;22(16):7659–66.

[78] Ramamurthi A, Vesely I. Ultraviolet light-induced modification of crosslinked hyaluronan gels. J Biomed Mater Res A 2003;66(2):317–29.

[79] Kuo CK, Ma PX. Ionically crosslinked alginate hydrogels as scaffolds for tissue engineering: part 1. Structure, gelation rate and mechanical properties. Biomaterials 2001;22(6):511–21.

[80] Kim IY, Seo SJ, Moon HS, Yoo MK, Park IY, Kim BC, et al. Chitosan and its derivatives for tissue engineering applications. Biotechnol Adv 2008;26(1):1–21.

[81] Sawhney AS, Pathak CP, Hubbell JA. Bioerodible hydrogels based on photopolymerized poly(ethylene glycol)-co-poly(.alpha-hydroxy acid) diacrylate macromers. Macromolecules 1993;26(4):581–7.

[82] Burdick JA, Anseth KS. Photoencapsulation of osteoblasts in injectable RGD-modified PEG hydrogels for bone tissue engineering. Biomaterials 2002;23(22): 4315–23.

[83] Bryant SJ, Anseth KS. Hydrogel properties influence ECM production by chondrocytes photoencapsulated in poly(ethylene glycol) hydrogels. J Biomed Mater Res 2002;59(1):63–72.

[84] Adelow C, Segura T, Hubbell JA, Frey P. The effect of enzymatically degradable poly(ethylene glycol) hydrogels on smooth muscle cell phenotype. Biomaterials 2008;29(3):314–26.

[85] Peyton SR, Raub CB, Keschrumrus VP, Putnam AJ. The use of poly(ethylene glycol) hydrogels to investigate the impact of ECM chemistry and mechanics on smooth muscle cells. Biomaterials 2006;27(28): 4881–93.

[86] Tessmar JK, Göpferich AM. Matrices and scaffolds for protein delivery in tissue engineering. Adv Drug Deliv Rev 2007;59(4–5):274–91.

[87] Sokolsky-Papkov M, Agashi K, Olaye A, Shakesheff K, Domb AJ. Polymer carriers for drug delivery in tissue engineering. Adv Drug Deliv Rev 2007;59(4–5): 187–206.

[88] Xie J, Schultz PG. Adding amino acids to the genetic repertoire. Curr Opin Chem Biol 2005;9(6):548–54.

[89] Wang L, Brock A, Herberich B, Schultz PG. Expanding the genetic code of *Escherichia coli*. Science 2001;292(5516):498–500.

[90] Sakamoto K, Hayashi A, Sakamoto A, Kiga D, Nakayama H, Soma A, et al. Site-specific incorporation of an unnatural amino acid into proteins in mammalian cells. Nucleic Acids Res 2002;30(21): 4692–9.

[91] Cretu A, Castagnino P, Assoian R. Studying the effects of matrix stiffness on cellular function using acrylamide-based hydrogels. J Vis Exp 2010;(42).

[92] Drumheller PD, Hubbell JA. Polymer networks with grafted cell adhesion peptides for highly biospecific cell adhesive substrates. Anal Biochem 1994;222 (2):380–8.

[93] Ghosh K, Ren XD, Shu XZ, Prestwich GD, Clark RA. Fibronectin functional domains coupled to hyaluronan stimulate adult human dermal fibroblast responses

critical for wound healing. Tissue Eng 2006;12(3): 601–13.
[94] DeLong SA, Moon JJ, West JL. Covalently immobilized gradients of bFGF on hydrogel scaffolds for directed cell migration. Biomaterials 2005;26(16): 3227–34.
[95] Phelps EA, Landázuri N, Thulé PM, Taylor WR, García AJ. Bioartificial matrices for therapeutic vascularization. Proc Natl Acad Sci USA 2010; 107(8):3323–8.
[96] Young JL, Engler AJ. Hydrogels with time-dependent material properties enhance cardiomyocyte differentiation *in vitro*. Biomaterials 2011;32(4):1002–9.
[97] Nuttelman CR, Mortisen DJ, Henry SM, Anseth KS. Attachment of fibronectin to poly(vinyl alcohol) hydrogels promotes NIH3T3 cell adhesion, proliferation, and migration. J Biomed Mater Res 2001;57 (2):217–23.
[98] Trappmann B, Gautrot JE, Connelly JT, Strange DG, Li Y, Oyen ML, et al. Extracellular-matrix tethering regulates stem-cell fate. Nat Mater 2012;11(7): 642–9.
[99] Stabenfeldt SE, Munglani G, Garcia AJ, LaPlaca MC. Biomimetic microenvironment modulates neural stem cell survival, migration, and differentiation. Tissue Eng Part A 2010;16(12):3747–58.
[100] Chan V, Collens MB, Jeong JH, Park K, Kong H, Bashir R. Directed cell growth and alignment on protein-patterned 3D hydrogels with stereolithography. Virtual Phys Prototyp 2012;7(3):219–28.
[101] Seidlits SK, Drinnan CT, Petersen RR, Shear JB, Suggs LJ, Schmidt CE. Fibronectin–hyaluronic acid composite hydrogels for three-dimensional endothelial cell culture. Acta Biomater 2011;7(6):2401–9.
[102] Underhill GH, Chen AA, Albrecht DR, Bhatia SN. Assessment of hepatocellular function within PEG hydrogels. Biomaterials 2007;28(2):256–70.
[103] Patel PN, Gobin AS, West JL, Patrick Jr. CW. Poly (ethylene glycol) hydrogel system supports preadipocyte viability, adhesion, and proliferation. Tissue Eng 2005;11(9–10):1498–505.
[104] Saik JE, Gould DJ, Watkins EM, Dickinson ME, West JL. Covalently immobilized platelet-derived growth factor-BB promotes angiogenesis in biomimetic poly (ethylene glycol) hydrogels. Acta Biomater 2011;7 (1):133–43.
[105] Owen SC, Fisher SA, Tam RY, Nimmo CM, Shoichet MS. Hyaluronic acid click hydrogels emulate the extracellular matrix. Langmuir 2013;29(24):7393–400.
[106] Hynd MR, Frampton JP, Dowell-Mesfin N, Turner JN, Shain W. Directed cell growth on protein-functionalized hydrogel surfaces. J Neurosci Methods 2007;162(1–2):255–63.
[107] Hynd MR, Frampton JP, Burnham MR, Martin DL, Dowell-Mesfin NM, Turner JN, et al. Functionalized hydrogel surfaces for the patterning of multiple biomolecules. J Biomed Mater Res A 2007;81(2):347–54.
[108] Cosson S, Kobel SA, Lutolf MP. Capturing complex protein gradients on biomimetic hydrogels for cell-based assays. Adv Funct Mater 2009;19(21):3411–9.
[109] Zhang C, Hekmatfer S, Karuri NW. A comparative study of polyethylene glycol hydrogels derivatized with the RGD peptide and the cell-binding domain of fibronectin. J Biomed Mater Res A 2014;102A: 170–9.
[110] Shu XZ, Ghosh K, Liu Y, Palumbo FS, Luo Y, Clark RA. Attachment and spreading of fibroblasts on an RGD peptide-modified injectable hyaluronan hydrogel. J Biomed Mater Res A 2004;68(2):365–75.
[111] Marklein RA, Burdick JA. Spatially controlled hydrogel mechanics to modulate stem cell interactions. Soft Matter 2010;6(1):136–43.
[112] McCall JD, Luoma JE, Anseth KS. Covalently tethered transforming growth factor beta in PEG hydrogels promotes chondrogenic differentiation of encapsulated human mesenchymal stem cells. Drug Deliv Transl Res 2012;2(5):305–12.
[113] Chatani S, Nair DP, Bowman CN. Relative reactivity and selectivity of vinyl sulfones and acrylates towards the thiol-Michael addition reaction and polymerization. Polym Chem 2013;4(4):1048–55.
[114] Wacker BK, Alford SK, Scott EA, Das Thakur M, Longmore GD, Elbert DL. Endothelial cell migration on RGD-peptide-containing PEG hydrogels in the presence of sphingosine 1-phosphate. Biophys J 2008; 94(1):273–85.
[115] Zisch AH, Lutolf MP, Ehrbar M, Raeber GP, Rizzi SC, Davies N, et al. Cell-demanded release of VEGF from synthetic, biointeractive cell-ingrowth matrices for vascularized tissue growth. FASEB J 2003;17(15): 2260–2.
[116] Phelps EA, Enemchukwu NO, Fiore VF, Sy JC, Murthy N, Sulchek TA, et al. Maleimide cross-linked bioactive PEG hydrogel exhibits improved reaction kinetics and cross-linking for cell encapsulation and *in situ* delivery. Adv Mater 2012;24(1):64–70, 2.
[117] Fu Y, Kao WJ. *In situ* forming poly(ethylene glycol)-based hydrogels via thiol-maleimide Michael-type addition. J Biomed Mater Res A 2011;98A(2):201–11.
[118] Nie T, Akins Jr. RE, Kiick KL. Production of heparin-containing hydrogels for modulating cell responses. Acta Biomater 2009;5(3):865–75.
[119] Lin CC, Raza A, Shih H. PEG hydrogels formed by thiol-ene photo-click chemistry and their effect on the formation and recovery of insulin-secreting cell spheroids. Biomaterials 2011;32(36):9685–95.

[120] Raza A, Ki CS, Lin CC. The influence of matrix properties on growth and morphogenesis of human pancreatic ductal epithelial cells in 3D. Biomaterials 2013;34(21):5117–27.

[121] Polizzotti BD, Fairbanks BD, Anseth KS. Three-dimensional biochemical patterning of click-based composite hydrogels via thiolene photopolymerization. Biomacromolecules 2008;9(4):1084–7.

[122] Deforest CA, Sims EA, Anseth KS. Peptide-functionalized click hydrogels with independently tunable mechanics and chemical functionality for 3D cell culture. Chem Mater 2010;22(16):4783–90.

[123] Capila I, Linhardt RJ. Heparin–protein interactions. Angew Chem Int Ed 2002;41(3):391–412.

[124] Thompson LD, Pantoliano MW, Springer BA. Energetic characterization of the basic fibroblast growth factor–heparin interaction: identification of the heparin binding domain. Biochemistry 1994;33(13):3831–40.

[125] Schultz GS, Wysocki A. Interactions between extracellular matrix and growth factors in wound healing. Wound Repair Regen 2009;17(2):153–62.

[126] Saksela O, Moscatelli D, Sommer A, Rifkin DB. Endothelial cell-derived heparan sulfate binds basic fibroblast growth factor and protects it from proteolytic degradation. J Cell Biol 1988;107(2):743–51.

[127] Sommer A, Rifkin DB. Interaction of heparin with human basic fibroblast growth factor: protection of the angiogenic protein from proteolytic degradation by a glycosaminoglycan. J Cell Physiol 1989;138(1):215–20.

[128] Pellegrini L. Role of heparan sulfate in fibroblast growth factor signalling: a structural view. Curr Opin Struct Biol 2001;11(5):629–34.

[129] Mohammadi M, Olsen SK, Ibrahimi OA. Structural basis for fibroblast growth factor receptor activation. Cytokine Growth Factor Rev 2005;16(2):107–37.

[130] Tae G, Scatena M, Stayton PS, Hoffman AS. PEG-cross-linked heparin is an affinity hydrogel for sustained release of vascular endothelial growth factor. J Biomater Sci Polym Ed 2006;17(1–2):187–97.

[131] Zieris A, Chwalek K, Prokoph S, Levental KR, Welzel PB, Freudenberg U, et al. Dual independent delivery of pro-angiogenic growth factors from starPEG-heparin hydrogels. J Control Release 2011; 156(1):28–36.

[132] Kim M, Lee JY, Jones CN, Revzin A, Tae G. Heparin-based hydrogel as a matrix for encapsulation and cultivation of primary hepatocytes. Biomaterials 2010;31(13):3596–603.

[133] Ohta M, Suzuki Y, Chou H, Ishikawa N, Suzuki S, Tanihara M, et al. Novel heparin/alginate gel combined with basic fibroblast growth factor promotes nerve regeneration in rat sciatic nerve. J Biomed Mater Res A 2004;71(4):661–8.

[134] Locci P, Marinucci L, Lilli C, Martinese D, Becchetti E. Transforming growth factor beta 1-hyaluronic acid interaction. Cell Tissue Res 1995;281(2):317–24.

[135] Lyon M, Deakin JA, Rahmoune H, Fernig DG, Nakamura T, Gallagher JT. Hepatocyte growth factor/scatter factor binds with high affinity to dermatan sulfate. J Biol Chem 1998;273(1):271–8.

[136] Catlow K, Deakin JA, Delehedde M, Fernig DG, Gallagher JT, Pavao MS, et al. Hepatocyte growth factor/scatter factor and its interaction with heparan sulphate and dermatan sulphate. Biochem Soc Trans 2003;31(2):352–3.

[137] Martino MM, Hubbell JA. The 12th–14th type III repeats of fibronectin function as a highly promiscuous growth factor-binding domain. FASEB J 2010; 24(12):4711–21.

[138] Martino MM, Briquez PS, Ranga A, Lutolf MP, Hubbell JA. Heparin-binding domain of fibrin(ogen) binds growth factors and promotes tissue repair when incorporated within a synthetic matrix. Proc Natl Acad Sci USA 2013;110(12):4563–8.

[139] Macri L, Silverstein D, Clark RAF. Growth factor binding to the pericellular matrix and its importance in tissue engineering. Adv Drug Deliv Rev 2007; 59(13):1366–81.

[140] Mooradian DL, Lucas RC, Weatherbee JA, Furcht LT. Transforming growth factor-beta 1 binds to immobilized fibronectin. J Cell Biochem 1989;41(4):189–200.

[141] Rahman S, Patel Y, Murray J, Patel KV, Sumathipala R, Sobel M, et al. Novel hepatocyte growth factor (HGF) binding domains on fibronectin and vitronectin coordinate a distinct and amplified Met-integrin induced signalling pathway in endothelial cells. BMC Cell Biol 2005;6(1):8.

[142] Hoshijima M, Hattori T, Inoue M, Araki D, Hanagata H, Miyauchi A, et al. CT domain of CCN2/CTGF directly interacts with fibronectin and enhances cell adhesion of chondrocytes through integrin $\alpha_5\beta_1$. FEBS Lett 2006;580(5):1376–82.

[143] Smith EM, Mitsi M, Nugent MA, Symes K. PDGF-A interactions with fibronectin reveal a critical role for heparan sulfate in directed cell migration during *Xenopus* gastrulation. Proc Natl Acad Sci USA 2009;106(51):21683–8.

[144] Wijelath ES, Murray J, Rahman S, Patel Y, Ishida A, Strand K, et al. Novel vascular endothelial growth factor binding domains of fibronectin enhance vascular endothelial growth factor biological activity. Circ Res 2002;91(1):25–31.

[145] Wijelath ES, Rahman S, Namekata M, Murray J, Nishimura T, Mostafavi-Pour Z, et al. Heparin-II domain of fibronectin is a vascular endothelial growth factor-binding domain: enhancement of

VEGF biological activity by a singular growth factor/matrix protein synergism. Circ Res 2006;99(8):853–60.

[146] Bossard C, Van den Berghe L, Laurell H, Castano C, Cerutti M, Prats AC, et al. Antiangiogenic properties of fibstatin, an extracellular FGF-2-binding polypeptide. Cancer Res 2004;64(20):7507–12.

[147] Martino MM, Tortelli F, Mochizuki M, Traub S, Ben-David D, Kuhn GA, et al. Engineering the growth factor microenvironment with fibronectin domains to promote wound and bone tissue healing. Sci Transl Med 2011;3(100):100ra89.

[148] Sahni A, Odrljin T, Francis CW. Binding of basic fibroblast growth factor to fibrinogen and fibrin. J Biol Chem 1998;273(13):7554–9.

[149] Sahni A, Francis CW. Vascular endothelial growth factor binds to fibrinogen and fibrin and stimulates endothelial cell proliferation. Blood 2000;96(12):3772–8.

[150] De Laporte L, Rice JJ, Tortelli F, Hubbell JA. Tenascin C promiscuously binds growth factors via its fifth fibronectin type III-like domain. PLoS One 2013;8(4):e62076.

[151] Somasundaram R, Schuppan D. Type I, II, III, IV, V, and VI collagens serve as extracellular ligands for the isoforms of platelet-derived growth factor (AA, BB, and AB). J Biol Chem 1996;271(43):26884–91.

[152] Schuppan D, Schmid M, Somasundaram R, Ackermann R, Ruehl M, Nakamura T, et al. Collagens in the liver extracellular matrix bind hepatocyte growth factor. Gastroenterology 1998;114(1):139–52.

[153] Paralkar VM, Vukicevic S, Reddi AH. Transforming growth factor β type 1 binds to collagen IV of basement membrane matrix: implications for development. Dev Biol 1991;143(2):303–8.

[154] Upton Z, Webb H, Hale K, Yandell CA, McMurtry JP, Francis GL, et al. Identification of vitronectin as a novel insulin-like growth factor-II binding protein. Endocrinology 1999;140(6):2928–31.

[155] Noble A, Towne C, Chopin L, Leavesley D, Upton Z. Insulin-like growth factor-II bound to vitronectin enhances MCF-7 breast cancer cell migration. Endocrinology 2003;144(6):2417–24.

[156] Schoppet M, Chavakis T, Al-Fakhri N, Kanse SM, Preissner KT. Molecular interactions and functional interference between vitronectin and transforming growth factor-beta. Lab Invest 2002;82(1):37–46.

[157] Tada S, Kitajima T, Ito Y. Design and synthesis of binding growth factors. Int J Mol Sci 2012;13(5):6053–72.

[158] Tuan TL, Cheung DT, Wu LT, Yee A, Gabriel S, Han B, et al. Engineering, expression and renaturation of targeted TGF-beta fusion proteins. Connect Tissue Res 1996;34(1):1–9.

[159] Andrades JA, Han B, Becerra J, Sorgente N, Hall FL, Nimni ME. A recombinant human TGF-beta1 fusion protein with collagen-binding domain promotes migration, growth, and differentiation of bone marrow mesenchymal cells. Exp Cell Res 1999;250(2):485–98.

[160] Hall FL, Kaiser A, Liu L, Chen ZH, Hu J, Nimni ME, et al. Design, expression, and renaturation of a lesion-targeted recombinant epidermal growth factor–von Willebrand factor fusion protein: efficacy in an animal model of experimental colitis. Int J Mol Med 2000;6(6):635–43.

[161] Andrades JA, Santamaria JA, Wu LT, Hall FL, Nimni ME, Becerra J. Production of a recombinant human basic fibroblast growth factor with a collagen binding domain. Protoplasma 2001;218(1–2):95–103.

[162] Chen B, Lin H, Zhao Y, Wang B, Liu Y, Liu Z, et al. Activation of demineralized bone matrix by genetically engineered human bone morphogenetic protein-2 with a collagen binding domain derived from von Willebrand factor propolypeptide. J Biomed Mater Res A 2007;80(2):428–34.

[163] Visser R, Arrabal PM, Becerra J, Rinas U, Cifuentes M. The effect of an rhBMP-2 absorbable collagen sponge-targeted system on bone formation *in vivo*. Biomaterials 2009;30(11):2032–7.

[164] Han B, Perelman N, Tang B, Hall F, Shors EC, Nimni ME. Collagen-targeted BMP3 fusion proteins arrayed on collagen matrices or porous ceramics impregnated with Type I collagen enhance osteogenesis in a rat cranial defect model. J Orthop Res 2002;20(4):747–55.

[165] Lin H, Chen B, Sun W, Zhao W, Zhao Y, Dai J. The effect of collagen-targeting platelet-derived growth factor on cellularization and vascularization of collagen scaffolds. Biomaterials 2006;27(33):5708–14.

[166] Sun W, Lin H, Xie H, Chen B, Zhao W, Han Q, et al. Collagen membranes loaded with collagen-binding human PDGF-BB accelerate wound healing in a rabbit dermal ischemic ulcer model. Growth Factors 2007;25(5):309–18.

[167] Sun W, Lin H, Chen B, Zhao W, Zhao Y, Dai J. Promotion of peripheral nerve growth by collagen scaffolds loaded with collagen-targeting human nerve growth factor-beta. J Biomed Mater Res A 2007;83(4):1054–61.

[168] Fan J, Xiao Z, Zhang H, Chen B, Tang G, Hou X, et al. Linear ordered collagen scaffolds loaded with collagen-binding neurotrophin-3 promote axonal regeneration and partial functional recovery after complete spinal cord transection. J Neurotrauma 2010;27(9):1671–83.

[169] Zhang J, Ding L, Zhao Y, Sun W, Chen B, Lin H, et al. Collagen-targeting vascular endothelial growth factor improves cardiac performance after myocardial infarction. Circulation 2009;119(13):1776–84.
[170] Lin N, Li X, Song T, Wang J, Meng K, Yang J, et al. The effect of collagen-binding vascular endothelial growth factor on the remodeling of scarred rat uterus following full-thickness injury. Biomaterials 2012;33(6):1801–7.
[171] Yang Y, Zhao Y, Chen B, Han Q, Sun W, Xiao Z, et al. Collagen-binding human epidermal growth factor promotes cellularization of collagen scaffolds. Tissue Eng Part A 2009;15(11):3589–96.
[172] Zhao W, Chen B, Li X, Lin H, Sun W, Zhao Y, et al. Vascularization and cellularization of collagen scaffolds incorporated with two different collagen-targeting human basic fibroblast growth factors. J Biomed Mater Res A 2007;82(3):630–6.
[173] Pang Y, Wang X, Ucuzian AA, Brey EM, Burgess WH, Jones KJ, et al. Local delivery of a collagen-binding FGF-1 chimera to smooth muscle cells in collagen scaffolds for vascular tissue engineering. Biomaterials 2010;31(5):878–85.
[174] Brewster LP, Washington C, Brey EM, Gassman A, Subramanian A, Calceterra J, et al. Construction and characterization of a thrombin-resistant designer FGF-based collagen binding domain angiogen. Biomaterials 2008;29(3):327–36.
[175] Nishi N, Matsushita O, Yuube K, Miyanaka H, Okabe A, Wada F. Collagen-binding growth factors: production and characterization of functional fusion proteins having a collagen-binding domain. Proc Natl Acad Sci USA 1998;95(12):7018–23.
[176] Akimoto M, Takeda A, Matsushita O, Inoue J, Sakamoto K, Hattori M, et al. Effects of CB-VEGF-A injection in rat flap models for improved survival. Plast Reconstr Surg 2013;131(4):717–25.
[177] Kitajima T, Terai H, Ito Y. A fusion protein of hepatocyte growth factor for immobilization to collagen. Biomaterials 2007;28(11):1989–97.
[178] Lu H, Kawazoe N, Kitajima T, Myoken Y, Tomita M, Umezawa A, et al. Spatial immobilization of bone morphogenetic protein-4 in a collagen-PLGA hybrid scaffold for enhanced osteoinductivity. Biomaterials 2012;33(26):6140–6.
[179] Ishikawa T, Terai H, Kitajima T. Production of a biologically active epidermal growth factor fusion protein with high collagen affinity. J Biochem 2001;129(4):627–33.
[180] Ishikawa T, Eguchi M, Wada M, Iwami Y, Tono K, Iwaguro H, et al. Establishment of a functionally active collagen-binding vascular endothelial growth factor fusion protein *in situ*. Arterioscler Thromb Vasc Biol 2006;26(9):1998–2004.
[181] Sakiyama SE, Schense JC, Hubbell JA. Incorporation of heparin-binding peptides into fibrin gels enhances neurite extension: an example of designer matrices in tissue engineering. FASEB J 1999;13(15):2214–24.
[182] Sakiyama-Elbert SE, Hubbell JA. Development of fibrin derivatives for controlled release of heparin-binding growth factors. J Control Release 2000;65(3):389–402.
[183] Zisch AH, Schenk U, Schense JC, Sakiyama-Elbert SE, Hubbell JA. Covalently conjugated VEGF—fibrin matrices for endothelialization. J Control Release 2001;72(1–3):101–13.
[184] Ichinose A, Tamaki T, Aoki N. Factor XIII-mediated cross-linking of NH2-terminal peptide of α2-plasmin inhibitor to fibrin. FEBS Lett 1983;153(2):369–71.
[185] Schense JC, Hubbell JA. Cross-linking exogenous bifunctional peptides into fibrin gels with factor XIIIa. Bioconjug Chem 1998;10(1):75–81.
[186] Sakiyama-Elbert SE, Panitch A, Hubbell JA. Development of growth factor fusion proteins for cell-triggered drug delivery. FASEB J 2001;15(7):1300–2.
[187] Ehrbar M, Djonov VG, Schnell C, Tschanz SA, Martiny-Baron G, Schenk U, et al. Cell-demanded liberation of VEGF121 from fibrin implants induces local and controlled blood vessel growth. Circ Res 2004; 94(8):1124–32.
[188] Schmoekel HG, Weber FE, Schense JC, Gratz KW, Schawalder P, Hubbell JA. Bone repair with a form of BMP-2 engineered for incorporation into fibrin cell ingrowth matrices. Biotechnol Bioeng 2005;89(3): 253–62.
[189] Lorentz KM, Yang L, Frey P, Hubbell JA. Engineered insulin-like growth factor-1 for improved smooth muscle regeneration. Biomaterials 2012;33(2):494–503.
[190] Geer DJ, Swartz DD, Andreadis ST. Biomimetic delivery of keratinocyte growth factor upon cellular demand for accelerated wound healing *in vitro* and *in vivo*. Am J Pathol 2005;167(6):1575–86.
[191] Hu B-H, Messersmith PB. Rational design of transglutaminase substrate peptides for rapid enzymatic formation of hydrogels. J Am Chem Soc 2003; 125(47):14298–9.
[192] Ehrbar M, Rizzi SC, Schoenmakers RG, Miguel BS, Hubbell JA, Weber FE, et al. Biomolecular hydrogels formed and degraded via site-specific enzymatic reactions. Biomacromolecules 2007;8(10):3000–7.
[193] Ehrbar M, Rizzi SC, Hlushchuk R, Djonov V, Zisch AH, Hubbell JA, et al. Enzymatic formation of modular cell-instructive fibrin analogs for tissue engineering. Biomaterials 2007;28(26):3856–66.

[194] Agrawal V, Kishan KV. Promiscuous binding nature of SH3 domains to their target proteins. Protein Pept Lett 2002;9(3):185–93.
[195] Vulic K, Shoichet MS. Tunable growth factor delivery from injectable hydrogels for tissue engineering. J Am Chem Soc 2012;134(2):882–5.
[196] Beckett D, Kovaleva E, Schatz PJ. A minimal peptide substrate in biotin holoenzyme synthetase-catalyzed biotinylation. Protein Sci 1999;8(4):921–9.
[197] Chen I, Howarth M, Lin W, Ting AY. Site-specific labeling of cell surface proteins with biophysical probes using biotin ligase. Nat Methods 2005;2(2):99–104.
[198] Howarth M, Takao K, Hayashi Y, Ting AY. Targeting quantum dots to surface proteins in living cells with biotin ligase. Proc Natl Acad Sci USA 2005;102(21):7583–8.
[199] Wylie RG, Ahsan S, Aizawa Y, Maxwell KL, Morshead CM, Shoichet MS. Spatially controlled simultaneous patterning of multiple growth factors in three-dimensional hydrogels. Nat Mater 2011;10(10):799–806.
[200] Leipzig ND, Wylie RG, Kim H, Shoichet MS. Differentiation of neural stem cells in three-dimensional growth factor-immobilized chitosan hydrogel scaffolds. Biomaterials 2011;32(1):57–64.
[201] Tam RY, Cooke MJ, Shoichet MS. A covalently modified hydrogel blend of hyaluronan–methyl cellulose with peptides and growth factors influences neural stem/progenitor cell fate. J Mater Chem 2012;22(37):19402–11.
[202] Hartley RW. Barnase and barstar: two small proteins to fold and fit together. Trends Biochem Sci 1989;14(11):450–4.
[203] Schreiber G, Fersht AR. Interaction of barnase with its polypeptide inhibitor barstar studied by protein engineering. Biochemistry 1993;32(19):5145–50.
[204] Ni X, Castanares M, Mukherjee A, Lupold SE. Nucleic acid aptamers: clinical applications and promising new horizons. Curr Med Chem 2011;18(27):4206–14.
[205] Jayasena SD. Aptamers: an emerging class of molecules that rival antibodies in diagnostics. Clin Chem 1999;45(9):1628–50.
[206] Ogawa A, Tomita N, Kikuchi N, Sando S, Aoyama Y. Aptamer selection for the inhibition of cell adhesion with fibronectin as target. Bioorg Med Chem Lett 2004;14(15):4001–4.
[207] Daniels DA, Chen H, Hicke BJ, Swiderek KM, Gold L. A tenascin-C aptamer identified by tumor cell SELEX: systematic evolution of ligands by exponential enrichment. Proc Natl Acad Sci USA 2003;100(26):15416–21.
[208] Hicke BJ, Marion C, Chang YF, Gould T, Lynott CK, Parma D. Tenascin-C aptamers are generated using tumor cells and purified protein. J Biol Chem 2001;276(52):48644–54.
[209] Mi Z, Guo H, Russell MB, Liu Y, Sullenger BA, Kuo PC. RNA aptamer blockade of osteopontin inhibits growth and metastasis of MDA-MB231 breast cancer cells. Mol Ther 2009;17(1):153–61.
[210] Green LS, Jellinek D, Jenison R, Östman A, Heldin C-H, Janjic N. Inhibitory DNA ligands to platelet-derived growth factor B-chain. Biochemistry 1996;35(45):14413–24.
[211] Potty ASR, Kourentzi K, Fang H, Jackson GW, Zhang X, Legge GB, et al. Biophysical characterization of DNA aptamer interactions with vascular endothelial growth factor. Biopolymers 2009;91(2):145–56.
[212] Jellinek D, Lynott CK, Rifkin DB, Janjić N. High-affinity RNA ligands to basic fibroblast growth factor inhibit receptor binding. Proc Natl Acad Sci USA 1993;90(23):11227–31.
[213] Ng EW, Shima DT, Calias P, Cunningham Jr. ET, Guyer DR, Adamis AP. Pegaptanib, a targeted anti-VEGF aptamer for ocular vascular disease. Nat Rev Drug Discov 2006;5(2):123–32.
[214] Pagratis NC, Bell C, Chang YF, Jennings S, Fitzwater T, Jellinek D, et al. Potent 2′-amino-, and 2′-fluoro-2′-deoxyribonucleotide RNA inhibitors of keratinocyte growth factor. Nat Biotechnol 1997;15(1):68–73.
[215] Jellinek D, Green LS, Bell C, Lynott CK, Gill N, Vargeese C, et al. Potent 2′-amino-2′-deoxypyrimidine RNA inhibitors of basic fibroblast growth factor. Biochemistry 1995;34(36):11363–72.
[216] Green LS, Jellinek D, Bell C, Beebe LA, Feistner BD, Gill SC, et al. Nuclease-resistant nucleic acid ligands to vascular permeability factor/vascular endothelial growth factor. Chem Biol 1995;2(10):683–95.
[217] Xiao J, Carter JA, Frederick KA, McGown LB. A genome-inspired DNA ligand for the affinity capture of insulin and insulin-like growth factor-2. J Sep Sci 2009;32(10):1654–64.
[218] Kang J, Lee MS, Copland 3rd JA, Luxon BA, Gorenstein DG. Combinatorial selection of a single stranded DNA thioaptamer targeting TGF-beta1 protein. Bioorg Med Chem Lett 2008;18(6):1835–9.
[219] McCauley TG, Kurz JC, Merlino PG, Lewis SD, Gilbert M, Epstein DM, et al. Pharmacologic and pharmacokinetic assessment of anti-TGFbeta2 aptamers in rabbit plasma and aqueous humor. Pharm Res 2006;23(2):303–11.
[220] White RR, Roy JA, Viles KD, Sullenger BA, Kontos CD. A nuclease-resistant RNA aptamer specifically

inhibits angiopoietin-1-mediated Tie2 activation and function. Angiogenesis 2008;11(4):395–401.

[221] White RR, Shan S, Rusconi CP, Shetty G, Dewhirst MW, Kontos CD. Inhibition of rat corneal angiogenesis by a nuclease-resistant RNA aptamer specific for angiopoietin-2. Proc Natl Acad Sci USA 2003; 100(9):5028–33.

[222] Zhang Z, Guo L, Tang J, Guo X, Xie J. An aptameric molecular beacon-based "signal-on" approach for rapid determination of rHuEPO-α. Talanta 2009;80 (2):985–90.

[223] Orava EW, Jarvik N, Shek YL, Sidhu SS, Gariepy J. A short DNA aptamer that recognizes TNFalpha and blocks its activity *in vitro*. ACS Chem Biol 2013; 8(1):170–8.

[224] Kubik MF, Bell C, Fitzwater T, Watson SR, Tasset DM. Isolation and characterization of 2′-fluoro-, 2′-amino-, and 2′-fluoro-/amino-modified RNA ligands to human IFN-gamma that inhibit receptor binding. J Immunol 1997;159(1):259–67.

[225] Ishiguro A, Akiyama T, Adachi H, Inoue J, Nakamura Y. Therapeutic potential of anti-interleukin-17A aptamer: suppression of interleukin-17A signaling and attenuation of autoimmunity in two mouse models. Arthritis Rheum 2011;63(2):455–66.

[226] Soontornworajit B, Zhou J, Shaw MT, Fan T-H, Wang Y. Hydrogel functionalization with DNA aptamers for sustained PDGF-BB release. Chem Commun 2010;46(11):1857–9.

[227] Soontornworajit B, Zhou J, Zhang Z, Wang Y. Aptamer-functionalized *in situ* injectable hydrogel for controlled protein release. Biomacromolecules 2010;11(10):2724–30.

[228] Battig MR, Soontornworajit B, Wang Y. Programmable release of multiple protein drugs from aptamer-functionalized hydrogels via nucleic acid hybridization. J Am Chem Soc 2012;134(30):12410–3.

[229] Shangguan D, Li Y, Tang Z, Cao ZC, Chen HW, Mallikaratchy P, et al. Aptamers evolved from live cells as effective molecular probes for cancer study. Proc Natl Acad Sci USA 2006;103(32):11838–43.

[230] Chen N, Zhang Z, Soontornworajit B, Zhou J, Wang Y. Cell adhesion on an artificial extracellular matrix using aptamer-functionalized PEG hydrogels. Biomaterials 2012;33(5):1353–62.

[231] Li S, Chen N, Zhang Z, Wang Y. Endonuclease-responsive aptamer-functionalized hydrogel coating for sequential catch and release of cancer cells. Biomaterials 2013;34(2):460–9.

[232] Zhang Z, Chen N, Li S, Battig MR, Wang Y. Programmable hydrogels for controlled cell catch and release using hybridized aptamers and complementary sequences. J Am Chem Soc 2012;134(38):15716–9.

[233] Burdick JA, Murphy WL. Moving from static to dynamic complexity in hydrogel design. Nat Commun 2012;3:1269.

[234] Wilkinson AE, McCormick AM, Leipzig ND. Central nervous system tissue engineering: current considerations and strategies. Synthesis Lect Tissue Eng 2011;3(2):1–120.

[235] Katti DS, Lakshmi S, Langer R, Laurencin CT. Toxicity, biodegradation and elimination of polyanhydrides. Adv Drug Deliv Rev 2002;54(7):933–61.

[236] Tayebjee MH, Lip GY, Blann AD, Macfadyen RJ. Effects of age, gender, ethnicity, diurnal variation and exercise on circulating levels of matrix metalloproteinases (MMP)-2 and -9, and their inhibitors, tissue inhibitors of matrix metalloproteinases (TIMP)-1 and -2. Thromb Res 2005;115(3):205–10.

[237] Sahoo S, Chung C, Khetan S, Burdick JA. Hydrolytically degradable hyaluronic acid hydrogels with controlled temporal structures. Biomacromolecules 2008;9(4):1088–92.

[238] Patterson J, Siew R, Herring SW, Lin AS, Guldberg R, Stayton PS. Hyaluronic acid hydrogels with controlled degradation properties for oriented bone regeneration. Biomaterials 2010;31(26):6772–81.

[239] West JL, Hubbell JA. Photopolymerized hydrogel materials for drug delivery applications. React Polym 1995;25(2–3):139–47.

[240] Zhao X, Harris JMilton. Novel degradable poly (ethylene glycol) hydrogels for controlled release of protein. J Pharm Sci 1998;87(11):1450–8.

[241] Elbert DL, Pratt AB, Lutolf MP, Halstenberg S, Hubbell JA. Protein delivery from materials formed by self-selective conjugate addition reactions. J Control Release 2001;76(1–2):11–25.

[242] Zustiak SP, Leach JB. Hydrolytically degradable poly (ethylene glycol) hydrogel scaffolds with tunable degradation and mechanical properties. Biomacromolecules 2010;11(5):1348–57.

[243] Cho E, Kutty JK, Datar K, Soo Lee J, Vyavahare NR, Webb K. A novel synthetic route for the preparation of hydrolytically degradable synthetic hydrogels. J Biomed Mater Res A 2009;90A(4):1073–82.

[244] Jo YS, Gantz J, Hubbell JA, Lutolf MP. Tailoring hydrogel degradation and drug release via neighboring amino acid controlled ester hydrolysis. Soft Matter 2009;5(2):440–6.

[245] Drury JL, Mooney DJ. Hydrogels for tissue engineering: scaffold design variables and applications. Biomaterials 2003;24(24):4337–51.

[246] Sargeant TD, Desai AP, Banerjee S, Agawu A, Stopek JB. An *in situ* forming collagen–PEG hydrogel for tissue regeneration. Acta Biomater 2012;8(1): 124–32.

[247] Ahmed TA, Griffith M, Hincke M. Characterization and inhibition of fibrin hydrogel-degrading enzymes during development of tissue engineering scaffolds. Tissue Eng 2007;13(7):1469–77.
[248] Afify AM, Stern M, Guntenhoner M, Stern R. Purification and characterization of human serum hyaluronidase. Arch Biochem Biophys 1993;305(2): 434–41.
[249] Oh EJ, Kang SW, Kim BS, Jiang G, Cho IH, Hahn SK. Control of the molecular degradation of hyaluronic acid hydrogels for tissue augmentation. J Biomed Mater Res A 2008;86(3):685–93.
[250] Nilasaroya A, Martens PJ, Whitelock JM. Enzymatic degradation of heparin-modified hydrogels and its effect on bioactivity. Biomaterials 2012;33(22):5534–40.
[251] West JL, Hubbell JA. Polymeric biomaterials with degradation sites for proteases involved in cell migration. Macromolecules 1998;32(1):241–4.
[252] Mann BK, Gobin AS, Tsai AT, Schmedlen RH, West JL. Smooth muscle cell growth in photopolymerized hydrogels with cell adhesive and proteolytically degradable domains: synthetic ECM analogs for tissue engineering. Biomaterials 2001;22(22):3045–51.
[253] Brandl FP, Seitz AK, Tessmar JK, Blunk T, Gopferich AM. Enzymatically degradable poly(ethylene glycol) based hydrogels for adipose tissue engineering. Biomaterials 2010;31(14):3957–66.
[254] Kraehenbuehl TP, Zammaretti P, Van der Vlies AJ, Schoenmakers RG, Lutolf MP, Jaconi ME, et al. Three-dimensional extracellular matrix-directed cardioprogenitor differentiation: systematic modulation of a synthetic cell-responsive PEG-hydrogel. Biomaterials 2008;29(18):2757–66.
[255] Ehrbar M, Sala A, Lienemann P, Ranga A, Mosiewicz K, Bittermann A, et al. Elucidating the role of matrix stiffness in 3D cell migration and remodeling. Biophys J 2011;100(2):284–93.
[256] Khetan S, Guvendiren M, Legant WR, Cohen DM, Chen CS, Burdick JA. Degradation-mediated cellular traction directs stem cell fate in covalently cross-linked three-dimensional hydrogels. Nat Mater 2013;12(5):458–65.
[257] Lévesque SG, Shoichet MS. Synthesis of enzyme-degradable, peptide-cross-linked dextran hydrogels. Bioconjug Chem 2007;18(3):874–85.
[258] Stephan MT, Moon JJ, Um SH, Bershteyn A, Irvine DJ. Therapeutic cell engineering with surface-conjugated synthetic nanoparticles. Nat Med 2010; 16(9):1035–41.
[259] Stephan MT, Stephan SB, Bak P, Chen J, Irvine DJ. Synapse-directed delivery of immunomodulators using T-cell-conjugated nanoparticles. Biomaterials 2012;33(23):5776–87.
[260] Holden CA, Yuan Q, Yeudall WA, Lebman DA, Yang H. Surface engineering of macrophages with nanoparticles to generate a cell-nanoparticle hybrid vehicle for hypoxia-targeted drug delivery. Int J Nanomedicine 2010;5:25–36.
[261] Christian DA, Hunter CA. Particle-mediated delivery of cytokines for immunotherapy. Immunotherapy 2012;4(4):425–41.
[262] Morachis JM, Mahmoud EA, Almutairi A. Physical and chemical strategies for therapeutic delivery by using polymeric nanoparticles. Pharmacol Rev 2012;64(3):505–19.
[263] Serda RE. Particle platforms for cancer immunotherapy. Int J Nanomed 2013;8:1683–96.
[264] Leong YS, Candau F. Inverse microemulsion polymerization. J Phys Chem 1982;86(13):2269–71.
[265] Mitra S, Gaur U, Ghosh PC, Maitra AN. Tumour targeted delivery of encapsulated dextran-doxorubicin conjugate using chitosan nanoparticles as carrier. J Control Release 2001;74(1–3):317–23.
[266] Lee H, Mok H, Lee S, Oh YK, Park TG. Target-specific intracellular delivery of siRNA using degradable hyaluronic acid nanogels. J Control Release 2007;119(2):245–52.
[267] Braun O, Selb J, Candau F. Synthesis in microemulsion and characterization of stimuli-responsive polyelectrolytes and polyampholytes based on *N*-isopropylacrylamide. Polymer (Guildf) 2001;42(21):8499–510.
[268] Fernandez VVA, Tepale N, Sánchez-Díaz JC, Mendizábal E, Puig JE, Soltero JFA. Thermoresponsive nanostructured poly(*N*-isopropylacrylamide) hydrogels made via inverse microemulsion polymerization. Colloid Polym Sci 2006;284(4): 387–95.
[269] Juraničová V, Kawamoto S, Fujimoto K, Kawaguchi H, Bartoň J. Inverse microemulsion polymerization of acrylamide in the presence of *N,N*-dimethylacrylamide. Die Angew Makromol Chem 1998;258(1):27–31.
[270] Bartoň J. Inverse microemulsion polymerization of oil-soluble monomers in the presence of hydrophilic polyacrylamide nanoparticles. Macromol Symp 2002;179(1):189–208.
[271] Kaneda I, Sogabe A, Nakajima H. Water-swellable polyelectrolyte microgels polymerized in an inverse microemulsion using a nonionic surfactant. J Colloid Interface Sci 2004;275(2):450–7.
[272] Gaur U, Sahoo SK, De TK, Ghosh PC, Maitra A, Ghosh PK. Biodistribution of fluoresceinated dextran using novel nanoparticles evading reticuloendothelial system. Int J Pharm 2000;202(1–2):1–10.
[273] Bharali DJ, Sahoo SK, Mozumdar S, Maitra A. Cross-linked polyvinylpyrrolidone nanoparticles: a potential

carrier for hydrophilic drugs. J Colloid Interface Sci 2003;258(2):415–23.
[274] Missirlis D, Tirelli N, Hubbell JA. Amphiphilic hydrogel nanoparticles. preparation, characterization, and preliminary assessment as new colloidal drug carriers. Langmuir 2005;21(6):2605–13.
[275] Oh JK, Perineau F, Matyjaszewski K. Preparation of nanoparticles of well-controlled water-soluble homopolymers and block copolymers using an inverse miniemulsion ATRP. Macromolecules 2006;39(23):8003–10.
[276] Oh JK, Tang C, Gao H, Tsarevsky NV, Matyjaszewski K. Inverse miniemulsion ATRP: a new method for synthesis and functionalization of well-defined water-soluble/cross-linked polymeric particles. J Am Chem Soc 2006;128(16):5578–84.
[277] Genggeng Q, Christopher WJ, Schork FJ. RAFT inverse miniemulsion polymerization of acrylamide. Macromol Rapid Commun 2007;28(9):1010–6.
[278] Oh JK, Dong H, Zhang R, Matyjaszewski K, Schlaad H. Preparation of nanoparticles of double-hydrophilic PEO-PHEMA block copolymers by AGET ATRP in inverse miniemulsion. J Polym Sci A Polym Chem 2007;45(21):4764–72.
[279] An Z, Shi Q, Tang W, Tsung CK, Hawker CJ, Stucky GD. Facile RAFT precipitation polymerization for the microwave-assisted synthesis of well-defined, double hydrophilic block copolymers and nanostructured hydrogels. J Am Chem Soc 2007;129(46):14493–9.
[280] An Z, Tang W, Wu M, Jiao Z, Stucky GD. Heterofunctional polymers and core-shell nanoparticles via cascade aminolysis/Michael addition and alkyne-azide click reaction of RAFT polymers. Chem Commun (Camb) 2008;(48):6501–3.
[281] Shen W, Chang Y, Liu G, Wang H, Cao A, An Z. Biocompatible, antifouling, and thermosensitive core–shell nanogels synthesized by RAFT aqueous dispersion polymerization. Macromolecules 2011;44(8): 2524–30.
[282] Rieger J, Grazon C, Charleux B, Alaimo D, Jérôme C. Pegylated thermally responsive block copolymer micelles and nanogels via *in situ* RAFT aqueous dispersion polymerization. J Polym Sci A Polym Chem 2009;47(9):2373–90.
[283] Rider CC. Heparin/heparan sulphate binding in the TGF-beta cytokine superfamily. Biochem Soc Trans 2006;34(Pt 3):458–60.
[284] Park JE, Keller GA, Ferrara N. The vascular endothelial growth factor (VEGF) isoforms: differential deposition into the subepithelial extracellular matrix and bioactivity of extracellular matrix-bound VEGF. Mol Biol Cell 1993;4(12):1317–26.
[285] Ferrara N. Binding to the extracellular matrix and proteolytic processing: two key mechanisms regulating vascular endothelial growth factor action. Mol Biol Cell 2010;21(5):687–90.
[286] Raines EW, Ross R. Compartmentalization of PDGF on extracellular binding sites dependent on exon-6-encoded sequences. J Cell Biol 1992; 116(2):533–43.
[287] Rolny C, Spillmann D, Lindahl U, Claesson-Welsh L. Heparin amplifies platelet-derived growth factor (PDGF)- BB-induced PDGF alpha -receptor but not PDGF beta-receptor tyrosine phosphorylation in heparan sulfate-deficient cells. Effects on signal transduction and biological responses. J Biol Chem 2002;277(22):19315–21.
[288] Lyon M, Deakin JA, Mizuno K, Nakamura T, Gallagher JT. Interaction of hepatocyte growth factor with heparan sulfate. Elucidation of the major heparan sulfate structural determinants. J Biol Chem 1994;269(15):11216–23.
[289] Ohkawara B, Iemura S, ten Dijke P, Ueno N. Action range of BMP is defined by its N-terminal basic amino acid core. Curr Biol 2002;12(3):205–9.
[290] Ruppert R, Hoffmann E, Sebald W. Human bone morphogenetic protein 2 contains a heparin-binding site which modifies its biological activity. Eur J Biochem 1996;237(1):295–302.
[291] Irie A, Habuchi H, Kimata K, Sanai Y. Heparan sulfate is required for bone morphogenetic protein-7 signaling. Biochem Biophys Res Commun 2003; 308(4):858–65.

CHAPTER

15

Probe and Control of Cell–Cell Interactions Using Bioengineered Tools

Feng Li[1,2], *Rangoli Aeran*[1], *Egest J. Pone*[1], *Dong-Ku Kang*[1], *Linan Liu*[1], *X. Chris Le*[2] *and Weian Zhao*[1]

[1]Department of Pharmaceutical Sciences, Chao and Family Comprehensive Cancer Center, Sue and Bill Gross Stem Cell Research Center, Edwards Lifesciences Center for Advanced Cardiovascular Technology, Department of Biomedical Engineering, University of California, Irvine, CA, USA, [2]Department of Laboratory Medicine and Pathology, University of Alberta, Edmonton, AB, Canada

OUTLINE

J.M. Karp & W. Zhao (Eds): Micro- and Nanoengineering of the Cell Surface.
DOI: http://dx.doi.org/10.1016/B978-1-45-573146-6.00015-5

15.1 INTRODUCTION

Cell–cell interactions broadly comprise the exchange of signals between various cells of multicellular organisms that underpin the diverse biological processes and responses [1,2]. The benefits of having groups of specialized cells working together for a supreme purpose—the survival, growth and reproduction of the organism—required continuous evolution of, on the one hand ever more specialized cells, and on the other, a more complex and robust means [3] of ensuring their cooperation with each other by means of communication through cellular and molecular signals [4]. The fundamental mechanism that ensures appropriate cooperation of individual cells and cell types is cellular communication: just like individuals within various groups each perform their specialized functions for common, overarching goals, so do cells of metazoa. Appropriate real-time communication ensures that each cell performs the function that most benefits the survival of the entire organism. This usually includes homeostasis and reproduction of individual cells (to replace aging and dying ones). Conversely, aberrant cell–cell communication is thought to give rise to devastating diseases such as cancer [5] (runaway cell proliferation) or autoimmunity [6] (immune attack on own healthy cells). Thus, understanding cell–cell communication is essential for understanding life and developing therapeutics.

Cellular interactions can be guided by physical contact between cells [7], by various secreted extracellular vesicles known as microvesicles and exosomes carrying protein or nucleic acid messengers [8,9], by soluble factors secreted by cells to act locally or distantly [1,10], electrical charges and pulses [11], etc. Physical cell–cell interactions comprise receptor–ligand pairs that tether cells together and to their underlying matrix. Besides giving rise to characteristic tissue tensile strength [12], such close cell–cell interactions ensure that some cell types can interact with each other by exchanging molecular messages through gaps in their membranes, referred to as gap junctions. These are permanent physical connections that provide a free passageway for ions and small molecules between adjacent cells but prevent the flow of larger proteins and nucleic acids [5,13]. As a result, they couple cellular metabolic status, signal transduction state, or electrical responses. Gap junctions are thought to be numerous and important particularly to cells of the same type within a tissue. In effect, a response initiated in one cell can spread to neighboring cells to synchronize their responses, as happens with calcium waves in cardiomyocytes during heart beating. Moreover, unlike most cells of a tissue that never migrate for their entire lifetime, cells of the immune system are motile, and therefore communicate primarily not through gap junctions but through receptor–ligand pairs [7]. The receptor and ligand could be expressed on the membranes (particularly on the plasma membrane), could be soluble intracellular, or could be soluble in the extracellular environment (interstitial fluids or blood) [14]. A case in point is the characteristic migration of leukocytes from bone marrow, lymph nodes, and spleen as they undergo various phases of maturation, and their subsequent recirculation and patrolling between lymphoid organs and periphery. Upon encountering signals for the presence of pathogen or cancer (released by the pathogen or cancer itself), various immunocytes of the innate and adaptive arms of the immune system migrate to the site of infection or cancer to eradicate the threat [15,16]. Clearly, these are complex scenarios that involve the action of a multitude of conserved microbial molecules, cytokines, chemokines, lipid mediators, other signaling molecules, and immune cells must constantly communicate with each other as well as with the surrounding tissue [17]. A decrease in cell–cell communication, or worse, its breakdown as in the case of ambiguous or conflicting signals, would have catastrophic consequences for the organism, as in the case of cancer, autoimmunity, or immunodeficiency. Likewise, integration and coordination of various responses through the endocrine system is essential for proper homeostasis and adaptive responses, whereas its breakdown results in various diseases, most notably diabetes. Finally, cross-system cellular communication is also critical for homeostasis and adaptation, and its breakdown has severe consequences: a misdirected immune attack on pancreatic insulin-producing β cells results in type 1 diabetes mellitus (T1DM) whereas a misdirected immune attack on the nervous system results in diseases such as multiple sclerosis (MS) [6].

Just as erroneous cellular interaction and communication leads to disease, many treatments rely on restoring cell–cell communication. Thus, in order to understand at a fundamental level the various types of cell–cell interaction and communication, we utilize appropriate molecular and cellular tools at our disposal. Traditionally, our need to investigate cell–cell interactions and communication has been limited by approaches of low resolution, the use of bulk cell population methods that obscures individual cellular state heterogeneity, and lack of ability to manipulate individual cells and their signaling molecules [18]. However, with the advent of molecular biology techniques on the one hand and the emergence of micro- and nano-scale engineering on the other, we now possess ever finer tools to probe cell–cell interaction and communication [19,20]. In particular, these new tools allow us to elucidate and manipulate cell–cell interactions at a single cell level at an unprecedented temporal and spatial resolution both *in vitro* and *in vivo* [21,22]. In this chapter, we summarize recent and emerging micro-, nano-, and molecular tools for the investigation of different types of cell–cell interactions. We will limit our discussion to selected examples that best illustrate recent advances in the cell–cell interaction methodology, and which have applications to diseases such as cancer, those of immune dysfunction, and endocrine disorders. We will first describe the engineered micro-scale tools to pattern cells in well-defined fashion and study cell–cell interactions at cellular and subcellular levels, particularly in a 3D context. We will then describe recent advances in engineered tools at the "molecular" level, including genetic and chemical tools, used to probe and control cellular interactions with other cells as well as their surrounding microenvironments. Finally, we will conclude with future directions in which our ever-growing list of high-resolution tools both informs, and is inspired, by cellular and subcellular behavior.

15.2 *IN VITRO* STUDY OF CELL–CELL INTERACTIONS USING ENGINEERED MICRODEVICES

Although *in vivo* studies of cell–cell interactions are ideal in maintaining the native microenvironment during the targeted biological process, the complexity of biological environment and limitations to the current probing techniques usually hinder the accurate and quantitative understanding of cell–cell interactions. As important complementary tools for the investigation of cell–cell interactions, *in vitro* techniques simplify the complicated *in vivo* microenvironment into well-controlled experimental conditions, leading to better understanding of many cell–cell interactions. Here, we emphasize the applications of microfabricated devices, due to their advantages in terms of control over the cell spatial and temporal arrangement, as well as their relatively high sensitivity and resolution in analyzing cell–cell interactions at single cell levels, while mimicking to a great extent the *in vivo* micro-environments where cell interactions occur.

15.2.1 Control Over Cell Spatial and Temporal Arrangements Using Engineered Microdevices

Cell–cell interactions depend on both physical contacts and soluble and particulate factors. Therefore, it is imperative to utilize devices that provide both defined spatial arrangements of each cell type and the spatial/temporal control of cellular secretions. Engineered microdevices are particularly advantageous due to their customizable size, shape, material composition, and design to introduce stimuli as well as monitor cellular responses [23,24]. Numerous microdevices have been engineered to study the roles of gap junctions in permanent cell–cell interactions [25–28]. Gap junctions are nano-scale protein channels that connect the cytosolic compartments of adjoining cells, allowing for the transfer of ions, metabolites, and other small molecules between cells below typical cutoff sizes [5,13,29]. Understanding the physiological roles of gap junctions is essential to uncover how cells synchronize and communicate signals among large cell populations; however, it remains challenging to achieve quantitative characterization of gap junctions in conventional macro-scale assays due to the lack over spatial control. To address this, Lee and coworkers [25] have developed a quantitative microfluidic assay to measure dye diffusion through gap junctions in connected cells (Figure 15.1). In this study, the microfluidic device was designed to allow cells to adhere to the surface as well as express functional gap junctions. Calcein dye was used as the diffusion tracer because its acetomethoxy (AM) form is nonfluorescent and membrane permeable, but when cleaved by intracellular esterases it becomes fluorescent and can freely move through cellular gap junctions (Figure 15.1D). To achieve spatially targeted delivery of calcein dye in the device, a noninvasive hydrodynamic focusing approach was developed to focus fluids at low Reynolds number into a column of C6 cells within a confluent cell culture chamber while leaving their neighboring cells unperturbed (Figure 15.1C). Subsequent time-lapse fluorescent microscopy

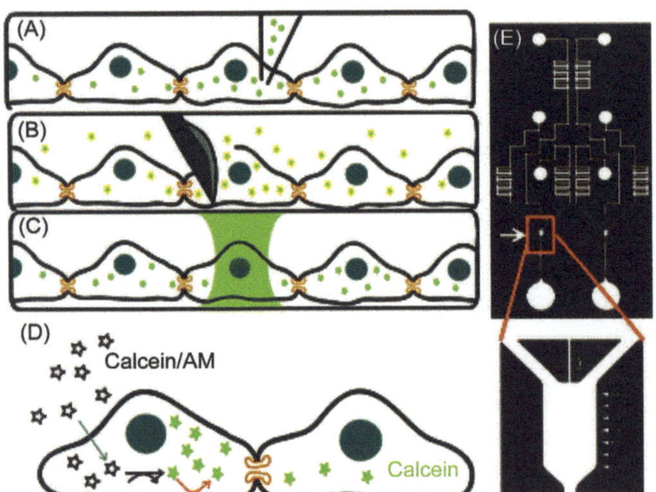

FIGURE 15.1 Microfluidic assay to visualize the extent of dye migration across connected cells and measurement of the effective diffusivities through gap junctions. (A–C) Methods for loading dye into a subset of contacting cells include microinjection (A), scrape leading (B), and microfluidic dye focusing (C). (D) Schematic showing dye transfer through gap junctions and conversion to fluorescent form inside cells. (E) Microfluidic chip design: cells were first seeded in the culture chamber (enlarged figure). Upon delivering dyes to target cells, the dye solution was loaded through the middle channel into the culture chamber and was then focused by applying high pressure (3.1 psi) into side channels and low pressure (0.5 psi) into the middle channel [25].

was used to quantitatively monitor the dynamics of dye transfer via gap junction to neighboring cells, yielding a calculated effective diffusivity of $3.4 \times 10^{-13}\,m^2/s$. In a similar study, Hua et al. [27] developed a tri-stream microfluidic system for multiplexed measurements of connexin-43 (Cx43) gap junctions with several dyes, including calcein AM, 5-(6)-carboxyfluorescein diacetate (CFDA), Oregon green 488, in the absence or presence of Cx43 inhibitors 1-heptanol, 2-APB, and MFA. The use of multistream laminar flow to selectively expose cultured cells with these dyes and inhibitors enabled measurement of the diffusion coefficient of the four dyes. This device could also be used to screen multiple inhibitors in parallel, demonstrating the potential of micro-engineered tools for high-throughput cell analysis.

Besides studying interactions between homogeneous cell types, engineered microdevices are also suitable for studying heterotypic cell interactions by arranging different cell types in spatially defined cocultures. The precise arrangements are achieved typically by controlling the micropatterned surfaces on engineered devices [30–33]. A number of fabrication strategies, including photolithography, microcontact printing, micromodeling, inkjet printing, and dip-pen spotting, have been used to create micropatterned surfaces to localize cells to defined regions. To construct patterned cocultures, selective adhesion of one cell type is commonly utilized. For example, hepatocyte–fibroblast cocultures can be fabricated on collagen-patterned substrates by allowing hepatocytes to adhere to the patterned collagen-coated regions with fibroblasts only adhering to the non-collagen-coated regions [34]. Layer-by-layer deposition of biopolymers, such as polysaccharides and proteins, is another commonly used strategy to create such cell patterning. For example, Langer et al. [35] developed a cell coculture system by using layer-by-layer deposition of three major extracellular matrix (ECM) components: hyaluronic acid (HA), fibronectin (FN), and collagen (Figure 15.2). HA was micropatterned on a glass substrate by capillary force lithography, and the regions of exposed glass were coated with FN to generate cell adhesive islands. To achieve the coculture of multiple cells on this micropatterned substrate, murine embryonic stem (ES) cells were first cultured onto FN-coated islands, forming cell islets with 100 μm diameters. Once ES cells were immobilized on FN islands, the HA-patterned surface and the seeded ES cells were treated with collagen. This treatment switched the non-adherent HA surface to adherent, thus allowing the seeding of the second NIH-3T3 fibroblasts. Such coculture systems of ES cells with other differentiated cell types could potentially be used to study the effect of heterotypic cell–cell interactions on ES cell fate decisions. In a separate study, Jiang et al. [36] developed a method for patterning multiple types of cells on the same substrate using electrochemical desorption of self-assembled monolayers (SAMs) in a microfluidic system. Importantly, this technique allows the patterning of different types of cells without the presence of physical constraints, therefore enabling control over cell motility. This method first employed alkanethiol (EG_6) to form an inert SAM on gold surface to resist protein adsorption and cell adhesion. By using a PDMS stamp with embedded microfluidic channels to selectively electrochemically desorb EG_6 from the gold substrate, each cell type could be patterned on the same substrate and confined to the activated regions. Because there is no physical barrier between these cells, the free exchange of substances between these different types of cells facilitates studies of cell–cell interactions. Subsequently, an electrochemical desorption step can be applied to convert the previously "inert" surface to an "active" that allows cells to move under the influence of each other. Having established this method, the same group patterned mammalian cells on a fabricated microfluidic device

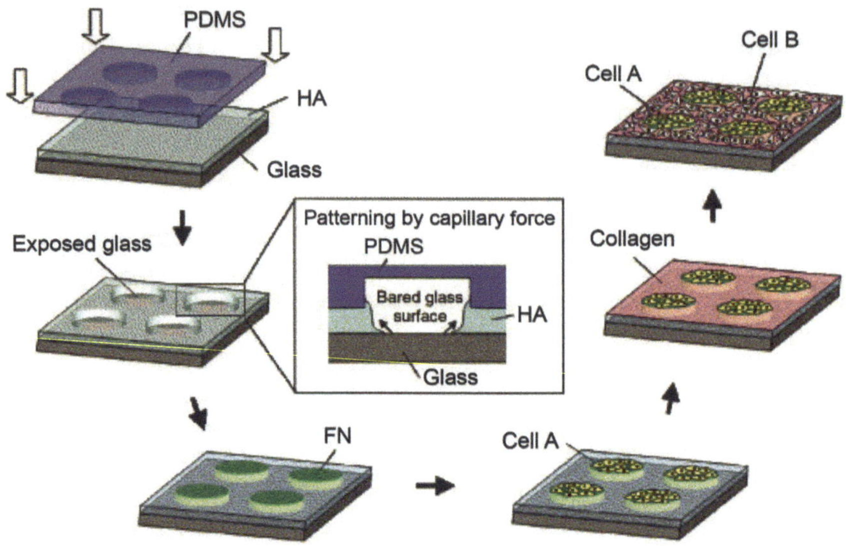

FIGURE 15.2 Schematic illustrating the fabrication of the coculture system using capillary lithography and layer-by-layer deposition. HA solution was spun-coated onto glass slides, and a PDMS (polydimethyl siloxane) mold was immediately placed on the thin layer of HA. HA under the void space of the PDMS mold receded until the glass surface became exposed. The exposed region of a glass substrate was coated with FN, where primary cells could be selectively adhered. Subsequently, the HA surface was complexed with collagen, allowing for adhesion of secondary cells [35].

for modeling three types of naturally occurring cell–cell interactions: (i) those between two types of immobilized cells, (ii) those between immobilized and freely moving cells, and (iii) those between two types of freely moving cells [37].

In addition to using chemical deposition to control cell patterning, microenvironments consisting of defined physical substrates have also been used to control cell–cell interactions. Typically, different cells are grown on different substrates, which can be rearranged to change the spatial organization of the culture. Therefore, such mechanically controlled microdevices allow the decoupling of cell–cell interaction processes into manipulations mimicking biological events such as cell differentiation based on physical topology and other cues. Hui and Bhatia [38] have developed such a device to study the intracellular communication between hepatocytes and supportive stromal cells in coculture. In their approach, cells were grown on an array of microfabricated plates that were physically separated or locked to change the spatial organization of the culture (Figure 15.3). By imposing a small micrometer-scale separation between the plates, cell–cell contact could be abrogated while soluble signaling is unperturbed. By increasing separation distances, the extent of soluble signaling could be further modulated. In addition, cells could be exchanged by removing one set of fingers and inserting in a replacement seeded with another cell type. Using this device, the maintenance of the hepatocellular phenotype by stroma required direct contact for a limited time followed by a sustained soluble signal that had an effective range of $<400\ \mu m$. This work demonstrated the power of using a mechanical devices to investigate dynamic cell–cell interactions in a multitude of applications, spanning embryogenesis, homeostasis, and when dysregulated, pathogenic processes. Likewise, Nishizawa et al. [39] developed a

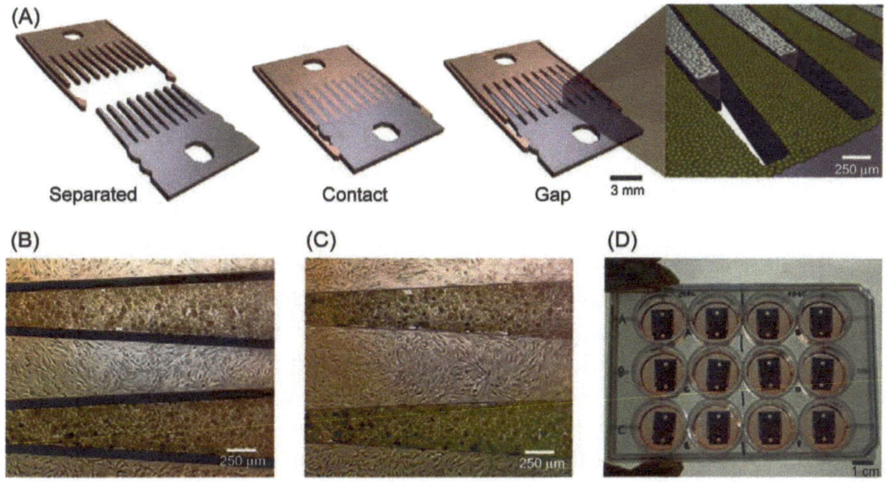

FIGURE 15.3 Schematic illustrating the fabricated mechanically detachable cell coculture device. (A) Microfabricated cell substrates can be fully separated (left), locked together with comb fingers in contact (center), or slightly separated (right). Cells were cultured on the top surfaces; manual scraping can be used to restrict cells to the comb fingers only (inset). (B and C) Bright-field images of hepatocytes (darker cells) and 3T3 fibroblasts cultured on the comb fingers. (D) Devices in a standard 12-well plate [38].

mechanically controlled device to study interactions between HeLa cells and human umbilical vein endothelial cells (HUVECs). HeLa cells and HUVECs were cultured separately on two detachable substrates, and coculturing was initiated by mechanically assembling the two complementary substrates. Using this coculture system as a tumor/endothelium model, the authors found that HeLa cells migrated toward HUVECs, whereas HUVECs retreated. Additionally, when direct contact between the two cell types was prevented, HUVECs initially migrated toward the HeLa cells and then retreated. In a separate study, the same group further integrated this mechanically controlled cell culture device with a microfluidic component to control the direction of medium flows [40]. By using this integrated device, they further observed that the retreat of HUVECs only happened when medium flow was directed from the HeLa side toward the HUVEC side, suggesting that the migration of HeLa cells and HUVECs was likely mediated by soluble factors produced by HeLa cells. Indeed, increased levels of VEGF produced by HeLa cells were observed in the presence of HUVECs. The increasing levels of VEGF could attract the HUVECs, leading to the initial HUVEC migration toward HeLa cells. However, since the tumor cells also produced repulsive signals, such as reactive oxygen species (ROS), the initial HUVECs approach halted and actually turned into a retreat. The characteristics of the cell movements including the direction and speed observed from the mechanically controlled cell culture devices revealed the significant role of cell–cell communication through secreted soluble factors, frequently with contrasting activities, in the processes of tumor cell intra- and extravasations.

15.2.2 Cell Interactions Studies at Single Cell Levels Using Engineered Microdevices

An additional advantage of using engineered microdevices to study cell–cell interactions is their adaptation for use in a high-throughput manner, i.e., to interrogate many cell interactions, one at a time, and then draw conclusions based on individual and rare events, as well as use statistical analysis to understand cellular responses at a population level. High-throughput single cell studies enabled by engineered microdevices give insights on the heterogeneity of intercellular interactions. One simple but effective approach is to segregate target cells into arrays of micro/nanoliter scaled wells [41]. One immune cell type that is amenable to high-throughput microfluidic analysis is the natural killer (NK) cell. The latter are granular lymphocytes lacking antigen receptors yet having the ability to recognize and kill tumor or virally infected cells and other non-self cells (e.g., transplanted or donor grafted cells) [42]. The efficacy of the NK cell-mediated immune response is heterogeneous with cell-to-cell variability in their cytolytic and secretory activities. To study NK cell responses at single and population levels, Guldevall et al. [43] measured immune surveillance by individual NK confined in microwell arrays. The authors designed dense arrays of microwells (100 × 100 square wells with 80 μm sides and a depth of 45 μm) to spatially confine NK cells, in effect allowing the analysis of many single NK cell events. Time-lapse imaging of NK cell cytotoxicity in these microwells revealed heterogeneity in the rate of 293T cell killing upon NK cell encounter, indicating heterogeneity in individual NK killing efficiency. In a separate study using a similar array of subnanoliter microwells, Love et al. [44] investigated the individual NK cell–target cell interactions and quantified the resulting cytolytic and secretory responses (Figure 15.4). As shown in Figure 15.4, NK cells and K562 target cells were co-deposited onto an array of 30 μm cubic nanowells to obtain small, isolated groups of cells (0–3 cells of each type). In these single cell cytolysis assays, measurements were performed in 1000–5000 wells per

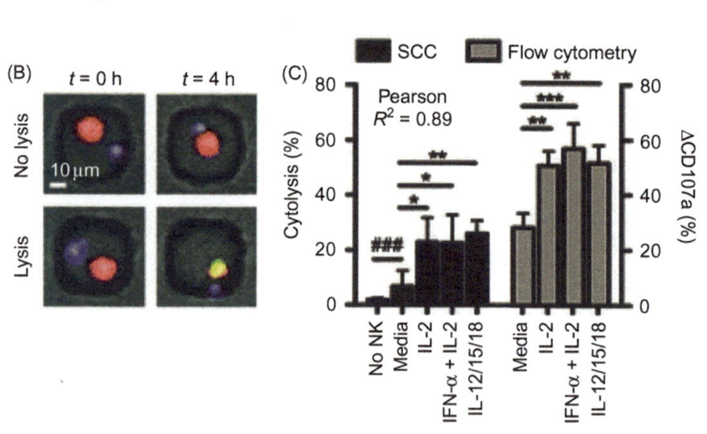

FIGURE 15.4 (A) Schematic illustrating the interaction between NK cells and K562 target cells in nanowells. (B) NK cells (blue) were co-incubated with target cells (red) in the nanowells, which when lysed become positive for SYTOX (green). (C) Frequencies of cytolytic events measured in nanowells, and degranulated NK cells measured by flow cytometry after 4 h of NK cell–target cell interactions [44].

array containing exactly one NK cell and one target cell. NK cells were shown to behave with individual variations when lysing single target cells and lysis was most probable during an NK cell's first encounter with target. Furthermore, the secretion of interferon-γ (IFN-γ), a key inflammatory immunomodulator that coordinates immune responses particularly during viral infection or cancer, occurred most often among NK cells that became the least motile upon contacting a target cell. However, the secretion of IFN-γ was unexpectedly largely independent of cytolysis. Furthermore, the frequencies of cytolytic events measured from thousands of individual NK cell–target cell pairs in the nanowells were strongly correlated with the expression levels of target cell-induced degranulation marker CD107a, as measured by flow cytometry analyses of bulk cultures (Figure 15.4C). Both cytolytic and degranulation frequencies are increased upon stimulation by exogenously added cytokines including IL-2, IFN-α plus IL-2, or a combination of IL-12, IL-15, and IL-18. This study represents a good example of how integrated analysis of cell–cell interaction parameters, cytolytic activity, and secretory activity enabled by micro-engineered tools can reveal new insights into the correlations of these complex functions in individual cells.

Microfluidic devices can also be engineered to trap single cells and study single cell interactions using hydrodynamic trappings. For example, Lee et al. developed a microfluidic device for selective trapping of cell pairs. Specifically, their microfluidic channel contains a single pair of trapping sites that are separated by 20 μm on opposite sides. Since the mouse fibroblast cell line used in their study averaged 12 μm in diameter and were rather homogenous in size distribution, the channel was wide enough to allow cells to flow through but narrow enough to ensure cell membrane contact when the cell pairs were trapped across from each other. This device enabled monitoring of cell–cell communications through gap junctions using the calcein transfer system as described previously [45]. Moreover, the same group developed a coculture device that can maintain and track single cell pairwise interactions [28]. As shown in Figure 15.5(A), by combining hydrodynamic trapping with a semi-isolated chamber that is accessible to continuous medium supplementation, this device can efficiently isolate cell pairs using minimal physical restraints to study cell migration, proliferation, and cell–cell interactions through multiple generations. Specifically, the first cell is trapped at the trapping junction, and thereafter fluidic resistance is increased such that subsequent cells will pass by the filled chamber due to self-variable fluidic resistance. Subsequently, the trapped cell migrates and opens the trapping junction, enabling the loading of second cells. Using this system, the intercellular communication between mouse embryonic stem cells (mESCs) and mouse embryonic fibroblast cells (MEFs) was investigated. Two distinctive cellular behaviors in migration and proliferation emerged: (i) the paired cells' migration patterns depended on their initial cell distance, and (ii) heterotypic pairing led to distinct proliferation patterns compared to homotypic, single cell cultures. Microfluidic-based hydrodynamic trapping also provides a powerful approach for cell–cell fusion that is important, for example, for the generation of hybridomas and somatic cell reprogramming. For example, Voldman et al. [46] developed a device to achieve cell paring and fusion using an array of weir-based passive hydrodynamic cell traps on a microfluidic platform. Specifically, thousands of PDMS cell traps were densely arrayed within a flow-through channel (Figure 15.5B). Each cell trap consisted of a weir structure that extended vertically into the channel and contained front-side and back-side capture cups. Using a unique three-step loading protocol, cells were trapped individually once they entered the capturing cup. In this manner, different cell types, including

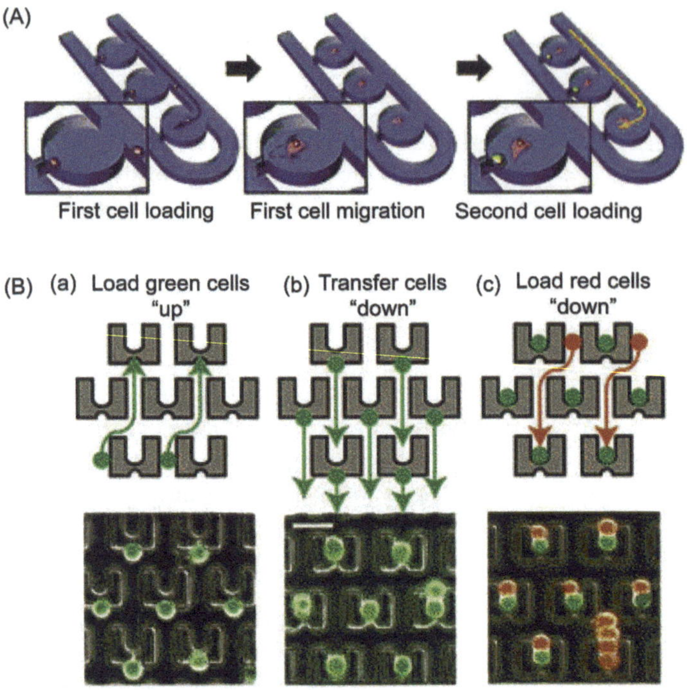

FIGURE 15.5 (A) A microfluidic system capable of creating a large number of heterotypic single cell pairs in isolated chambers by hydrodynamic cell trappings [28]. (B) Schematic illustrating the high-throughput cell pairing device and three-step loading protocol. (a) Cells are first loaded "up" toward the smaller back-side capture cup. (b) The direction of the flow is reversed, and the cells are transferred "down" into the larger front-side capture cup two rows below. (c) The second cell type is loaded from the top, and cells are captured in front of the first cell type [46].

fibroblasts, mESCs, and myeloma cells, were paired with efficiencies up to 70%. Furthermore, this device is compatible with both chemical and electrical fusion protocols, providing flexible cell fusions with high efficiency of 89%, or a fivefold increase compared to available commercial electrofusion chambers.

In addition to microwell devices, microdroplet devices have also been used to study cell–cell interactions at single cell levels. Yarmush et al. [47] recently developed a device for monitoring interactions between T cells and dendritic cells (DCs) at single cell levels. Specifically, monodisperse nanoliter droplets that encapsulated individual cell pairs were generated through a nozzle formed with three inlet microchannels containing an oil phase, an aqueous suspension of naïve T cells, and an aqueous suspension of LPS-stimulated DCs, respectively. Tubulin marker assays were performed in each droplet to monitor the polymerized tubulin in live cells. This system allowed monitoring the remodeling of the cytoskeleton and microtubule polymerization in DCs at the immunological synapse (IS) formed between co-encapsulated pairs of DCs and T cells. Importantly, it was revealed that not all DC and T cells were able to interact and create IS, demonstrating heterogeneity in cell–cell interaction at a single cell level that may be missed in traditional bulk population measurements.

15.2.3 Mimicking the *In Vivo* Microenvironment for Cellular Interactions Using Engineered Microdevices

Cells integrate and respond to cues in their *in vivo* microenvironments, including cell organizations, contacts with neighboring cells,

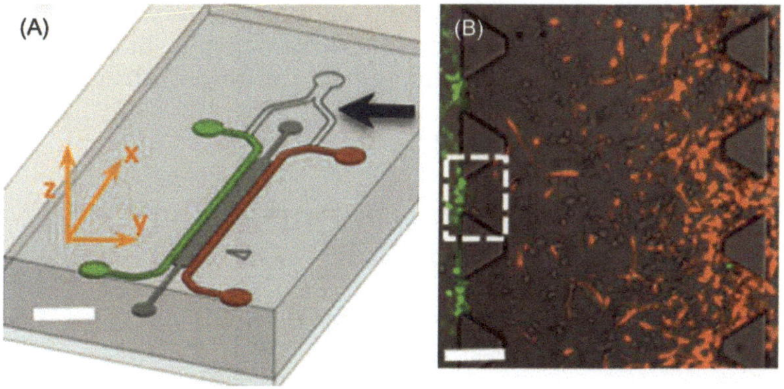

FIGURE 15.6 Microfluidic device to mimic tumor–vascular interface. (A) Endothelial channel (green), tumor channel (red), and 3D ECM (dark gray) between the two channels. (B) Phase contrast image showing the fibrosarcoma cells (HT1080, red) invading through the ECM (gray) toward the endothelium. A single 3D ECM hydrogel matrix region is illustrated with the white dashed square [54].

soluble factors, as well as mechanics of the surrounding fluid and ECM. In cancer, in particular, the microenvironment, or the niche, has been shown to play critical roles in cancer development, metastasis, and drug resistance. Therefore, microdevices that offer 3D controls over multiple cues of cancer microenvironment have attracted great attention to exploit cell–cell interactions under a more physiologically relevant condition [48–53]. For example, to examine the role of tumor–endothelial cell interactions during cancer metastasis, Zervantonakis et al. [54] developed an *in vitro* 3D microfluidic model of the tumor–vascular interface that integrates live imaging, precise control of microenvironmental factors, and endothelial barrier measurement. In this system, the relationship between tumor cell intravasation and endothelial activation and permeability as a function of cytokines present was shown to also depend on paracrine signaling involving macrophages interacting with tumor cells. As shown in Figure 15.6, this microdevice consisted of two independently addressable microchannels where tumor and endothelial cells are seeded. The two channels were connected through a 3D ECM hydrogel. The tumor cells invaded in 3D hydrogel in response to externally applied growth factor gradients or paracrine signals produced by the endothelial cells or macrophages. Macrophages were shown to regulate endothelial barrier function and paradoxically increase tumor cell intravasation through secreted tumor necrosis factor alpha (TNF-α).

In a related approach, the Bhatia group developed a microfluidic platform for high-throughput analysis of microtissues that allows systematic probing of tumor microenvironmental cues, including ECM, soluble factors, and stromal cells, in a 3D context [55]. As shown in Figure 15.7, the workflow could be divided into three phases. First, fluorescently labeled tumor cells were encapsulated with the desired combination of stromal cells or ECM into synthetic 3D microtissue using a droplet microfluidic device (Figure 15.7A). In the second phase, a cytometry-like flow analyzer was used to sort out defined populations of freshly generated microtissues by various parameters, including cell proliferation and cell density (Figure 15.7C). This system therefore allowed selection of ECM molecules and soluble factors that lead to a range of cell proliferation profiles. In the third phase, sorted microtissues were cultured in the presence of drugs to examine tumor response in a 3D-defined niche. Importantly, TGFβ, a benchmark immunomodulatory factor used in their study, had an antiproliferative effect that is unique to the 3D tumor microtissue compared to the 2D monolayer culture. Such 3D microtissues that more closely resemble *in vivo* systems

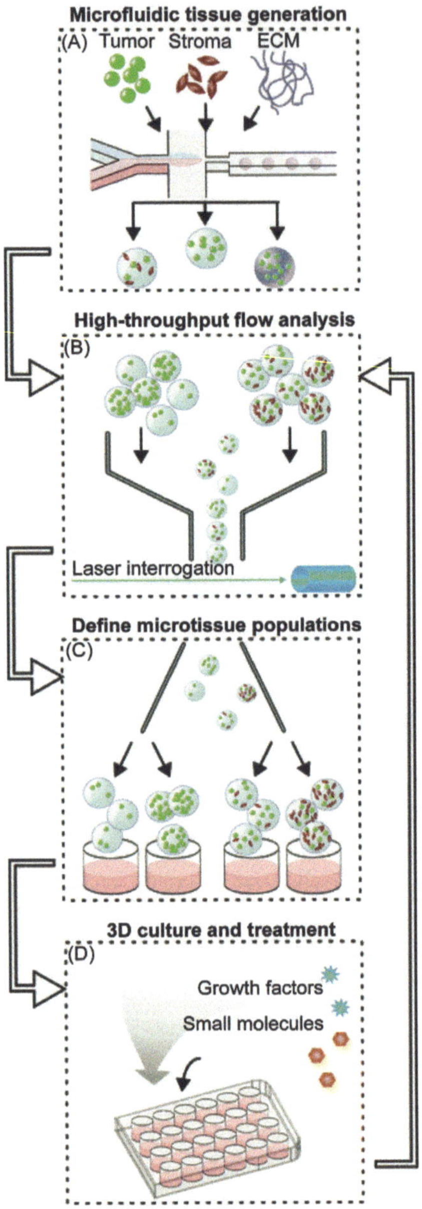

FIGURE 15.7 Microfluidic 3D tumor microenvironment screening platform. (A) Microfluidic droplet-based encapsulation of tumor cells into microtissues that can be tuned with co-encapsulated stromal cells or entrapped ECM molecules. (B) Microtissues are analyzed in multiple fluorescent channels using large particle flow analysis. (C) Flow sorting separates and defines microtissues by cell proliferation or homotypic and heterotypic interactions. (D) Microtissues in 3D cellular microenvironment are used to screen cancer drugs [55].

compared to the traditional 2D monolayer cultures are therefore advantageous in studying new and emerging anticancer drugs and approaches.

15.3 PROBING AND MANIPULATION OF CELL–CELL INTERACTIONS USING ENGINEERED MOLECULAR TOOLS

In addition to above-mentioned microfabricated tools, new genetic approaches allow the study of cells at unprecedented temporal and spatial resolution. In cell therapy, in particular, engineering of cells with molecular sensors allows tracking and monitoring of transplanted cells to elucidate cell–cell and cell–niche interactions *in vivo* using imaging techniques such as intravital two-photon fluorescent microscopy (IVM). Furthermore, the ability to engineer novel and enhanced cell–cell interactions makes it possible to manipulate and control the fate and function (e.g., homing) of transplanted cells. In this section, we will summarize the recent advances in genetically and chemically engineered tools that have opened up possibilities to explore and control cell–cell interactions *in vitro* and *in vivo*.

15.3.1 Genetically Encoded Proteins for Probing Cell–Cell Interactions

Genetic engineering, the direct manipulation of the genetic composition of cells and organisms, is a powerful technique to generate reporters for probing cell functions and interactions *in vivo*. By genetically engineering cells with fluorogenic or catalytic proteins, cells can be modified to produce optical signals inside a living system in controlled, frequently inducible, manner. Along with powerful *in vivo* optical imaging systems such as IVM, cell migration, trafficking, biodistribution, and cell–cell interactions can now be monitored in real time *in vivo*. Through these approaches, previously unknown intercellular interactions have been revealed with prominent examples including stem cell homing, stem cell–tumor cell interactions [56], immune cell–tumor cell interactions, immune cell interactions with each other or with stromal cells, endothelial cells or pathogens [57]. In particular, cell pairs can be engineered with fluorescent proteins or luciferase to report signals only upon cell–cell communication or if interactions occur, therefore minimizing nonspecific signals from complex biological systems. For example, Prescher et al. [57] recently developed a proximity reporter strategy that produced luminescent signals only when two different types of cells are physically in proximity to each other *in vivo*. As shown in Figure 15.8, the proximity reporter strategy was achieved by genetically engineering target cell pairs with β-galactosidase (β-gal) and luciferase to generate "activator cells" and "reporter cells," respectively. "Caged" luciferins (lugal), i.e., a luciferin molecule attached to a steric appendage that precludes the binding of the molecule to luciferase, were used as substrates that are released only upon β-gal activity. Next, intracellular enzymes catalyze the cleavage of Lugal to release luciferin, which diffuses near neighboring reporter cells and is processed by the luciferase to produce bioluminescence. Thus, reporter cells closest to activator cells would consume the most substrate, resulting in a correlation between the signal intensity and reporter cell and activator cell proximity. This proximity reporter system was used to illuminate metastatic lesions and tumor–immune cell interactions *in vivo*. Specifically, mouse bone marrow from β-gal expressing transgenic mice was transplanted into irradiated immunodeficient mice. Three weeks following bone marrow transplantation, mouse breast carcinoma 4T1-luc2 reporter tumor cells were orthotopically allografted into the mammary fat pads. Metastatic sites were visualized after injection of Lugal and the resulting bioluminescence as described earlier. Potential sites of metastatic–immune cell interactions, especially at submandibular lymph nodes, could be observed in a living animal. This represents a good example in the use of genetically engineered tools to visualize cellular proximity and interactions in a variety of biological settings in the minimally perturbed *in vivo* microenvironments, and thus confirming their potential to better understand physiological processes such as immune responses or stem cell niche development.

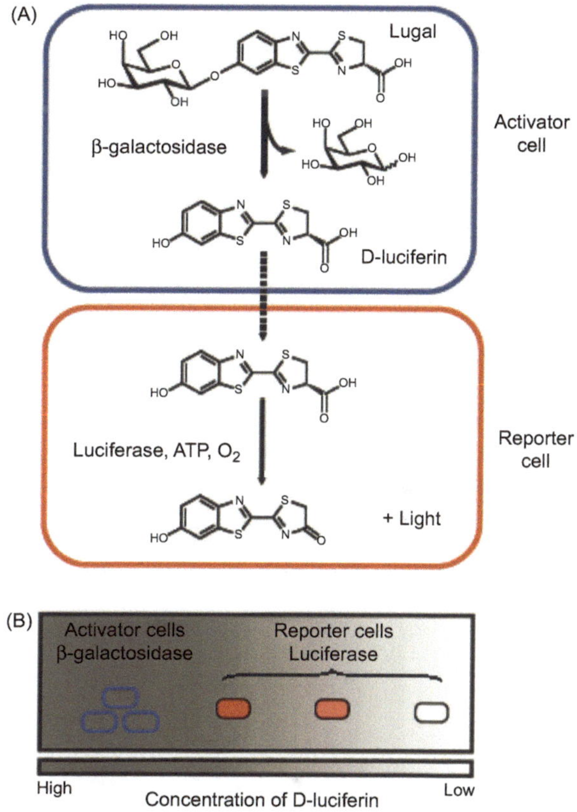

FIGURE 15.8 Proximity reporter assay design. (A) Activator cells (expressing β-gal) catalyze the cleavage of Lugal, releasing D-luciferin. (B) The concentration of D-luciferin is highest near the activator cells (indicated by gray-scale shading). As the distance between the reporter and activator cell decreases, light emission increases (indicated by the red color) [57].

15.3.2 Cell Surface Aptamer Sensors to Probe Cell–Cell Communications

Dynamic cell–cell interactions within an *in vivo* microenvironment largely rely on communications through ligand–receptor interactions at cell surface. The ability of chemical modifications to alter cell surfaces with synthetic ligands and sensors opens up a new venue to probe cell–cell interactions and communications. In this section, we describe representative approaches that make use of chemical engineering strategies to facilitate the study of cell interactions for emerging biological processes such as stem cell trafficking and chemical transmitter dynamics.

Although the ability to explore cellular interactions and communication is essential for understanding stem cell trafficking *in vivo*, it remains challenging to monitor the interaction of cells in their natural environments and niches in real time. To overcome these technical challenges, Zhao et al. [22] developed fluorescent cell surface sensors that quantitatively detect surrounding soluble factors with high spatial and temporal resolution (Figures 15.9 and 15.10). The sensor consists of a single strand DNA oligonucleotide aptamer modified with a pair of quenchable fluorescent dyes, frequently referred to as molecular beacons (Figure 15.9B and C). In the presence of the target analyte platelet-derived growth factor (PDGF), the aptamer binds to PDGF, triggering

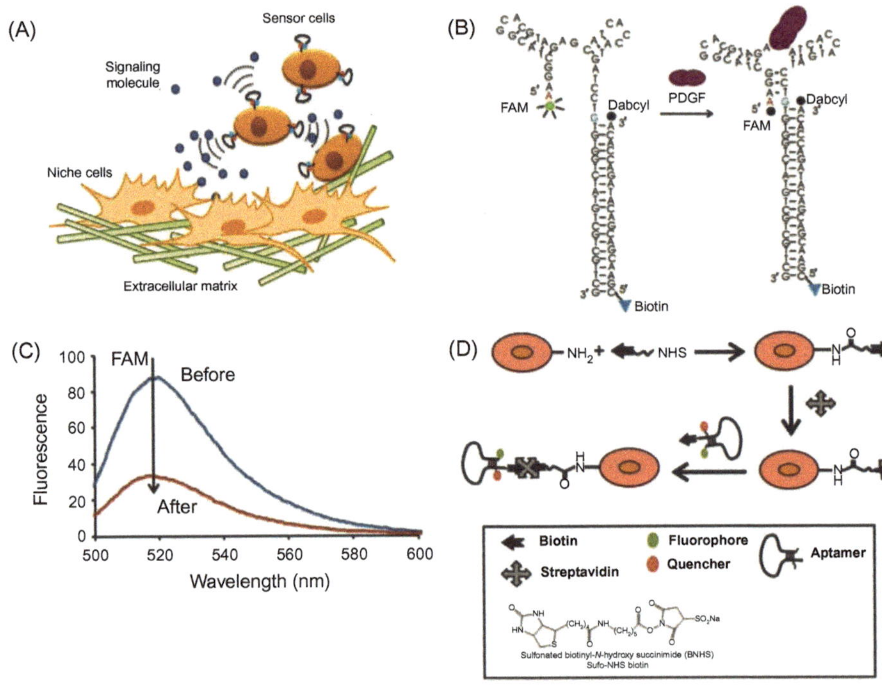

FIGURE 15.9 (A) Probing the interaction between MSC and the cellular niche microenvironment using aptamer sensor-engineered MSC. (B) PDGF DNA aptamer modified with FAM dye and Dabcyl quencher. Upon binding PDGF, a conformational change results in the two dyes coming together, leading to a decrease in fluorescent signal. (C) Quantified decrease in fluorescent signal upon aptamer binding to PDGF. (D) Schematic showing the chemical approach to attach PDGF aptamer sensors on MSC [22].

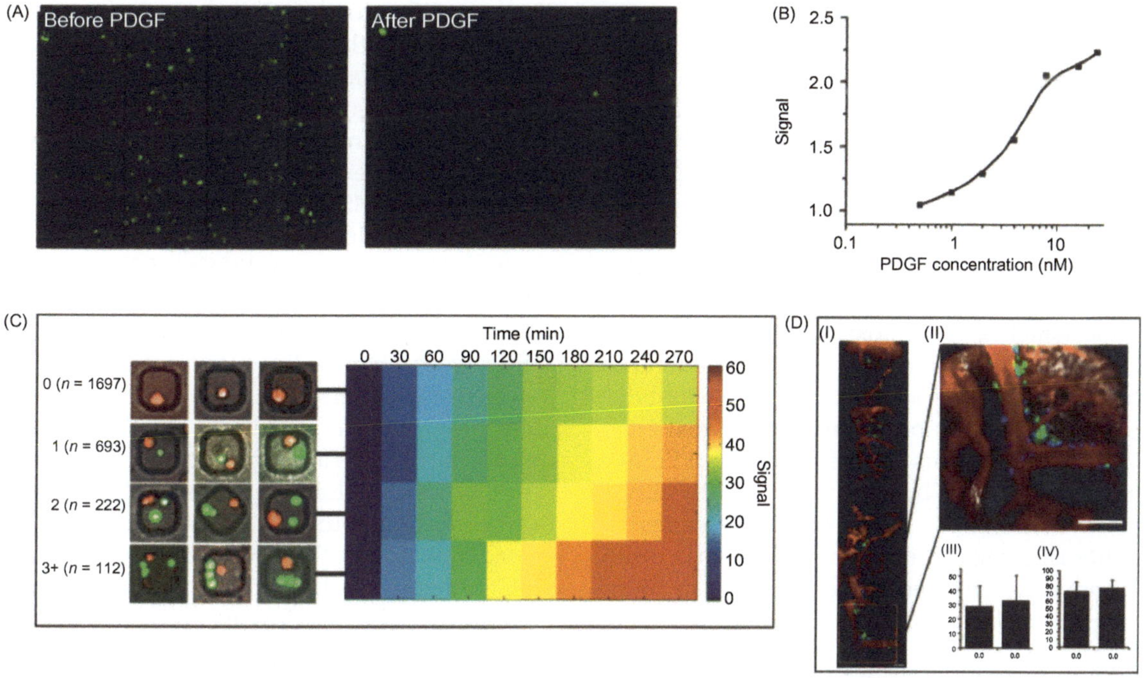

FIGURE 15.10 (A) Representative fluorescent microscopy images of sensor cell before and immediately after addition of 10 nM PDGF. (B) Sensor signal (defined as the ratio of geometric mean before and after addition of PDGF based on flow cytometry data) as a function of PDGF concentration. (C) Left: Representative images of microwells containing different numbers of PDGF-producing MDA-MB-231 cells (green) in the same well with sensor MSC (red) at time $t = 0$ (n is the number of MSCs used in the analysis). Right: Fluorescence of sensor MSC declining during the course of PDGF production. Signal is defined as the percentage of MSCs that have fluorescence intensity <50% of their initial value at the indicated time. (D) (I) Large area map of right parietal bone marrow compartments in a Balb/c mouse 24 h after injection of MSC and aptamer MSC. (II) Zoomed-in image of the area in the white box in (I) shows a similar distribution of MSC (green) and aptamer MSC (blue) in the vicinity of a large venule (red). (III) Quantification of the average number of cells per z-stack. (IV) Quantification of percentage of cells positioned outside blood vessels shows no difference between MSC and aptamer MSC. MSC: Green, aptamer MSC: blue, blood vessels: red, bone: white, DiO: Native MSC, DiD: aptamer MSC. Scale bar: 100 μm [22].

conformational changes that bring the two attached dyes in close proximity, thereby yielding a real-time fluorescent signal. This PDGF sensor was first attached on the mesenchymal stem cell (MSC) cell membrane by treating cell surface amines with sulfonated biotinyl-*N*-hydroxy-succinimide (NHS-Biotin), followed by streptavidin–biotin conjugation (Figure 15.9D). The engineered sensor MSCs are highly specific and able to quantitatively detect low local concentrations of PDGF that is either added into the medium or, more importantly, secreted by neighboring cells (Figure 15.10A–C). These cell surface sensors have unmatched spatiotemporal resolution at cellular and subcellular dimensions, opening new avenues to study cell–cell interactions by locally acting, i.e., autocrine, paracrine, or juxtacrine signaling factors. Furthermore, the engineered sensor MSCs retain their ability to home to the bone marrow and can be monitored *in vivo* at the single cell level using IVM (Figure 15.10D). The cell surface sensor approach makes use of functional aptamer sensors that dynamically monitor intercellular signaling in real time and therefore represents a new tool to elucidate unaddressed cell biology questions in the native cellular microenvironment.

In a related approach, Tokunaga et al. [58] used a fluorescent aptamer immobilized on the surface of brain astrocytes to monitor in real time the release of adenosine triphosphate (ATP) as a gliotransmitter. The structure-switching fluorescent aptamer was conjugated to astrocyte cell surface through a tocopherol anchor which is a nontoxic lipophilic molecule and incorporates directly to cellular membranes. Upon mechanical stimulation of astrocytes, fluorescence could be monitored in real time, giving novel insights on the action of ATP as a neurotransmitter. Importantly, ATP release was synchronized with calcium waves in neighboring cells, confirming the unique utility of chemically engineered sensor cells in monitoring intercellular communications.

15.3.3 Chemical Modification of the Cell Surfaces with Adhesion Molecules to Control Cell–Cell Interactions

In addition to probes for cellular microenvironments and communications, cell surfaces engineered with exogenous ligands can alter cellular interactions with homotypic or heterotypic cells and the surrounding extracellular scaffolds. In cell-based therapies, in particular, one of the greatest challenges is to deliver modified cells to the injured sites in way that is minimally invasive and does not perturb normal host functions, especially immune functions [59,60]. For example, the inefficient homing of injected MSC is thought to be a major limitation to existing MSC-based therapy. In contrast, leukocytes can home efficiently to site of inflammation using a variety of homing and adhesion molecules. By mimicking this leukocyte homing cascade, Karp and coworkers chemically engineered MSCs with Sialyl Lewis^{X} (SLeX), a cell rolling ligand that is involved in leukocyte rolling and homing. They demonstrated that SLeX modification promotes a robust cell rolling response both *in vitro* on a P-selectin-coated dish and on inflamed endothelium *in vivo* [61]. In addition to naturally occurring ligands, synthetic adhesion molecules, including aptamers, have also been exploited to control the interactions between MSCs and endothelial cells. Our laboratory has developed a simple approach to study the interactions of MSCs with endothelial cells under dynamic flow conditions using cell surface-attached DNA aptamers [62]. MSCs were chemically engineered with P- or L-selectin aptamers using the NHS-Biotin protocol. As shown in Figure 15.11, aptamer-displaying MSC can replicate the three major types of cell–cell interactions under dynamic flow condition found in the leukocyte homing cascade including: (i) flowing P-selectin aptamer MSC tethers to adherent P-selectin expressing endothelial cells, (ii) flowing L-selectin aptamer MSC tethers to L-selectin expressing leukocyte that is first tethered on a P-selectin substrate, and (iii) flowing L-selectin aptamer MSC and L-selectin expressing leukocyte first complex in the flowing stream and then tether to a P-selectin-coated substrate. The concept of engineering specific cell–cell interactions by chemically engineering cell surface ligands has potential applications in cell delivery, tissue engineering, and construction of *in vitro* microtissues to study cellular interactions. Likewise, Tan et al. [63] showed that chemical modification of cell surfaces with artificial aptamer-lipid receptors could be used to alter cellular functions, including protein binding, enzymatic activity, and intercellular interactions.

15.4 CONCLUSIONS AND PERSPECTIVES

Appropriate intercellular interactions and communication are essential to homeostasis and recovery, whereas erroneous signals

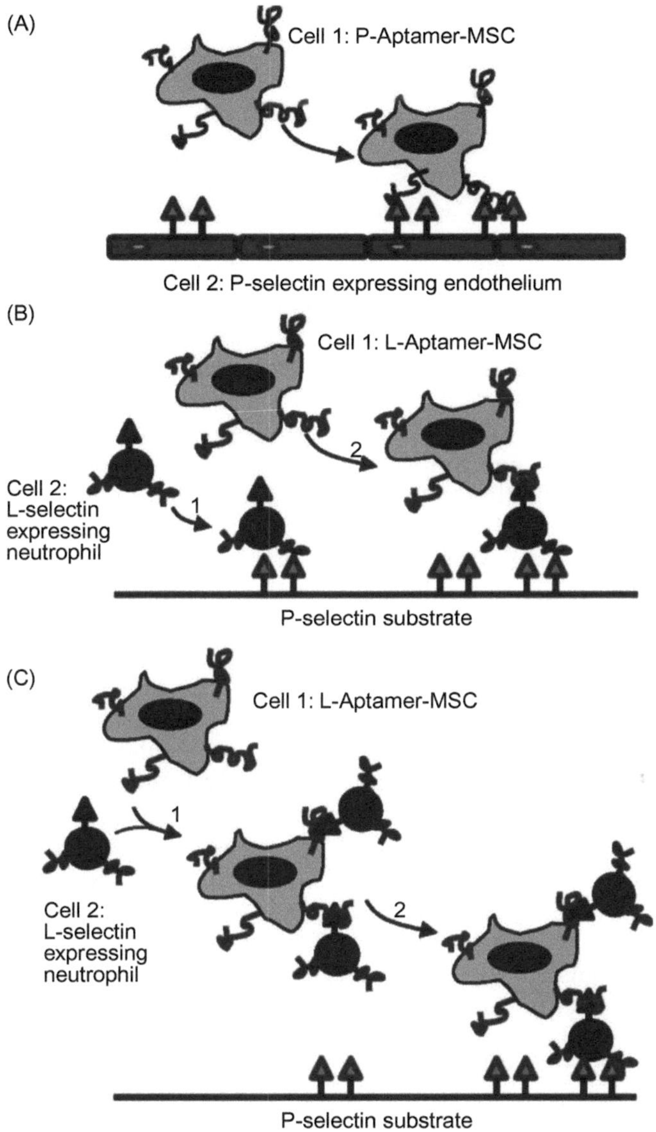

FIGURE 15.11 MSC chemically engineered with P-selectin or L-selectin aptamers to mimic three types of cell–cell interactions in leukocyte homing under dynamic flow conditions. (A) Flowing cell 1 (P-selectin aptamer MSC) tethers to adherent cell 2 (P-selectin expressing endothelial cell). (B) Flowing cell 2 (L-selectin expressing leukocyte) tethers to a P-selectin-coated substrate using the native leukocyte–P-selectin interaction and then captures flowing cell 1 (L-selectin aptamer MSC). (C) Cell 1 (L-selectin aptamer MSC) and cell 2 (L-selectin expressing leukocyte) first complex in the flowing stream and then tether to a P-selectin-coated substrate [62].

would have catastrophic consequences for the organism, as in the case of cancer, diabetes, autoimmunity, or immunodeficiency. Traditional approaches that investigate parameters based on cell population averages are gradually being supplanted by emerging approaches with higher spatial and temporal resolution that take into account and can distinguish the heterogeneity of cellular and molecular systems. With recent advances in various fronts of the natural sciences, our tools for studying and manipulating life at the organismal, systems, cellular, subcellular, and molecular levels have sharpened considerably. This has enabled ever greater insight and control over cellular behavior, including cell–cell interactions and communication.

In particular, microengineering advances in the emerging field of biological microfluidics have greatly enhanced our ability to peer into the world of intercellular interactions in controlled and defined microenvironments that closely mimic their *in vivo* counterparts [19,20]. In addition, enhanced genetic, chemical, and micro- and nanotechnological tools have enabled the measurement of various membrane and soluble signals during homotypic and heterotypic intercellular interactions and communication. For example, the interactions of different types of immune cells with each other [64] or with pathogens [65] and cancer cells [66,67], the reciprocal communication of neural cells [68] with heterotypic cells, or the interaction of stem cells with immune cells and stressed cells during repair of tissue injury [69,70] are now being analyzed at cellular and subcellular resolution in defined *in vitro*, as well as *in vivo*, microenvironments. In particular, the micro-engineered tools and molecular-engineered tools are complementary and can be used in combination to explore cell–cell interactions. For example, cells that have been molecular engineered with fluorescent proteins have been applied to microdevices to generate signals *in situ* for real-time analysis of cell interactions [55]. Furthermore, micro-engineered microwell devices have been used to evaluate chemically modified sensor MSCs before injecting those cells into living systems [22]. Importantly, cells are not only being studied using advanced methods, but their own unique properties, including their sensing of external stimuli, communication with other cells and trafficking and homing to defined microenvironments—in short their adaptive behaviors—are in turn making possible their use as robust and customizable platforms for applications ranging from biological research to diagnostic biosensors and therapeutic delivery agents [21].

Challenges remain toward more clearly and comprehensively understanding cell–cell interactions. First, there are many types of signals that make up intercellular interactions and communications, including physical contact, soluble factors, electrical signal transmission, and mechanical cues. Therefore, better micro- and molecular-engineered tools that can precisely isolate or sense specific and a combination of signals will be required. As we have described in previous sections, several techniques have begun to show their potential in this endeavor. For example, the mechanically detachable microdevices are able to distinguish interactions through soluble factors from physical contacts [38]. Second, cell–cell interactions are highly dynamic *in vivo*, thus requiring better molecular-engineered tools to capture the dynamics of cell–cell interactions [71]. Genetic and chemical modifications are advantageous to light up target cells with fluorescence or luminescence; however, it will be beneficial to generate interaction-specific signals that can monitor dynamic biological and pathological processes in real time. The "caged" luciferin-based proximity reporter assay represents such a trend [57]. Third, for *in vivo* applications, novel nanomaterials have been synthesized and demonstrated to be useful in numerous *in vivo* imaging techniques, including MRI, ultrasound, PET, and optical imaging [72]. By chemically modifying target cells with novel nanomaterials, it is possible to further expand the current *in vivo* imaging techniques to probe cell–cell interactions *in vivo*.

In conclusion, in the coming years, it is anticipated that advances on several fronts—micro- and nano-scale engineering and fluidics, genetic and chemical approaches—will not only continue to enhance the resolution and information content of biological experiments but likely will reveal novel paradigms of cellular interactions and communication. For example, novel genetically encoded reporters, such as robust ratiometric genetically encoded calcium indicators (GECIs) [73], will likely give unique and novel time-resolved information on the *in vivo* interaction and communication of immune cells with each other and other cell types during immune responses. Similarly, novel indicators used to map neuronal interactions will greatly enhance our understanding of the nervous system and its interaction with other physiological systems and the environment [74]. Likewise, advances in stem cell research will likely merge genetic and materials engineering approaches to control the complex, multistep intercellular interactions and communication giving rise to tissue and organ formation for regenerative medicine. Synthetic biology is also expected to benefit from, as well as inform, the above-mentioned trends in experimental biology, bioengineering, and nanobiotechnology [75,76]. Advances in the related field of systems biology are likely to play an increasingly active and important role in uncovering previously unknown or obscure links between diverse and seemingly unrelated cellular and physiological processes [77]. Coupled to these advances in fundamental experimental biology and bioengineered tools, it is anticipated that our ability to diagnose, and therefore treat, diseases caused by aberrant intercellular interactions and communication, including cancer, immune, and endocrine disorders, will also continue to increase in efficacy and selectivity.

References

[1] Purvis JE, Lahav G. Encoding and decoding cellular information through signaling dynamics. Cell 2013;152:945–56.
[2] Hynes NE, Ingham PW, Lim WA, Marshall CJ, Massague J, Pawson T. Signalling change: signal transduction through the decades. Nat Rev Mol Cell Biol 2013;14:393–8.
[3] Kitano H. Biological robustness. Nat Rev Genet 2004; 5:826–37.
[4] Richter DJ, King N. The genomic and cellular foundations of animal origins. Annu Rev Genet 2013;47: 509–37.
[5] Naus CC, Laird DW. Implications and challenges of connexin connections to cancer. Nat Rev Cancer 2010; 10:435–41.
[6] Cho JH, Gregersen PK. Genomics and the multifactorial nature of human autoimmune disease. N Engl J Med 2011;365:1612–23.
[7] Batista FD, Dustin ML. Cell: cell interactions in the immune system. Immunol Rev 2013;251:7–12.
[8] Mittelbrunn M, Sanchez-Madrid F. Intercellular communication: diverse structures for exchange of genetic information. Nat Rev Mol Cell Biol 2012;13: 328–35.
[9] El Andaloussi S, Mager I, Breakefield XO, Wood MJ. Extracellular vesicles: biology and emerging therapeutic opportunities. Nat Rev Drug Discov 2013;12:347–57.
[10] Kholodenko BN, Hancock JF, Kolch W. Signalling ballet in space and time. Nat Rev Mol Cell Biol 2010;11:414–26.
[11] Debanne D, Campanac E, Bialowas A, Carlier E, Alcaraz G. Axon physiology. Physiol Rev 2011; 91:555–602.
[12] Brooke MA, Nitoiu D, Kelsell DP. Cell–cell connectivity: desmosomes and disease. J Pathol 2012;226: 158–71.
[13] Mese G, Richard G, White TW. Gap junctions: basic structure and function. J Invest Dermatol 2007;127: 2516–24.
[14] Iwasaki A, Medzhitov R. Regulation of adaptive immunity by the innate immune system. Science 2010;327:291–5.

[15] Masopust D, Schenkel JM. The integration of T cell migration, differentiation and function. Nat Rev Immunol 2013;13:309–20.
[16] Ley K, Laudanna C, Cybulsky MI, Nourshargh S. Getting to the site of inflammation: the leukocyte adhesion cascade updated. Nat Rev Immunol 2007; 7:678–89.
[17] Matzinger P, Kamala T. Tissue-based class control: the other side of tolerance. Nat Rev Immunol 2011;11:221–30.
[18] Guo F, French JB, Li P, Zhao H, Chan CY, Fick JR, et al. Probing cell–cell communication with microfluidic devices. Lab Chip 2013;13:3152–62.
[19] El-Ali J, Sorger PK, Jensen KF. Cells on chips. Nature 2006;442:403–11.
[20] Primiceri E, Chiriaco MS, Rinaldi R, Maruccio G. Cell chips as new tools for cell biology—results, perspectives and opportunities. Lab Chip 2013;13: 3789–802.
[21] Ali MM, Kang DK, Tsang K, Fu M, Karp JM, Zhao W. Cell-surface sensors: lighting the cellular environment. Wiley Interdiscip Rev Nanomed Nanobiotechnol 2012; 4:547–61.
[22] Zhao W, Schafer S, Choi J, Yamanaka YJ, Lombardi ML, Bose S, et al. Cell-surface sensors for real-time probing of cellular environments. Nat Nanotechnol 2011;6: 524–31.
[23] Unger MA, Chou HP, Thorsen T, Scherer A, Quake SR. Monolithic microfabricated valves and pumps by multilayer soft lithography. Science 2000;288:113–6.
[24] Gomez-Sjoberg R, Leyrat AA, Pirone DM, Chen CS, Quake SR. Versatile, fully automated, microfluidic cell culture system. Anal Chem 2007;79:8557–63.
[25] Chen S, Lee LP. Non-invasive microfluidic gap junction assay. Integr Biol (Camb) 2010;2:130–8.
[26] Xu T, Yue W, Li CW, Yao X, Cai G, Yang M. Real-time monitoring of suspension cell–cell communication using an integrated microfluidics. Lab Chip 2010;10:2271–8.
[27] Bathany C, Beahm D, Felske JD, Sachs F, Hua SZ. High throughput assay of diffusion through Cx43 gap junction channels with a microfluidic chip. Anal Chem 2011;83:933–9.
[28] Hong S, Pan Q, Lee LP. Single-cell level co-culture platform for intercellular communication. IntegrBiol (Camb) 2012;4:374–80.
[29] Kumar NM, Gilula NB. The gap junction communication channel. Cell 1996;84:381–8.
[30] Ma H, Liu T, Qin J, Lin B. Characterization of the interaction between fibroblasts and tumor cells on a microfluidic co-culture device. Electrophoresis 2010;31: 1599–605.
[31] Liu W, Li L, Wang X, Ren L, Wang X, Wang J, et al. An integrated microfluidic system for studying cell–microenvironmental interactions versatilely and dynamically. Lab Chip 2010;10:1717–24.
[32] Zheng C, Zhao L, Chen G, Zhou Y, Pang Y, Huang Y. Quantitative study of the dynamic tumor–endothelial cell interactions through an integrated microfluidic coculture system. Anal Chem 2012;84:2088–93.
[33] Huang JW, Pan HJ, Yao WY, Tsao YW, Liao WY, Wu CW, et al. Interaction between lung cancer cell and myofibroblast influenced by cyclic tensile strain. Lab Chip 2013;13:1114–20.
[34] Bhatia SN, Yarmush ML, Toner M. Controlling cell interactions by micropatterning in co-cultures: hepatocytes and 3T3 fibroblasts. J Biomed Mater Res 1997; 34:189–99.
[35] Fukuda J, Khademhosseini A, Yeh J, Eng G, Cheng J, Farokhzad Omid C, et al. Micropatterned cell co-cultures using layer-by-layer deposition of extracellular matrix components. Biomaterials 2005;27:1479–86.
[36] Li Y, Yuan B, Ji H, Han D, Chen S, Tian F, et al. A method for patterning multiple types of cells by using electrochemical desorption of self-assembled monolayers within microfluidic channels. Angew Chem Int Ed Engl 2007;46:1094–6.
[37] Chen Z, Li Y, Liu W, Zhang D, Zhao Y, Yuan B, et al. Patterning mammalian cells for modeling three types of naturally occurring cell–cell interactions. Angew Chem Int Ed Engl 2009;48:8303–5.
[38] Hui EE, Bhatia SN. Micromechanical control of cell–cell interactions. Proc Natl Acad Sci USA 2007;104:5722–6.
[39] Kaji H, Yokoi T, Kawashima T, Nishizawa M. Controlled cocultures of HeLa cells and human umbilical vein endothelial cells on detachable substrates. Lab Chip 2009;9:427–32.
[40] Kaji H, Yokoi T, Kawashima T, Nishizawa M. Directing the flow of medium in controlled cocultures of HeLa cells and human umbilical vein endothelial cells with a microfluidic device. Lab Chip 2010;10:2374–9.
[41] Lindstrom S, Andersson-Svahn H. Miniaturization of biological assays—overview on microwell devices for single-cell analyses. Biochim Biophys Acta 2011; 1810:308–16.
[42] Sun JC, Lanier LL. NK cell development, homeostasis and function: parallels with $CD8^+$ T cells. Nat Rev Immunol 2011;11:645–57.
[43] Guldevall K, Vanherberghen B, Frisk T, Hurtig J, Christakou AE, Manneberg O, et al. Imaging immune surveillance of individual natural killer cells confined in microwell arrays. PLoS ONE 2010;5:e15453.
[44] Yamanaka YJ, Berger CT, Sips M, Cheney PC, Alter G, Love JC. Single-cell analysis of the dynamics and functional outcomes of interactions between human

natural killer cells and target cells. Integr Biol (Camb) 2012;4:1175–84.

[45] Lee PJ, Hung PJ, Shaw R, Jan L, Lee LP. Microfluidic application-specific integrated device for monitoring direct cell–cell communication via gap junctions between individual cell pairs. Appl Phys Lett 2005;86 223902–223902(3).

[46] Skelley AM, Kairak O, Suh H, Jaenisch R, Voldman J. Microfluidic control of cell pairing and fusion. Nat Methods 2009;6:147–52.

[47] Konry T, Golberg A, Yarmush M. Live single cell functional phenotyping in droplet nano-liter reactors. Sci Rep 2013;3:3179.

[48] Lii J, Hsu WJ, Parsa H, Das A, Rouse R, Sia SK. Real-time microfluidic system for studying mammalian cells in 3D microenvironments. Anal Chem 2008;80:3640–7.

[49] Chung S, Sudo R, Mack PJ, Wan CR, Vickerman V, Kamm RD. Cell migration into scaffolds under co-culture conditions in a microfluidic platform. Lab Chip 2009;9:269–75.

[50] Danino T, Volfson D, Bhatia SN, Tsimring L, Hasty J. *In silico* patterning of vascular mesenchymal cells in three dimensions. PLoS ONE 2011;6:e20182.

[51] Zervantonakis IK, Kothapalli CR, Chung S, Sudo R, Kamm RD. Microfluidic devices for studying heterotypic cell–cell interactions and tissue specimen cultures under controlled microenvironments. Biomicrofluidics 2011;5:13406.

[52] Shin Y, Han S, Jeon JS, Yamamoto K, Zervantonakis IK, Sudo R, et al. Microfluidic assay for simultaneous culture of multiple cell types on surfaces or within hydrogels. Nat Protoc 2012;7:1247–59.

[53] Chen MB, Srigunapalan S, Wheeler AR, Simmons CA. A 3D microfluidic platform incorporating methacrylated gelatin hydrogels to study physiological cardiovascular cell–cell interactions. Lab Chip 2013;13:2591–8.

[54] Zervantonakis IK, Hughes-Alford SK, Charest JL, Condeelis JS, Gertler FB, Kamm RD. Three-dimensional microfluidic model for tumor cell intravasation and endothelial barrier function. Proc Natl Acad Sci USA 2012;109:13515–20.

[55] Li CY, Wood DK, Huang JH, Bhatia SN. Flow-based pipeline for systematic modulation and analysis of 3D tumor microenvironments. Lab Chip 2013;13:1969–78.

[56] Sasportas LS, Kasmieh R, Wakimoto H, Hingtgen S, Van De Water JA, Mohapatra G, et al. Assessment of therapeutic efficacy and fate of engineered human mesenchymal stem cells for cancer therapy. Proc Natl Acad Sci USA 2009;106:4822–7.

[57] Sellmyer MA, Bronsart L, Imoto H, Contag CH, Wandless TJ, Prescher JA. Visualizing cellular interactions with a generalized proximity reporter. Proc Natl Acad Sci USA 2013;110:8567–72.

[58] Tokunaga T, Namiki S, Yamada K, Imaishi T, Nonaka H, Hirose K, et al. Cell surface-anchored fluorescent aptamer sensor enables imaging of chemical transmitter dynamics. J Am Chem Soc 2012;134:9561–4.

[59] Herberts CA, Kwa MS, Hermsen HP. Risk factors in the development of stem cell therapy. J Transl Med 2011;9:29.

[60] Cherry AB, Daley GQ. Reprogrammed cells for disease modeling and regenerative medicine. Annu Rev Med 2013;64:277–90.

[61] Sarkar D, Spencer JA, Phillips JA, Zhao W, Schafer S, Spelke DP, et al. Engineered cell homing. Blood 2011; 118:e184–91.

[62] Zhao W, Loh W, Droujinine IA, Teo W, Kumar N, Schafer S, et al. Mimicking the inflammatory cell adhesion cascade by nucleic acid aptamer programmed cell–cell interactions. FASEB J 2011;25:3045–56.

[63] Altman MO, Chang YM, Xiong X, Tan W. Modifying cellular properties using artificial aptamer-lipid receptors. Sci Rep 2013;3:3343.

[64] Baratchi S, Khoshmanesh K, Sacristan C, Depoil D, Wlodkowic D, McIntyre P, et al. Immunology on chip: promises and opportunities. Biotechnol Adv 2014;32(2): 333–46.

[65] Banerjee P, Bhunia AK. Mammalian cell-based biosensors for pathogens and toxins. Trends Biotechnol 2009;27:179–88.

[66] Chen J, Li J, Sun Y. Microfluidic approaches for cancer cell detection, characterization, and separation. Lab Chip 2012;12:1753–67.

[67] Businaro L, De Ninno A, Schiavoni G, Lucarini V, Ciasca G, Gerardino A, et al. Cross talk between cancer and immune cells: exploring complex dynamics in a microfluidic environment. Lab Chip 2013;13: 229–39.

[68] Soe AK, Nahavandi S, Khoshmanesh K. Neuroscience goes on a chip. Biosens Bioelectron 2012;35:1–13.

[69] Gupta K, Kim DH, Ellison D, Smith C, Kundu A, Tuan J, et al. Lab-on-a-chip devices as an emerging platform for stem cell biology. Lab Chip 2010;10:2019–31.

[70] Keating A. Mesenchymal stromal cells: new directions. Cell Stem Cell 2012;10:709–16.

[71] Germain RN, Robey EA, Cahalan MD. A decade of imaging cellular motility and interaction dynamics in the immune system. Science 2012;336:1676–81.

[72] Eckert MA, Vu PQ, Zhang K, Kang D, Ali MM, Xu C, et al. Novel molecular and nanosensors for *in vivo* sensing. Theranostics 2013;3:583–94.

[73] Mues M, Bartholomaus I, Thestrup T, Griesbeck O, Wekerle H, Kawakami N, et al. Real-time *in vivo* analysis of T cell activation in the central nervous system using a genetically encoded calcium indicator. Nat Med 2013;19:778–83.

[74] Cai D, Cohen KB, Luo T, Lichtman JW, Sanes JR. Improved tools for the brainbow toolbox. Nat Methods 2013;10:540–7.

[75] Wieland M, Fussenegger M. Engineering molecular circuits using synthetic biology in mammalian cells. Annu Rev Chem Biomol Eng 2012;3:209–34.

[76] Cheng AA, Lu TK. Synthetic biology: an emerging engineering discipline. Annu Rev Biomed Eng 2012; 14:155–78.

[77] Walpole J, Papin JA, Peirce SM. Multiscale computational models of complex biological systems. Annu Rev Biomed Eng 2013;15:137–54.

Index

Note: Page numbers followed by "*f*", "*t*", and "*b*" refers to figures, tables and boxes, respectively.

FSC
www.fsc.org
MIX
Papier | Fördert gute Waldnutzung
FSC® C083411

Kolibri 360 GmbH
Ferdinand-Jühlke-Straße 7
99095 Erfurt, Deutschland
produktsicherheit@kolibri360.de